URBAN NIGHTSCAPES

Youth Cultures, Pleasure Spaces and Corporate Power.

Despite unprecedented growth in the entertainment economy, in many western cities, urban nightlife is increasingly experiencing a form of 'McDonaldisation', with big brands taking over large parts of downtown areas, leaving consumers with an increasingly standardised experience and a lack of alternative, creative provision. *Urban Nightscapes* takes a new look at this rapidly changing aspect of urban life, examining the relationships between young adults, nightlife and city spaces.

The first part of the book explores three inter-related aspects of these nightscapes: *production* and the role of large-scale corporate entertainment operators, who provide branded, themed and stylised experiences; *regulation* through practices which aid capital accumulation and city 'image-building'; and *consumption*, where a night out is characterised by not only pleasure-seeking hedonism, but by a segmentation of youth identity and activity as well. *Urban Nightscapes* highlights who owns and controls the night-time economy and, in particular, the increasing amount of mergers and concentration of ownership; the pervasive use of surveillance (both technological and social); and how mainstream, commercial nightlife is squeezing out both historic and alternative/independent forms of enjoyment. The second part of the book then colourfully explores these ideas through detailed ethnographic case studies of young professionals, students, women and gay consumers, excluded youth groups and also alternative nightlife activity, such as squats and free parties.

Throughout the book the authors explore pockets of resistance to this standardised experience and suggest a number of potential future scenarios for cities at night beyond the corporate nightlife machine. *Urban Nightscapes* presents a theoretical and lively ethnographic account for understanding contemporary youth cultures, political urban change and city nightlife.

Paul Chatterton is Lecturer in the Department of Geography, and **Robert Hollands** is a Senior Lecturer in the Department of Sociology and Social Policy, both at the University of Newcastle upon Tyne in the UK.

CRITICAL GEOGRAPHIES

Tracey Skelton

Lecturer in Geography, Loughborough University

Gill Valentine

Professor of Geography, University of Sheffield

This series offers cutting-edge research organised into three themes of concepts, scale and transformations. It is aimed at upper-level undergraduates, research students and academics and will facilitate inter-disciplinary engagement between geography and other social sciences. It provides a forum for the innovative and vibrant debates which span the broad spectrum of this discipline.

1 MIND AND BODY SPACES

Geographies of illness, impairment and disability

Edited by Ruth Butler and Hester Parr

2 EMBODIED GEOGRAPHIES

Spaces, bodies and rites of passage

Edited by Elizabeth Kenworthy Teather

3 LEISURE/TOURISM GEOGRAPHIES

Practices and geographical knowledge

Edited by David Crouch

4 CLUBBING

Dancing, ecstasy, vitality

Ben Malbon

5 ENTANGLEMENTS OF POWER

Geographies of domination/resistance

Edited by Joanne Sharp, Paul Routledge, Chris Philo and Ronan Paddison

6 DE-CENTRING SEXUALITIES

Politics and representations beyond the metropolis

Edited by Richard Phillips, Diane Watt and David Shuttleton

7 GEOPOLITICAL TRADITIONS

A century of geopolitical thought

Edited by Klaus Dodds and David Atkinson

8 CHILDREN'S GEOGRAPHIES
Playing, living, learning
Edited by Sarah L. Holloway and Gill Valentine

9 THINKING SPACE
Edited by Mike Crang and Nigel Thrift

10 ANIMAL SPACES, BEASTLY PLACES
New geographies of human–animal relations
Edited by Chris Philo and Chris Wilbert

11 BODIES
Exploring fluid boundaries
Robyn Longhurst

12 CLOSET SPACE
Geographies of metaphor from the body to the globe
Michael P. Brown

13 TIMESPACE
Geographies of temporality
Edited by Jon May and Nigel Thrift

14 GEOGRAPHIES OF YOUNG PEOPLE
The morally contested spaces of identity
Stuart C. Aitken

15 CELTIC GEOGRAPHIES
Old culture, new times
Edited by David C. Harvey, Rhys Jones, Neil McInroy and Christine Milligan

16 CULTURE/PLACE/HEALTH
Wilbert M. Gesler and Robin A. Kearns

17 SOUND TRACKS
Popular music, identity and place
John Connell and Chris Gibson

18 URBAN NIGHTSCAPES
Youth cultures, pleasure spaces and corporate power
Paul Chatterton and Robert Hollands

URBAN NIGHTSCAPES

Youth Cultures, Pleasure Spaces and Corporate Power

Paul Chatterton and Robert Hollands

LONDON AND NEW YORK

First published 2003 by Routledge
2 Park Square, Milton Park, Abingdon, Oxon, OX14 4RN

Simultaneously published in the USA and Canada
by Routledge
270 Madison Ave, New York, NY 10016

Transferred to Digital Printing 2006

Routledge is an imprint of the Taylor & Francis Group

Typeset in Perpetua by Taylor & Francis Books Ltd
Printed and bound in Great Britain by TJI Digital, Padstow, Cornwall

British Library Cataloguing in Publication Data
A catalogue record for this book is available from the British Library

Library of Congress Cataloging in Publication Data
A catalog record for this book has been requested

ISBN 0–415–28345–0 (hbk)
ISBN 0–415–28346–9 (pbk)

CONTENTS

ILLUSTRATIONS

Plates

Figures

Tables

PREFACE

This book is a truly collectively written and produced endeavour in terms of workload and intellectual input which emerged from extended discussions and debates over many hours (and the occasional beer). It goes without saying, then, that it is the product of a longstanding set of personal and academic interests we both have held around youth cultures, nightlife and cities. As with many collaborative works, the initial idea came through a meeting of two people who were working and thinking in similar areas, theoretically, methodologically and politically. One of us, a sociologist, has been researching and teaching in the area of youth studies for many years, writing books on working-class leisure (Cantelon and Hollands, 1988), youth transitions (Hollands, 1990) and youth cultures and nightlife (Hollands, 1995, 1997, 2000). The other, a geographer whose PhD was on student cultures and their impact on cities (Chatterton, 1998), continues to teach and research around urban cultures, youth identities and political activism (Chatterton, 1999, 2000, 2002). Both of us share an outlook which combines the study of political-economic forces, and in particular a concern about the increasing power of corporate and global capital in our daily lives, with critical ethnographies sensitive to the nuances of locality, agency and political resistance.

These personal and academic biographies came together most fruitfully between 2000 and 2002 at the University of Newcastle upon Tyne, when we jointly managed a research project funded by the Economic and Social Research Council (ESRC) entitled 'Youth cultures, identities and the consumption of nightlife city spaces in three English cities' (award number R000238288). Much of the empirical data presented here is derived from this research project, which looked at the production, regulation and consumption of urban nightlife in several UK cities. This book, then, while partly emanating from our UK research project, also draws upon examples outside the UK and represents a significant extension of our thinking reported elsewhere (see Hollands and Chatterton, 2003; Chatterton and Hollands, 2001, 2002; Hollands, 2002). A more detailed methodological note appears in the introduction.

As the research progressed, a story of corporate power, greed, domination and marginalisation, not to mention hedonism/pleasure, dissatisfaction and resistance across city streets at night, unravelled itself. The book is driven by a concern for who loses and who wins in the constant 'merry-go-round' of urban change, renewal and gentrification. While there are many readings of urban nightlife, and hence many different books which could have been written, this book aims to be a powerful reminder of the consequences of letting corporate power, profit and the pro-growth entrepreneurial state go unchecked. As such, it is not only an invitation to critically engage with theory and detailed empirical findings about the making of urban nightlife – it is also a call for radical thinking and praxis about possible new urban worlds.

ACKNOWLEDGEMENTS

During the course of this research we received a great deal of support from a wide variety of people and organisations. First and foremost, a big thanks to all those who took part in the research, including all the young (and not so young!) adults involved in our focus groups, as well as the large number of people we interviewed who are involved in regulating, owning or managing nightlife. Second, we would especially like to thank Bernie Byrnes, Cait Reid, Meg Aubrey and Emer O'Sullivan, who offered valuable research support, and the Economic and Social Research Council in the UK, who funded the research we draw on here. Colleagues at the University of Newcastle upon Tyne who gave us their views, time and enthusiasm include Richard Collier, John Vail, Anoop Nayak, Les Gofton, Colin Clark and James Cornford, while further afield we acknowledge the support of Rob MacDonald, Steve Redhead and Justin O'Connor. A number of colleagues working in the area of the entertainment economy, namely Mark Gottdiener, John Hannigan and Will Straw, offered encouraging yet critical advice. Thanks also to Pau Sarracant from the Youth Observatory in Barcelona for hosting us, and Roger Martinez and David Tximmi for showing us around this city. And thanks to 'Uncle Jimmy' for putting us up in Montreal when we presented our first paper based on the ideas in this book at McGill University at the 'Night and the City' Conference in March 2001.

In terms of the production of the book, thanks to the Routledge team, including Andrew Mould, Vanessa Winch and Ann Michael, and also Gill Valentine and Tracey Skelton as enthusiastic series editors. Thanks also to all those who provided visual material for the book; in particular we appreciate the efforts of Ann Rooke, Louise Hepworth and Michael Duckett, and of Tim Wray, who provided material for the cover photo. Over the years a number of people have been our social inspiration and have been part of some great nights out, which have given us some useful reality checks for our work. Thanks here, in no particular order, to Tash, Jen and Watson, Bon, James, James, Shaun, Matt, Giles, Jim, Harriet, Paul, John, C, Aki, Haydn, Paula, Joey C-H, Carole (Wearsy) and the

'Heaton Widows', the secretaries at CURDS and all the folk from the Brighton Palais and Tyneside Action for People and Planet. Thanks also, of course, to our parents and families. Finally, *A Tale of Two Cities* (1996) by Ian Taylor and his colleagues was a book which both of us read prior to embarking on this writing project. Ian's untimely death in 2001 was a shock to many, and we are indebted to the critical spirit of much of his thinking.

Robert and Paul
Newcastle upon Tyne, UK
July, 2002

The authors and publishers would like to thank the following for granting permission to reproduce material in this work:
University of Newcastle upon Tyne for Figure 2.1 and Figure 2.2
Carlos Aparicio for Plate 2.2
The *Newcastle Evening Chronicle* for Plate 3.3 and Plate 10.1
Louise Hepworth for Plate 4.1
Scottish & Newcastle Retail for Plate 5.3 and Figure 5.1
North News & Pictures for Plate 7.1
The trustees of Mass Observation Archive, University of Sussex, for Figure 8.1
Mike Duckett for Plate 8.1
Arcova Studio for Plate 9.1
All other photographs courtesy of the authors.

1

INTRODUCTION

Making urban nightscapes

This book is about the making and re-making of urban nightlife. While a night out is a common and widespread experience, many of us do not spend much time thinking about what exactly goes into its making, which is not surprising as most of the time we are too busy enjoying ourselves. For example, rarely do we pose questions such as: who owns and profits from the nightlife venues we socialise in? Who develops, designs and promotes nights out and for which groups? What laws and legislations govern nightlife and how it is policed? What implicit and explicit codes structure a night out, which people go to which venues and why, and what do young people actually think about their nightlife experiences? What follows in this book is an attempt to unpack these related sets of issues.

What this book is not is a celebration of the diversity and the countless experiences which a night out can offer. Such a venture is clearly another project in itself. This is also not another book solely about clubbing or rave culture, or indeed drug and Ecstasy cultures (of which there have been numerous examples: see, Malbon, 1999; Saunders, 1995; Redhead *et al.*, 1997), but more broadly 'urban nightscapes', which entails a variety of youth cultural spaces and groups, including those who are excluded from, or challenge, what's on offer downtown at night. Instead, in what follows we pursue a more political-economy perspective of a night out, which not only looks at consumer experiences and draws upon our real experiences with young people during a night out, but also explores issues of production and regulation. Hence, our work takes a critical look at the role of the entrepreneurial local state, the various effects of corporate capital, and the increasing uniformity and standardisation of many modern-day consumer practices. To give the reader some idea of what to expect in the following pages, our main concern relates to the growing dominance of large corporate operators and their nightlife brands, and what this means for the future character and liveability of urban areas.

We start this book within a context of change and transformation. Social and economic restructuring over the last three decades has resulted in the development of a new urban 'brand' which has reshaped many parts of city landscapes into corporate entertainment and leisure hubs (Gottdiener, 2001; Hannigan, 1998). While urban areas have always been sites of pleasure-seeking, a central focus of recent rebranding has been the promotion of the night-time economy, much of which is characterised by the ritual descent of young adults into city-centre bars, pubs and clubs, especially during the weekend (Hollands, 1995; Thornton, 1995; Chan, 1999; Andersson, 2002). One stark example which provides a glimpse into the widespread nature of the growth of night-time entertainment, and gives us a way into thinking about the making, remaking and unmaking of urban nightscapes, comes from Barcelona in the Catalunya region of Spain, and more specifically the Port Vell area on the city's water-front. This area has been unrecognisably transformed since the mid 1990s, from a decaying old dock area into one of the city's foremost and most fashionable party zones. The main quay is dominated by the Maremagnum complex, an entertainment and shopping area opened in 1995 containing around fifty shops, twenty-five restaurants and dozens of bars and clubs (see Plate 1.1). The 2001 *Time Out* guide to Barcelona states: 'Maremagnum gleams temptingly in the middle of the Port Vell and draws in huge crowds both night and day.'

An initial and indeed understandable response is to applaud such a development as a positive example of 'urban cultural revival' (Landry, 2000; Comedia and Demos 1997). Indeed, the night-time economy in many cities has now become an accepted part of wider urban renewal strategies and is seen as a significant source of income, employment and civic 'image-building' (Chatterton and Hollands, 2001, 2002).

Plate 1.1 Maremagnum Entertainment Complex; Port Vell, Barcelona, Spain, 2002

However, there is another largely untold story here. Port Vell, for example, the once-glistening jewel in the gentrified regeneration of Barcelona's waterfront has recently lost some of its lustre. The reality by 2002 was that the city's wealthier population, who once danced and drank here with style, have moved on to the latest nightlife venues elsewhere. As a result, Maremagnum has become a more middle-ground 'mainstream' nightlife space for young, often under-age, tourists and the city's immigrant and working-class communities. After forty incidents involving the police at the leisure complex between 2000 and 2002, complete with security guards armed with batons, the problem finally came to a head. In January 2002, it was alleged that four door staff from Bar Caipirinha in Maremagnum beat and then threw a young Ecuadorian immigrant, Wilson Pacheco, into the sea when he became aggressive after he was denied entry to the bar. Pacheco drowned, creating a media frenzy calling for a new set of nightlife regulations, and Bar Caipirinha was closed. Subsequently, the Catalan Interior Ministry, along with the Barcelona city government and the port authority, developed a plan to restrict late-night bar and alcohol-led activity in the area and promote a greater diversity of more cultural and family-oriented activities.[1]

Although only a single example, what exactly does this 'other' story reveal about the making and remaking of urban nightscapes generally? First, it demonstrates that while there is a growing popularity and domination of large-scale, glossy corporate nightlife developments in many cities around the world, they often contain their own set of contradictions. Constant attempts to upgrade and gentrify urban waterfronts and central areas, which include nightlife facilities, invariably result in a tail-chasing game of 'cool-hunting' (Klein, 2000), as young professionals go in search of the latest cool, chic, fashionable bar or club, leaving yesterday's stylish haunt in their wake. Indeed, much of the new nightlife economy is all about being 'cool'. In their attempt to define this rather tricky term, Pountain and Robins (2000) point out that coolness, as a consumption strategy, is largely an individual identity strategy rather than a collective political response. As the authors suggest, 'no-one wants to be good any more, they want to be Cool' (*ibid.*: 10). Yet, more importantly for this book, coolness has also become a vehicle for big business, the media and advertisers to push their way further into the wallets of young consumers.

Second, while such upgrading initially appears to be enforced through pricing and various stylistic codes, it often requires more 'direct' and violent forms of regulation (i.e. bouncers, security staff; see Hobbs *et al.*, 2000) which, as the Maremagnum example shows, can quickly spiral out of control. Finally, attempts to gentrify leisure and the night-time economy in the city have resulted in various nightlife consumption groups jockeying for position and territory, leading to a socially segregated, conflictual and increasingly polarised use of space (Hollands, 2002).

The above example from Barcelona, then, reflects some of the main concerns of *Urban Nightscapes*. Our focus is 'urban' – and in particular downtown – areas, which continue to represent the most visible manifestation of these trends, especially in Europe. However, we are cognisant that our discussion reflects broader social and cultural changes rather than mere 'city-based' phenomena (see Gottdiener, 2001). In this sense, many of the processes we emphasise, including economic concentration, corporatisation, branding and theming, segmentation of consumer markets and pandering to middle-class tastes, are occurring across central, suburban and complex polycentric city-regions (Soja, 2000) across the western world, and clearly here North America is leading the way. This is not to say that there are not important variations in nightlife cultures between countries and cities and that wider processes do not work themselves out at different rates in specific local conditions. For example, there are clear differences in nightlife infrastructures between larger global cities, established metropolitan centres and smaller cities. However, the phenomena we are studying are increasingly global, or at least western, in their orientation, and hence it is possible to generalise findings to some degree, as the night-time economies of many post-industrial countries continue to converge and follow similar trends.

Our second focus is on the term 'nightscapes', which refers specifically to young adults' varied nightlife activities in licensed premises such as bars, pubs, nightclubs and music venues, as well as the streets and spaces in-between. We recognise that other activities such as cinema, theatre, restaurants, casinos, cafés and sporting events also combine to make up these nightscapes, but these are not our concern here as they are not primarily the preserve of young people in cities.[2] Our use of the term 'nightscapes' also refers to issues raised by Zukin (1992) about the aestheticisation and commodification of urban landscapes, but also to the increased use of the city as a place of consumption, play and hedonism in the evening (Featherstone, 1991).

The framework

Studying nightscapes entails unravelling certain inherent difficulties, contradictions and dichotomies involved in the actual 'experience' of nightlife. We are acutely aware that framing it through academic discourse and theoretical concepts erodes much of the fluidity, excitement and sociability of a night out. It is quite impossible to capture what the city is about at night within the stiff pages of an academic text. In fact, any attempt to represent nightlife will automatically make certain groups and actions more visible than others. The city, especially at night, contains many contradictory elements that cannot always be resolved, understood or related. Nightlife is simultaneously conflictual and transgressive, at the same time as being segregated, commodified and sanitised. It also has emotional (enhanced through alcohol, drugs, dance, sex, encounter) and rational elements (planning, surveillance and policing), which are not always easy to understand and reconcile. Nevertheless, it is worth highlighting, to quote Thornton (1995: 91), that 'the seemingly chaotic paths along which people move through the city are really remarkably routine'.

Our perspective is to stress the active making and remaking of urban nightscapes, an approach which is sensitive to processes as well as possibilities. As such, the book operates on two levels (see Table 1.1). First, we present an understanding of nightscapes through an *integrated 'circuit of culture' which comprise the three processes of production, regulation and consumption* (Du Gay, 1997). By this we mean that, to fully understand an area of activity such as nightlife, it is imperative to simultaneously explore who and what is involved in producing nightlife spaces (i.e. designing, marketing, selling, property markets, corporate strategies, etc.), who and what is involved in regulating them (i.e. laws and legislations, surveillance, entrance requirements, codes of conduct), and who and what is involved in consuming them (i.e. lived experience, perceptions, stereotypes, etc.). Hence, while nightlife venues are clearly commercially manufactured by a range of multinational, national, regional and local operators, and regulated by various legislative frameworks and formal and informal surveillance mechanisms, it is also necessary to explore the lived consumer experience and the role young adults play in shaping such spaces.

Second, *urban nightscapes can be understood as a mixture of mainstream, residual and alternative nightlife spaces*. Mainstream spaces are the well-recognised commercially provided bars, pubs and nightclubs that exist in most large urban centres. While there are a range of venue types here, the unifying feature of the mainstream is that it is driven by commercial gain and the profit-motive, rather than the other concerns such as access, equality or creativity. The mainstream is also characterised by ownership by large national and international corporate players who are increasingly using strategies like branding and theming to target and segment certain cash-rich groups such as professionals and service sector workers (including professional women and the gentrifying gay population) and higher education students. These spaces cater for much of the hedonistic rituals and raucous behaviour one normally associates with a night out. Residual community spaces such as traditional pubs, ale-houses and saloons, as well as the purview of the street, which were a common feature of most industrial city centres have been left to decline or are disappearing altogether, due to the changing priorities of nightlife operators and consumer tastes. Finally, there is a range of independently run and alternative nightlife spaces which cater for more specific youth cultures, identities and tastes, some of which are self-organised, such as free parties, unofficial raves and squatted social centres. Clearly, spaces such as the mainstream, the residual and the alternative and resistant margins are constantly shifting entities, with rather nebulous boundaries. Today's fringe cultures become tomorrow's mainstream fashions. Hence, we have tried to avoid over-literal interpretations which regard the mainstream as mere commercial blandness while the underground is teeming with resistance and creativity. Instead, we have focused upon how different spaces and boundaries are made and remade, regulated and experienced.

Table 1.1 Mapping out urban nightscapes

Mode of analysis	Type of nightlife space: Mainstream	Residual	Alternative
Production	Corporate brand	Community	Individual
	Profit-oriented	Need-oriented	Experimental
	Global/national	National/regional	Local (and global networks)
Regulation	Entrepreneurial	Stigmatised	Criminalised
	Formal (CCTV, bouncers) and informal (style, price)	Formal (police)	Informal (self- regulated)
Consumption	Profit-oriented	Community-oriented	Creative-oriented
	Divided consumer–producer relations (brand/lifestyle)	Traditional consumer–producer relations (product)	Interactive consumer–producer relations
	Gentrified/up-market	Down-market	Alternative/resistant
Spatial location	Dominant centre	Under-developed centre	Margins

Source: Chatterton and Hollands, 2002

The central argument of *Urban Nightscapes* is that urban nightlife is increasingly characterised by dominant regimes of: mainstream production, through the corporatisation of ownership via processes of branding and theming (Klein, 2000; Gottdiener, 2001); regulation, through practices which increasingly aid capital accumulation and urban image-building (Zukin, 1995; Harvey, 1989b) yet increase surveillance (Davis, 1992); and consumption, through new forms of segmented nightlife activity driven by processes of gentrification and the adoption of 'up-market' lifestyle identities among groups of young adults (Butler, 1997; Wynne and O'Connor, 1998; Savage and Butler, 1995). In this sense, although many city centres have achieved a 'cool' status through branded and upgraded nightlife, they are also increasingly becoming more exclusive, segmented and conflictual arenas (Smith, 1996; Sibley, 1995). And while we stress that new opportunities have opened up, especially for young women (McRobbie, 2000), ethnic cultures and music (Forman, 2002) students (Chatterton, 1999), and gay nightlife in particular (Knopp, 1992), these have often been sanitised and commercially incorporated into the mainstream. Moreover, historic, residual and alternative forms of nightlife are increasingly marginalised to the geographic periphery of the urban core, over-regulated until they simply disappear, transformed by the changing corporate priorities of their owners, or are bought out under the weight of urban renewal and gentrified leisure.

Our concern, then, is how to make sense of production, regulation and consumption patterns within urban nightlife infrastructures, with a focus on young adults' experience and use of particular spaces. Our emphasis on processes and the 'making' of nightlife circumvents, we hope, some of the rather unhelpful dichotomies used in understanding youth lifestyles, such as culturalism versus structuralism, objective versus subjective and material versus symbolic constructions of society (see Miles, 2000 for a discussion of some of these approaches). As such, throughout the book we use examples of how young people actively talk about, and make sense of, their social and spatial world and that of others in the night-time economy.

In this regard, first, it is fair to say that much youth cultural analysis has been implicitly aspatial in its orientation (for an exception, see Skelton and Valentine, 1998; Massey, 1998). Therefore, we seek to locate nightlife provision and youth experiences in a spatial context, both in our use of notions like mainstream, alternative and residual spaces, and also in terms of different national and local conditions. Second, with regard to understanding the relationship between 'circuits of culture', there are many ways to unpack consumer experiences, including nightlife. 'Horizontal' readings explore the relations and meanings circulated between consumers, while 'vertical' readings consider consumers as part of a commodity chain including production as well as consumption (Williams *et al.*, 2001). While we attempt to utilise both of these approaches, we give particular weight to the relationship between the production and regulation of nightlife, and its consumption by young adults. In this sense, we adopt a spatialised 'political economy' of youth cultural activity in the night-time economy, combined with a neo-gramscian perspective which stresses the interplay of dominant, residual and emergent tendencies (Williams, 1977). Political economy has been sorely neglected within the study of popular and youth culture generally (although see Fine and Leopold, 1993; Hollands, 1998), and traditionally much youth cultural analysis has focused on cultural resistance (Hall and Jefferson, 1976; Willis, 1990), postmodern hybrids (Muggleton, 1998) and the active making of lifestyles (Miles, 2000), without exploring this wider context of the changing role of the state and corporate strategies.

So, while we are sympathetic to elements of an 'agency-based' or 'experiential' approach (see Malbon, 1999), we also feel that a preoccupation solely with cultural creativity underestimates the material constraints in which consumers operate, as well as ignoring ongoing fundamental inequalities within the youth population. Clearly, young adults are actively involved in meaning-making in the night-time economy (see Malbon, 1999); however, at the same time many nightlife premises have become, to paraphrase Le Corbusier, 'gentrified machines for drinking in'. Similarly, it is important to stress the endurance of significant consumption divides within youth populations (Furlong and Cartmel, 1997) – between, for example, unemployed young people or those dependent on unstable employment, university students and those in high-level training, and young professionals in stable, well-paid and mobile employment – as well as to understand how social divisions have become more complex today. Our approach, then, is that wider processes of capital accumulation and restructuring, especially through the globalisation and corporatisation of the cultural industries (Klein, 2000; Monbiot, 2000), enduring social inequalities, and the changing role of the state, all are extremely influential in shaping modern-day nightlife experiences. Hence, to borrow from Marx: 'Young adults make their own nightlife, but not under conditions of their own choosing.'

Key concepts in understanding change and transformation within youthful nightlife spaces

Within this book, we draw upon a number of key concepts within social science research which will help the reader understand some of the processes of change and transformation shaping urban nightscapes and the lives of young people. The three chapters which comprise Part I of the book discuss these ideas and concepts in depth and illustrate them with examples. Here, in this first chapter, we briefly introduce and outline various key terms to help the reader locate this book in some ongoing and longstanding conceptual debates.

Our first context explores some of the productive forces underlying the emergence of a new entertainment economy. Service employment, and especially activity in the 'cultural economy', has grown rapidly in many cities to offset manufacturing loss. As a result of this shift, the entertainment and nightlife economies have become a central rather than add-on part of urban economic growth and employment. This transformation has been partly explained through ideas of *post-Fordism, in which there has been a saturation of Fordist mass markets, and subsequent changes in consumer preferences towards more individualised, reflexive and globally oriented forms of consumption*. These changes can also be set within a context of *economic and cultural globalisation, and a rationalisation and concentration of the organisation of production* across many sectors of the economy (Held *et al.*, 1999). Hence, in what follows we utilise a more *neo-Fordist approach which stresses the continuance of mass markets combined with a concentration of ownership and shift in control away from national ownership towards a small number of global corporate entities*. Economies of scale, standardisation and homogeneity, then, still play a key role for many global firms in terms of ordering consumer markets.

Branding and the use of themes have also become a central element for today's global firm, with value generated not solely from price or product differentiation but also through symbolic, lifestyle or brand differentiation (Gottdiener, 2001). Along with the use of branding has come the heavy stylising and scripting of consumer spaces around certain accepted, sanitised and safe norms and codes of behaviour. Further, branding and themes are used to create not only product loyalty but also consumer identity, social status and differentiation.

These changes in production and the economic organisation of entertainment and nightlife have taken place against a backdrop of *rapid urban change and restructuring over the last three decades*. After decades of decay and neglect created through economic restructuring (Harvey, 2000; Sugrue, 1998; Hudson and Williams, 1994; Taylor *et al.*, 1996), many traditional metropolitan and industrial centres have slowly been remodelled as places to live, work and be entertained. The specific characteristics of this 'return to the centre' are a renewed emphasis on the so-called knowledge-based economy, city-centre living and the idea of the 24-hour city, a greater economic role for corporately organised leisure and retail, and consumption-based (cultural) rather than production-based activities (Zukin, 1995; Hall, 1999). Clearly, each urban area has steered its own course through this reinvention process and has borrowed differentially from both the excesses of the North American model of casinos, multiplexes and malls (Davis, 1992; Hannigan, 1998) and the continental European model associated with 'café culture' and socially inclusive city-centre living.

These various urban reinventions, although successful in terms of reanimating and transforming the physical aesthetics of city centres, have done little to address questions of equality and access. Hence, *the contemporary urban entertainment economy is marked by social and spatial inequality and segmentation of consumer markets*. While both niche and mass brands are developed within entertainment and consumer markets, the key point is that both are increasingly controlled by a small number of global players who develop a portfolio of brands to dominate markets. Within both these markets there is a general pandering towards cash-rich groups of consumers and a tendency to create safe entertainment which offers 'riskless risk' (Hannigan, 1998; Hubbard, 2002). Here, *concepts of gentrification, stylisation and sanitisation are key* – that is to say, there has been a displacement of lower-order activities and working-class communities by higher-order activities aimed at cash-rich groups. Hence, within such urban transformations, it is important to note who gains and who loses, who is guiding urban nightscapes and to what ends, and who, literally, has been invited to the 'party'.

Our second context concerns the regulation and governance of nightlife. *Nightlife has been subject to much legal, political and indeed moral regulation*, fuelled largely by a longstanding anti-urbanism and a fear of crime and disorder, especially at night. Currently, governing the night involves a number of formal and informal dimensions which include legal (laws and legislation), technical (closed-circuit television (CCTV) and radio-nets), economic (drinks and door entry prices) and social–cultural (musical taste, youth cultural styles and dress codes) elements.

While there has been an erosion of the strict times and spaces of industrial work, where nightlife was carefully regulated to ensure that workers' leisure did not interfere with their productivity, contemporary nightlife is still subject to moral panics, regulation, rationalisation and planning. While there is a demand for a greater variety of nightlife compared to fifty years ago, *current nightlife developments point to the increased use of formal and informal surveillance and control techniques* (including CCTV, door entry policies, design, pricing) aimed at sanitising and controlling access to certain consumer groups. Hence, *urban nightlife contains a number of contradictory tendencies towards both deregulation and (re)regulation, and fun and disorder.*

Urban nightlife has also become an arena for a more complex set of negotiations between a range of groups. Traditional bodies (the judiciary and police), primarily concerned with social order and public safety, now increasingly have to negotiate with private capital and the more entrepreneurial local state with their imperatives of capital accumulation. Current nightlife developments backed by increasingly cash-strapped local urban authorities, have a tendency to benefit private capital rather than meet older notions within city planning such as civic pride, cultural diversity and universal access.

Our final context concerns the changing nature of consumption experiences, and in particular stresses that the social meaning of what it means to be young is constantly changing. *One of the fundamental shifts in the last two decades has been the extension of a youthful phase, as evidenced by terms like 'post-adolescence' and 'middle youth'* (Irwin, 1995; Goldscheider and Goldscheider, 1999; Roberts 1997), characterised by liminality and experimentation with youthful cultural activity for an extended period of time. By this we are referring to well-recognised delayed transitions into adulthood, marriage or full-time work due to staying on in education and training, increased dependency on the parental household, erosion of income benefits or student grants, and a changing labour market. The extended suspension of adult roles has meant that increasing numbers of youth are remaining at home into their twenties and even early thirties, which also implies that many have more disposable income for consumer spending. Marketing agents such as Mintel (1998) use phrases such as 'young adults' and 'pre-family adults' to reflect this extended period. In this book, we use such extended definitions of young people, and hence ethnographic material is drawn from young people aged between their late-teens and mid-thirties.

Moreover, the so-called process of *individualisation (where there is a presumed greater level of individual choice in terms of style and identity) and a greater global reach of consumer goods and media forms has fuelled a seemingly complex array of youthful, and not so young, lifestyles and identities* (Miles, 2000; Bennett, 2000; Epstein, 1998). More specifically, as traditional social relations and sites of identity for young adults are affected by social and economic change (Wilkinson, 1994a), *consumption and leisure, especially in cities, have become more central elements of youth identity* (Willis, 1990; White, 1999; Hollands, 1995). Re-imagined urban centres, then, have become important contexts in which young adults continue to deal with changes in their lifeworlds and forge roles and identities. Gender issues (McRobbie, 2000; Henderson, 1997; Wilkinson, 1994a), sexual orientation (Knopp, 1992; Whittle, 1994) and ethnic identities (Back, 1996) have also become more prominent within urban popular culture and nightlife.

The contemporary city at night, then, is often regarded as a 'stage' which acts as a backdrop for a diverse and varied collection of 'mix and match' youth styles, cultures and lifestyles (Redhead, 1997). Hence, many young people appear to be able to choose from a greater range of consumer goods and services and images than in the past. In this sense, *the city can be seen to offer abundant resources for experimentation and play, and opens up liminal and carnivalesque social spaces* (Featherstone and Lash, 1999; Shields, 1991). Many such postmodern readings of the urban explore the metaphors of play and hedonism rather than work and order.

However enticing such readings can be, it is important not to uncritically accept postmodern analyses of either youth consumption or urban change. Behind the seemingly fragmented and individualised patterns of consumption and underneath the 'free-floating' array of consumer goods and urban lifestyles, *differential transitions, inequalities and exclusions continue to assert their influence in both social and spatial terms on young people* (Hollands, 2002; MacDonald and Marsh, 2001). Hence, large numbers of young people remain excluded from pleasurable consumer spaces, with unemployment and poverty continuing to be a reality for many. The significant aspect for what follows in this book is that while urban nightlife is a popular pursuit for many young people, *some remain socially and economically excluded from new downtown entertainment playgrounds, while others choose to openly reject and contest it and build alternatives.*

Organisation of the book

The book explores the making of a 'night out' in two main parts. Part I develops in more detail the conceptual framework and context alluded to above for understanding urban nightscapes, and young people's involvement in them, by examining the production, regulation and consumption of different types of nightlife. Chapter 2 focuses specifically on ownership in the night-time economy, and looks at the role of corporatisation, branding and theming in creating a dominant nightlife infrastructure, while Chapter 3 examines the changing regulatory context behind nightlife activity. Chapter 4 focuses on a critical appraisal of theories of youth identities, cultures, lifestyles and transitions, and looks at how they aid an understanding of young adults' segmented nightlife consumption patterns.

Part II of the book empirically and ethnographically explores in detail how different youth groupings experience the three general spatial forms of nightlife activity – mainstream, residual and alternative – and looks at processes of production, regulation and consumption within each sphere. Chapter 5 explores some of the variations and complexities within the dominant mainstream, while Chapters 6 and 7 look at how specific sub-groups like students, young women and gay consumers are increasingly being drawn into the commercial mainstream. Chapter 8 examines the fate of residual youth groups and the demise of traditional nightlife spaces, while Chapter 9 looks at examples of alternative and resistant forms of nightlife on the margins. Finally, in the conclusion (Chapter 10) we explore the changing relationship between these processes and spaces, and suggest a number of potential future scenarios for youth and nightlife development beyond the corporate nightlife machine – including visions based on inclusion, diversity and creativity, rather than more limited notions of exclusion, social control and commodification.

Methodological note

At the outset, it is important to note that although this book attempts to provide a general theoretical treatment and analysis of urban nightlife trends, and seeks, wherever possible, to draw on a range of international examples and cases, like all empirically grounded social science research it reflects particular contexts, places and spaces. Further, due to the rapidly evolving nature of urban nightlife, the reader needs to keep in mind that many things will have changed since this book was researched and written, such as ownership patterns, regulatory laws, consumer trends, brands and venue styles. Nevertheless, we are confident that the themes and ideas we have drawn out can be used as tools for understanding some of the more general trends concerning the past, present and future of a night out.

The material presented in this book is the outcome of a number of years of thinking, researching and writing on youth cultures and cities by the authors, and a number of specific research projects including an examination of youth culture in a post-industrial city (Hollands, 1995), a PhD looking at youth and student cultures (Chatterton, 1998) and, most recently, a two-year research project undertaken jointly by the two authors looking at youth and urban nightlife (Chatterton and Hollands, 2001). This later project, funded by the UK government's Economic and Social Research Council, was based upon intensive case studies in several city centres in the UK, undertaken between 1999 and 2002 (principally Leeds, Newcastle upon Tyne and Bristol, with a smaller amount of additional material drawn from Edinburgh, Manchester and Liverpool).

Three principal cities in the UK (Leeds, Bristol, Newcastle) were chosen because they offered different backdrops against which to examine changes in youth culture and nightlife, including: diversity and reputation of cultural and nightlife infrastructure; character of urban and regional economic base; nature of past, current and future development trajectories; class and occupational structures; and local cultural traits and identity. In this sense, the former industrial city of Leeds represents a move towards a post-industrial corporate city which has been successful in attracting high-level business services and transforming its city core into a high value-added consumption, housing and leisure zone (Haughton and Whitney, 1994). It has developed a strong creative, independent nightlife alongside a growing corporate branded sector, which is increasingly utilised by a young affluent population who have benefited from the city's economic prosperity. In contrast, Newcastle represents a city struggling with the post-industrial transition (Tomaney and Ward, 2001) in terms of capturing new investment and encouraging service employment and high-level retail and leisure activities, despite having a rash of recent arts projects and attracting accolades for its bid to become the European Capital of Culture in 2008. As a result, Newcastle's nightlife, although changing (Chatterton and Hollands, 2001), continues to be marked more by established gender roles, local embedded working-class customs and a stronger regional ownership structure. Finally, Bristol represents one of the UK's 'sunbelt' cities (Boddy *et al.*, 1986) which has not been scarred to the same extent by industrial decline. Quality of life and cultural amenity is high, with medium income 25 per cent above the national average, a varied nightlife infrastructure, albeit with a strong corporate presence in the centre, and a substantial underground and subversive nightlife scene in the periphery.

During the course of this research into youth and urban nightlife in the UK, a number of techniques were used to gather material. First, an extensive survey was conducted in our three main UK cities – Leeds, Bristol and Newcastle – to gather base-line data on nightlife venues (defined here as pubs, bars, nightclubs and music venues) in the city centre area delineated by metropolitan police boundaries. Data gathered in these three cities included number and capacities of venues (pubs, bars, clubs, music venues), number of Section 77 special hours certificates (allowing bars to open past 11 p.m.); ownership of venues (whether international, national, regional or local); and style of venues (broken down into seven types: style bar, café-bar, traditional pub, ale-house, theme pub/bar, disco bar, alternative venue). This quantitative material is drawn upon at several points in the book in the form of charts and graphs.

More intensive, in-depth fieldwork was also conducted based around a number of largely ethnographic and qualitative methods. Most central here were eighteen focus groups in which over eighty young adults participated, ranging in age from 16 to 32. Here we use broader notions of youth, where, due to extended adolescence and delayed transitions into adulthood, being young includes those from their mid teens to their mid thirties. The focus groups represented a 'purposive' (Hammersley, 1983) rather than random sample, and were chosen to reflect a number of 'commonly discussed' nightlife groups, which emerged from pilot studies, previous research, conversations with those involved in nightlife, and current labels used by nightlife operators and marketing agencies. These groups included: gay men, lesbians and their friends; higher education students; young professionals and graduates; alternative/subcultural groups including Goths, rockers, skaters, squatters and participants in warehouse parties and raves; local working-class youths who had not attended further or higher education; and 'all-women' groups on a night out. Groups were contacted through a variety of means in order to avoid a straightforward 'snowballing' or 'known contacts' effect (May, 1993), and these included placing adverts in local listings press and contacts through venue managers, DJs, promoters and local music journalists, as well as through workplace managers of large firms and universities. Focus groups were generally held on neutral territory – in bars, cafés and pubs – where the participants felt more at ease to share their views. Following the initial focus groups, particular individuals were subsequently contacted and accompanied on a night out to help follow up research themes in a less formal setting.

Each focus group was taped and transcribed, providing over thirty hours of material. Semi-structured and open-ended questions pursued during the sessions included biographical information; information on types of nightlife participation; views on identities and groups in the city's nightlife; opinions on nightlife infrastructures, the role of the city council, police, venues; and areas of satisfaction and dissatisfaction. Our research aims during the focus groups were overtly stated, and confidentiality was preserved by changing all names.

Further, one-to-one interviews with forty-one producers of nightlife (owners, company directors, area and regional managers, bar managers, bar staff, journalists, DJs, musicians) and thirty-two regulators (various city council personnel, magistrates, police, doormen, residents' associations) were undertaken across the three cities. These were selected on a 'representative' basis – with, for example, all city-centre managers and police inspectors with responsibility for the city centre across the three locations interviewed (for comparative purposes) – and were chosen strategically to represent 'types' of producers (i.e. a cross-section of large corporate firms with a national and international scope, smaller independent micro-businesses and regional firms).

All this interview material, along with previous research and other written and overseas sources, was read, reread and analysed and used as the basis to formulate ideas and approaches and explore a number of nightlife groups and spaces which are set out in this book. Hence, quotations appear in the following chapters which were directly taken from the focus groups and interview transcripts. When we have used this material, names have been changed to preserve anonymity, but we have provided a small amount of biographical information for context. Hence, for quotes from the focus groups with young adults we have provided an assumed name, age and the city where the focus group took place; for the more formal one-to-one interviews we have labelled them sequentially, such as 'Bar manager 1', and provided the person's job position and the city, but for venue owners we have omitted the name of the venue to ensure anonymity. We have also included a number of boxed texts to provide more in-depth case-study material. Three of these boxes (5.3, 8.2, 9.1) recount our real experiences from a selection of nights out with our focus group participants, and here we use a more informal writing style. These portraits are intended to provide more ethnographic material, and reflect the pleasures, fears and aspirations of those we shared moments with during their actual nightlives. We locate these stories in the cities in which they occurred, but all names of people and venues and some of the details have been changed to ensure confidentiality.

While much of the empirical material and quotations in this book stem from this UK research project, we also felt it was important to provide a glimpse into how these processes we identified in the UK were working out in other parts of the western world. Hence, we use several examples of nightlife developments from cities in continental Europe and North America. Here, additional material was collected for this book through brief visits and fieldwork in North America and Spain (in Barcelona and the nearby renowned gay tourist destination of Sitges). The choice of locations largely reflected our existing research links and hence ease of access to material. In terms of Spain, the inclusion of Barcelona was part of an ongoing collaboration with the Youth Observatory of Catalunya, which provided opportunities and research material for a comparative glance at a European city much heralded as a first-class nightlife and cultural destination.

It is not the intention that this material should form the basis of in-depth comparative portraits, but it is offered so that readers in other geographical contexts can formulate their own comparisons with the situation and trends we outline in the UK. In particular, we invite students and researchers in North America, continental Europe and the Asia-Pacific region to use the material in this book to look at nightlife in their own locality and explore the similarities and differences they find compared with the UK situation to guide their own work and thoughts. We have also drawn upon our previous research and academic and journalistic writings, as well as various internet sources, to look at what was happening in other contexts. Hence, while our stories are flavoured by a particular national context and indeed specific localities, it is clear to us that some of the processes discussed in this book – such as increasing corporate control, economic mergers, and the standardisation and globalisation of much youth cultural activity – are fairly endemic to cities throughout the west.

Clearly, pursuing questions such as young people's habits, opinions and use of nightlife spaces raised a number of tensions and conflicts during our fieldwork. In particular, many research participants revealed sensitive and personal information about certain people, friends, venue owners or members of the police, judiciary and the local state during focus groups and interviews. Many personal grudges, past arguments, likes and dislikes also emerged. In places where a number of research participants clearly knew one another, we had to use and report such research findings very carefully to avoid inflaming tensions. Nevertheless, all of our findings were disseminated in our case-study cities through seminars, written reports and on the web (www.ncl.ac.uk/youthnightlife). Additionally, while researching the night may initially seem inviting and exciting, there were a number of pitfalls, not least the unpredictability of making arrangements with those directly involved in nightlife (young consumers, DJs, bar owners, etc.), gaining trust and access with groups such as the police, and issues of security for the research team.

Finally, while this is an academic project, the ideas in this book have also been developed through our own lived experiences, participation and perceptions of what makes a night out, as well as through direct experience and interaction with the many people who took part in our research (ranging from producers to regulators and consumers). We have been keen to find a balance between letting our nightlife participants speak for themselves – using, for example, direct quotes and recounting experiences from nights out – and drawing upon theory to make interpretations about what they have said. In sum, what follows in this book reveals some important questions to be addressed which relate to people's daily and nightly experiences of the city. We realise that, in translation, something of the excitement and intangibility of a night out will be lost. However, we hope that we have adequately captured something of that experience, and the urgency of the issues presented, in the many voices expressed here.

Part I

UNDERSTANDING NIGHTLIFE PROCESSES AND SPACES

Producing, regulating and consuming urban nightscapes

2

PRODUCING NIGHTLIFE

Corporatisation, branding and market segmentation in the urban entertainment economy[1]

The Ministry of Sound, which started life as a London club in the early 1990s, now has the largest global dance record label and the most popular music website in the world, and publishes the UK's fastest growing music magazine, *Ministry*. It promotes club events around the globe and broadcasts its radio show on fifty stations in thirty-two countries, thus becoming the world's most famous dance and clubbing lifestyle brand for young people. Universal Studios Japan, which opened in Osaka in 1999, combines theme park rides with media studios creating a modern-day 'symbiotic' media production and consumption entertainment destination (Davis, 1999: 439). On 57th Street West Manhattan, NikeTown, Nike Corporation's flagship retail outlet, is described by Klein (2000: 56) as a hallowed shrine to the heroic ideals of athleticism rather than simply a sporting goods shop, with its three-storey-high screens and famed sports memorabilia. And Mythos, a theme park based on Greek mythology which opens in 2004 to coincide with the Olympic Games in Athens, comes complete with rides, mythological figures and wandering minstrels (Emmons, 2000).

These are just a few examples typifying the world-wide spread of the entertainment and leisure industries (Gottdiener, 2001; Sorkin, 1992) and the emergence of an economy rooted in an infrastructure of themed restaurants and bars, nightclubs, casinos, sport stadia, arenas, concert hall/music venues, multiplex cinemas and various types of virtual arcades, rides and theatres (Hannigan, 1998). While it is clear that popular culture has long played an important role in cities, we would argue that the current urban entertainment economy is distinguishable by a concentration of corporate ownership, increased use of branding and theming, and conscious attempts to segment its markets, especially through the gentrification and sanitisation of leisure activities. The night-time economy, especially through the growth of up-market style and café-bars and nightclubs, has a key part to play in contemporary entertainment infrastructures.

Our approach in this book is concerned with how corporate control in the urban entertainment and nightlife economies is usurping and commercialising public space, segmenting and gentrifying markets and marginalising historic, alternative and creative local development. In this first chapter, we outline the emergence of a dominant mode of urban entertainment and night-time production, and situate it within critical discussions concerning the transformation from Fordist[2] to post-Fordist production and the related shift from mass to more segmented and varied forms of consumption (Kumar, 1995; Amin, 1994), the move from the welfarist to the entrepreneurial state and city (Harvey, 1989b) and the growing globalisation and corporatisation of economic activity (Held *et al.*, 1999; Monbiot, 2000; Klein, 2000). Beyond the rather seductive argument surrounding 'flexibility' and post-modern consumption, we stress a more neo-Fordist interpretation of the nightlife industry, characterised by some novel features but also by a continuation and intensification of concentration and conglomeration of ownership, a lack of real consumer choice and diversity in spite of increases in designs and branding, and continued social and spatial segregation due to market segmentation. This dominant mode of production is displacing older, historic modes of nightlife based around the community bar and pub (Mass Observation, 1970) connected largely to Fordist forms of collective consumption in the working-class industrial city, and marginalising more independent modes of nightlife associated with various alternative youth and subcultures (McKay, 1998). What we highlight throughout this book are the implications of this shifting balance between different modes of nightlife production.

Understanding the urban entertainment and night-time economy

The boom in the urban entertainment economy is well documented beyond the obvious visual transformations of city landscapes. Scott (2000), for example, claims that over three million Americans work in the 'cultural economy',[3] Hannigan (1998: 2) argues that jobs in the entertainment industry in California now surpass those in the aerospace industry, while it has been calculated that 'fun' services grew by over 7 per cent in the USA between 1960 and 1984 (Esping-Andersen, 1990). Davis (1999: 437) meanwhile states that entertainment is one of the hottest sectors in real-estate circles, and Gottdiener *et al.* (1999: 256) have referred to the spread of its ethos and architecture in the USA as 'the Las Vegasization of city downtowns'.

In the UK, the leisure sector has become an important job creator, employing nearly 1.8 million people, or 8 per cent of the workforce, a figure which has more than doubled since the 1930s (Gershuny and Fisher, 2000: 50). There has also been a steady influx of Urban Entertainment Destinations (UEDs) in downtown areas in the UK bringing together cinema, retail, eating and nightlife and drawing upon anchor tenants such as Warner Brothers, TGI Fridays, Starbucks coffee shops, Hard Rock Cafés, Planet Hollywoods and Disney Stores. The most famous among them is the Trocadero in London's Piccadilly Circus, an entertainment and retail destination comprising global brands such as UGC Cinemas, Planet Hollywood, Bar Rumba and the Rainforest Café. DLG Architects have built a new wave of urban entertainment complexes including the Light in Leeds, heralded as a 'whole new city-centre experience', the Great Northern Experience in Manchester and Broadway Plaza in Birmingham, all comprising multiplex, family entertainment centres, health and fitness suites, bars, restaurants and residential and retail uses. In Spain, our opening discussion of Port Vell in Barcelona is one of the most well-known examples among the growing number of 'waterside leisure areas' composed of shops, bars/clubs, restaurants, hotels, IMAX theatres and marinas/aquariums, while China has seen the building of forty-one theme parks over the last decade (Hannigan, 1998:2).

While there are a host of examples and strong empirical support documenting the growth and development of these new urban 'landscapes of consumption', especially shopping malls (Wrigley and Lowe, 1996; Shields, 1992; Goss, 1993; Connell, 1999), the night-time economy has received far less attention. Despite the fact that much of the current entertainment economy is being fuelled by the growth of night-time activity, very little work has analysed the transformation of many cities into 'nightlife hotspots' (although see Chatterton and Hollands, 2001; Chatterton and Hollands, 2002). To aid us here, the current development of the entertainment and nightlife economy can be theoretically situated with reference to wider economic, political and socio-cultural changes characterised generally under the rubrics of Fordism, post-Fordism and neo-Fordism (Kumar, 1995; Amin, 1994; Harvey, 1989b; Lash and Urry, 1987). More specific discussions about flexible specialisation and accumulation (see Piore and Sabel, 1984; Harvey, 1989c), a growing literature on (anti) globalisation and corporatisation (Held *et al.*, 1999; Monbiot, 2000; Klein, 2000), the move towards a service-based, cultural and 'symbolic' economy (Lash and Urry, 1994), changes in the local/welfare state and the rise of the entrepreneurial city (Harvey, 1989b; Jessop, 1997; Burrows and Loader, 1994), and critiques of post-modern consumption (Hollands, 2002; Warde, 1994), especially in relation to market segmentation, gentrification and branding: all these are useful components of this wider debate, which we discuss below.

Fundamentally, the post-Fordist transition refers to changes in the production process, although this clearly implies broader political and cultural transformations (Kumar, 1995: 37). Rooted in the development of new types of small-scale, flexible, specialised, integrated and high-tech production units clustered in industrial districts (epitomised through the 'Third Italy' phenomenon of the 1970s and 1980s; see Goodman *et al.*, 1989; also Piore and Sabel, 1984), it is viewed by some as a new stage of capitalist and political organisation (Lash and Urry, 1987). Responding to the inflexibility and saturation of national mass production markets, post-Fordism is supposedly characterised not only by more flexible production techniques, but also by organisational changes like the decentralisation and globalisation of capital, outsourcing and subcontracting, a decline in the function of national welfare states, and changes in consumer preferences towards more individualised forms of global consumption (Kumar, 1995; Urry, 1990; 1995). In effect, crisis-ridden western capitalist economies since the 1970s, faced with declining growth, dis-investment of material production and 'manufacturing flight' to lower-cost locations (Held *et al.*, 1999; Massey and Allen, 1988), have sought new avenues of wealth generation. Service employment, especially business and financial services, and increasingly activity in the 'cultural economy' (Scott, 2000), have grown rapidly to offset manufacturing loss. In its constant search for new profit areas, then, 'capitalism itself is moving into a phase in which the cultural forms and meanings of its outputs are becoming critical if not dominating elements of productive strategies' (*ibid.*: 2).

Flexible accumulation, one particular take on the post-Fordist transition (Harvey, 1989a), is based upon the assumption of increased flexibility, not only in relation to the labour process but also with respect to types of products, services and markets. In this sense, Kumar (1995) points out that saturation of markets for mass goods, the exhaustion of groups of mass consumers and the dictates of new styles of life, along with ceaseless technological innovation, have all resulted in a rapid turnover and swift changes in production. Such flexibility suits firms, too, as they eagerly search for ways to exploit and expand new markets. Not surprisingly, there has been a shift towards investment in, and the marketing of, different types of products and services in the cultural economy. To quote Harvey (1989c: 285): 'If there are limits to the accumulation and turnover of physical goods ... then it makes sense for capitalists to turn towards the provision of very ephemeral services in consumption.' Perhaps more accurately in terms of the production of entertainment, Lash and Urry (1994) point to the emergence of 'reflexive accumulation' in which the accumulation process is based around more knowledge- and service-intensive activities and a concentration on signs, symbols and lifestyles, rather than just material goods.

In general terms, under this model, production is more knowledge-intensive and involves small-batch tasks undertaken within dense networks of vertically disintegrated units. Capital flows towards the production of goods and services that are more ephemeral and spectacular (a live music concert, a casino), disposable (beer, fast food), lifestyle-based (branded venues and products, including 'premium' brands) and even 'virtual' (internet, virtual reality parks, computer games). In other words, 'fast-moving consumer goods and services', as they are known in the business world (du Chernatony and Malcolm, 1998) – epitomised by entertainment, popular culture and nightlife activity – require constant replenishment and are a particularly effective tool for speeding up capital accumulation. Moreover, while cultural and entertainment products might initially involve high start-up costs, reproduction and distribution can generate economies of scale and almost ceaseless profits (Scott, 2000). Entertainment, and especially nightlife, for example, involves a temporal expansion of capital accumulation past the typical retail 'flight', encouraging late-night activities catering for pre-family young adults, students and tourists. Finally, multifunctionality is now the cornerstone of many leisure developments (Gottdiener, 2001: 101), developing synergies between retail, media, real estate, sports, nightlife, dancing, eating and other pursuits. Such multifunctionality requires new spaces for profit-making, like UEDs (see Plate 2.1), which often combine theme bars and restaurants, cinemas, arcades, internet cafés, retail outlets and licensed merchandise shops, generating entertainment hybrids such as edutainment, eatertainment and shopertainment (Hannigan, 1998).

While the Fordist/post-Fordist typology is a useful 'ideal type' for an analysis of the entertainment and night-time industries, there are a number of important caveats and reservations around the idea of a linear and unfettered transition. First and foremost, it might be argued that some versions of the transition have overstated the flexibility argument and have mistakenly assumed that standardised mass markets have indeed been exhausted under capitalism (Fine, 1995: 136). There is clearly a degree of continuity within this restructuring, whereby post-Fordism is a reworking of earlier mass systems of Fordism (Aglietta, 1979). Hasse and Leiulfsrud (2001: 111) have recently written that 'flexible modes of production are predominately integrated into established forms of mass production', while Kumar (1995: 58) notes that for the transnational globalisers, the 'global standardisation of Dallas and McDonald's can co-exist quite happily with the artificial diversity of Disneyland and the manufactured localism of the heritage industry'. Present-day consumer markets are characterised as much by the non-differentiated mass production of standardised

Plate 2.1 Urban Entertainment Destination at Fountain Park, on the fringe of Edinburgh's city centre; Scotland, 2002

goods by certain global producers as they are by a preference for non-mass, specialist goods, new consumer lifestyles and greater aesthetic rather than functional consumption patterns (Urry, 1990: 14; 1995: 151). As such, it might be argued that elements of Fordist production remain alongside more differentiated, post-Fordist forms in many parts of the entertainment economy, including the alcohol/brewing and nightlife sectors.

In this regard, one notable trend in the current urban entertainment economy has been the continuing shift in ownership and control away from national entities and more locally grounded collections of self-made entrepreneurs towards a small number of global corporate entities (Hannigan, 1998). Clearly, this trend should not imply that monopolies did not exist historically under Fordist mass entertainment production. Although the contemporary processes of globalisation and market concentration vary across space and are tempered by national and international regulatory frameworks, their impact on everyday culture and entertainment activities is increasingly visible. Some commentators have recognised that large corporations themselves have begun to take on aspects of post-Fordism, including decentralisation and flexible specialisation, alongside standardisation and market domination (Kumar, 1995: 44–5). As Held *et al.* (1999: 158) comment: 'in the post-war era every sector of the communications and cultural industries has seen the rise of larger and larger corporations, which have become increasingly multinational in terms of their sales, products and organisation'. For example, while a small number of major global entertainment entities such as Time Warner, Sony, Viacom, Disney, Bertelsmann and NewsCorp have come to play a key role in production, more importantly they now also play a central role in the distribution of cultural forms (Held *et al.*, 1999; Scott, 2000) and so dominate everything from the conception to the consumption of cultural goods and services. Ultimately, then, the goal for most global corporations is market expansion and domination, and hence global mergers, synergies and cross-promotions abound. With the erosion of anti-trust and anti-monopoly laws across the USA and Europe, such large companies increasingly have a free hand in directing and controlling entertainment across the world.

Beneath the spectacle and carnivalesque atmosphere of entertainment and the production of individualised niche markets, then, lurks an increased concentration and conglomeration of ownership by a small number of large corporate firms. As Klein (2000: 130) suggests, 'despite the embrace of the polyethnic imagery, market-driven globalisation doesn't want diversity; quite the opposite. Its enemies are national habits, local brands and distinctive regional tastes.' The globalisation of the entertainment industries and products is a reminder that the logic of capital is still based at least partly on economies of scale, standardisation and homogeneity (Kumar, 1995: 188–9; Ritzer, 1993). De-nationalisation and de-localisation of entertainment, in conjunction with concentration of ownership, is thus a central feature of this transformation. While global players seek to create the impression that they are sensitive to local and national contexts with their language of 'global localities', the overall effect, however, is often one of 'serial' monotony or reproduction (Harvey, 1989b), with the majority of cities and regions around the world adopting a familiar approach in their creation of an entertainment economy infrastructure.

One of the most obvious features of the current entertainment and nightlife economies, then, is that they have become highly branded, theme-centric and stylised (Hannigan, 1998; Gottdiener, 2001; Chatterton and Hollands, 2002). 'To brand' originally meant 'to burn', or to mark a product in some way to distinguish it from other similar items (Stobart, 1994: 1; Murphy, 1998: 1), while 'theming' refers to entertainment venues being rolled out across the globe accorded to a scripted idea, such as the all-pervasive Irish pub or the Las Vegas-style casino. While branding is not a new phenomenon – Bass beer is often considered to be one of the first product brands, created in 1876 (Stobart, 1994: 3) – it has developed over the last hundred or so years into a 'business process' which is designed to exemplify a company's core essence. There have been crucial changes in the nature of branding over the past fifty years which are important for our later discussion of nightlife. For example, with the move from a manufacturing to a service economy, there has been a corresponding shift towards the branding of services and images, and not just products. In this sense, the physical elements of commodities have increasingly given way to 'intangible' or 'product-surround' qualities – i.e. aesthetic and emotional elements (Hart, 1998).

Branding has also become an international phenomenon, with numerous successful 'power brands' (Murphy, 1998) emerging across the world aided by the impact of global marketing and advertising. A whole host of corporate strategies are used to manage the ever-increasing global portfolio of consumer brands. Globally recognised brands are bought and sold daily on international stock markets, often signalling significant modifications to, or the end of, well-known brands, but at other times equating to few changes. What is clear is that the world's largest branders have busily been divesting from material production and shifting it overseas to cheap-cost locations to concentrate on high-value added activities such as marketing, advertising and branding rather than making products. Brand value, not necessarily the financial stability of a company, is increasingly important in today's climate of corporate mergers and takeovers, as evidenced by Interbrew's recent purchase of Beck's Beer for £1 billion because it was seen as a 'good brand' (Clark, 2001). In late capitalist economies awash with consumer goods, then, surplus value is generated not from price or product differentiation, but rather through symbolic or brand differentiation (Gottdiener, 2001).

Branded entertainment spaces draw heavily upon design and stylisation. As Julier (2000) points out, a 'culture of design' pervades not just leisure and retail spaces but the fabric and image of whole localities. Design involves a complex mesh of symbolic, material and textual factors, with symbolic aspects taking on a particularly important role in today's ephemeral culture (Lash and Urry, 1994). Designers now play a key role in the development of branded, stylised spaces with design geared increasingly towards attracting desirable consumers, repelling undesirable ones and maximising consumer spending. Mainstream commercial spaces are designed environments which connect with widely held social and ideological values and the desire of particular social groups to distinguish themselves through not only material but symbolic or positional goods (Veblen 1994; Bourdieu 1984). Branding, then, is designed to create not only product loyalty but consumer identity, social status and differentiation (Klein, 2000), especially in relationship to other style groups (Julier, 2000: 98). In this sense, there are variations in the form of design within the mainstream, correlating to different taste groups within the city which cluster around particular environments, each with their own ambience, design and social codes.

The current entertainment economy has also flourished due to political processes and regulatory responses by the national (Burrows and Loader, 1994) and local state to changes in the global economy and shifts in production (Harvey, 1989b). The 'return to the urban centre' or 'downtown' (O'Connor and Wynne, 1995; Harvey, 2000) is underpinned by a belief that the revitalisation of core areas of old industrial cities is crucial for economic renewal. This has resulted in a fundamental rethink of the role of the local state, in particular, chronicling a shift in its historic managerial and welfarist functions towards aiding urban regeneration via property development, deregulation and encouraging corporate inward investment (Jessop, 1997). Along with a renaissance of city-centre employment and housing markets, cultural or creative industries have been used in the economic and symbolic rejuvenation of local economies throughout the west in the wake of manufacturing decline (Williams, 1997; Hall, 1999; Pratt, 1997; Scott, 2000).

Entertainment and nightlife activities have become central components of this economic restructuring process and have provided many localities assumed escape routes to offset decline in the local economy (Chatterton and Hollands, 2002). While the city at night has historically been regarded as the shadowy 'other' of the working day as a place for marginal, crime-ridden and liminal pleasures (Lovatt, 1995), since the 1980s nightlife and a host of popular cultural activities, often promoted through the idea of the '24-hour city', have become an accepted part of urban growth (Lovatt, 1995; Bianchini, 1995; Heath and Stickland, 1997). As a result, a raft of public subsidies, public–private partnerships and regulatory changes (see Chapter 3) have emerged, not only to help kick-start the urban housing and office markets, but also to develop cultural, night-time and entertainment facilities (Harvey, 2000). Numerous commentators have noted that such developments, as well as having had a tendency to aid private capital, are stylistically partial to catering for cash-rich groups at the expense of more locally grounded economic development and the needs of the urban poor (Smith, 1996; Harvey, 1989b).

The rise of the entertainment economy also parallels changes in traditional sites of identity formation such as the home, work and the church and the rise of new consumer identities in the mall, stadium, nightclub and bar (Sennett, 1998; Lash and Urry, 1994; Rojek, 1995). Young adults have a particular role to play here, as they are often identified most strongly with the changing relationship between work and leisure, and the growing demand for specialised lifestyle goods and services (Miles, 2000; Roberts, 1997). Part of this relates to changes in the economy which have resulted in extended youth transitions – exemplified by higher rates of unemployment and terms like 'post-adolescence' and 'middle youth' – and involvement in nightlife and entertainment for much longer periods of time (Hollands, 1995).

Additionally, elements of the post-Fordism paradigm have drawn attention to the move away from mass to more individualised forms of consumption (Hall and Jacques, 1989; Urry, 1990), as well as the rise of new social identities in relation to gender, sexuality, ethnicity and more specific youth cultural and hybrid forms (Muggleton, 1997). However, in this regard it is also important not to overstate the flexibility thesis in cultural terms (i.e. the 'cultural turn'), and simply read off a particular set of more differentiated postmodern consumption practices from a supposed more flexible mode of production (see Warde, 1994, for example). As we argue in Chapter 4 (also see Hollands, 2002), there remain significant cleavages in the youth population between highly paid young professionals, those in lower-level service and manual work, and a section either permanently unemployed or in unstable employment (Ball *et al.*, 2000), not to mention significant gender, sexual (see Chapter 7) and ethnic divisions (see Chapter 8), which continue to contour nightlife destinations and spaces. In fact, despite an apparent opening up of markets to new groups of consumers (women, students, gay and ethnic groups) it is increasingly the cash-rich, middle-class factions of these populations that are the industry's favoured consumers of entertainment and nightlife facilities (Hannigan, 1998; Wynne and O'Connor, 1998; Chatterton and Hollands, 2002).

This brings us to our final feature of the contemporary urban entertainment economy – the social and spatial segmentation of consumer markets. As Christopherson (1994: 409) suggests:

> The signal qualities of the contemporary urban landscape are not playfulness but control, not spontaneity but manipulation, not interaction but separation. The need to manage urban space and particularly to separate different kinds of people in space is a pre-eminent consideration in contemporary urban design.

Market segmentation (both socially and spatially) exists in tandem with the emergence of standardised globally branded products and services, exploiting economies of scale as well as scope (Kumar, 1995: 190). The development and co-existence of niche and mass brands is particularly important for understanding tendencies towards segregation in the entertainment economy. Urry's (1990) discussion of the travel industry is a good initial model here. For example, he outlines how differences in historic ownership patterns in the industry – between, for example, smallholders and the landed classes – have determined the social tone of consumption destinations, differentiating mass/cheap from elite/niche travel destinations. Urry (1990) further elaborates on these different consumer markets through the duality of the 'collective gaze' of the sociable working classes and the 'romantic gaze' of the more detached middle classes. Hiding major social divisions beneath trendy lifestyle categories, the entertainment sector, then, is in reality carved up between mainstream and premium lifestyle provision, with a general pandering towards middle-class taste, 'riskless risk' (Hannigan, 1998), and conscious attempts to sanitise and exclude the poor and disenfranchised (Toon, 2000; Ruddick, 1998; MacDonald, 1997; Sibley, 1995).

City-centre gentrification (see Smith 1996; Ley, 1996), traditionally conceived through changes in the housing market, has become increasingly concerned with the production and consumption of urban social and spatial differentiation.[4] Smith (1996: 114) argues: 'Gentrification is a redifferentiation of the cultural, social and economic landscape, and to that extent one can see in the very patterns of consumption clear attempts at social differentiation.' In his terms, it is the 'class remake of the central urban landscape' (Smith, 1996: 39). Gentrification is also tied up with economic, social and cultural restructuring, broadening its focus to include the cultural and aesthetic infrastructure necessary to support different lifestyles and identities in the new urban economy (Zukin, 1988).

Hence, the rise of incomes among wealthy city-livers and the urban service professional class (Ley, 1996; Savage and Butler, 1995; Butler, 1997) has stimulated demand for gentrified and 'safe' entertainment, on top of mainstream and commercially oriented provision for those in more routine and lower-order service jobs seeking weekend escapism and 'hedonism in hard times' (Redhead, 1993). Additionally, when many previously marginalised groups such as women, gay and ethnic populations are brought into the arena of entertainment consumption, they are either absorbed into mainstream or gentrified sectors (see Chapter 7; also Whittle, 1994; Chasin, 2001) or simultaneously separated into entertainment ghettos such as gay villages, ethnic entertainment zones, women-only nights, etc. In contrast, unemployed, low-income and welfare-dependent groups literally have no space here. Gentrified spaces, then, not only reaffirm existing structures in the labour market (Smith, 1996), they also hide the 'dirty' back regions of entertainment production by constructing the illusion of a wealthy urban oasis (Zukin, 1995).

These branded and segmented entertainment markets have spatially encroached into the everyday urban public realm of 'the street' (Klein, 2000). One of the most significant markers here is the transformation of abandoned, ageing architecture into leisure and consumption destinations, a notable recent trend being the conversion of banks, churches, schools and hospitals into restaurants or large chain bars (see Plate 5.4). Industrial buildings once rooted in the fabric of working-class and community life have become the infrastructure for a new class of high-income pleasure-seekers and city-livers (Zukin, 1992). Increasingly, the shapers of these new urban spaces are multinational media and entertainment conglomerates who have shifted their emphasis to making 'places' as much as making products (Davis, 1999) – so much so that some commentators have begun to talk about the creation of urban 'brandscapes' rather than community landscapes (Hart, 1998), subsequently involving the squeezing out of what one might refer to as 'unmarketed cultural spaces' and dispossessed groups in cities (Klein, 2000: 45; Sibley, 1995). The dominance of this urban entertainment economy, and how it specifically relates to the nightlife industries, is the focus of the next section.

Corporatisation, branding and market segmentation in the nightlife industries

In the first section of this chapter, we have sought to generally theorise the emergence of the entertainment and night-time economy and reveal some of its general features, including transformations in its mode of production, increased concentration of ownership, and use of branding/theming and market segmentation. The following section empirically examines the emergence of this dominant pattern with specific reference to urban nightlife in various contexts, and examines some of the implications for older historic and newer alternative forms of production.

Nightlife restructuring and corporate concentration

One of the most striking features of the current mode of nightlife production is the shift towards the concentration of corporate activity across a number of areas such as alcohol manufacturing, venue ownership and product distribution. Despite being a global trend, concentration and restructuring takes on varied forms in different locations (see Held *et al.*, 1999). Of particular importance here are national regimes of production and regulation which have historically shaped economic sectors such as brewing and the ownership of nightlife venues.

In terms of brewing, the trend towards global concentration continues at a rapid pace. World-wide, around thirty big brewery companies currently account for two-thirds of the beer produced (European Commission, 2001: 1). Moreover, trade in hops has become concentrated in the hands of two major groups over the last four years, accounting for 40 per cent and 30 per cent respectively of the total world market in hops (*ibid.*: 3). A number of large firms have grown from their home markets to dominate large geographical regions. These include Anhäuser–Busch

(A–B) the world's largest brewer, Adolph Coors Co. and Miller Brewing Co. (owned by Philip Morris) in the USA, AmBev in South America, Kirin/Lion Nathan in the Asia-Pacific, South African Breweries in Africa, and Scottish & Newcastle (S&N), Heineken, Carlsberg and Interbrew in Europe (see Table 2.1).

Interbrew, the second-largest brewer in the world after a spate of acquisitions, is particularly intent on market domination and has an unashamed goal of beating America's Anhäuser–Busch to be the world's largest brewer. Interbrew's products are sold in 110 countries, and in the highly competitive UK market it is already the market leader. Over the 1990s, Interbrew entered a phase of rapid expansion, and completed thirty acquisitions and strategic joint ventures, the largest of which were Labatts (Canada), Oriental Breweries (South Korea), SUN Interbrew (Russia) and Bass Brewers and Whitbread Beer Company (UK). Interbrew's recent acquisition of the German family-owned Becks company for £1.1 billion signals its ongoing commitment to gaining a foothold and expanding in the world's most prosperous alcohol markets. The company motto, 'the world's local brewer', demonstrates its desire to be both global and local and its commitment to the rather awkward goal of

Table 2.1 World's biggest brewers: by sales volume, 1999

Rank	*Company (country HQ)*	*Million hecto litres*
1	Anhäuser–Busch (USA)	154.7
2	Interbrew (Belgium)	79.6
3	Heineken (Netherlands)	74.0
4	Ambev (Brazil)	59.0
5	South African Breweries (South Africa)	55.2
6	Miller (USA)	51.8
7	Kirin/Lion Nathan (Japan/New Zealand)	39.7
8	Carlsberg (Denmark)	37.0
9	S & N/Kronenburg (UK/France)	33.0
10	Coors (USA)	30.0

Source: Bilefsky, 2000

the 'glocalisation' of its markets. Although the beer sector is increasingly concentrated, what is also evident are highly intricate webs of interrelationships, cross-investments, collaboration and competition which have been woven by the leading international beer companies in their fight for geographical domination (Bellas, 2001). Recent mergers between, for example, Antarctica and Brahma in South America and Interbrew, Bass and Whitbread in Europe are evidence of this.

In terms of both brewing and the ownership of venues, levels of concentration vary considerably between countries. In the USA, the production of beer is highly concentrated, with A–B, Miller and Coors accounting for 80 per cent of the industry's shipments in 1997. A progressive tax is in place in the USA which has helped to create a microbrewing industry with a $1 billion turnover, and 'brew pubs' are a distinctive part of the North American mode of nightlife production (see Box 2.1). However, the 1,610 microbreweries, brew pubs and regional speciality brewers (known together as 'craft brewers') hold only 3.0 per cent of the market share in the USA and face intense competition from large national brewers (Institute for Brewing Studies, 2001). In particular, the US Department of Justice has looked into allegations that A–B is engaging in unfair sales and distributions practices, when three California micro-breweries allegedly filed a class action suit against the company for using its large market share to coerce independent wholesalers into dropping smaller brands (All About Beer, 1997). A–B is also engaged in micro-brewing in order to gain a share in the growing independent market.

Box 2.1: North American nightlife ownership patterns

The ownership of nightlife venues in North America differs somewhat from the UK situation where, historically, brewers controlled the majority of pubs, either through direct ownership or through a 'tied' system whereby the tenant leased the premises and had to stock a certain percentage of beers and ales made by their brewery landlord (see Mason and McNally, 1997, and the subsequent discussion below). In contrast, control of liquor licensing in the USA and Canada rests with individual states and provinces, and various examples here suggest that regulations generally prohibit drinks manufacturers from gaining access to a licence to sell to the public, with some exceptions (i.e. brewpubs). Indeed, in New York state, the Alcohol Beverage Control Law says that 'no brewer shall sell any beer, wine or liquor at retail',[5] while the Liquor License Act of Ontario states that a licence to sell liquor shall not be issued to a manufacturer, or any person likely to promote the sale of liquor or to sell the liquor of a manufacturer exclusive of any other manufacturer.[6] As such, nightlife ownership in the USA is more diffuse and is characterised by a mix of: corporate chain 'bar-restaurants' in which a number of activities such as seated drinking, food and entertainment are mixed (see Gottdiener, 2001; Hannigan, 1998); independent operators who have promotional deals with the major breweries; and brew pubs, which combine making and selling their own beer but also offer a full range of commercial beers as well. All three types invariably offer food as well as drink. Despite this mix, the general pattern of dominance by large chain bar- restaurants, like Hard Rock Café, Planet Hollywood, ESPN Zone, Hooters and Dave & Buster's, in many cities around North America (and the activities of the Firkin Group of Pubs and Prime Restaurants in Canada) is comparable to some of the branding and theming patterns which we note in the UK. Similarly, as we outline in Chapter 5, parts of North America are also currently experiencing a growth in themed English and Irish pubs in both the independent and corporate chain sector of the industry. (see also Box 5.2)

Germany, the world's third largest beer market and a country internationally renowned for its beer and drinking cultures, still has a diffuse pattern of ownership and an unconsolidated beer market. This is underpinned by dynastic family control of beer production and a system of 'progressive beer taxation' which enables small companies to survive alongside larger ones. As a result of this tax, Germany is home to 1,270 breweries, accounting for three-quarters of all the beer production sites in the European Union. However, the centuries-old traditions associated with small-scale family brewing are being eroded by the imperatives of the global market. In 2001, the family-dominated board of the Becks company decided that the company was too small to compete internationally. The subsequent sale of Becks to Interbrew for £1 billion signals the beginning of the end of Germany's position as the last bastion of independent local brewers (Clark, 2001). Similarly, in Belgium the number of breweries has fallen from 3,223 at the start of the twentieth century to 115 by 2000. These fewer remaining companies, rather than focusing upon the domestic market, are focusing on export. For example, in 1960 Belgian breweries exported 2 per cent of their beer production, whereas by 2000 Belgian brewers exported 37.5 per cent, nearly five and a half million hecto litres (The Confederation of Belgian Breweries, 2001).

The British story is interesting due to the considerable ongoing consolidation of ownership in both the brewing sector and the ownership of nightlife venues. The watershed event was the 1989 Monopolies and Mergers Commission Report, which concluded that a complex monopoly existed in the brewing industry largely as a result of high levels of vertical integration, in which brewers owned everything from production to the point of sale (Mason and McNally, 1997). At this time, 88 per cent of public houses were either managed or tied as tenanted houses to a small number of large breweries. This report led to the Supply of Beer Orders Act, which aimed to break the monopoly ownership of the national brewers by restricting the 'tied house' system so that no brewer could own, lease or have any other interest in more than 2,000 pubs; in addition, at least one guest beer should be sold, and loan tying should be abolished (Mason and McNally, 1997: 412). As a result of this legislation, most large national brewers sold off large stocks of public houses to come within these limits or divested from brewing altogether to get around the limits on pub ownership imposed upon them. However, the Act was never fully implemented, as breweries only had to release ties on half the pubs held over the 2,000 limit, and the loan ties were never completely abolished (*ibid.*). Since then, the brewing and pub-owning sectors have grown increasingly functionally separated, and there has been an acceleration of mergers, concentration and rationalisation within both. In terms of brewing, while in 1930 there were 559 brewery companies in Britain, by 1998 there were only 59 (BLRA, 1999). By 2000, Scottish–Courage remained the only national-level brewer with annual beer sales in excess of £2 billion (Ritchie, 1999) and alongside Interbrew, Carlsberg–Tetley and Guinness, these four super-brewers continue to control 81 per cent of beer sales in the UK.

More significant has been the restructuring of the ownership of nightlife venues in the UK. Traditional operators, such as local or regional brewers or independent entrepreneurs, declined in importance over the twentieth century as they were acquired by and merged with a small number of large national brewers. Over the last decade of the twentieth century the monopoly of these national brewers was broken up by the Beer Orders Act, and the ownership and production of nightlife spaces now represents a complex hierarchy between a number of types of operators. First, many well-established historic brewers, such as Scottish & Newcastle, Bass (now Six Continents) and Whitbread, have grown into large national and multinational entertainment conglomerates. Their retail wings continue to own large pub estates and have the resources to manage a wide portfolio of venues, including premium-branded bar venues and unbranded tenanted pubs. However, many of them are increasingly divesting their unbranded, smaller and older stock, which includes traditional community pubs, and are concentrating on branded mixed-use lifestyle venues, restaurants, health centres and hotels (Leisure and Hospitality Business, 2000).

Second, an emerging breed of highly profitable 'pubcos' also play a dominate role in the high-street nightlife market, which is a rapidly growing sector worth an estimated £2.5 billion (*The Publican*, 5 February 2001: 17). These companies are highly acquisitive, are usually backed by international corporate financial houses, and are profiting greatly as former brewers continue to sell off pub estates. For example, 70 per cent of the Punch Group is owned by the US investment firm Texas Pacific Group, Pubmaster is backed by WestLB, one of the largest German banks, and Morgan Grenfell Investment Company has acquired much of Whitbread's pub estate. Venture capitalists such as Alchemy and 3i are also getting in on the act and are buying up nightlife venues (Leisure and Hospitality Business, 2000). Over the last decade, such 'pubcos' have expanded, with around seventy such companies existing across the UK owning nightlife estates of thirty or more venues. Most of these companies are undergoing internal restructuring in preparation for floating on the stock market. While the number of pubs has stayed roughly static at about 62,000, the number owned by national brewers has fallen from 32,000 to 3,300 over the last ten years (now accounting for 5.3 per cent of the pub market). In contrast, 'pubcos', who owned 16,000 outlets in 1989, owned around 48,000 in 2000 (accounting for nearly 80 per cent of the market). In particular, the growth of multi-site 'pubcos' has been dramatic, accounting for nearly 50 per cent of all pubs in the UK in the same year (Table 2.2). Many of these 'pubcos' have shown remarkable levels of growth: for example, Nomura Principal Investment Group has prospered by buying up premises from brewers or former brewers (see Box 2.2). Similarly, the JD Wetherspoon pub chain, which started from a single premises in London, was touted as the fastest growing company in the UK and the ninth in Europe in early 2002 (JD Wetherspoon, 2002).

Table 2.2 Change in pub ownership in the UK, 1989–2000

	1989	*2000 January*	*2000 July*
National brewers			
Tenanted	22.000	2,724	1,000
Managed	10,000	7,336	2,300
Sub-total	32,000	10,060[a]	3,300[b]
Regional brewers			
Tenanted	9,000	5,939	5,939
Managed	3,000	3,498	3,498
Sub-total	12,000	9,437	9,437
Non-brewer operators			
Single/independent	16,000	18,098	18,098
Multi-site pubcos	–	24,196	30,956
Sub-total	16,000	42,294	49,054
Total	60,000	61,791	61,791

Source: The Publican newspaper, 2000

[a] *Bass, Scottish & Newcastle, Whitbread*

[b] *Scottish & Newcastle only*

Box 2.2: Nomura: a global–local landlord

One of the UK's biggest pub landlords, Nomura is also the largest securities firm in Japan, playing a significant role in many key markets around the world through its banking, investment and venture capital divisions. The firm has operations in some thirty countries around the world, 12,310 employees and total assets of some ¥20,529,135 in 2001. While its main business is providing individual and corporate trading services in its home market, it has also sought to revive its fortunes with a range of mergers and acquisitions outside Japan. Under the direction of the former managing director of its Principal Finance Group, Guy Hands, reportedly said to personally earn in the region of £40 million a year, Nomura embarked on a £10 billion buying spree, acquiring some of the UK's best-known brand names. The Nomura portfolio includes betting chains, off-licences, international hotel chain Le Meridien, the Ministry of Defence married quarters, and joint ownership of Boxclever (an amalgamation of Granada and Radio Rentals). It also paid £700 million for one third of British Rail's rolling stock, and won the bidding contest to redevelop the Millennium Dome after attempts to buy it failed. Nomura is well known for buying companies, turning them around and selling them at a profit. Nomura's recent acquisition of nearly a thousand pubs from Bass plc and 1,800 from GrandMet vaulted it to number one among pub owners in the UK, with around 5,500 pubs. Through such mergers it aims to move away from a centrally managed estate to 'tenanted' outlets, leased to local entrepreneurs who pay rent. Nomura continues to get even more dominant as market leader, as it bids for more pubs recently put on the market.

By 2000, the ten largest pub operators owned nearly 50 per cent of all pubs and bars in the UK; only three still have a connection with brewing. The biggest included Nomura, the Punch Group, Whitbread, Six Continents Leisure, and Scottish & Newcastle Retail, who each owned over four thousand venues (*The Publican*, 2000). Smaller, independent pub companies owning only a handful of venues do still exist and have introduced innovative new nightlife venue concepts, and there are also a number of regional brewers with sizeable pub estates, such as Greene King, Wolverhampton and Dudley and Young & Co. However, both independent operators and regional brewers have been extremely susceptible to buy-outs from larger predatory operators, eager to buy successful bar brands to expand and be able to float on the stock exchange. As one independent bar owner from the UK commented:

> If you look around in cities you will find a handful of people involved in setting up bars and the rest of it is just the corporates. But then the corporates come straight in afterwards and if you're setting up a bar and struggling to make a living and somebody comes along and says, well, we'll give you half a million, you take it and run.
>
> (Bar manager 1, Leeds)

This ongoing restructuring has significant implications for the ways in which pubs and bars are operated in the UK, with a shifting balance between managed or tenanted/leased outlets. Up until the massive changes in pub ownership in the 1990s, most traditional pubs owned by the brewers were operated as tenancies. However, the number of tenanted premises fell dramatically from nearly 45,000 to just under 10,000 between 1967 and 1998, while the number run as managed houses dramatically increased (BLRA, 1999). The recent growth of super-pubs, style bars and branded restaurants has shifted ownership in favour of managed rather than tenanted outlets, which is indicative of a resurgence of more 'Fordist' centre-branch plant management structures (see Piore and Sabel, 1984).

There are some signs that tenanted outlets were enjoying a limited renaissance by 2000, as they offered stable rental income and reduced overhead costs for pub operators, with less need for area managers, head-office staff, personnel and marketing departments. Moreover, these operators are aware that tenancies can offer a differentiated product, in contrast to the large glut of branded pubs and bars which continue to fill Britain's high streets. This counter-trend is more indicative of post-Fordist notions of subcontracting and outsourcing (Kumar, 1995: 60–1). However, there are limits to diversity even here: companies like Nomura tie their tenants into particular buying agreements which inevitably lead to a standardisation of product availability – arrangements which resonate with the idea of 'flexible mass production' (Piore and Sabel, 1984). Nomura has even set up a website for its tenants, listing potential suppliers to buy from. The Nomura-owned 'Inntrepreneur pubco' is currently facing a number of legal challenges from tenants for illegally 'tying' them into above-market-price buying arrangements with suppliers, a practice which was outlawed under the Beer Orders Act (Clark, 2001).

Clearly, such dramatic restructuring has implications for older (residual), independent and alternative modes of nightlife production. Many small independent operators were pessimistic about the encroaching influence of large corporate operators. As one independent owner commented:

> With corporate enterprise taking over more and more, they have a game plan that they will follow, which is domination of city-centre sites ... but I think the long-term view is that corporate rape and pillage will continue. You know they're all gobbling each other up because they've got to grow.
>
> (Bar owner 1, Newcastle)

Large, corporately backed 'pubcos' are able to put up large sums of money to transform high-value listed city-centre buildings in prime locations into new premises, spatially squeezing out independent entrepreneurs and dominating the urban landscape. Urban nightlife, then, has become a competitive arena with only the strongest, or wealthiest, able to survive (Zukin, 1995).

As Figure 2.1 shows, across a number of older industrial cities in the UK by 2000 ownership of bars and pubs is concentrated in the hands of a small number of national/multinational operators, notably S&N and Bass, plus a number of growing 'pubcos'. While there are also a number of regional brewers, recent indications

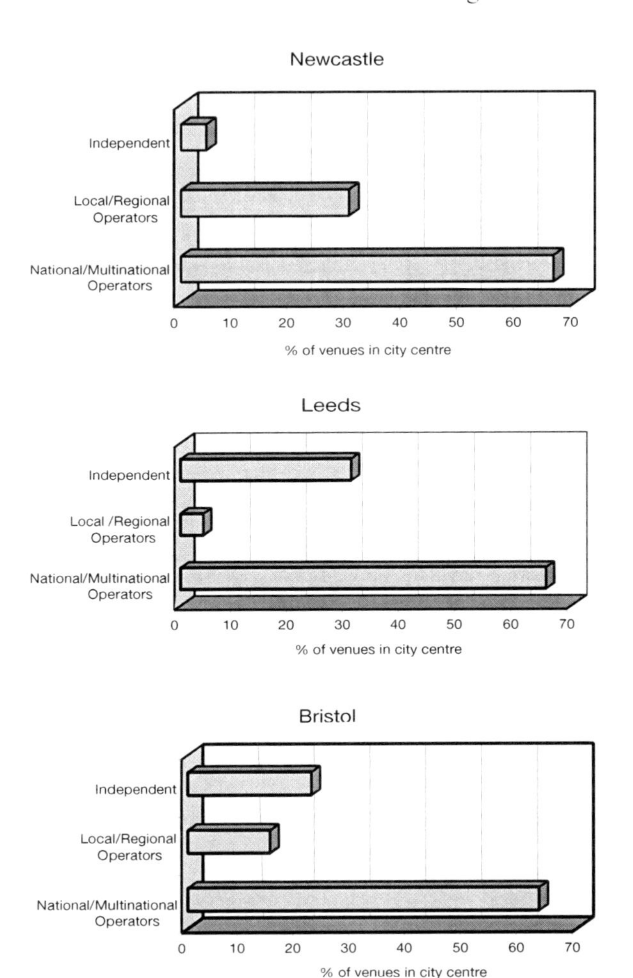

Figure 2.1 Ownership in the night-time economy – Newcastle, Leeds and Bristol, UK, 2000
Source: Chatterton and Hollands, 2001

suggest they are likely either to divest from brewing or to sell off their successful bar brands to expanding 'pubcos'. The revocation of the Beer Orders Act in 2001 however, has opened up opportunities for companies, especially former brewers, to make bulk purchases of nightlife venues. The prospect of such acquisitions could lead to a re-integration of the industry. Market concentration is likely to increase further as many traditional operators also slowly divest from alcohol-oriented nightlife into higher-profit areas such as pub-restaurants, fitness centres and hotels, due to perceived shifts in demographic and consumption patterns.

Nightclubs are currently experiencing similar levels of market concentration, although overall less so than bars and pubs. In the UK, for example, the nightclub industry had sales of over £2 billion in 1997, and admitted around 185 million people through their doors (Mintel, 1998: 15). However, many nightclub operators are facing new challenges due to falling audiences and the blurring of divisions between pubs and clubs, which has made the nightclub sector particularly difficult for small operators. These tight margins opened the way for large operators in the UK, such as the Po Na Na Group and Luminar Leisure. Luminar, for example, grew through the £360 million acquisition of Northern Leisure and Rank Leisure, to become one of the most established entertainment operators in the UK. Luminar runs 250 late-night venues, including brands such as Chicago Rock Café, Jumpin Jacks and Life Café Bars, and by 2002 had established itself as the largest nightclub operator in the UK, owning 15 per cent of all nightclubs. British dance clubs, in particular, are eager to play on their reputation as the birthplace of dance culture to expand their global reach. As a result, established and world-recognised dance clubs such as Cream and Ministry of Sound have been able to expand their operations by staging huge festivals which have commodified some of the original elements of rave culture. Such a winning formula has been extended across the globe from India to South Africa.

In sum, the nighlife sector is a highly volatile and unstable, with a significant proportion of the country's pubs up for sale, and takeovers, mergers and divestments continue apace in both the club and bar sectors. In particular, the British pub and bar sector has virtually been severed from its historical association with national and place-based brewers and pub retailers and, more recently, its monopolisation by the big breweries since the 1950s. However, deregulation has not produced a post-Fordist panacea of small companies emerging to drive forward nightlife production. Rather, there has simply been a carving up of the different wings of the industry, with the overwhelming proportion of urban nightlife venues now owned by a handful of corporate operators. Many of these are backed by global equity and finance houses, while the UK beer market is dominated by two multinationals, Interbrew and Scottish–Courage. Brewers and former brewers staying in the pub business are increasingly concentrating company efforts on their branded premises, while competing with 'pubcos' and corporately financed companies who are pursuing both branding and theming strategies, with both busy developing more 'premium' markets. It is to this aspect that we now turn.

Branding and market segmentation in the production of urban nightlife

While branding is far from a new phenomenon in urban nightlife, it has grown from its origins in alcoholic products (Tennant, 1994) to apply to whole/multiple retail outlets and has become a central part of the expansion strategies of many pub and bar-restaurant operators. Well-known global brands such as Hard Rock Café (forty-one outlets world-wide) and Planet Hollywood (sixty-eight outlets world-wide) combine drinking with eating, while in the USA many of the branded bars and clubs are combined with restaurants or other entertainment packages such as sport, virtual arcades, the sex industry or live music, and include ESPN Zone, Dave & Buster's, Hooters, Spearmint Rhino and Billboard Live (for a more detailed discussion of some of these examples, see Box 5.2). So, while the branding and theming of nightlife destinations is a global phenomenon, it has become particularly strong in the UK over the last decade. Eight per cent of all pubs in the UK are now branded using one of 206 brands, with the top five pub operators controlling 63 per cent of branded pubs (*The Publican*, 2000). In city centres the branding process is much higher, with about 30 per cent of premises branded. In particular, out of its 3,300 outlets, Scottish & Newcastle Retail claim that '50 per cent of the estate is currently branded. This will rise to 70 per cent by April 2002' (Scottish & Newcastle, 2001). All nation-wide operators are now organised around branded divisions rather than geographical areas.

Branding has also become a key driver for the nightclub sector, especially with the growth of 'super-clubs' such as Gatecrasher, Ministry of Sound and Cream. Pacha, with a 34-year history and over eighty venues world-wide marked by the distinctive two-cherries logo, is the biggest global nightclubbing brand. First opened by Ricardo Urgell in Sitges, near Barcelona, in 1966, and then on the party island of Ibiza in 1973, the brand has expanded through franchises across the world in places such as Buenos Aires, Munich and Budapest. Nineteen more opened in 2002 with the prospect of Pacha restaurant franchises, and the brand has developed out of nightclubs and into the music sector through Pacha Records, launched in 2000. Meanwhile, www.pacha.com offers live TV and radio link-ups to nightclubs, and an on-line 'storePacha' sells clubbing accessories from bikinis and t-shirts to wallets and jewellery. Pacha opened its first UK venue in London in 2001; as the manager Bill Reilly explained, the venue isn't driven by the DJs or the music, but by the Pacha brand (*Night Magazine*, February 2002: 19). (See Plate 2.2.)

The branding and theming of nightlife venues has numerous benefits for large corporate operators. Its attractiveness as a strategy stems from its ability to

Plate 2.2 The Pacha brand goes global. Flier for Pacha nightclub; Barcelona, Spain, 2002
Source: Carlos Aparicio

increase rational production techniques, and hence reduce costs and overheads, and tap into sacred consumer principles such as consumer choice, quality through reputation, safety, convenience and reliability (du Chernatony and Malcolm, 1998; Ritzer, 2001; Gottdiener, 2001). Branding has become an imperative for most large entertainment conglomerates as a way of minimising risk, maximising profits for shareholders and gaining the trust of stock-market investors. As a representative from a large nightlife operator in the UK claimed:

> As far as the City [London Stock Exchange] is concerned, half a dozen pubs in one town means nothing to them. Whether they make, you know, good money or not, it is not something. I mean the City loves brands, they love things that you can roll out and you can have 20/30/40.
>
> (Regional manager for a multinational leisure operator 1, Newcastle)

Additionally, hiding the reality of corporate ownership behind lifestyle brands is also a way for operators to detract attention away from their market domination and to encourage consumers to believe that they are making a discerning nightlife choice. As Nick Tamblyn, the managing director of the Chorion Group, owner of the Tiger Tiger brand in the UK, has said: 'There has to be a bigger difference between each Tiger Tiger club than between Burger King and McDonald's' (Doward, 1999: 1). And despite being an obvious franchised chain in Canada, the Firkin Group of Pubs logo is 'Everyone's a little different. One's just right for you' (www.firkinpubs.com). Moreover, developing a portfolio of brands allows companies to develop a number of distinct identities, target several audiences and operate at several venues in one location without competing with themselves for customers.

Whitbread Beer Company in the UK, for example, has broken down drinkers into seven categories, which include: the Breezer, defined as 'the most common style', where 'drinking is just part of a good night out'; the Steamer, 'the rowdy type, aims to drink as much as possible and quite possibly raise a little hell'; the Poser, for whom 'drinking is a fashion statement'; and the Adapter, who has 'just turned 18, has a little money but no confidence and no experience. Just goes with the flow.' Similarly, a study completed by the Carat Media Group (Carat Insight, 2001) analysed 15–34 year olds and categorised them in terms of how they respond to consumer goods and advertising: L Plate Lads, Disillusioned Young Mums, Cross Roaders, Progressive Leaders, City Boys, Survivors, New Traditionalists and Confident Introverts. This process of deconstructing the market and reconstructing it around branded identities, while often involving young adults through market research, is also based upon the work of company directors and marketers intent on formulating brands and dominating markets rather than responding to consumer tastes.

The current mode of nightlife production based upon brand development is a purposeful attempt to shape new consumer identities in the night-time economy, and can be understood as part of the wider restructuring of entertainment production. On the one hand, niche branding can be seen as evidence of the industry moving away from the declining Fordist model associated with a mass consumption experience in the largely male-working-class-dominated traditional pub (Gofton, 1983; Harring, 1983; Harrison, 1971) towards a more differentiated yet segmented set of markets. Typically, stereotypical profiles of social groups are conflated into lifestyle categories, which then form the basis for a number of supposed niche markets (Goss, 1993). As a result, in the UK, Firkin and It's A Scream brands are associated with students, All Bar One and Quo Vadis target professional office workers, Bar 38 is allegedly 'women- and gay-friendly', while Bar Oz, Walkabout, OutBack Bar and SpringBok target sports fans. Similarly, in the USA, Spearmint Rhino (a lap-dancing chain) courts corporate business groups, while ESPN Zone and Dave & Buster's target sports and virtual arcade fans respectively.

At the same time, even niche branding can be viewed as an extension of Fordist principles in that it can represent simply a more 'flexible' type of mass production (Piore and Sabel, 1984). Despite the fact that some of these new pub and bar concepts have promoted themselves as developing new types of licensing arrangements and different attitudes to dress codes and gender relations, and encouraging a diversity of uses generally mixing eating, drinking and entertainment, many brands are still recognised as 'much the same' or serially reproduced. For example, theming has come under heavy criticism from consumer groups and publicans alike for its damaging effects on the identity of the traditional British pub and its clientele (Everitt and Bowler, 1996), not to mention the fact that some consumers recognise that, rather than being unique, such premises are often both artificial and homogenous (see Chatterton and Hollands, 2001). Some newer 'pubcos' have even gone as far as branding their premises as a 'traditional' nightlife experience, with Wetherspoons' 'just a pub' philosophy being one example.

Nightlife production around branded lifestyle niches is a highly reflexive process which requires continual adjustment. Many first-generation brands are now tired and unprofitable, and new brand 'roll-outs' are constantly required to reinvigorate consumer demand. Larger national chains have taken the branding concept a step further. Nightlife venues are increasingly disconnected from their placed-based and brewing legacies and refer instead to a wider lifestyle experience. Freed from the chains of the mundane production of beer, corporate pub companies now have the time and extra financial resources to develop brand images (Klein, 2000), and attempt to draw on wider synergies and lifestyle experiences linking food, fashion and sport, and based around certain dress codes and social mores. The alcoholic drinks themselves have become more brand than product. Hence, 'Can I have a beer?' has been replaced by 'Can I have a Becks?' Further, top brands draw upon 'aspirational advertising', which sells not merely alcoholic products but a series of packaged consumer experiences based around emotive feelings, such as success, glamour, sex, risk, youth and social status.

A central trend is towards branding up-market premises, which target cash-rich, high-disposable-income groups, or those perceived to be older and less rowdy. Smarter up-market exclusive style and café-bars have emerged in order for certain social classes to redistinguish themselves from the mass nightlife market (Chatterton and Hollands, 2001). Business tourism and corporate hospitality are also significant influences in the creation of segmented nightlife markets (Doward, 2001). There are numerous examples of the gentrification of nightlife to draw on here, and we explore a number of case studies in Chapter 5. As housing, office, leisure and nightlife markets are recast through successive waves of property redevelopments, attached to them is a strong narrative of the 'public' who should (young professional service workers, trendy urbanites) and should not (younger teens, the homeless) inhabit these city spaces. At the same time, many corporate nightlife operators continue to provide more standardised and mass mainstream brands for those consumers in more routine, lower-order service jobs seeking weekend escapism and 'hedonism in hard times'. These themed and chain mainstream spaces offer predictable environments and familiar pathways through consumption choices for the mass-market nightlife consumer (Gottdiener, 2001: 148) and exist alongside a developing trend towards more premium premises.

Drawing on the situation in the UK, this process of upgrading nightlife spaces is readily apparent. In particular, the growth of style, themed and branded venues has been dramatic and to the detriment of traditional, alternative and residual older pubs. Cities such as Bristol and Leeds, for example, have witnessed significant high-level service-sector growth and repopulation in central areas which has fuelled upgraded, branded nightlife expansion over the 1990s. As Figure 2.2 shows, style and café-bars account for about 40 per cent of all venues in these cities. Conversely, alternative pubs and ale-houses account for a small and rapidly falling amount (around 10 per cent), and while traditional pubs still account for around one third of venues, this is likely to fall over the coming years as city-centre operators shift their focus to branded operations.

What is evident, then, is that older/historic and independent/alternative modes of nightlife are being quickly displaced by a post-industrial mode of corporately driven nightlife production in the consumption-led city. In the shadows exists the 'residue' of near-forgotten groups, community spaces and traditional drinking establishments marginalised by new city brandscapes. These residual spaces and people of the industrial city are now no longer required in the newly emerging and redeveloping corporate landscape. And while many aspects of nightlife remain seedbeds of resistance, often located on the margins, they find fewer opportunities to exist in the corporate city.

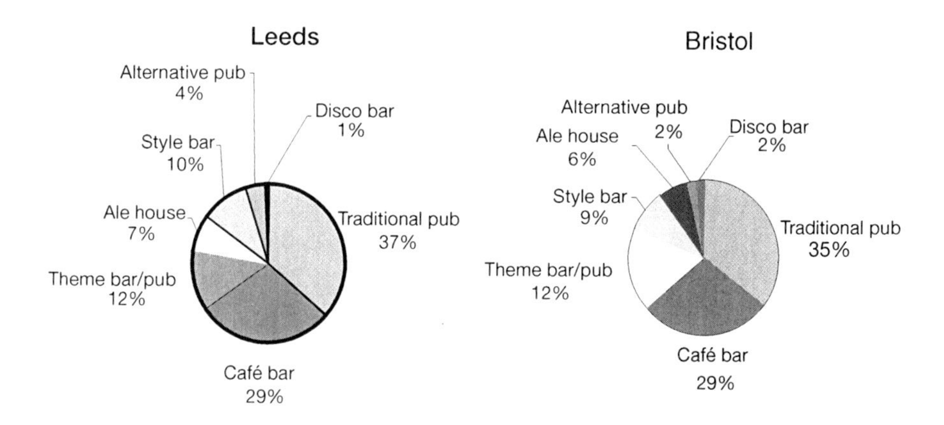

Figure 2.2 Venue styles in city-centre nightlife; Leeds and Bristol, UK, 2000
Source: Chatterton and Hollands, 2001

The interaction between these different modes of nightlife production varies between localities. In many places, older or resistant modes of production are still evident and sit uncomfortably alongside newer, stylish nightlife, and there is some evidence of resistance to branding and also brand failure. Many peripheral industrial cities, struggling to reinvent themselves into post-industrial consumption destinations, have a remaining legacy of strong local nightlife cultures based around local companies and brewers, often run by individual, family and self-made business entrepreneurs. They also have stronger and tight-knit regional identities, masculine working cultures and a kind of hedonism in hard times through a desire for escapism for many young locals in the face of continuing economic constraints. In such a context, many global operators have found it difficult to enter local markets and brand penetration has been lower. However, even such transitional cities are entering into the global corporate fray, and the encroachment of nightlife brands and the creation of new stylish downtown identities continues to unfold (Chatterton and Hollands, 2001).

Many independent operators remain pessimistic about the encroaching influence of large corporately backed 'pubcos' and their chains, who are able to put up large sums of money to transform high-value buildings in prime locations into new premises. Moreover, large operators are able to gain cost advantages through rational techniques of production such as bulk buying arrangements and 'synergies' between products (Ritzer, 2001), and exploit their greater influence over the cash-strapped local state (Monbiot, 2000), whereas smaller operators suffer from a lack of access to start-up capital, business and marketing skills and a lack of affordable property, and face complex and expensive licensing laws. Left to the market, many smaller-scale, locally based nightlife producers continue to be closed down, pushed to the margins or simply bought out. These competing modes of nightlife are explored in greater detail in Part II of the book. In the next chapter, however, we turn to the regulation of the night-time economy.

3

REGULATING NIGHTLIFE

Profit, fun and (dis)order[1]

In February 2000, the doors opened to the exclusive Rock Club on London's Victoria Embankment. The venue was owned by Piers Adams, longstanding entrepreneur behind London nightlife chains K Bar and Po Na Na and long-term friend of Guy Ritchie and Madonna, and a glut of paparazzi and stars including Kate Moss, Robbie Williams, All Saints, George Clooney and Jude Law graced its opening night. Advertising itself as a members-only table club 'offering an intimate atmosphere with an emphasis on exemplary standards of service which has helped make Rock a magnet for the UK's smartest people' and 'London's Beautiful It Girls and It Boys', Rock is not for the average reveller. Corporate VIP Privilege Cards start at £1,000, with bottles of spirits served at your table for £140 (Rock Nightclub, 2001).

In the same year, Fred Braugthon, chair of the Police Federation in the UK, commented that there was a 'sense of disorder and anarchy' in many city centres due to the drunken, yobbish and often violent behaviour of many weekend revellers. Calls were made for new legislation to shut down 'thug pubs' and introduce 'on-the-spot' fines for drunken behaviour. These laws have been rushed through the UK parliament under the Criminal Justice and Police Act (2001). In May of the following year, 160 police in full riot gear raided the Gatecrasher nightclub in Sheffield due to concerns over drug dealing. The club was closed for over a month and is now in negotiations with the police to increase club security (*Night Magazine*, July 2001: 13). The super-club Home, in London's Leicester Square, was also recently closed and its licence revoked by Westminster City Council using emergency procedures under the Public Entertainment Licence (Misuse of Drugs) Act 1997, after an undercover police operation led to several arrests of alleged drugs suppliers. Owners of the club, the Big Beat Group, went into liquidation.

How are we to make sense of these seemingly contradictory tendencies in which the night-time economy is associated with both the good times of stylish, exclusive activity and the bad times of violence, disorder and criminality? To unravel these contradictions, it is useful to think about how the night-time economy is governed. In a general sense, Miller *et al.* (2000) have outlined a shift from 'government' towards a wider process of 'governance' involving a range of actors from the business world, the voluntary sector and citizen groups. However, shifts to a more governance-style approach can also be seen to reflect a wider restructuring of state, capital and consumer practices. In particular, some commentators have suggested that there is a decline of public accountability among local institutions and an intrusion of a business-led quangocracy into the local–regional economic development process (Imrie *et al.*, 1995; Geddes, 2000; Jessop *et al.*, 1999).

Viewed in this context, significant changes have occurred within nightlife over the last few decades. In the industrial era, traditional regulators such as the police and the licensing judiciary largely focused upon rigid ordering, control and restraint so as not to allow entertainment and leisure to interfere with the world of work (Harring, 1983; Cunningham, 1980). Aspects of this more Fordist mode of regulation continue, albeit in different forms, via various licensing controls and laws, not to mention through surveillance and policing (Hobbs *et al.*, 2000). Yet more recently, governing the night appears to have become a more fluid, differentiated and complex yet pervasive process. As many cities seek to rebuild themselves around a post-industrial, service-based economy, nightlife has become an important economic entity in its own right. New coalitions of interest groups – including real-estate companies, property developers and entertainment conglomerates keen to profit from the new boom in the cultural economy, in conjunction with increasingly entrepreneurial and cash-strapped city councils and local governments – have formed, and have been effective in building a 'new consensus' for how the night-time economy should develop (Chatterton and Hollands, 2001).

Urban nightlife within this framework is largely based around profit generation and selling the city through developing up-market, exclusive leisure spaces, while marginalising local, independent and alternative nightlife and sanitising historic residual groups and spaces (Harvey, 2000). Despite the considerable success of this new consensus, problems and contradictions remain, including the stifling of local economic creativity in nightlife, a lack of consultation with the consumers and workers in the industry, and continuing problems of disorder, crime and noise. The night-time economy, then, continues to be framed through a number of – often contradictory – discourses such as law and order, economic development, creativity and inclusion.

As we explore in Part II of the book, regulatory changes appear to largely favour the development of urban spaces aimed at the needs of highly mobile, cash-rich youth groups such as business professionals, tourists, service workers and particular sections of the student, gay and female markets, while working against alternative/oppositional and older, community-based forms of nightlife. In the conclusion to the book, we return to some of the questions raised here, and in particular the possibilities for regaining ground for more liberatory, locally grounded and creative nightlife practices.

Regulating urban nightlife: history and context

The regulation of nightlife is far from straightforward. As such, it is necessary to appreciate that it has a number of different dimensions – legal (laws and legislation), technical (CCTV and radio-nets), economic (drinks and door entry prices) and social–cultural (musical taste, youth cultural styles and dress codes). In this sense, regulation entails both formal strategies and mechanisms such as policing, CCTV and door prices, alongside more subtle and informal elements like norms, habits, dress, language, style and demeanour. Yet attempts to control open spaces such as pubs, bars and nightclubs are always partial and contested. Nightlife is an example of an ambivalent space: 'a space in which there is a desire both to accommodate a pluralistic public and to control it through rational strategies of surveillance and discipline' (Lees, 1999: 245). Nightlife, then, represents a constant renegotiation and subversion of codes, styles and rules.

In spite of this fluidity, nightlife has been subject to much legal, political and indeed moral regulation (Dorn, 1983), fuelled by a longstanding anti-urbanism and a fear of crime and disorder, especially at night. While the control of licensed premises such as ale-houses, saloons and taverns dates back several centuries, the industrial city, through its connotations of immorality, vice and overcrowding, represented the heyday for the formalised control of entertainment and nightlife (Harring, 1983; Evans, 2002). In particular, there were strong, often contradictory, beliefs from the bourgeois classes that recreation time both demoralised and radicalised the working classes (Harring, 1983). While orchestras, theatre companies, professional associations and opera emerged for high-brow tastes, and variety halls, pleasure gardens, picture palaces, popular theatres and vaudeville catered for the increasingly educated, more middle-brow consumers (DiMaggio, 1991; Hannigan, 1998; Evans, 2002), the bawdy dance and music halls, burlesque houses, variety theatres, saloons and gin palaces of the industrial working classes were looked down upon, policed and heavily surveyed (Harring, 1983; Harrison, 1971).

Over the course of the twentieth century, leisure and entertainment has been subject to pervasive regulation and has been increasingly rationalised and planned through greater state involvement, censorship, licensing, planning guidelines and more formalised policing. The modern police force, which emerged out of the crisis of urban administration in the industrial city (Cohen, 1997: 113; Harring, 1983), played a key role in this regulation, especially in city centres where social classes, elsewhere residentially segregated, congregated in large numbers (*ibid.*, 129). The last hundred years have witnessed the criminalisation of numerous traditional street pastimes and working-class pursuits (Pearson, 1983). Regulation, then, was often mainly targeted at working-class entertainment, as they were seen to be the main source of social vice and moral decline (Cunningham, 1980).

The heyday of organised industrial society, epitomised through the system of Fordist production, formalised and extended the rational social control of leisure (Sherman, 1986). This entailed a regimentation of the times and places of work along the strict lines of scientific Taylorism, which was mirrored in the non-work sphere through the emergence of a distinct leisure time (Rojek, 1995; Thompson, 1967). The notion of instrumental rationality and the creation of the rational person (Weber, 1976; Marcuse, 1964; Sennett, 1977) is central in understanding the creation and regulation of mass consumption patterns. For instance, under Fordist work patterns, the night-time drinking economy was carefully regulated through the curtailment of entertainment and opening hours to ensure that workers' leisure did not interfere with their productivity (Harrison, 1971; Dorn, 1983; Gofton, 1990).

Neat links between consumption and production were never so clear-cut, even under a Fordist regime, especially in the rather messy and unstructured times and places of a night out. However, in the contemporary period, the connection here has become more blurred, complex and multifaceted (Lash and Urry, 1987). With the decline of the predominantly industrial/productionist society and the rise of a more service-based, consumption-oriented society (Zukin, 1991), many urban areas have sought new avenues for wealth creation in the entertainment, night-time and pleasure industries. In this context, the ideal of the '24-hour cultural economy' is part of a move away from the older industrial city, with its emphasis on manufacturing production and its visible temporal and spatial ordering. Pleasure-seeking and a broad range of nightlife activities now have a legitimate stage within the urban economy, which has created the need for more complicated and differential forms of governance.

Urban nightlife, then, contains a number of contradictory tendencies towards both deregulation and (re)regulation, and fun and disorder (Bannister and Fyfe, 2001). On the one hand, during weekend evenings city streets host tens of thousands of young revellers intent on fun, spending, drug-taking, dancing, encountering and subversion. It is now well accepted that this 'economy of pleasure' (Lovatt, 1995) and the '24-hour city' (Bianchini, 1995; Heath and Stickland, 1997) are vehicles for economic growth, profit generation and entrepreneurialism. The financial success of this after-dark economy has stimulated demand for its further deregulation. The other side of the coin is that, as Lovatt (1995) observes, regulation of the night-time economy has been slow to change, due to its perceived peripheral status to the daytime economy and a historical suspicion of it as a site of excess, vice and crime. In many ways, then, the night continues to be heavily influenced by Fordist concerns for tighter regulation, social control and zoning, due to lingering moral panics about lawlessness and disorder.

Young people, in particular, have a long history of being the focus for night-time moral panics and social control (Pearson, 1983), and the image of 'youth as trouble' continues to the present day (Wyn and White, 1997; Griffin, 1993; Males, 1996).

Historically they provided various 'folk devils' (Cohen, 1980) for respectable society, be it in the guise of street hooligans, mods, rockers, teds, football fans, lager louts, ravers or joy-riders (Hollands, 2000; White, 1999). More recently, in the UK there is growing concern over drink-fuelled violence and vandalism among young adults (Hobbs *et al.*, 2000; Lister *et al.*, 2000). Similar moral panics have arisen in the USA in relation to street gangs and hip-hop culture (Giroux, 1996), and there is also concern in Australia about excessive drinking, often based around youth, tourist and surf cultures (Homel and Clark, 1994). Such representations of youth continue to fuel a whole raft of restrictive regulations ranging from CCTV surveillance (Toon, 2000) to curfews (Collins and Kearns, 2001) and attempts to curb under-age drinking. So, in spite of efforts to create a profit-making 24-hour night-time culture, substantial obstacles remain, especially in relation to what is seen as an 'exclusionary' youth-dominated pub and club culture (Thomas and Bromley, 2000). Curiously, calls to speed up economic development and deregulate the night-time economy in the UK are made alongside new legislation to crack down on violence and crime on nights out.

As a result of these contradictions and various conflicts of interest, urban nightlife has become an arena for a more complex set of negotiations between a range of groups. As Table 3.1 highlights, there are a number of groups involved in the governance and regulation of nightlife, each of which has a different set of concerns and parameters. The rest of this chapter explores some of these tensions within and between these groups and, in particular, charts the clear shift in power from traditional bodies (the judiciary and police) primarily concerned with social order and public safety, towards groups such as business interests and the local state, with their imperatives of capital accumulation and desire to expand the cultural and night-time economy. At the same time, door security is caught between issues of social control and profit-making. While some citizens' groups (largely middle-class residents) are also part of this equation, consumers and workers in the night-time economy are largely silent here. Such regulatory transformations are also applied differentially, aiding corporate investment and profits while ignoring and often criminalising alternative/oppositional and residual nightlife groups and spaces.

Legislating the night

Traditions, cultural norms and habits within nightlife vary considerably between and within national contexts. The UK, for example, remains an island apart at the beginning of the twenty-first century with respect to an 11 p.m. watershed for the closing of bars and pubs, in contrast to other parts of the western world where nightlife patterns are later and staged between a variety of activities such as eating, drinking and going to a club. Drinking habits also vary. Sharing pitchers of beer is more widespread across North America, pints of lager are the preferred option in

Table 3.1 Parameters among nightlife groups

Group	*Parameters*
Licensing judiciary	Implement national laws Respond to need for growth
Police	Restrict activity and maintain law and order Accept and manage growth of nightlife
Local state	Manage nightlife problems and promote equality and access Promote inward investment and economic development
Door security	Control access and stop disorder, often through use of violence Legitimate, professionalise and upgrade door security provision
Nightlife operators	Small-scale operators – creative motive Large-scale operators – profit motive backed by trade associations
Residents	Reduce nightlife to protect quality of life Seek fun in nightlife activities
Consumers	Distinction, creativity, difference Fun, hedonism, escapism
Workers	Low pay, long hours, poor conditions (bar staff) Financial and creative opportunities in night-time economy (managers, DJs, promoters, etc.)

northern Europe, while small glasses of beer and wine are drunk in Mediterranean Europe. However, designer beers, strong 'psychoactive' alcopops, spirits and wines are now universally popular, and there is evidence of a convergence of drinking trends among young people (Beccaria and Sande, 2002).

Approaches to regulating the night, especially alcohol consumption, are framed within a variety of moral, social and political concerns which vary between countries. The USA, for example, has a long history of strict regulation of alcohol, most emphatically represented through Prohibition, which became law in January 1920 and lasted for thirteen years under the Volstead Act. Although the intention was to reduce criminal behaviour, the reality was an increase in illegal smuggling, sales and organised crime. 'Speakeasies' (illegal saloons), for example, proliferated, and by 1929 there were 32,000 in New York alone, twice the number of official bar saloons which had existed before 1920 (Miller, 2000). Strong anti-drinking sentiments have a long history in the USA, especially through right-political discourses on morality, the family and personal control. Efforts to curb, if not eliminate, alcohol sales historically rested with the religious-based temperance societies, for whom collective enrollment reached more than 1 million by 1840 (*ibid.*). The saloon and the dance hall, centrepieces for working-class life in industrial cities such as Chicago, Milwaukee and Buffalo, were singled out as the dens of vice and moral decline in nineteenth-century America (Harring, 1983), and the anti-saloon movement went to great lengths to raise taxes to price them out of existence.

Remarkably similar moral concerns about alcohol have emerged again in contemporary American society (Engs, 1991). However, current attitudes towards alcohol are difficult to discern, reflecting both hedonism and puritanism and contemporary influences of a more globalised consumer society. Nightlife in the USA is also more decentralised, car- and home-based, and alcohol consumption is framed through strong moralistic undertones and a higher legal drinking age (21) than in most other western countries.[2] Drink-driving is a particular concern, and groups such as Mothers Against Drunk Driving (MADD) have played a key role in pushing for nation-wide legislation in terms of raising drinking ages and increasing penalties for drunk drivers. However, many young people have found creative ways around higher drinking ages. In Tijuana on the north Mexican border, lower legal drinking-age limits and inexpensive drinks have given rise to a nightclub district frequented by thousands of young Southern Californians; on weekend nights, more than 6,500 people cross back into the United States between 12 a.m. and 4 a.m. (Lange and Voas, 2000). Alcohol control remains high on the priority list for both state and federal governments. Each state has strong measures over the night-time economy through dedicated departments such as Alcoholic Beverage Control Departments and the State Liquor Authorities, which regulate the sale and consumption of alcohol and the issue, suspension and revocation of liquor licences, while the Federal Bureau of Alcohol, Tobacco and Firearms (ATF) regulates alcohol at the federal level.

Attitudes and regulations towards nightlife and alcohol vary widely across Europe. Northern European countries, especially the UK, are plagued by images of 'lager-fuelled yobs'. Here, drinking cultures are distinctive. Heavy binge drinking occurs on weekends and special occasions rather than with meals; many people drink for the purposes of getting drunk, and public drunkenness is more or less accepted. Age limitations are often established for legal alcohol consumption, and alcohol is generally prohibited for children (Engs, 1991). Nordic countries exhibit strict laws over personal consumption through state monopolies which control the production and distribution of alcohol through state-run stores.

In contrast, southern European countries have more relaxed drinking cultures often based around wines, generally consumed with meals. Drunkenness is less accepted, even at celebrations, and children are often given diluted wine with meals as part of a rite of passage into adult drinking cultures. In these cultures there are fewer perceived psycho-social problems and few strict control policies regarding alcohol use (*ibid.*). Nevertheless, regulations are hardening. In Spain in 2002, for example, the government introduced a law, backed by heavy fines, banning drinking in the streets and the sale of alcohol to people under 18 years old. Such laws reflect a growing moral panic towards rowdy youth street drinking, known as the *Botellón*, (see Box 8.1) and concern over recent figures, which showed that 76 per cent of people between the age of 14 and 18 consume alcohol. This country, which has pursued a rather liberal approach to social control in the post-Franco era where children have been allowed freedom to drink, is now showing evidence of turning towards the USA model of banning alcohol to those under 21 years of age (De Rituerto, 2002).

In the UK, the government has been involved with the regulation of the sale and distribution of alcoholic drinks since the thirteenth century, due to concerns about beer purity, price and public order. Towards the end of the nineteenth century, licensing magistrates were granted absolute powers to refuse or grant new licences for pubs, taverns and hostelries where there was deemed to be sufficient 'need', in order to control what was seen as the 'odious and loathsome sin of drunkenness' (Justice Clerks Society, 1999: 29). These archaic laws have remained, with very few amendments, for almost two centuries, and licensing magistrates still wield power in the control and development of the night-time economy despite their non-elected and non-representative status. They are often highly dependent on police information and intelligence about particular persons, places and premises, and are criticised for being out of touch with current trends in youth popular cultures and urban development. Many magistrates rely on stereotypes of young people and nightlife and have little direct experience of the activities for which they are legislating.

However, this seems set to change in the near future. By the 1990s, there was a growing awareness that licensing magistrates were interpreting 'need' in a way which was protecting the interests of existing licence-holders and restricting competition (Justice Clerks Society, 1999: 77). In this context, the judiciary were encouraged to balance the needs of the free market and the demand for urban regeneration with those of controlling potential disorder and disruption. More fundamentally, the whole licensing system has been reviewed through the White Paper *Time for Reform* (Home Office, 2000), much of which is aimed at simplifying procedures for the sale of alcohol and entertainment and encouraging more and later nightlife activity, while at the same time granting tough new powers to the police and transferring more responsibility to the local state. While the judiciary has often upheld the status quo and controlled competition, many groups have voiced concerns over this shift in power. There is a recognition that magistrates at least provide independent judgement, anchored within a legal framework, while conflicts of interest may arise within the local state as it tries to balance economic development with public need (Chatterton and Hollands, 2001).

Legislating the night, then, is an arena of conflict between established and emerging styles of governance. In many cities with tough working-class industrial images, the authorities have kept a tight control over the development of nightlife through concerns over violence, guns, drugs and under-age drinking. Such fears seemed partially founded in the case of Manchester in the UK. Experiencing a rapid growth of nightlife in the 1990s, it was also widely dubbed 'Gunchester' due to gang violence in its clubland (Haslam, 1999). However, most large urban areas have actively transformed themselves from their industrial pasts through a business-led quangocracy, and in such places established regulatory groups have succumbed to a more deregulated and pro-growth approach to the night-time economy. In European countries, European Union legislation is encouraging a standardisation of regulatory arrangements which are more flexible, simple and market-responsive.

Policing the night

Any discussion of the role the police play in nightlife needs to be situated historically. Harring (1983) suggests that the police largely emerged from the class struggle of industrial capitalism and are part of the broader development of capitalist social institutions. In particular, from the point of view of commercial employers, middle-class residents and religious and temperance groups, the police were seen as essential in upholding morality in the emerging industrial city, including nightlife. While policing only became formalised in the nineteenth century, in countries such as the UK their role in regulating the night dates back to the fourteenth century, when parish constables were given duties to regulate ale-houses and taverns (Rawlings, 2002). In the USA, historically, police attitudes to nightlife activity were mixed, as constables could often be found in saloons having a beer and simultaneously monitoring the behaviour of locals (Harring, 1983).

Within the contemporary night-time economy, the police have adopted a more systematic, sober and professionalised approach, and largely fulfil a dual role: advising on the development of licensed premises and directly policing nightlife. In effect, they simultaneously pursue a moral and a coercive role (Cohen, 1997). This latter role has come under stark relief due to concerns over public disorder from increases in nightlife activity. In the UK, for example, the image of 'lager-fuelled youth' has become commonplace in the media and has led to legislation aimed at shutting down 'thug pubs' and curtailing drink-fuelled violence and vandalism.[3] Framing the night through such discourses of disorder has led the police to embark upon a crusade to crack down on perpetrators and 'clean up' the night.

The USA has its own set of issues in terms of policing the night, relating to more pronounced urban–suburban divides and car-based nightlife cultures, the race issue, gang violence and higher gun use. Policing in the United States is decentralised, with county and local police forces having significant levels of autonomy; hence, variations exist in terms of approaches. The USA has also witnessed a rise in nightlife violence, especially in clubland, which many believe relates to the reluctance of some nightclubs to summon police through concerns that they will be shut down. Leslie Ayers of San Francisco Late Night Coalition, a group of citizens, DJs and record producers, suggested that frequent emergency calls by venues increase the likelihood that they may be closed, which 'makes people think twice about calling 911' (quoted in Myers, 2001: 3). More importantly, as we see in Chapter 8, certain music genres, especially rap and hip hop, have been blamed for creating a culture of violence (Giroux, 1996).

However, these impending law and order crises and concerns with moral decline overlook a long history of street violence and crime (Pearson, 1983). Additionally, the extent to which the police can claim absolute control over the night is always open to contestation (Herbert, 1999). Night-time spaces are inherently difficult to police, monitor and control as they are subject to flux, instability and constant renegotiation by the fluid movement of consumers. Police investigations here often have to unravel encounters which are framed through a cocktail of drugs (both legal and illegal) and emotionally charged behaviour. Nightlife is evasive, fleeting and fast-paced, which repels the order that modern-day policing relies upon. Many nightlife spaces are indeed 'no go' areas for police.

Police attitudes to nightlife regulation vary and, in general, views on liberalisation and deregulation often depend upon the nature of their relationship with groups such as the local state, the business sector and religious groups. Young (1995) describes how police in Newcastle in the north of England have historically dealt with drunkenness in highly gender-specific ways, much of which relates to time-honoured institutionally imbued cultures of drinking among male police officers themselves. In localities which have embraced a more pro-business approach to development and the 24-hour city, the police have moved away from narrow law and order discourses and have taken on board more liberal interpretations. A central part of such moves has been the acceptance of staggered closing hours as a method for dissipating late-night flashpoints and violence. The police are also cognisant of the influence of design on nightlife violence and are keen to back café and style-based venues, where people are seated and alcohol is mixed with food.

However, problems including anti-social behaviour, outbursts of violence, excessive and under-age alcohol consumption, urinating and vomiting in the streets remain, whatever the type of nightlife. Many police forces have been unprepared for the scale of growth in nightlife, and, compared to other large sports and music events, street nightlife receives comparatively few resources. This has led them to seek complementary methods to police the city at night. In particular, there has been the emergence of what Newburn (2001) has called 'new security networks' which involve hybrid, and increasingly privatised, policing networks. Private door-security firms and doormen (or bouncers) play a key role in such networks, who in general outnumber police by a ratio of ten to one in many downtown areas at peak times. There has been a departure from the old-style tactics of containment and confinement, towards focused use of officers in conjunction with wider urban surveillance networks.

CCTV plays a crucial role here, yet its status as an effective tool for policing the city is hotly contested (Norris and Armstrong, 1998; Fyfe and Bannister, 1998; Toon, 2000). The effects of CCTV are beyond those of mere crime prevention and there is little doubt that it has changed the individual's experience of the street. While this so-called 'silver bullet' of crime prevention has brought cost savings and reductions in crime, it has wider implications as a '1984'-style Big Brother tool which induces conformity and abolishes the potential for deviance (Norris and Armstrong, 1998: 6). In this sense, it assumes deviance is a taken-for-granted part of urban life, and seeks to manage it and appoint blame rather then looking at its causes. Although reported support for CCTV is debatable and is often based upon dubious surveys (see Ditton, 1998), as Bannister *et al.* (1998: 27) point out, CCTV is popular as it feeds off a 'fear of difference' and the unpredictability of collective behaviour. Hence, it is often used to further the privatisation and purification of public space. Those opposed to CCTV outline a number of alternative strategies involving self-policing by repopulating streets, coupled with a greater diversity of downtown activities and a sense of civic responsibility.

Police are also increasingly channelling their work through multi-agency teams and partnerships, comprising local councils, licensees and door security, which aim to tackle some of the root causes of late-night disorder. One of the remaining issues for the police is the reconciliation of their agenda of maintaining law and order with the agenda of larger nightlife operators who have a legally binding and fiduciary obligation to maintain commercial profits for shareholders. Here, there is a growing recognition that nightlife companies must take some responsibility for late-night disorder, through, for example, financial contributions to policing, especially the larger branded venues which sell alcohol in significant amounts.

As we discuss in more detail in Part II of the book, alternative independent venues, rather than using formal policing methods and relying on door security, draw upon self-regulation through customer identification with the ethos of the premises, which includes a more liberal approach to dress codes and a blurring of the consumer–producer divide. Yet the police often view alternative venues as 'deviant', particularly in terms of illegal drug use, despite their better record in terms of lack of violence. Nightlife wedded more to the working-class industrial city, as in market taverns, ale-houses and saloons, is perceived by the police to be inhabited by what has been described as the urban 'underclass' (Campbell, 1993); these spaces are hence often regarded as sites of criminality, violence and debauchery, worthy only of containment or surveillance.

Policing styles in the contemporary nightlife economy, then, reflect different nightlife contexts. While alternative and oppositional nightlife such as squats and free parties are often policed out of existence for being 'illegal', rough working-class places are treated with suspicion and interventionist policing. However, in other contexts, especially the world of downtown corporate-led branded nightlife, the police no longer treat it just as a source of vice and crime, but have acknowledged its role in profit, growth and employment in the post-industrial city. Policing roles here are not so much about controlling the morality of the industrial working classes, but supervising the pleasure-seeking of the young and the wealthy. However, in general the police find it difficult to grasp the wider social significances of drinking and drunkenness for young people beyond those of disorder and moral decline (Tomsen, 1997; Warrell, 1994). Policing urban nightlife, then, is caught between competing discourses of law and order and the imperatives of growth.

And while there is some evidence that the police at a local level show some willingness to understand socio-cultural phenomena such as drug and dance cultures, new legislation is also emerging which is constraining the parameters for acceptable nightlife activity across the board. In the UK, for example, the 1997 Public Entertainments Licences (Misuse of Drugs) Act was introduced to tackle drug use at dance events and allows local authorities to revoke the licences of clubs that have a 'drug problem', while the Criminal Justice and Police Act (2001) gives police powers to shut down problem venues and issue on-the-spot penalties for disorderly behaviour as well as ushering in powers to clamp down on public and under-age drinking. Moreover, legislation such as the Criminal Justice and Public Order Act (1994) and the Terrorism Act (2000) has severely restricted certain forms of activity, especially the right to assemble, protest and party. Similarly, in Australia the New South Wales Ministry of Police issued a 'Code of Practice for Dance Parties' (1998) which, although not as restrictive as anti-rave legislation in UK, has sought to contain dance culture within legitimate sites, and hence curtail smaller-scale, illegal rave spaces (St John, 2001a).

Changing times on the door? Bashers, bouncers and style selectors

> 'Bouncer' is a very old-fashioned term for somebody who used to be a big gorilla. Originally that's all the job was in the 1960s. You just had to be able to bash people. And now obviously it's changed an awful lot and it's a highly skilled job.
>
> (Door staff manager 1, Bristol)

Door supervisors, otherwise known as 'bouncers', play a key role in regulating the night. As Hobbs *et al.* (2000) have outlined, the 'culture of the door' has long been pervaded by violence, physical force and intimidation, and this culture is still very much alive. However, the actual operation of door staff has begun to change over the last few decades, supplementing the still pervasive 'hard man', with 'door pickers' and 'style selectors'. Door supervision in general has become more professionalised. In the UK, this trend is being encouraged through local authority Door Registration

Schemes and also the National Security Industry Authority, which has established a register of approved providers of security industry services. Similarly, in the USA, more than a million people have undertaken Training for Intervention Procedures (TIPS), a nationally certified programme organised by Health Communications Inc. which is designed to teach bartenders, managers, security personnel and consumers of alcohol how to prevent intoxication, drink-driving, under-age drinking, and alcohol abuse. TIPS certification lowers the premium bars and restaurants pay on liquor liability insurance, and hence in most licensed venues it is a requirement for employment. Due to the high number of alcohol-related deaths on college campuses, a particular focus of TIPS has been educating university students about responsible drinking.

The basic job of door staff, however, remains deciding upon the suitability of customers in order to maintain order and the commercial viability of licensed venues, using both violent and non-violent tactics (Lister *et al.*, 2000). They are the definitive gatekeepers of the night-time economy, who ensure a connection between venue ambience and clientele. As one door security member interviewed commented,

> It's the constant problem of trying to ram square pegs in round holes, isn't it? We know that there are certain people that are comfortable and right for a dance-based venue and there are certain people where you put them in a café-based place. So, yeah, we don't just slam people in at all, it's got to be thought about.
>
> (Door security personnel 1, Leeds)

Door cultures still vary significantly. Many old industrial cities which have a lingering tradition of tough, male-dominated working-class nightlife are often regarded as behind the times (Winlow, 2001). The link between door cultures and criminal cultures is still clear enough, especially in localities flavoured by a hard working-class history. In Newcastle in the UK, for example, thirty-eight door staff had police files, while Morris (1998: 11) outlined that in both Tyne and Wear and Merseyside criminal groups forced 'existing door supervisors, through intimidation and extreme violence, to "pay" them a "tax" for running a door, whilst also requiring them to allow "approved" drug dealers to operate in the premises under their supervision'. Lister *et al.* (2000) outline how many door staff operate with an ambivalent relationship towards the formal law. When complaint cases of assault arise, very rarely are bouncers successfully prosecuted, thanks to collusion among door staff, police empathy, the victim's perceived risk of intimidation, and problems of drunkenness which lead to poor-quality evidence (*ibid.*).

Clearly, different types of nightlife venues have their own set of entry requirements, expectations and subtle forms of discrimination at the door based on age, appearance, social class, gender, ethnicity and sexuality (Chatterton and Hollands, 2002). One of the most visible differences in door cultures is between more mainstream branded and alternative independent venues. While many busy downtown mainstream venues view strong-arm tactics as necessary due to problems such as under-age drinking, excessive alcohol consumption and violence, many alternative venues do not use door supervisors at all, relying more on self-regulation. Images associated with alternative or fringe venues based around particular musical styles, sexual preferences or ethnicities act as effective forms of self-policing, and many of these premises form an 'extended family' or 'community' which literally helps the venue to police itself and detract unwelcome clientele. Such forms of self-regulation create subtle forms of 'autosurveillance' (Atkinson, 2001) in which consumers internalise a set of codes, assumptions and expected behaviours. The peripheral location of many alternative venues also creates a kind of self-policing and reduces the chances of infiltration and disruption by unexpected groups of consumers.

With the rapid growth and diversification in nightlife, door staff have to respond to the introduction of new venue concepts and wider shifts in music and youth cultural styles. In particular, as young people express a more eclectic 'mix and match' approach to style and appearance, it is more difficult for bouncers to make simple judgements about clientele based simply on their initial appearance. However, in some localities and types of venues, many door staff still adhere to established nightlife style conventions such as 'no jeans', 'no trainers', 'no skinheads', 'no visible tattoos' policies, which in such eclectic times, where links between style and social structure are more complex, raises a whole host of problems for identifying 'the right sort of people'.

Many door policies in central areas are encouraging an 'upgrading' of styles and appearances (Chatterton and Hollands, 2002). Much of this upward drift is due to a number of perceived, yet extremely problematic, links between style and behaviour. As one city-centre bar owner commented: 'We do really push for reasonably smart dress purely because if people have made the effort to get dressed up they're not going to be causing trouble. They don't want to wreck their clothes' (Bar owner 2, Bristol). Here, more exclusive venues use 'door pickers' in conjunction with bouncers to implement 'hyper-selective' style barriers. Such upgrading of door policies is generally an attempt to sanitise consumer markets and price out 'trouble' from the market. This has a number of implications for diversity and access, not least in relation to provision for poorer groups. At the same time, basic criteria for non-entry, such as the 'wrong' style or excessive drunkenness, are often disregarded by managers of large busy corporate nightlife venues, who are under tremendous pressure to fill the venue and maximise beer sales.

Door cultures, then, have to reflect a complex interplay of styles and aspirations of consumer cultures, a need for order, and the dictates of corporate owners. As a result, 'informal' door cultures where access is maintained through community and networks of trust, 'rational' door cultures based around strict delineations between types of consumers and types of venues, and 'hyper-selective' door cultures which are more subtle yet equally coercive, all co exist within contemporary urban nightlife. As housing, labour and leisure markets continue to be upgraded in central areas, the 'door' will increasingly becoming a mechanism for distinction and exclusion.

The makers and rakers of urban nightlife

A number of institutional players, who come together through a complex set of interrelationships, profit from urban nightlife developments and thus have a keen interest in how it is governed. First, as we outlined in the previous chapter, a small number of leisure merchants (Hannigan, 1998) have emerged to dominate the ownership, distribution and consumption of nightlife. Second, an equally small number of land, property and real-estate developers and managers back such large entertainment conglomerates, especially through complexes drawing together a number of corporate tenants. In some cases, these companies are one and the same, with entertainment giants such as Sony, Disney and Warner becoming shrewd real-estate developers. In the UK in the 1990s, property developers Urban Splash were among the first to recognise that redundant buildings could be adapted for new residential and entertainment uses, and have spearheaded a new wave of developments which have brought wealthy professionals back downtown looking for the buzz of city living.

Further, Scottish & Newcastle have taken a leading role in developing sizeable urban entertainment destinations. They have been closely involved in Birmingham's £75 million, 25-acre Star City leisure complex, which features four of its branded bar outlets among an assortment of multiplexes, bowling alleys and restaurants, and are planning other Star City developments across Spain. Similarly, by 2000 the Heron Corporation was developing a new generation of urban entertainment centres called Heron City across Europe, in Madrid, Barcelona, Valencia and Stockholm. Also spearheading this spate of new leisure–retail–nightlife complexes are companies such as Land Securities, redevelopers of Birmingham's notorious Bull Ring, who with fixed assets in excess of £8.3 billion is one of Britain's leading real-estate companies. Third, a whole set of corporate financiers, venture-capitalist and pension fund operators, have recognised the potential gains to be made from lifestyle nightlife destinations, especially those which are branded and hence risk-averse. Finally, as we discuss next in this chapter, the local state has come to profit from this area, mainly through selling land and collecting tax revenue from these middle-class consumption ghettos.

National regulations entail different opportunities for these makers and rakers of urban nightlife. As we highlighted in the previous chapter, in the UK remonopolisation and reconcentration has occurred with a small number of large, often multinational, companies dominating ownership, and independent operators and local brewers increasingly squeezed. However, regulatory mechanisms in countries such as Germany and Belgium are more supportive of local producers, and in North America the established system of brew pubs promotes a slightly more deconcentrated nightlife market to a certain degree.

Well-organised and vocal trade associations and lobbying groups have emerged to influence the current development of nightlife. One of the most powerful of these is the Portman Group in the UK, an independent company established in 1989 comprising the world's largest alcohol producers, such as Bacardi, Scottish & Newcastle and Interbrew. While the company's stated aim is to 'reduce the misuse of alcohol', the Portman Group have been an effective vehicle for ensuring that large alcohol providers are portrayed as responsible corporate citizens and that their needs are taken on board by government. Further, the Association of Licensed Multiple Retailers (ALMR) was set up in 1994 in the UK to promote the growing number of independent multiple retailers. ALMR now exists as a strong lobby group for licensed retail companies (tenanted and leased pub estates, the retail divisions of brewers) and key suppliers of goods and services (brewers, distributors and support services). A similar group exists in the nightclub sector through the British Entertainment and Discotheque Association (BEDA). The net effect of such groups has been more effective lobbying for large established capital interests in terms of influencing the national regulatory terrain for nightlife.

Small-scale independent operators have been less successful in mobilising and creating a sectoral voice in such a regulatory landscape, and as a result their views are not as readily heard. Nevertheless, the independent sector has created its own organisations. In the UK, longstanding lobby and advocacy groups such as the Licensed Victuallers' Associations, the British Beer and Pub Association (BBPA) and the Campaign for Real Ale (CAMRA) campaign to maintain the traditional nature of British pub and beer culture. Many clusters of small operators have also joined forces to lobby for their rights. In Leeds, for example, the growing number of independent bars formed the Leeds Café Bar Association, while a number of cutting edge nightclubs formed the Leeds Nightclub Association, both of which felt a need to voice their concerns about encroaching corporate influence in the city's nightlife.

Finally, there are those who actually work in the nightlife industry, producing it night after night, including managers, bar staff, cleaners, promoters, DJs, etc. In the UK it has been estimated that the nightlife sector (including brewing, bars and clubs) directly and indirectly employs 830,000 people (BLRA, 1999). The growth of nightlife has opened up financial opportunities for many people, especially young people looking to 'double-job' or supplement their income. This is not just in terms of bar work, but also in terms of more creative jobs such as interior design and music.

However, the Low Pay Commission (1998) in the UK outlined that 40 per cent of people employed in the hospitality sector are paid below the minimum wage, the highest of any sector in the economy. On a more personal level, work here is often mundane and disempowering, and very rarely are the voices of many people working within the industry heard at all in the debate about nightlife. The disconnection which is evident between the head offices of large nightlife conglomerates and their individual venues is often reflected in customer relations. As one young reveller commented to us: 'Go to most bars and staff will be on £3.50 an hour. Their bosses are somewhere in the Shetland Islands that don't know their name and they're just on a payroll, and you can see that in somebody's face' (Jason, 21 years old, Newcastle). The realities of working in the nightlife sector, then, for all but a few successful entrepreneurs equate to long hours, few entitlements, low pay and little or no say in how the industry is run.

The local state and the entrepreneurial nightlife city

It is now common parlance to suggest that the local state has added a more entrepreneurial, promotional and partnership role to its more mundane task of 'managing' social welfare (Harvey, 1989a; Cochrane, 1987). Such shifts are part of a wider restructuring of institutional arrangements across the west, and in conjunction with the transformation from mass production to one of 'flexible accumulation' (Harvey, 1989b) there has been a rolling back of the 'welfarist' state in terms of its powers of economic intervention and in its style of governance. In both the UK and the USA, cash-strapped city councils have increasingly been supplanted by various QUANGOs (quasi-autonomous non-governmental organisations) (O'Toole, 1996) and public–private partnership schemes and corporations (Zukin, 1995) to stimulate economic development, or have become increasingly dependent on attracting mobile corporate capital investment.

This shift towards a more pro-active business-led entrepreneurial local state is now a common feature not just of mainstream economic development but also of the cultural and night-time economies. Many cities have sought to reinvent themselves as places of consumption dependant on a diverse and vibrant 'after dark' economy, partly as a response to a rapidly changing post-industrial populace (Savage and Butler, 1995; Wynne and O'Connor, 1998) and shifting patterns of investments towards the 'symbolic' economy (Lash and Urry, 1994) and lifestyle brands (Klein, 2000). Localities which have been at the forefront of the emergence of 'cool' economies have been those which have promoted a liberal, business-led partnership and market-driven approach. In places such as Glasgow, Melbourne and Barcelona, nightlife activity is now heralded as an integral part of the new post-industrial urban economy as much as the business or retail park. However, many older industrial localities have neither the infrastructure nor the clientele to fuel a 24-hour cultural economy and have found it difficult to create a more cosmopolitan image because of a strong tradition of highly gendered masculine nightlife cultures and the lack of a critical mass of professional classes (Hollands, 1997; Gofton, 1990).

This 'entrepreneurial turn' signifies a shifting balance of power between the local state and capital interests. The erosion of national or local government's ability to control and regulate the activity of capital interests is well documented (Monbiot, 2000; Hertz, 2000). In the UK, for example, the government used to have powers to dismantle any commercial enterprise 'tending to the common grievance, prejudice and inconvenience of His Majesty's subjects'; however, the state has largely renounced this historic role and refuses to interfere with the operation of the free market (Monbiot, 2000: 314). In particular, the emergence of public–private partnerships encourages the local state to come into line with the needs of business and creates platforms from which business elites can exercise political influence (Jessop *et al.*, 1999). Fundamentally, then, the deregulation of nightlife is part of the reassertion of capital and the renewal of new forms of capital accumulation (Jessop *et al.*, 1991).

Urban nightlife has become a visible example of this process in which capitalist enterprises, aided by a new business-friendly state, can seek out new profit arenas. Many large property developers, landowners and nightlife operators receive public subsidies for renovating and expanding buildings for entertainment and nightlife use. As one city council employee in the UK stated:

> The role of the local authority is to create the conditions in which developers can invest so as to stack up fairly major sites ... inevitably plc's [public limited companies] are going to be the ones who come forward ... So I think the effort of the city at the moment is behind large-scale development proposals.
>
> (Local authority representative 1, Newcastle)

The fact is that many urban governments have little room for manoeuvre due to their own declining financial position in relation to central government funding, restrictions on raising local revenue, and various protocols ensuring they get value for money whenever development opportunities involve the sale of public land. The local state is increasingly using the rhetoric of access, creativity and diversity, associated with the idea of the 24- or 18-hour economy, to 'court' big national corporate operators, getting the best deal on public land sites and, understandably, is happy to fill up what were once derelict and empty buildings.

While it is not the remit of local authorities to inhibit or encourage certain types of activity, this is not to suggest that city councils are somehow entirely powerless to influence the nature of nightlife development. Local planning policies and guidelines are critical in deciding types of uses, and the local state is responsible for granting a variety of entertainment and liquor permits. Yet planning powers are extremely blunt instruments and in many cases can be easily overturned. The local state does, however, have to balance its entrepreneurial role with the more mundane management of the side-effects of the night-time economy, such as noise and litter. Many councils have established legislation to restrict the growth of late-night venues in so-called 'stress areas' and have introduced bylaws to curb drinking alcohol in the streets (see Plate 3.1). Westminster City Council in London, for example, with the unenviable job of managing Soho, has a tough reputation for restricting night venues to reduce noise and disturbances. However, such legislation is not set in law and several large leisure groups have challenged and overturned decisions in High Court appeals.

Many small-scale local entrepreneurs find it difficult to find a place within this new night-time economy geared towards meeting the needs of large-scale corporate capital. Such operators face further problems as they are regarded as unknown or 'risky' entities, while national/international operators are seen as a 'safe bet' in terms of credibility, financial situation and policing methods such as mandatory use of door staff. Moreover, ale-houses, taverns and saloons which provide leisure options for working-class groups do not feature in the priorities of large corporate operators or the entrepreneurial local state, who are both eager to change the image of downtown areas away from their industrial past and instead court the wealthier post-industrial service classes.

Plate 3.1 Tough regulations to curb public alcohol consumption; Liverpool city centre, UK, 2002

Waking up the neighbourhood

Ultimately, one stumbling block for the deregulation of the night-time economy is the clashes which emerge between night-time revellers and local residents. Where nightlife activity has grown in central and suburban areas, residents have become more vocal participants within the governance debate. In particular, as suburbanisation and the growth of decentred polycentric areas continue apace, places far from downtown areas are becoming nightlife destinations in their own right. These areas are often more attractive to developers due to the saturation of city-centre markets, cheaper property and less restrictive licensing, and a side-stepping of the disbenefits of central areas such as overcrowding, violence and lack of late-night travel.

Many traditional residential and suburban areas are becoming saturated with late-night bars, pubs, restaurants and multiplexes. This is partly associated with the clustered growth of nightlife-active groups such as university students, young professionals and single couples in more cosmopolitan and transitional parts of large cities (Chatterton, 1999). Growth usually occurs along established arterial routes, many of which are near university campuses or halls of residence, and examples in the UK include Clifton in Bristol, Jesmond in Newcastle, Headingly in Leeds, Selly Oak in Birmingham and Chorlton in Manchester. In many places, growth of this nature has provoked a strong response from established, older and wealthy residents. One paper in Newcastle ran the headline 'Drinkers turn upmarket suburb into "a hell hole"' (*The Journal*, 5 October 2000: 12). Meanwhile, in Bristol, vocal residents' associations successfully opposed several nightlife developments at public inquiries, drawing upon letter-writing campaigns, testimonies and covert video footage, winning a landmark decision from the local authority to restrict the granting of new licences to places which also serve substantial meals along with alcohol.

Many cities across the west have also encouraged a rapid, if selective, repopulation of downtown areas which has created tensions between partying and living. Initially, this trend is often spearheaded by gentrifying pioneers such as artists, writers and students. As such groups increase the bohemian feel and amenity value of an area, higher-income groups quickly move in. This continued growth of wealthy professional classes in central areas displaces many of the traditional lower-income residents, and stimulates demand for a variety of central amenities, including exclusive stylish nightlife activity (Solnit and Schwartzenberg, 2000). Ironically, although these new gentrifiers are often extensive users of city-centre nightlife, they are also the most articulate and vocal in asserting their objections to its negative aspects such as noise, disorder and vandalism.

The Ribera district in the heart of old Barcelona has experienced such change. An area near the city's waterfront formerly populated by merchants and sailors, in the twentieth century it experienced decline as the city grew outwards. Over the last few decades, Ribera has slowly been colonised by more bohemian groups of artists, students and musicians, due to its central location, its ramshackle medieval architecture and labyrinthine network of streets. As more affluent groups move in, the area has become one of the city's newest nightlife destinations, with the once quiet Passeig de Born in particular now packed with bars and restaurants and wandering crowds on weekend nights. A concerted campaign by local residents has ensued, with banners, lobbying the mayor for restrictions on the growth of nightlife, draped from numerous balconies and windows in the area (see Plate 3.2), and a dedicated website has been set up, linking up with similar campaigns in cities throughout Spain.

In spite of small victories which residents might gain, their perception is that licensing procedures are heavily weighted in favour of the trade, its legal advisors, statutory agencies and the court, who all use the system regularly. The public in

Plate 3.2 '*Ssst ... nens dormint* (Shhhh ... children asleep).' Residents voice their concerns over the growth in Barcelona's nightlife; Ribera district, Barcelona, Spain, 2002

whose name the whole process is said to be necessary are seldom mentioned and are uncertain how to participate effectively. Nor are the laws framed to take into account the cumulative effect of granting a large number of licences in one area on the rest of that environment.

Consumers, self-regulation and consumer democracy

Finally, consumers of nightlife are rarely, if ever, included in the governance equation. This is not surprising considering the complex array of motivations for a night out, which range from quiet socialising to hedonism, escapism and creative engagement. Nevertheless, one of the problems is that many regulators such as the police, local authorities and licensing magistrates have little understanding of the range of social groups, styles, identities and divisions within the night-time economy. It seems fair that those being regulated must be left some freedom to decide how nightlife is governed. Yet how this is to be achieved is unclear, and there are few examples of regulators consulting consumers as to their views on solving problems, let alone defining them. Moreover, the mainstream press is also susceptible to villainising young people for some of the excesses within nightlife (Hollands, 2000).

One concern is that the police and the local state often restrict nightlife options as they adopt a paternalistic, and often patronising, approach towards certain consumers. In particular, there are few opportunities for young people in city centres outside the narrowly defined 'consumption experiences' (Toon, 2000). Specific groups of young people such as Goths, punks and skaters are generally stereotyped and subject to police harassment, mainly due to their outwardly different appearance and perceptions that their presence will have a negative effect on retail (see Plate 3.3 and Borden, 2001; Snow, 1999). Younger people have fewer and fewer reasons to be in cities if they are not 'consumers'. Many parts of city centres at night, then, are largely alcohol-fuelled consumption ghettos with few public, flexible, mixed-age places. As we discuss in Part II, oppositional forms of nightlife (squats, free parties, raves) open up avenues for more democratic and participatory forms of regulation, while more historic forms of nightlife, like community pubs and bars and even the street itself which cater for lower-income groups, are increasingly being redeveloped or sanitised out of existence.

What we have highlighted in this chapter are the substantial changes over the last few decades in the regulation of nightlife. In particular, while the organisation and control of urban nightlife has shifted from a rather straightforward control-oriented set of mechanisms towards a seemingly more complex 'governance' approach, the interests of capital accumulation and the more well-off are clearly dominant. In particular, the needs of acquisitive nightlife corporations and developers have increasingly come first, aided by a more compliant and entrepreneurial local state. Similarly, while there have been voices of dissent, including more vocal residents'

Unruly kids are damaging our businesses, claim city traders

They have just Goth to go!

SQUARING UP – gangs of Goths gathering in Newcastle's Old Eldon Square have been causing a nuisance, hitting takings at shops there

Plate 3.3 Young kids villainised for hanging out downtown; Newcastle upon Tyne, UK
Source: *Newcastle Evening Chronicle*, 23 February 2001: 27

groups and overstretched police forces, most regulators have begun to embrace the 'new consensus' of profit-making first and dealing with social problems second. In this sense, there is evidence of a legacy of more historical modes co-existing with newer sets of priorities, and hence impulses towards both re-regulation (to maintain law and order, especially in terms of certain groups regarded as more troublesome) and de-regulation (to stimulate economic development). In Part II of the book, we explore in greater detail the changing balance between more entrepreneurial/pro-business, alternative/resistant, and historic/working-class nightlife.

4

CONSUMING NIGHTLIFE

Youth cultural identities, transitions and lifestyle divisions

Nightlife activity is an integral part of many young people's consumption lives (Hollands, 1995; Chatterton and Hollands, 2001; Malbon, 1999). It encompasses a complex array of youth cultural styles, experiences, identities and spaces – from free warehouse and street parties to sports bars filled with rugby players, corporate Christmas parties in lap-dancing clubs, rock bars, gay villages, young office workers in wine bars, ladies' nights out, student unions and underground clubs. Understanding the plethora of young adults' experiences in these various nightlife consumption spaces is further complicated by the fact that youth as an age category is cross-cut by a range of overlapping social identities like class, gender, race, ethnicity, geography and sexual orientation. Additionally, the development of a prolonged 'post-adolescent' phase and rapidly changing labour market transitions has meant that young people are continuing to engage in youth cultural activity for much longer periods of time.

With a significant proportion of urban-livers composed of young adults (Mintel, 2000: 14), many cities around the world are reasserting themselves to meet their consumption and entertainment needs (Chatterton and Hollands, 2002). Visiting bars, pubs and clubs is a core element of many young people's lifestyles. In the UK, 80 per cent visited pubs and clubs in 1999, an increase of 12 per cent over the previous five years (Mintel, 2000: 15). The 15–24 age group is also ten times more likely than the general population to be a frequent visitor to a club, with 52 per cent going once a month or more (Mintel, 1998: 22). The over-25 'rave generation' in the UK, Europe and North America continue to visit clubs, and as a result 'clubbing will remain as popular as it is now, and more sophisticated nightclubs will cater for die-hard party animals in their thirties' (Mintel, 2000: 45). There has also been a move away from traditional nightclubs, and their association with seediness, violence and excess, in the wake of the phenomenon of 'clubbing', which emerged from the 'one-nation' dance, rave and – to a certain extent – drug cultures of the late 1980s and early 1990s (Collins, 1997). While the club scene has diversified, grown and fractured along the lines of a number of smaller consumer groups, since the mid 1990s it has also been commercialised and a distinction between underground and mainstream clubs has been clearly drawn (Thornton, 1995). The experience of going to bars and pubs has also been transformed as traditional ale-houses, bars and taverns have given way to the emergence of style and café-bar venues and hybrid bar/clubs (Chatterton and Hollands, 2001).

Motivations for engaging in nightlife activity have also changed. While immensely varied, changes in the nightclub and pub/bar sectors mean that music, socialising, atmosphere, dancing and lifestyle performance and distinction are now among the main motivations for a night out (Hollands, 1995; Chatterton and Hollands, 2001), alongside more traditional reasons such as letting go, courtship or seeking casual sex. On the surface, more fragmented experimental 'mix and match' behavioural patterns are evident, in which different styles and types of venues are woven together to create a night out. However, young people's labour market position and social identities continue to contour their experiences of nightlife (Hollands, 2002).

Alcohol and, increasingly, drugs play key roles in shaping young people's nightlife activities. In more industrial times, drinking was associated with masculinity and the rituals and relationships of the workplace (Gofton, 1983; Brain, 2000). The changes which have been wrought over the last few decades through the advent of lagers, ciders, spirits and the presence of more women have altered the role of drinking in pubs and bars from a largely male ritual to a broader phenomena associated with fun, hedonism and lifestyle. Going out in urban centres has more and more become the preserve of the young, who in contrast to their more mature predecessors sometimes have problems handling their drink (Coffield and Gofton, 1994). In terms of young people's drinking habits, in the 1930s 18–24 year olds were the lightest drinkers in the UK population. By the 1980s this situation had reversed (Institute for Alcohol Studies, 1999; General Household Survey, 1999). In the USA, by the time they reach high school, over 60 per cent of young people have been drunk, while almost half of college students binge drink (Wechsler *et al.*, 1994). There is also evidence that many young people over the last ten years have turned towards illegal drugs such as Ecstasy, especially through the growth of club cultures (Henderson, 1997), or have chosen to smoke cannabis at home (Coffield and Gofton, 1994). In the UK, among 16–29 year olds, 49 per cent admitted having ever taken any drug, with 42 per cent consuming cannabis, 20 per cent amphetamines, 11 per cent LSD, 10 per cent Ecstasy and 16 per cent poppers (Mirrlees-Black *et al.*, 1998). Such changing habits have been cause for considerable concern from the brewing industry. Drinks producers have responded by 're-commodifying' alcohol products and creating what Brain (2000: 2) has called a 'postmodern alcohol market', in which the range of alcohol products is more extensive and product strengths are higher in an attempt to win back the 'rave' generation who are eager to find greater highs, and where products are based more around marketing and lifestyle advertising.

In Chapters 2 and 3 we have already outlined the varied ways in which production and regulation differentially contour young adults' consumption of the night-time economy. In this chapter we go on to provide a context for understanding youth cultural identities and lifestyles in the night-time economy. Slater (1997: 4), for example, reminds us that the consumption of goods and services is 'always carried out under specific social arrangements of productive organisation, technological abilities, relations of labour, property and distribution'. Furthermore, we also need to recognise that youth consumption is increasingly being framed by contradictory regulatory regimes involving the expansion of the night-time economy as part of urban regeneration strategies, more informal and 'privatised' forms of policing style and appearance, and a general sanitisation of activity through increased surveillance. While it is essential to see consumption as a symbolically meaningful and active relationship through which experiences, identities and feelings are created (Campbell, 1995; Bocock, 1993; Slater, 1997) and young people as active 'reflexive' participants in nightlife, our approach also stresses how the production and regulation of nightlife provision actively creates 'leisure divisions' within the youth population.

First, we look briefly at the historical role young people have played as consumers, including a discussion of subcultural theory (Hall and Jefferson, 1976). We then turn to an analysis and critique of some postmodern approaches to youth culture through the guise of terms such as 'club cultures' and ideas about 'neo-tribes' (Redhead, 1993, 1997; Muggleton, 1997; Bennett, 2000). Here, one suggestion has been that identities for many young people may be as likely to develop around the consumption of commodities, experiences and lifestyles as through engaging in economic production itself (Wilkinson, 1995; Willis, 1990). This notion, combined with the fact that nightlife cultures are changing so quickly and frequently through increased individualisation, fragmentation and globalisation, has supposedly fuelled a complex array of youthful lifestyles in the postmodern period (Miles, 2000).

However, our approach suggests a much closer link between the work and production sphere (Fine and Leopold, 1993), lifestyles and social divisions (Crompton, 1996; Miles, 2000), and the creation of a hierarchy of taste cultures (Bourdieu, 1984; Thornton, 1995), resulting in a more socially segmented set of youth consumption spaces and groupings (Chatterton and Hollands, 2001, 2002). We develop a more complex typology of youth identities in the night-time economy which reveals the formation of a number of consumption spaces, encompassing various social groups among young people. While minority elements of 'neo-tribal' and hybrid forms of youth identity and consumption clearly exist in the night-time economy, we argue that the main focus for the development of downtown nightlife is a more 'mainstream' form which exploits existing social cleavages in the population, segregating young adults into particular spaces and places. This is an active process, fuelled partly through an internal competition among youth groups to maintain social and status distinctions (Bourdieu, 1984). Finally, we expand on how the mainstream also works to marginalise older residual nightlife spaces and groups and alternative and oppositional youth cultural forms.

Consuming youth: from the 'affluent' teenager to subcultural analysis

As we have argued, young people today, particularly in the realm of nightlife, are heavily coveted members of consumer society. Yet historically, this was not always the case. Young people have only recently emerged as an important consumer entity in the twentieth century, mainly through the creation of a distinct period of adolescence. This does not mean that youth cultures did not exist previously – evidence of urban youth street cultures can be traced back to the early twentieth century, although these were often small and localised, as well as being almost exclusively (male) deviant/criminal and non-commercial (Pearson, 1983; Davis, 1992). Fowler (1995) provides evidence of the beginnings of a teenage market in the UK in the 1930s, with magazines, films and dance halls increasingly orienting themselves to young people. Palladino's (1995) work on the history of the American teenager relates the beginning of youth consumption in the USA with the development of mainstream, largely middle-class consumer tastes. Interestingly, Miles (2000: 108) locates these earlier forms of consumption as being 'in training for adulthood', rather than as youth cultures functioning to maintain some kind of independent status between childhood and adulthood.

The Second World War is often viewed as an important turning point in consumer society in general, with post-war economic growth bringing employment, increased production of goods and services and higher disposable incomes and consumer expectations among young people.[1] Notions of youngsters as 'affluent teenagers' (Abrams, 1959) having more money to spend, then, were characteristic of this period (Osgerby, 1998). This era, which was seen generally as the start of a mature consumer society, and a developing teenage market in the late 1950s and 1960s, around fashion, music, food, clothes and drink (see Bocock, 1993), led to ideas that youth was at the forefront of a conspicuous, fast-paced, leisure-oriented consumption phase (Stewart, 1992). Discussions of the younger generation as a 'class in itself' (Berger, 1963), functionalist analysis of the 'generation gap' (Parson, 1942), combined with the views of advertisers and manufacturers that there was a homogenous teen market, fuelled a somewhat Fordist approach to youth consumption. In other words, it was assumed that all young people shared primarily the same values and ideals and hence all would want to consume roughly the same types and styles of goods.

The main critique of the 'disappearance of class' and 'affluent teenager' thesis came from the Marxist-inspired class-cultural perspective developed by Stuart Hall and his colleagues at the Centre for Contemporary Cultural Studies (CCCS) in the early 1970s in the book *Resistance through Rituals* (Hall and Jefferson, 1976; also see Cohen 1997, chapter 2). Although they did not claim that their subcultural paradigm was intended to be a model which explained youth cultures in their entirety, Hall and Jefferson's (1976) work clearly demonstrated that there were very different sub-sets of youth subcultural styles, expressions and meanings, rather than a general and homogenous 'youth culture'. While it was true that CCCS concentrated almost exclusively on looking at male white working-class cultures, this early work contained one of the initial discussions of the absence of young girls (McRobbie and Garber, 1976) and the impact race and ethnicity had on white subcultures (also see Hebdige, 1979). A more careful reading of *Resistance through Rituals* also reveals some discussion of middle-class youth cultures and how they might be more differentially understood (pp. 57–71). In particular, Hall and Jefferson (1976: 60) prophetically hinted at the 'individualisation thesis', a dominant theme in 1990s youth sociology (see Furlong and Cartmel, 1997), by arguing that middle-class youth cultures 'are diffuse, less group-centred, more individualistic'. In fact, many of the youth subcultures CCCS looked at involved working-class youths borrowing from media-based middle-class cultures, punk being the most exemplary (see Hebdige, 1979), debunking the fact that the 1990s alone were characteristic of a 'pick and mix of youth styles'. Teddy boy and mod culture are other classic examples here (see Jefferson in Hall and Jefferson, 1976).

Yet one of the continuing problems with the subcultural perspective is that it remained somewhat within a 'correspondence' framework, with working-class youth leisure paralleling elements of their 'parent' or wider working-class culture (i.e. encompassing the values of masculinity, toughness, solidarity, territoriality – see Clarke's (in Hall and Jefferson, 1976) study of skinhead culture in particular), with close homologies existing between the cultures adopted and social position (Willis, 1978; although for an exception see Hebdige, 1979). Additionally, issues around gender and ethnicity, although raised, were not sufficiently well developed theoretically to deal with the expanding role young women and black youth were beginning to play in adolescent cultures of the 1980s (Griffin, 1985; McRobbie and Nava, 1984; CCCS, 1982; Pryce, 1979). Finally, the economic crisis of the late 1970s and early 1980s brought about by the decline of manufacturing, political conservatism represented by Thatcherism (Hall and Jacques, 1989) and Reaganism, and increasing concerns about policing youth through disciplining them in the labour market (Hollands, 1990) meant that the youth resistance paradigm of CCCS appeared far less appropriate as a form of analysis.

Ironically, as young people's declining economic position in the 1980s appeared to distance many of them from the rapidly growing consumer market, Miles (2000) argues that youth consumerism during this period actually became more of an accepted 'way of life'. In the USA, the work of Gaines (1991) unearthed a nihilistic culture of 'no future' among many young people coping with economic downsizing and industrial decline, while others spoke pessimistically about 'slackers' (Kellner, 1994) and 'Generation X' (Howe and Strauss, 1993; Coupland, 1992). Stewart (1992: 224) also suggests that youth consumption in this period became increasingly more incorporated into the mainstream, representing core values rather than rebellion, and becoming more individualistic yet more homogenous in many aspects. Of course, this was also the era of the 'yuppie', not to mention the expansion of higher education, fuelling the beginning of both a gentrified and student leisure and consumption market. Yet it also represented a time of increasing concern about 'race' and young ethnic identities (Hall *et al.*, 1978; CCCS, 1982) and changes in the relationship between gender and generation (McRobbie and Nava, 1984). By the end of the decade, youth culture had returned as the 'second summer of love', with the rise of acid house and rave, cultures described as decidedly fragmented and postmodern, encompassing elements of punk DiY, gay culture, disco and New-Ageism (Redhead, 1993). Youth subcultures were viewed as in terminal decline in the 1990s, replaced by more loose, fragmented, hybrid and transitory global 'cultures of avoidance' (Redhead, 1997). Similarly, as youth cultural styles continue to explode up the age hierarchy, the teenage consumer is proclaimed dead.

Consuming youth: postmodern club cultures and neo-tribes

One of the major impacts on social theory in the last decade has been postmodernism (Harvey, 1989b) and more specifically the development of such a perspective in the analyses of culture and consumption (Wynne and O'Connor, 1998; Slater, 1997; Featherstone, 1991). This general approach has had a huge impact on a range of fields, including youth studies. As Cohen and Ainley (2000: 86) argue, by 'the early 1990s the textualisation of cultural studies was complete' and, by implication, the study of youth cultures was decidedly influenced by postmodern thinking exemplified by a somewhat diverse set of writings around 'club cultures' (Redhead, 1993, 1997; McRobbie, 1993), and 'neo-tribes' (Bennett, 2000; Malbon, 1999; Maffesoli, 1996).

Here, symbolic, hybrid forms of consumption are seen as crucial for an understanding of contemporary youth cultures, as is an assumed loosening of traditional sources of identity and a blurring of traditional social relations like class, gender, ethnicity, sexual orientations and national identities (Roberts, 1997). Moreover, the past two decades have seen an ongoing shift from collectivised to more individualised and privatised forms of consumption (Clarke and Bradford, 1998: 878). The difficulty for youth today, according to Willis (1990: 13), is that many 'of the traditional resources of, and inherited bases for, social meaning, membership, security and psychic certainty have lost their legitimacy for a good proportion of young people'. In this regard, he goes on to use the term 'symbolic work' to express young people's everyday creation of meaning and identity in the social world. Willis (1990: 10) defines the term as:

> the application of human capacities to and through, on and with symbolic resources and raw materials (collections of signs and symbols, for instance the language we inherit as well as texts, songs, films, images and artefacts of all kinds) to produce meaning.

According to Willis, symbolic work is as 'necessary' and as fundamental to the production and reproduction of daily life as material sustenance itself. The importance of this notion in the debate about the changing relation between work and leisure for young people is central. For instance, if youth is a social group more likely to be excluded from productive work, meaning and identity for them, it might be argued, is as likely to be undertaken on and around the products of material production as it is through engaging in economic labour itself.

Hence, a proportion of young people may experience an ambivalence in defining themselves in relation to work and occupation. As Ball *et al.* (2000: 7) state: 'occupational status and work may not be that important in the lives of the young ... other sources of identity and identification deriving from music, fashion and leisure may be more central to how they think about themselves' (also see Hollands, 1998). Some young adults are seeking to redress the balance between work and play, rejecting low-paid low-skilled jobs and reacting against the constraints of the work ethic (Kane, 2000; Wilkinson, 1995), while a minority may have become dissatisfied with consumer and corporate culture altogether (Klein, 2000). Additionally, the changing economic and social position of young women, and their greater insertion into youth cultural activity, consumption and nightlife (see Chapter 7) has led to the questionning traditional gender roles, while the mixing of black and white youth cultures has worked to create some examples of ethnic hybrids (Jones, 1998; Back, 1996). Finally, the expansion of higher education and the creation of student consumption identities, with the ascendance of more middle-class forms of youth culture (Roberts and Parsell, 1994), have also blurred traditional leisure demarcations.

Further processes at work here are the globalisation and internationalisation of youth cultures (Miles, 2000; Bennett, 2000). It is argued that there has been a move away from nationally based youth cultures to more global and eclectic formations. The appropriation of American rap and hip hop by youth across the world is one example here (Mitchell, 2001). Unfortunately, much of the discussion surrounding this globalisation process often fails to give enough emphasis to the growing role that international capital and corporatisation play in the production of increasingly

Plate 4.1 Clubbing has transformed many aspects of nightlife and has become an important 'way of life' for many young people; Rockshots nightclub, Newcastle upon Tyne, UK, 2001

Source: Louise Hepworth

standardised and homogenised youth cultural provision. The contribution made recently by youth geographers is more suggestive here, hinting that despite an increased globalisation of youth cultural activity, locality and 'socio-temporal' contexts remain important (Skelton and Valentine, 1998).

The first real alternative youth cultural paradigm to challenge the notion of subcultures from CCCS came out of Steve Redhead's postmodern-inspired work (see Redhead, 1993, 1997). Redhead's position can best be summed up by the assertion that class subcultures – if indeed they ever existed – have now been surpassed by 'club cultures' (Redhead, 1997): loose, globally based youth formations grounded in the media/market niches of contemporary dance music (see Plate 4.1), and the regulation and indeed criminalisation of youth cultures characterised by a kind of 'hedonism in hard times' (Redhead, 1993). Redhead's work clearly captures important transformations in youth cultures over the last thirty years, and is a useful benchmark in any analysis of contemporary youth identity and nightlife. For example, his contribution to ideas about fragmentation and the 'cultural mixing' of styles, highlighting the global nature of contemporary youth cultures and the state's obsession with regulating them, is exemplary.

Yet, as with many postmodern-inspired analyses, there are also some fundamental shortcomings and weaknesses in his approach. First, Redhead's use of the term 'club cultures' is vague and ill-defined, not to mention being somewhat all-encompassing. In essence, despite its importance, clubbing does not encapsulate all types of youth leisure and is actually only one form of nightlife activity. Second, postmodern conceptual frameworks often lack an underlying economic and spatial context. While it might be argued that club culture has become somewhat of a global phenomenon due to the role of the media and the increased mobility of young people, nightlife experiences for many remain largely rooted in specific localities (see Chatterton and Hollands, 2001). Third, Redhead's (1993, 1997) analysis of contemporary youth cultures is somewhat unidimensional, with too much emphasis on regulation, criminological and media-based forces and determinations, rather than examining the whole political economy of popular culture (such as ownership patterns, corporatisation, commercialisation, etc.).

Other postmodernists have been more explicit in arguing that there is no relationship between contemporary youth styles and their articulation with social relationship. Muggleton (1997: 198), for example, argues: 'Post-subculturalists no longer have any sense of subcultural authenticity, where inception is rooted in particular socio-temporal contexts and tied to underlying structural relations.' As such, style is seen as constituted solely through consumption, and is 'no longer articulated around the modernist structuring relations of class, gender, ethnicity or even the age span of youth' (*ibid.*: 199). Yet one might legitimately ask: how can youth cultural activity not be rooted in a socio-temporal context, as it is there that it has to be 'performed'? Similarly, how can style no longer be influenced by existing social structures and relations, even if they are changing? In an attempt to distance themselves from the class-based analysis of CCCS, many postmodern accounts might be accused of overlooking the whole issue of youth inequalities altogether.

Others, meanwhile, utilise the postmodern-inspired idea of tribes or neo-tribes to theorise aspects of contemporary youth cultures (Bennett, 2000; Malbon, 1999). The suggestion here is that today it is virtually impossible to tie youth cultural styles and musical taste down to any strict notion of social class (although see Martinez, 1999, for a counter-argument). This is particularly the case when it comes to looking at dance music, which itself has identifiable tribal elements. For example, dance events involve quite fleeting and loose forms of sociation, rather than any rigid structure or internal logic. Clubs and club events become temporary sites of 'social centrality' (Hetherington, K., 1996), providing intense but short-lived moments of collectivity and indeed 'ecstasy'. As such, young people, rather than remaining committed to a static and formal organisation, move easily in and out of particular styles, forms of music and temporary communities. What we do not appear to know about these tribes from available research is how common they are, or whether or not they represent existing fragments of older social divisions or the creation of new forms of social hierarchy and distinction (although see Henderson [1997] on gender).

New concerns with hybridity, tribes and post-subculturalist forms, although interesting, have often meant that 'En route questions of class trans/formation were rather left to one side' (Cohen and Ainley, 2000: 84). While postmodern analysis appears on the surface more amenable to examining how fragments of social identity like race, gender and sexuality combine and articulate with one another in leisure and consumption, often their analysis of youth simply collapses such distinctions through its emphases on decentred subjectivities. Clearly, there are examples of contemporary postmodern youth cultural activity to be found. Yet the same question that often plagued earlier theorists of youth subcultures remains: are postmodern examples any more representative or empirically demonstrable among the young than minority subcultures were? The problem with postmodern theorising and methodology here is that they do not appear to find inequalities or stratified youth cultures, partly because they are not looking for them. Additionally, they do not have a theoretical framework that allows them to see social division within the leisure sphere, differentiate young people on the basis of their economic or domestic situation, or account for the role locality plays in youth culture.

There is a sense in which much postmodern analysis of youth ends up either as a discussion about endless chains of signification, competing generational multi-cultures or decentred subjectivities (Cohen and Ainley, 2000). At best, attempts are made to show what these hybrid youth cultures say about so-called postmodern society and culture – usually in the form of reaffirming its existence, thereby masquerading as both 'evidence' and 'cause' simultaneously. Postmodern approaches have also curiously lacked the tools of analyses for understanding the 'economisation' of the cultural industries – in other words, looking at the impact of capital investment and disinvestment on lifestyle, consumer choice and the standardisation (Ritzer, 1993) and corporatisation of youth cultures. Consumption, as Edwards (2000: 3) argues, is not 'simply a matter of style, it is also a matter of money and economics, social practice and social division'. Existing social divisions and transitions, locality and corporate investment patterns, then, continue to be important contexts for understanding consumption 'choices'. Next, we take a closer look at a range of work on youth transitions, lifestyles and social divisions, to help us unpack how youth cultures in the night-time economy are not just fragmented but are also segmented.

Youth transitions, lifestyles and consumption

The various weaknesses of postmodern theorising around youth cultures have lead others to reassert how social divisions continue to influence consumption decisions and lifestyle choices. This has been partly accomplished through studies of youth transitions, in addition to a number of approaches concerned with theorising youth lifestyles. Transition studies of youth in the labour market (Roberts, 1995), education (Griffin, 1985; Willis, 1977), training (Mizen, 1995; Hollands, 1990) and the family household (Morrow and Richards, 1996), have continued to emphasise the structural aspects of young people's experience and highlight the persistence of social inequalities and divisions that exist around class, gender, ethnicity and sexual orientation. For example, studies of the youth labour market have shown both the disadvantages that labour market restructuring has brought to many young people (Roberts, 1995), and the persistence of intra-class and gender social divisions in youth transitions (Banks, 1991). However, with exceptions, very few of these studies have been helpful in determining how such transitions connect up to youth cultures, leisure and lifestyles.

Indeed, Miles (2000) has argued that many transition sociologists have 'exaggerated the impact of structural influences on young people's lives' (p. 9), treating them as 'troubled victims of social and economic restructuring' (p. 10). Such work, he suggests, blinkers us from important concerns with lifestyle, culture and identity. Age transition studies have often assumed a predetermined and linear sequence of events and transformations that are no longer applicable to the lives of many young people. They have also often been narrow, economically driven and positivistic in their orientation (Cohen and Ainley, 2000: 80). This is not to suggest that the economy is unimportant, only that there has been a too easy 'reading off' of identity from young people's economic or educational pathways (Banks, 1991). Feminists have argued that transition studies have also been limited to the 'masculine' realm of employment rather than thinking more widely about household transitions (Griffin, 1993), while others note an absence of a discussion of skills and knowledge and sexual or leisure transitions (Cohen and Ainley, 2000).

MacDonald *et al.*, (2001) suggest that, increasingly, transition studies are focusing on more complex, broken, interrupted, unpredictable, contingent transitions and 'critical moment' events and active biography making, rather than assuming linear pathways. A recent example is a study of youth choices and transitions by Ball *et al.*, (2000), which takes on much wider notions of transition by looking at the combined interaction of education, training, labour market, household and leisure pathways, while maintaining a notion of the impact of social divisions. MacDonald *et al.*, (2001) also cite work concerned with youth from a deprived northern estate (Johnson *et al.*, 2000), which provides biographies including discussions of drug use, anti-school cultures, and involvement in dance cultures and peer groups in 'creating transitions' (also see MacDonald and Marsh, 2002). One might, however, argue that much of the recent transitions literature, while increasingly interested in 'cultural' aspects, has tended to focus largely on disadvantaged youth while ignoring other whole categories of young people, like service workers, further and higher education students, privileged professional workers and studies of middle-class youth in general. The result is that there is very uneven information about the transitional experiences of a range of young people and how this might impact on youth cultural activity or lifestyle.

Another contribution to thinking about social divisions in the leisure sphere is through a general concern with theorising lifestyle. While the notion of lifestyle can be traced back to the work of some of the classical theorists including Weber, Simmel and Veblen (Bennett, 2000: 25), it has also been utilised by postmodernism writers such as Featherstone (1991) as well as in more sociological analyses concerned with exploring its relationship to social class (Crompton, 1998). As such, part of the argument here revolves around whether culture and lifestyle, rather than the economy or economic position, gives rise to identity. Pakulski and Waters' (1996) notion of 'status conventionalism' suggests that occupation is only one marker of stratification and that status is as likely to emerge from cultural consumption in today's society, while Crompton (1996, 1998) continues to maintain that identity formation is connected to social divisions and economic relations. Some of the most relevant theorising and research for our work on nightlife (see Chapter 5) has come from studies of lifestyle concerned with the middle class and gentrification (Butler, 1997; Savage and Butler, 1995; Savage *et al.*, 1992; Wynne and O'Connor, 1998), while Miles (2000) has attempted to apply the concept to youth.

Roberts (1997: 3) suggests that young people are often commonly cited as being especially associated with the rise of lifestyle-based identities, while Miles (2000: 106) argues that if young people are valued as anything in contemporary society, it is through their role as consumers. As Campbell (1995: 114) elaborates: 'Not only is youth ... necessarily a life-cycle stage in which experimentation with identity is a central concern, but it is also a stage when individuals generally have little in the way of regular, fixed financial commitments.' Miles (2000) defines lifestyle as an 'active expression of a "way of life"' (p. 16) and 'as the outward expression of an identity' (p. 26), and discusses the concept in relation to a range of theoretical approaches and conceptual debates. One of the advantages of using the term is that it is broader than either 'club culture' or 'subculture', and potentially accounts for a diversity of youth cultures and styles – mainstream as well as alternative. Additionally, one can think of lifestyle as being at least 'contoured' by social divisions and transitions (also see Crompton, 1998; Savage *et al.*, 1992).

Similarly, Bourdieu's method of social inquiry provides a useful framework for understanding the relationship between social space and lifestyles. His work locates:

> the production of lifestyle tastes within a structured social space in which various groups, classes and class factions struggle and compete to impose their own particular tastes as the legitimate tastes, and thereby, where necessary, name and rename, classify and reclassify, order and reorder the field. This points us towards an examination of the economy of cultural goods and lifestyles.
>
> (Featherstone, 1991: 87)

Bourdieu's central concepts of field, habitus and different forms of capital are useful here. A field, in Bourdieu's words, is 'a space in which a game takes place (*espace de jeu*), a field of objective social relations between individuals or institutions who are competing for the same stake' (quoted in Moi, 1991: 1021). Each field requires a practical mastery of its rules (Wacquant and Bourdieu, 1992), and once this is achieved it is possible to confer or withdraw legitimacy from others in that field without this being recognised as an act of power. Habitus is the collection of durable dispositions which allow people to know and recognise the laws of the field through practical experience within it. It can be thought of as the mediating mechanism between the objectivity of social reality and the subjectivity of personal experience.

Bourdieu's social spaces (fields) are also mediated by various types of 'capital': economic – access to various monetary resources, social – resources which one accrues through durable networks of acquaintance or recognition, and cultural or informational – competence and ability to appreciate legitimate culture related, in particular, to level of education. People actively invest social and cultural capital to realise economic capital, and vice versa (Savage *et al.*, 1992: 100). The interplay between different types of capital leads to the emergence of different social groups such as industrialists, new petite bourgeoisie and artists. Bourdieu, then, constructs a social space of tastes on the basis of the possession of different types of capital, not just income. Therefore, groups can be distinguished according to cultural practices, tastes and consumption preferences in many areas such as art, food, clothing and sport.

Thornton's (1995) work on youth and club cultures represents a reworking of various aspects of Bourdieu's model and offers a useful way of thinking about nightlife and lifestyles. Club cultures for her are the 'colloquial expression of youth cultures from whom dance clubs and their eighties offshoot, raves, are the symbolic axis and working social hub' (Thornton, 1995: 3). For Thornton, these club cultures are essentially taste cultures, and utilising the work of Bourdieu (1984), she develops the notion of 'subcultural capital' to explain how hierarchies of taste and a diversity of dance styles/cultures evolve. In her view these are based around notions of authenticity versus inauthenticity, hipness versus mainstream, and underground versus the media/commercialisation.

Thornton's approach, with its emphasis on young people defining themselves in relation to their peers and the role the media plays in constructing such subcultures, again forms an important contribution towards comprehending youth nightlife, despite once again being confined to clubbing *per se*. Yet there are problems and inconsistencies with her approach. For instance, the mainstream/underground divide is rather dichotomous and overlooks residual cultures, and it is not clear from her empirical work what subgroups go towards making up the mainstream. Her analysis also begs the question of how cultural capital, and indeed her own concept of subcultural capital, relates to social categories like class, gender, ethnicity and, importantly, place. With regard to class, for instance, Thornton (1995: 13) rather ambivalently argues that 'subcultural capital clouds class background', while stating that 'class is not irrelevant' (p. 12). She is somewhat clearer with respect to gender when she argues that young men are more likely to possess larger amounts of subcultural capital than young women (*ibid.*: 13), and she also mentions that 'race' is a 'conspicuous divider'.

Cohen and Ainley (2000) argue that new inequalities organised around the post-Fordist labour market, education cleavages, spatial inequalities of neighbourhoods and social networks, and new segmentations in the leisure and consumption spheres, can begin to show how youth transitions can be related to cultures without slipping into either a 'structureless' postmodernism or a structurally 'determinist' sociology. In terms of labour market restructuring, as Harvey (1989c) argues, the general impact has been to create not only a polarisation between core secure jobs and insecure peripheral ones, but greater flexibility in the labour process and more differentiated transitions and consumption patterns. For youth in particular, it has meant both an extension and fragmentation of transitions into the labour market and a delay in transitions into adulthood. The post-Fordist labour market has not only worked to delay and interrupt traditional youth transitions, but it has also worked to complexify them. Young people today make a bewildering array of labour market transitions, including moving through various training and educational routes through to temporary, contract and part-time work, and in some cases to secure full-time employment. As some commentators have argued, this fragmentation has resulted in greater 'individualisation' among young people, where they are more prone to take personal responsibility for both their economic fate and the construction of their own identity (Roberts and Parsell, 1992; Furlong and Cartmel, 1997).

Such fragmentation partly obscures the continuation of stark social divisions within both the general and the youth population. Bauman (1997), for example, refers passionately to the dichotomy between a relatively affluent and secure group tied to the work ethic but seduced by postmodern consumption enclaves and an endless stream of new products (which includes nightlife), and the swelling ranks of the dispossessed, the redundant and the criminalised. These latter groups, the 'flawed consumers' (*ibid.*: 14) who are time-rich but cash-poor, are no use to today's consumer-oriented markets. In a similar way, Hutton (1995) refers to a three-fold segmentation of the labour market – 40 per cent well-off core permanent workers; 30 per cent in unstable employment; and 30 per cent excluded. What is being stressed here, then, is the endurance of significant segments within youth populations (Furlong and Cartmel, 1997) – between, for example, the unemployed or those dependent on welfare benefits or unstable employment, university students and those in high-level training, and young professionals in stable, well-paid and mobile employment. Differences between and within these categories of young people are underpinned by a host of factors such as educational background, parental income, household situation, ethnicity, gender, sexual orientation and geographical location. For example, as MacDonald and Marsh's (2002) work shows, there is a growing section of young people exposed to unstable labour market conditions or welfare dependency, whose participation in certain forms of youth cultural activity is often extremely curtailed to certain activities and certain places. For a whole host of reasons such as price, geographical marginality, racism or merely feelings of disenfranchisement, significant groups of young people are largely restricted to leisure in their homes and estates, or community pubs and social clubs.

Other sections of working-class youth have adapted to the new post-Fordist service labour market, and while inherently insecure and underpaid, many attempt to engage with mainstream commercial lifestyles. Yet even here there are subtle leisure inequalities. As Ball *et al.*, (2000: 6) argue, 'as always, some are more able than others to participate in the experiential commodities of youth consumption. Going clubbing, drinking, smoking, recreational drugs, fashionable clothing and other lifestyle accessories do not come cheap.' They go on to mention the purchase of cheaper clothing imitations and 'look-alike' labels as a way of dealing with such inequalities. Others may save, beg and borrow in order to transcend what is in reality unaffordable, in order to purchase lifestyle accessories which are literally beyond their economic means.

Students have come to be an identifiable consumer group. Students in further and higher education have substantially increased, with the bulk of this group made up of 'non-traditional' students who are often older, locally based and living at home, sometimes working class and increasingly female (Chatterton, 1999). While the diversification of the student body has served to broaden the nature of cultural activity somewhat, traditional wealthy adolescent students remain strongly targeted consumers, with identifiable swathes of all British cities devoted to meeting their educational, housing and nightlife needs (Hollands, 1995; Chatterton, 1999).

Finally, many young people emerging from universities and professional qualifications are able to enter into a world of relatively stable employment and consumer lifestyles. In metropolitan centres which have benefited from the spoils of professional and business service decentralisation, entertainment and cultural provision for this relatively privileged young middle-class group is plentiful. These young urban service workers, knowledge professionals and cultural intermediaries – the denizens of the re-imagined urban landscape (Featherstone, 1991) – as well as accumulating economic capital also seek symbolic capital and status through consumption, and hence are implicated in a virtuous cycle of growth. Bauman (1997: 180) has called such groups 'postmodern sensation gatherers' who constantly search for new exhilarating 'peak' experiences. Hence, nightlife, like other forms of consumer practices, offers the potential for the consumption of positional goods for this group of 'white-collar' service workers to distinguish themselves from other social groups (Bourdieu, 1984). Such distinction requires constant effort and the continual reworking and upgrading of new nightlife concepts through which participants can maintain social 'distance' and social status. Numerous studies have examined these new class factions in urban contexts (Wynne and O'Connor, 1998; Savage *et al.*, 1992; Bourdieu, 1984), suggesting that they have stimulated an explosion of cultural goods and services.

Of course, these rather general categories represent only an analytical starting point. The task remains to develop a more sophisticated theory of social division beyond old dichotomies or correspondence class theories (Crompton, 1998) by taking into account how labour market conditions and structures have changed (i.e. class displacement, realignment and recomposition) and lifestyles have expanded. One should not assume a completely homogenous working-class youth cohort, either occupationally or in terms of lifestyle (Roberts, 1997). Cohen and Ainley (2000), for example, point to differences in the way some working-class kids are able to utilise cultural power and knowledge to achieve economic power, while Ball *et al.*, (2000) provide empirical evidence of a wide spectrum of working transitions based on educational, occupational, domestic and leisure factors. For instance, work and training in style occupations and the new cultural economies (hairdressing, fashion, nightlife, tourism, etc.) may help produce very different youth cultural identities from those produced through traditional service employment like retail and office work.

Even those employed in similar occupational strata may differ in their leisure choices, with some content with mainstream choices while others try to achieve higher social status through consumption of a more premium lifestyle. There are also documented distinctions evident within the young middle classes in terms of occupational and lifestyle identities (see Ball *et al.*, 2000; Wynne and O'Connor, 1998). Savage *et al.*,'s (1992) three-fold distinction of ascetic, postmodern and undistinctive lifestyles within the middle classes is useful. However, such work does not really deal well with the youth dimension, although Ball *et al.*,'s (2000) work is suggestive of different lifestyle patterns within middle-class youth in some instances. Higher and further education also provide many young people with a degree of latitude in which to forge more distinct identities within the general category of studenthood, even though they retain quite strong class connotations. And, of course, all of these above groups are made more complicated by gender, ethnicity and sexual orientation. Nevertheless, there remain clear transitional differences between the young poor literally unable to make the transition to work, adulthood and high-consumption lifestyles (MacDonald *et al.*, 2001), and those more middle-class patterns of avid consumerism, delayed or postponed employment, or a rejection of traditional definitions of work (Gorz, 1999). In the end, most young people come to recognise that their eventual status will be crucially influenced by their relationship (either negative or positive) to the labour market, and other key resources such as credit, housing, education and leisure (see Crompton, 1996).

As we go on to suggest in Part II, in the context of young people's use of urban nightlife there is a continuing polarisation between highly mobile, 'cash-rich, time-poor' groups of young people who can access a variety of entertainment choices, a large mainstream middle ground, and those experiencing unemployment, unstable employment, low wages, high debt and restricted leisure opportunities. Additionally, we outline how nightlife experiences are still contoured by higher education (Chapter 6), gender and sexual identity (Chapter 7) and ethnic differences (Chapter 8). In the next section we outline our own typology of consumption spaces in the night-time economy and briefly sketch out how they are inhabited by different segments of the youth population.

Consuming youth nightlife: mainstream, residual and alternative spaces and groups

Contrary to either a free-floating, individualised, 'pick and mix' story of post-modern youth cultures or a simple 'class correspondence' model of leisure, nightlife activity is instead characterised by hierarchically segmented consumption groupings and spaces in cities, which are highly structured around drinking circuits or areas, each with their own set of codes, dress styles, language and tastes. Despite post-modern assertions that hybrid eclectic styles today make it more problematic for young people to distinguish between themselves and other youth cultures (Muggleton, 1997: 199), many young people have no such difficulty in identifying varied nightlife spaces inhabited by different social groups. Below, we reveal three main types of nightlife consumption spaces – mainstream, residual and alternative – that are inhabited, made and remade by a range of competing youth cultural groupings characterised by various cultural tastes and polarised transitions. Clearly, such consumption groups and their spaces are fluid and overlapping rather than rigid categories. For example, students, gays and groups of women are identified at a 'common-sense' level on a night out, but clearly all have dominant, residual and alternative variants. Finally, our research reveals an increasing domination of gentrified mainstream nightlife space and its marginalisation and usurpation of residual and alternative places and groupings.

Primary among these segmented spaces is one that we have labelled 'the mainstream', as it stands out as the dominant mode of young adults' consumption of urban nightlife culture (also see Chatterton and Hollands, 2002). This is despite the fact that it proves rather hard to define precisely, partly because it often functions as 'the other' (what other less stylish young people consume) and partly because of its complex make-up (Thornton, 1995). Although at first the mainstream may seem a rather blunt instrument for understanding the diversity of groups and places downtown at night, we use it here to encompass the well-recognised weekend commercial provision of chain and theme pubs and traditional nightclubs which are characterised by smart attire, chart music, commercial circuit drinking, pleasure-seeking and hedonistic behaviour. Venues here include large, somewhat tacky (what many refer to as 'cheesy') established nightclubs, theme and chain super-pubs and bars owned by multinational entities, encompassing both dated 'chrome and mirror' styles and more contemporary minimalist café-bar styles. Hence a range of social groups and identities can be considered part of this large social space of the commercial mainstream, and it is inhabited differentially by a range of youth groups from across the middle-ground social spectrum, including the young working class, students and a range of young professionals. As such, it brings together a variety of traditional nightlife behaviours, alongside emerging middle-class, upgraded, sanitised and gentrified styles. The important point for our analysis here is that the unifying factor of the mainstream is commercially viable and profit-oriented provision. Further among all those who consume the mainstream there is a desire for ease of access and familiar environments, and a rather uncritical – or perhaps undeclared – stance towards issues such as profit, ownership and exclusion within the city at night. This is not to say that people are cultural dupes in their leisure time, but rather, obviously, that the majority of us choose to focus on socialising, pleasure and status-seeking during a night out, rather than critiquing and academically deconstructing the 'ins and outs' of what makes nightlife.

The mainstream, then, is consumed by numerous overlapping groups, entails a range of types of provision and is internally divided in terms of behaviours and styles. However, rather than a random and chaotic mix of postmodern styles, as we go on to show in Chapter 5 the mainstream has a number of internal social divisions and is a segmented social space reflecting, for example, intra-class, gender and local stereotypes (see Hollands, 2000). Differentiation here is constructed at two levels around labour market position and various taste communities based upon style, music, fashion and argot, leading to a preference for particular nightlife social spaces. As the boundaries of these two levels shift over time, youth groupings within the mainstream compete for new positions along the nightlife hierarchy and constantly seek out new fashions and spaces to re-differentiate themselves.

In this regard, one of the main transformations within the mainstream is a clear attempt to introduce more up-market and sanitised environments (Chatterton and Hollands, 2001). Traditional nightclubs, with their rather seedy images of violence, drunkenness, chart music and reputations as 'cattle-markets', have been joined downtown by smaller, safer and more niche-oriented specialist clubs, and larger super-pubs owned by multinational entertainment operators. Similarly, many of the new bars, characterised by terms like 'café culture' and 'style venues', have attempted to create an up-market feel, with polished floors, minimalist or branded décor and a greater selection of designer drinks (see Plate 4.2). As such, many of these new venues are perceived and experienced by young adults as more cosmopolitan (Chatterton and Hollands, 2002).

This cosmopolitanisation process has been partly driven by two important influences on youth nightlife spaces over the last thirty years – the increased presence of young women and the overlap between mainstream and gay cultures. As we go on to argue in Chapter 7, while many nightlife spaces remain heavily gendered and 'sexed', a loosening of sexualised identities has paradoxically opened up new opportunities for young women and gay youth to express themselves, while drawing them into the mainstream particularly through the

Plate 4.2 The wealthy and fashionable at play; the Fine Line Bar, Bristol, UK, 2000

gentrification process (for gender, see Chatterton and Hollands, 2001; Difford, 2000a; for gay gentrification see Whittle, 1994; Chasin, 2001).

The gentrifying mainstream also attracts elements of the wealthier end of the elite student market and an increasingly older, more mature and upwardly mobile section of the local working-class population, who clearly view such places as sites to express their perceived mobility and status. Part of the explanation for the trend towards a more 'exclusive' mainstream can be found in theorisations about 'subcultural capital', an emphasis on the importance of 'peer distinction' (Thornton, 1995) and the acquisition of positional goods and hierarchy within nightlife youth cultures. The idea that these cultures are essentially 'taste cultures' (Bourdieu, 1984) – in this case the acquisition of knowledge about what is 'trendy' – partly explains how more specific hierarchies of style evolve and constantly change. Differences between youth groupings and cultures fuel a never-ending and tail-chasing game of 'cool-hunting' (Klein, 2000). To sum up, the commercial mainstream is a differentiated playground for the active production and reproduction of various social groupings of young people, keen to refashion their consumption identities in relation to their peers and in light of their own socio-economic and labour market positions. In essence, it is the fluidity of the mainstream, and its constant trend towards gentrification and stylisation, which drives the system forward in a constant search for new consumption experiences.

However, while the mainstream succeeds in meeting the style aspirations of white-collar workers including young professionals, graduates and service employees, ironically it works to produce a rather predictable, serially reproduced and rationalised environment. Longhurst and Savage (1996) argue that it is actually often commonality and conformity, as well as a desire for difference and distinction, which marks out this large middle-class, middle-brow faction. As we consider in more detail in Chapter 5, such aspirations signify an increasing desire for safe, risk-free consumption environments. What Wynne and O'Connor's (1998) study of middle-class urban-livers in Manchester found was that this affluent, mobile and largely childless group were not particularly experimental or postmodern, but represented a large 'open middle' of consumer tastes who avoid environments where access is unfamiliar. Much of this lack of experimentation reflects balancing work, family and social commitments for such 'cash-rich, time-poor' groups, but it also reflects wider personal insecurities and a retreat to the familiar, in what is regarded for many as an increasingly complex and dangerous world (Beck, 1992). Hannigan (1998: 70) comments that to reduce the concerns of the middle classes over the safety of downtown areas, consumption experiences have become more sanitised and programmed. By regulating access, branding and rationalising product sales and controlling movement through design of interiors, music and furniture, urban playscapes more and more reflect this dystopic, standardised world of McDonaldisation (Ritzer, 1993). In this sense, Clarke and Bradford (1998: 875) point out that those consumers in stable employment are 'seduced' by the commercial mainstream and its marketing machine, which generates ever more sophisticated needs and desires and an all-important escape route from some of the less palatable realities of urban life.

Smith (1996) in his work on the gentrified city suggests that up-market developments represent a business and middle-class backlash against the urban working-class population. In light of this dominance of the mainstream, and more specifically the growth of up-market style bars, the fate of more 'residual' forms of urban nightlife consumption have rapidly diminished in many cities. By 'residual nightlife' we are referring to those more traditional spaces of the city inhabited by increasingly excluded sections of the youth population, including community bars, pubs and ale-houses, as well as the general purview of the street, the neighbourhood ghetto or council estate. These spaces are inhabited, predominately by the young urban poor or Bauman's (1997) 'flawed consumers' – the unemployed, welfare-dependent and criminalised who represent the 'other' city of dirt, poverty, dereliction, violence and crime (Illich, 2000), in contrast to the stylish gentrifying mainstream. Connected to the industrial city, residual spaces are now surplus to requirements in the newly emerging post-industrial corporate landscape, replete with its themed fantasy world and expanding consumer power. Similarly, residual youth groupings, including the young unemployed, homeless, poor and often those from ethnic backgrounds, are excluded, segregated, incorporated, policed and in some cases literally 'swept off the streets'.

As such, the younger generation of this 'socially excluded' section of the population have a limited number of choices of nightlife. They can steer clear of the increasingly 'middle-class' city (Ley, 1996) and socialise within what is left of their rapidly declining communities, and in some cases ghettos. Community pubs on estates, social clubs, house parties and the street (Toon, 2000) may remain their only choice of venue. Others continue to inhabit those declining traditional city spaces made up of 'rough' market pubs, ale-houses, saloons and taverns, earmarked for redevelopment, while local police often view them as sites of petty crime. Finally, some venture out and attempt to consume the 'bottom end' of mainstream commercial provision. Clearly, groups such as these will continue to be further maligned and increasingly excluded from city-centre nightlife, as gentrification and urban rebranding for a wealthy elite continues apace.

Finally, we found that a number of young adults were less sanguine about the new outward stylisation of much mainstream nightlife. As we highlighted in Chapter 2, increasing corporate control of much of the style revolution means that, for some young consumers, it is all style and no substance (Klein, 2000). The degree to which this stylisation represents choice or a 'real' step up the social mobility ladder is highly debatable for some young people. A disenchantment with what is on offer in the labour market and mainstream nightlife has led many young people into forms of active resistance and the search for alternatives. Urban nightlife, as such, contains numerous oppositional places, and hence retains elements of transgression. While many alternative spaces are simply more bohemian versions of mainstream culture – the British dance music industry (see Hesmondhalgh, 1998) and its associated infrastructure in clubs, hybrid bar-clubs, independent clothes and record shops is a good example of this model – others openly identify themselves as actively oppositional against the mainstream (Chatterton, 2002a).

'Alternative' nightlife spaces, usually independently owned or managed, in the form of unique single-site music, club and bar venues, or one-off squats and/or house or free parties, form the basis of more localised nightlife production–consumption clusters. Such places exist to meet the needs of particular youth identity groups, and styles here can be quite specific and related to certain genres of music, dance cultures, clothing styles and ethnicity (such as Goths, post-punk, grunge, indie, hip hop, garage, nu-metal, etc.; see Bennett, 2000), politics or sexual identity (Whittle, 1994), or are generally more 'casual' in relation to the formal regulation of dress than in the mainstream sphere. Because they are less likely to be corporately owned, such places are typically found on the lower cost margins of city centres, and consumption here is usually related to a conscious identity, style or lifestyle rather than a passing consumer whim. Marginal spaces are often distinguished by shocking, out of place bodily appearances, often displaying an anti-aesthetic, setting them apart from the respectable, fashionable mainstream. Consumption here can also be driven through musical appreciation (i.e. live music, or specialist DJs or particular underground club nights) or being with like-minded people, and can sometimes combine arts, culture, performance and politics.

The important point is that within such sites there is more of a blurring of the division between producers and consumers, through the exchange of music, shared ideas and values, business deals and networks of trust and reciprocity. Examples of more underground alternative provision include illegal warehouse, house parties or squats, where the link between production and consumption is literally indistinguishable (Chatterton, 2002b). It is here and in the more underground club scene that the more fleeting and loose forms of tribal sociation as suggested by Bennett (2000) and Maffesoli (1996) are identifiable.

However, minority cultural identities in the guise of lifestyles or 'neo-tribes' do not negate the idea that social and spatial divisions, inequalities and hierarchies continue to exist within urban nightlife. Within nightlife, internal cultural diversification in the industry is as much about social hierarchies and ensuring continued profitability as it is about 'tribes' and hybridity. Urban youth nightlife consumption cultures, then, remain segmented around a dominant commercial mainstream, with its various subdivisions, and diminishing opportunities for alternative and residual experiences. It is to these nightlife experiences we now turn in Part 2.

Part II

URBAN NIGHTLIFE STORIES

Experiencing mainstream, residual and alternative spaces

5

PLEASURE, PROFIT AND YOUTH IN THE CORPORATE PLAYGROUND

Branding and gentrification in mainstream nightlife

This chapter is about making and experiencing what we call mainstream commercial nightlife and here we discuss a number of aspects. First, and foremost the commercial mainstream is a place of capital accumulation and only coincidently does it have anything to do with creativity, diversity and access. Hence it is a battleground for profit-hungry entertainment conglomerates eager to attract both mass and niche audiences, but also equally concerned to segment them. Second, it is also an area where city councils are active in terms of 'place marketing', hence the emphasis on large-scale development showcases filled with corporate pubs and restaurant chains, rather than addressing the entertainment needs of a broad public and opportunities for local entrepreneurs. Finally, the commercial mainstream is a differentiated 'playground' which offers a number of goods and spaces for the active production and reproduction of social groupings of young people, keen to refashion their night-time consumption identities in relation to their peers and their own labour market positions.

Our story here considers the interaction between these three elements. Our argument is that a mode of mainstream commercial nightlife dominates downtown urban nightlife, characterised by growing corporate activity aided by the entrepreneurial city, resulting in a segmented and increasingly gentrified consumption experience. This should not imply that this mode is somehow either fixed, uncomplicated or monolithic. In fact, it is the fluidity of mainstream nightlife, with its changing styles and constant upgrading, which both drives it forward in a constant search for new consumption experiences and, ironically, works to produce a rather predictable and rationalised environment (what Harvey (1989b) refers to as 'serial reproduction'). In this sense, what was stylish for pioneering producers and consumers a few years ago becomes the fodder for today's less differentiated and mass nightlife market.

There are some general themes within mainstream nightlife which distinguish it from both residual and alternative nightlife (see Chapters 8 and 9 respectively). Foremost, it is characterised by corporately owned themed/branded or stylised environments and strict regulatory practices, including mandatory doormen and smart attire, and typified by the consumption of commercial chart music, circuit drinking, pleasure-seeking and often over-indulgence. It is the well-recognised young adults' Friday and Saturday night out in bars, pubs and clubs in most major cities around the world. Yet within the mainstream one can also discern different levels, such as the well-publicised mass market of hedonistic and sometimes over-exuberant drink-fuelled behaviours (Hollands, 1995), as well as a growing tendency towards stylisation and exclusivity, particularly in terms of prices and certainly dress sense (Chatterton and Hollands, 2001). Rather than condemning mainstream activity through moral panics associated with unruly yob cultures, or assuming that these spaces are part of a brave new world for post-industrial urban centres, it is important to place such nightlife within a broader economic, regulatory and consumption-based framework.

Second, difficulties defining mainstream nightlife arise from the fact that it functions as an imprecise 'other' for both young people and youth researchers. As a result of this othering, and its changing nature, it is often an imprecise and inconsistent category (Thornton, 1995). Echoing the Frankfurt School, the mainstream is denigrated to the level of unsophisticated, commercial mass culture, representing a creeping sameness which is evident across a number of cultural styles such as music, fashion, film and TV (Swingewood, 1977). However, it is important to recognise that nightlife here is internally divided and contains numerous overlapping and segregated groups of consumers. All forms of nightlife consumption require a range of knowledge and competencies and, in a Bourdieuian sense, different forms of social, cultural and economic capital (or subcultural capital) will circulate in different nightlife consumption arenas, even within the mainstream (Thornton, 1995). There is no single 'mainstream', then, but a variety of mainstream scenes.

However, as discussed in Chapter 2, a key trend within the mainstream is a growing tendency towards gentrification, the pandering towards middle-ground tastes, and associated effects such as commercialisation purification and privatisation of urban space. Gentrification of housing markets, schools, shops, leisure provision, entertainment and nightlife is increasing social and spatial differentiation in central areas. The move 'back into the urban centre' (Wynne and O'Connor, 1998) continues to be highly selective, creating hermetically sealed living–working–playing environments for a new group of mobile, wealthy, young urban-livers who are increasingly driving the development of mainstream nightlife. Along waterfront locations, this process has become all too obvious, with cities like Baltimore, Cleveland, Barcelona and Melbourne all creating leisure nightscapes from the residue of their industrial infrastructures. This has had the effect of creating certain demarcations and a hierarchy of nightlife spaces within the mainstream, suggestive of who is welcome and who is not. Yet, at the same time, in many of these now post-industrial cities, young people in less stable and lower-level service and manufacturing jobs also consume these spaces, resulting in a continual game of upgrading and downgrading of premises. Below, we outline an understanding of mainstream nightlife within such a framework.

Serial (re)production, corporate 'brandscapes' and the gentrification of mainstream urban nightlife spaces

As outlined in Chapter 2, the mainstream is defined through a particular mode of production characterised by the disproportionate role corporate operators play in nightlife in central areas and their use of branding/theming and market segmentation. With regard to ownership, one independent bar owner in the UK (Bristol), reflecting on the dominance of large national brewers, commented: 'What horrified me was their influence on the way cities operate, and that influence is still totally endemic.' Others referred to the impact of corporates through various 'disease' metaphors. The danger of corporate buy-out, as expansion-hungry corporations hunt for easy ways to adopt winning strategies and enter established markets, is particularly felt by smaller independent nightlife retailers.

The impact of corporate dominance within mainstream nightlife on cultural diversity and consumer experiences is considerable. For instance, large publicly quoted companies have a legal fiduciary duty to shareholders and directors, not the cities they reside in. As one bar manager interviewed intimated: 'At the end of the day, breweries and major companies are run by accountants. All they're interested in is how much profit is being made' (Bar manager 2, Newcastle upon Tyne). One of the quickest and safest routes to profit is branding, and hence branded nightlife ideas that work locally are rolled out nationally. Such a focus on profitable brands has implications for wider cultural development. As one music producer interviewed reflected: 'The people who are controlling the investment aren't looking at the cultural aspects, they're just looking at the business – selling pints' (DJ 1, Newcastle upon Tyne). Finally, the mainstream mode of production creates a particular set of relations between those who profit (owners), labour (bar staff) and consume (customers). In general, this relationship is more 'producer-led' and is often manifested by a division of command between the corporate centre and nightlife venues, with control over what happens in the venue in terms of style, music, lighting and products residing at external head offices. One bar manager, from a large national bar chain, commented: 'At the end of the day Head Office stipulates what we can do, what we can't do. We're not even allowed to promote things unless Head Office gives us consent' (Bar manager 3, Bristol). All of these factors result in a more standardised nightlife experience, with a serial reproduction of similar premises emerging across cities.

Clearly, large corporate operators understand that dominating mainstream nightlife markets is not simply a case of 'one size fits all' (see also Chapter 2). They are also cognisant that they have a relationship with smaller, more creative operators, whether about learning new bar concepts or in terms of potential buy-outs of successful independent brands (see Chapter 9 for an extended discussion of the independent mode of nightlife production). For example, attempts by smaller independent operators to create diverse, cosmopolitan and more café-bar atmospheres, drawing upon motifs associated with 'Europeanisation', have sought to encourage different types of licensing, new attitudes to dress codes and gender relations, a diversity of uses mixing eating and drinking, a 'chameleon' approach by appealing to different audiences throughout the day and a broader range of alcohol preferences, such as wine, spirits, bottled designer beers and alcopops (Difford, 2000a). Many new nightlife venues are also keen to place themselves as wider lifestyle venues and one-stop-shops for nightlife. Such multi-functionality means that the physical boundaries between types of premises are blurring. In particular, hybrid half-club/half-bars have emerged which have eroded the distinction between a 'bar' and a 'club' experience. While originating mainly from small-scale local entrepreneurs, many of these concepts have subsequently been usurped by corporate chains. Figure 5.1 shows the layout of S&N Retail's award-winning Bar 38 branded venue, based around a number of drinking, eating and socialising spaces between which similar cash-rich consumers are encouraged to circulate. Note in particular the TV screens, cash point and fourteen CCTV cameras. (Contrast this to the layout of a traditional pub in Figure 8.1, Chapter 8, which catered for a number of different social groups through distinct rooms such as the tap room or lounge.)

The 'brand' has become a key motif for success for operators within mainstream nightlife, which is of little surprise in our brand-obsessed (Klein, 2000) and symbol-ridden environment (Gottdiener, 2001). While urban areas have always had elements of a 'symbolic economy', what is new is the symbiosis of image and product, the scope and scale of selling global images, and the way in which branded spaces increasingly speak for social places and groups. The symbolic economy within branded nightlife works on a number of levels. Semiotically speaking, signifiers such as bar or club 'denote' certain meanings such as 'beer' or 'dancing' but, beyond this, branded and themed nightlife spaces 'connote' wider social meanings. Such connotations may refer to particular social and spatial settings and a myriad of historical or geographical reference points such as sports bars, sushi bars, Moroccan bazaars, Aussie and Irish pubs and beach bars, which invite consumers to step outside their immediate environment and participate in structured fantasies. The logic of such easy-to-read themed environments is that their legibility stimulates our propensity to spend. Themed environments, then, offer 'perfectly engineered enticements, directing our behaviour toward the spending of more money, and making us like it' (Gottdiener, 2001: 152). In short, people like to be entertained while they spend money (*ibid.*).

Rather than representing a homogenous mass, successful mainstream corporate operators have developed strategies to attract both niche and mass audiences through developing a portfolio of brands, with some following themes such as Europeanisation–cosmopolitanisation, while others adopt a more McDonaldisation–Americanisation model. On the one hand, smaller-scale more design-intensive venues for status-conscious customers involve higher start-up costs and lower revenue from a restricted consumer base, but are able to maintain profits by charging higher prices at the bar and at the door. The design of such specialised, one-off stylised environments appeals to small, niche-market segments, usually based upon highly individual, esoteric and avant-garde design and minimalist features – stripped wooden benches, metallic walls, glass frontages. Such places often become well-used leisure spaces for those involved in the creative industries. While these venues are more bohemian variations of the mainstream, they still share many characteristics of the commercial sector such as traditional gender roles, patterns of alcohol consumption, ownership, pricing policy and producer–consumer relations. Moreover, such premises can quickly become colonised by more mainstream audiences, eroding their creative impulses. Such is the fate of the latest chic style bar.

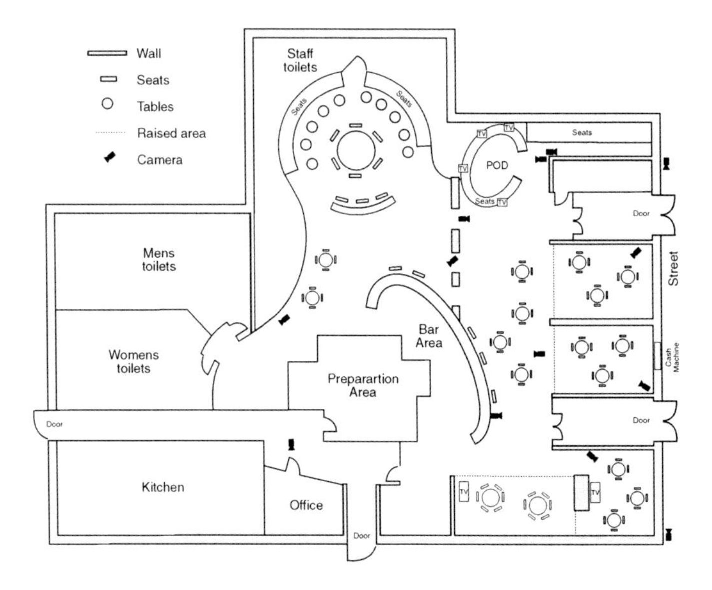

Figure 5.1 Plan of a style bar; Bar 38, Newcastle upon Tyne, UK, 2002
Source: Scottish & Newcastle Retail

On the other hand, it is the large-capacity (1,000 plus), themed, mass-oriented mainstream venues based around well-known high-street brands which have come to dominate nightlife, particularly in countries like the UK, as they provide the best opportunities for economies of scale and profit maximisation. While initial outlays on centrally located property may be costly, profit returns from downtown locations are high and multi-site brands offer production and maintenance savings, while the parent company can negotiate large bulk discounts from suppliers. Such massified venues attempt to appeal to the widest possible audience, and design features and themes are constructed to attract a broad, inoffensive middle ground comprising bright walls, rock and pop music, legible signage and well-known drink brands, in which many different groups can feel at home. Globally, operators such as Hard Rock Café (see Plate 5.1) and Planet Hollywood are the best-known examples of this latter strategy.

In the UK, the JD Wetherspoon pub chain perhaps best epitomises such an approach. It is actively engaged in the serial reproduction of nightlife, with its cheap beer, no music or TV, 'just a pub' philosophy. Established in the 1970s, the company rapidly reached a pre-tax profit of £15.6 million in 2000 and hopes to boost its estate from 400 to 2,000 pubs over the next few years. The company has been able to dominate the mainstream high street through bulk buying and

Plate 5.1 Hard Rock Café, Barcelona, Plaça de Catalunya, Barcelona, Spain, 2002

selling heavily discounted drinks and food. Similarly, Brannigans, one of the fastest growing UK late-night entertainment brands with plans to double the size of the business over the next three years (*UBS Capital*, press release, 16 November 2001) has a brash and no-nonsense approach to nightlife provision. Unlike many bars who aim to hide alcohol drinking behind a veneer of style, it has an upfront approach to the brand's philosophy of 'Eating, Drinking and Cavorting', placing this slogan in large bright neon letters outside all of its venues (see Plate 5.2). Its 'fun pub' formula combines music and dancing in a pub setting with a music policy which concentrates on pop classics and live tribute bands.

Other companies have used approaches to move towards the upper end of mass nightlife markets to attract more affluent, stylish, middle-class audiences (see Box 5.1 on Six Continents Retail, for example). Chorion Leisure's Tiger Tiger brand, which started in London and plans to roll out premises across the UK, is described as an 'up-market nightclub' or themed 'super-club' (comprising bars, restaurant and club in one) catering for a 25-plus age group. Moreover, Luminar Leisure, the UK's largest nightclub operators, owners of the Slug and Lettuce chain, describe these venues as 'yuppie' bars 'aimed at affluent urban professionals, offering quality food and premium beers'.

Plate 5.2 'Eating, Drinking and Cavorting' on the mainstream; Brannigans bar, Bristol, UK, 2002

Box 5.1: Six Continents Retail: from working-class beer to middle-class hotels

Bass Breweries, established in Burton-on-Trent in 1777, has recently moved out of beer production after selling its brewing division to Interbrew and now is slowly divesting from many of its traditional pubs. Subsequently, Bass has relaunched itself as Six Continents Retail because the Bass 'brand' was seen as too localised for its global ambitions. Six Continents now sees itself as a global leisure group concentrating upon up-market hotels such as Inter-continental in the USA, SPHC in Asia, Holiday Inns in Europe and branded restaurants and nightlife venues such as All Bar One, O'Neills and Harvester. Tellingly, the company has been divesting from what have traditionally been lower-profit working-class and community-based activities such as traditional tenanted pubs, Gala bingo halls and Coral bookmakers. A further 988 smaller pubs were sold in February 2001 for £625 million as they were deemed not suitable for rebranding. As a result, Six Continents Retail is now concentrating on 2,000 larger nightlife venues in the UK, with a very high proportion either branded or due to be branded.

Scottish & Newcastle Retail have also sought to utilise middle-class values as a yardstick for the development of their various brands. In a discussion of brands they argue: 'Today's middle-class customers are more experienced, better travelled, have higher aspirations, and higher standards' and 'bars have become the norm for increasing numbers of mainstream young consumers, whose levels of sophistication have risen' (Scottish & Newcastle, 2001; also see Plate 5.3). Additionally, JD Wetherspoon has introduced a more style and café-type venue called Lloyds No. 1 to its established portfolio. A review of its Manchester premises stated that: 'Lloyds is Wetherspoon's slightly more ambitious cousin – tastefully decked out with comfier seating, music to fill the pregnant pauses, nicer toilets, but the same hugely competitive beer, wine and food prices' (JD Wetherspoon, 2002). Finally, Wolverhampton and Dudley's own up-market chain, Pitcher and Piano, is aimed at 20–30-year-old young professionals and has attracted enough attention as a brand to be considered an attractive 'package' for potential buyers. Branding and the move up-market, then, is not only designed to tap into that cash-rich and allegedly more 'trouble-free' clientele, but is also driven by the need to increase the value of brands (Batchelor, 1998) in light of continued concentration and conglomeration within the industry, and the desire from growing companies to float on the stock exchange and raise money for further expansion.

Most large chains focus upon rolling out branded environments across several localities, which gives consumers an all-important sense of familiarity and connected-

Plate 5.3 Putting on the style: S&N Retail's award-winning Bar 38 brand; Newcastle upon Tyne, UK, 2002

ness across geographical places. At the same time, large branded nightlife venues are also trying to create a far-reaching corporate strategy in which the venue is not just a place for drinking, but offers a wider lifestyle experience, with international recognition, cross-merchandising and potential to expand in other locations. Hard Rock Café, Pacha and Ministry of Sound are some of the most visible examples here. As one UK nightclub owner told us: 'It's a brand that you buy into as a customer. When you go to Ministry of Sound it's the whole ethos, you get the magazine and the CDs and the t-shirts and the record bag' (Nightclub owner 1, Bristol).

As mentioned in Chapter 2 (see Box 2.1), different traditions of licensing, the sale of alcoholic beverages with food and/or in conjunction with other leisure activities, and a greater suburbanisation of entertainment, have resulted in a more varied pattern of mainstream nightlife ownership and branding in North America. The majority of branded premises here tend to be 'bar-restaurants', serving food and drink together, as well as being themed around other forms of entertainment whether it be sports, video and arcade games, music, sexuality or geographical 'place spaces' (i.e. Irish, Aussie and English pubs; see Box 5.2 for some examples). Sports bars, taverns and bar-restaurants in particular are one of the largest growth areas in North America, decked out with endless banks of TV screens and offering burger and beer deals served by often scantily clad waitresses. Whole web pages (such as www.sportstavern.com) are devoted to them. Similarly, Irish theme pubs have also grown tremendously, providing musical entertainment as well as 'authentic cuisine'.

Many of these themed environments tend to be casual and even family oriented, rather than being gentrified spaces targeting young adults specifically (although the cost of drinking and dining together may be prohibitive, and some places not suitable for children). Finally, such branded spaces in North America tend to be suburban, being located in strip malls and in out-of-town leisure complexes, while also increasing their presence in downtown core areas.

Box 5.2: Examples of North American themed 'bar-restaurants'

ESPN Zone

The ESPN cable sports network and sports bars and restaurant chain has become a well-known brand across the USA and has attracted the attention of global corporations like Disney, which has secured a majority shareholding in the company. ESPN Zone now has footholds in major entertainment sites such as Times Square, located on 42nd and Broadway (Hannigan, 1998).

Hooters

Described as having a 'casual beach theme', the bar-restaurant is primarily characterised by its Hooters Girls (cheerleader-like waitresses; see also Box 7.2), 1950s and 1960s jukebox music, and sports on television. From a single location opened in 1983 in Florida, the brand now has over 310 locations in forty-three states and in Asia, Aruba, Canada, England, Mexico, Singapore, Taiwan, Argentina, Brazil, Austria, Switzerland and Puerto Rico, with its website stating the brand 'has proven successful in small-town America, major metropolitan areas and internationally'. Around 75 per cent of its sales come from food and merchandising, with 25 per cent coming from its sale of wine and beer (Hooters does not sell liquor). Today, the company is involved in cross-marketing itself through its own magazine, billboards, TV and sports partnerships with the National Golf Association Hooters Tour and the United Speed Alliance Hooters Pro Cup racing series.

Firkin Group of Pubs

Following the opening of their first premises in downtown Toronto in 1987, the Firkin Group of pubs added three more in five years, before deciding to franchise the operation. Based on the idea of a traditional English neighbourhood pub, encompassing good beer, good food and good company, the company trades on the logo 'Everyone's a little different. One's just right for you.' There are now thirty-six franchises operating in Canada, the majority in Ontario.

Dave & Buster's

Dave & Buster's launched their concept of combing food, drink and interactive games back in 1982 in Dallas, Texas, and today are a nation-wide institution operating in fourteen states, Canada and Taiwan. In addition to their bars and restaurants, they offer interactive entertainment attractions such as pocket billiards, shuffleboard, state-of-the-art simulators, virtual reality and traditional carnival-style amusements and games of skill. Guests under the age of 21 (legal drinking age) must be accompanied by a parent or guardian, and are restricted after certain hours. As well as expanding in the USA, Dave & Buster's have granted several international licensing agreements for future expansion abroad. In 1999, they began trading on the New York stock exchange.

Prime Restaurants

Prime Restaurants develops, manages, buys and sells franchise restaurant concepts, and with 130 restaurants in Canada and six in the USA and revenues of nearly $300 million (Canadian), it is a going concern in the bar-restaurant sector. From its early beginnings in opening Casey's Bar and Grill (there are now thirty across Ontario and Quebec), the company is perhaps best known for developing the East Side Mario franchise. In 1996 it engaged in a joint venture to open a series of Irish pubs in Canada (with brand names such as Fionn MacCool's, D'arcy McGee's, Tir nan Og, Paddy O'Flaherty's). In 1999 they launched Esplanade Bier Market in Toronto, selling imported beers and upscale Belgium food.[1]

In a bid for some form of authenticity and history, branded and gentrified urban nightlife has also increasingly colonised abandoned and decaying industrial architecture. While many of these buildings were originally reused by bohemian artists, intellectuals and students (Zukin, 1988), not to mention small nightlife operators, in the face of increasing property values a vast collection of ageing buildings have more recently been transformed by large entertainment conglomerates. Old banking halls, schools and hospitals are now garish fun pubs, grain stores and warehouses have become wealthy riverside lofts or art galleries, bingo halls and working men's clubs are transformed into stylish nightclubs and restaurants, while mills and churches now stand as new offices, apartments or mixed-use arts venues. In many cases, older and industrial architecture rooted in the very fabric of working-class and community life now serves as an infrastructure for a new class of high-income pleasure-seekers, tourists, office workers and city-livers (see Plate 5.4).

Hallmarks of nightlife brandscapes include a fleeting and fast-moving experience, a preoccupation with the new, and a rapid turnover of styles and fads. Such new nightlife styles, no doubt, have radically changed the look and feel of many urban areas. In Bristol in the UK, for example:

> In 1982 there were pubs and a smattering of (God help us) cocktail bars. The middle-aged middle classes drank in wine bars. By 1992 there were theme pubs and theme bars, many of them dumping their old traditional names in favour of 'humorous' names like the Slug and Lettuce, the Spaceman and Chips or the Pestilence and Sausages (actually we've made the last two up). In 2001 we have a fair few pubs left, but the big news is

Plate 5.4 New nightlife mixes with the old city; Beluga, Edinburgh, Scotland, 2002

> bars, bright shiny chic places which are designed to appeal to women rather more than blokes with swelling guts. In 1982 they shut in the afternoons and at 11 p.m. weekdays and 10.30 p.m. Sundays. In 2001 most drinking places open all day and many late into the night as well.
>
> (Byrne, 2001: 23)

This branding process also works at the level of the city (Julier, 2000) through the creation of urban brandscapes, linking nightlife and the wider city identity. For example, Leeds, a former textile city in England's northern industrial heartland, has gone to great lengths to rebrand itself and create a new visual culture. As one newspaper commented: 'After years of being saddled with the image of flat caps and Tetley Bittermen, Renaissance Leeds has been largely modelled on continental 24-hour cities ... groups of chic revellers throng in the streets, Leeds is the only place to be' (Simpson, M., 1999). With European flair, Leeds has developed a bustling café-bar culture and claims to have 'waved goodbye to post-industrial decline and welcomed, with open arms, caffe-latte society' (Difford, 2000b: 4). There is some truth in this. For example, between 1994 and 2001, licensed premises increased by 53 per cent and special hours certificates (post 11 p.m. licences for bars) increased by 155 per cent (West Yorkshire Police, 2001). There is, however, another side to this story of prosperity, especially in the areas surrounding the city centre. The UK government's Index of Local Deprivation states that twelve of the thirty-three wards in Leeds are within the worst 10 per cent of wards nationally. Much of this rebranding process, then, is about creating place myths to reposition urban areas within an international hierarchy (and cater for wealthy city-liver groups), rather than a concern with the everyday life of cities and their citizens.

The development of mainstream branded nightlife is linked to more general social and spatial processes and hence is a visible feature of suburban and exurban areas (Gottdiener, 2001). Over the past thirty years, there has been a recognition of significant changes in urban form through a move away from a simple centre–suburb model towards outer or edge cities (Garreau, 1992), and a development of 'polycentric' or 'peripheral' urbanisation. Soja (2000) has described this through the term 'postmetropolis' in which the experiences, forms and functions of the city, suburbs and wider metropolitan region are blurred, and the notion of 'exopolis' – a city turned 'inside out' and 'outside in', in which spatial incoherence and polarisation are norms. The spread of gentrified living and branded nightlife beyond the urban core is part of these spatial transformations. Commercial developer interest has increasingly focused upon multifunctional themed complexes comprising strip malls, casinos, multiplexes, bars and nightclubs which serve expanding metropolitan regions. These multi-functional, exurban and suburban entertainment and nightlife ghettos have been established for many decades in suburban locations, especially in North America and Australia. They have appeal to groups who have significant demands on their daily lives from families and work, and hence demand an efficient use of leisure time. Moreover, by offering the option of private car travel, family conviviality and easy-to-navigate consumption environments, they reduce the risk factor of downtown nightlife and are removed from aspects of urban life such as poverty, dirt, homelessness and crime. While European cities continue to have a stronger tradition of more economically buoyant and cultural vibrant city cores which has resulted in a greater focus of nightlife in central rather than suburban or edge locations, nevertheless Urban Entertainment Destinations (UEDs) have markedly grown on urban fringes and in suburbs across the UK and across Europe, bringing together cinema, retail, eating and nightlife.

Mainstream branded and gentrified nightlife spaces, then, are colonising many parts of the urban fabric and redefining urban space. At the most basic level they are eroding more traditional historic forms of nightlife (as discussed in Chapter 8) and pushing alternative nightlife to the margins (Chapter 9). One can no longer linger without consuming, and traditional, older communities within nightlife venues are rarely present in the city today. The key point about any themed space is that it is part of the 'powerful forces that define the symbolic value of commodities to our society', which are essentially tools for business competition and place marketing rather than generating a wider social value or use (Gottdiener, 2001: 10). Success in the night-time economy results from using themes and symbols to differentiate what are essentially similar serially reproduced products (bars, pubs, clubs). Many mainstream nightlife venues are little more than 'machines for drinking in', employing lifestyle strategies and symbolic identification to increase alcohol consumption. Branded nightscapes, then, are part of 'the merry-go-round of consumer culture that circles and circles over the same territory, but requires ever new fantasies and modes of desire in order to maintain a high level of spending' (Gottdiener, 2001: 67).

Upgrading downtown: promoting and policing mainstream styles

As we outlined in Chapter 3, corporate dominance of mainstream nightlife is facilitated by a more entrepreneurial, business-led and growth-oriented local state, eager to attract large-scale developments and major inward investors. One small-scale bar owner outlined why local politicians and planners were likely to join forces with large corporate developers and nightlife entertainment conglomerates: 'They [the local state] just see that leisure-driven development is the easy way out for them ... they're not going to go bust, the big plcs [public limited companies]' (Bar owner 2, Leeds). This approach ensures that city authorities are getting the best deal on public land-sites, securing tenants for derelict buildings and decaying infrastructure, and raising the profile of urban areas. Further, simply to favour one type of nightlife developer over another would attract the prying eye of scrutiny councils and district auditors, not to mention law suits from aggrieved developers. As one planner told us: 'I mean, you can't say that I'm turning this down because it's corporate' (Local authority representative 2, Leeds).

As long as the local state remains tempted by the short-term gains of corporate investment, then, the impetus behind nightlife developments is likely to remain with large developers rather than the public interest. In particular, much of the development of the night-time economy is geared towards raising the external profile of the city. As one councillor in Bristol in south-west England suggested about their recent initiative to develop a '24-hour city':

> That was an attempt to promote the city and strengthen the local economy. Because by doing this we are obviously attracting more major club funding, leisure funding, brewery funding and that people are now being bussed from the Midlands, Birmingham, Bath, Cardiff, Devon.
>
> (Local authority representative 1, Bristol)

Such an approach, oriented towards meeting the needs of larger, externally located, publicly quoted companies, inevitably leads urban regeneration in a particular direction. It calls into question the use of public money to subsidise entertainment and nightlife facilities for the profit-hungry corporate sector (Harvey, 2000). Independent nightlife continues to face a number of barriers in such a context, such as rising property values, bulk buying by corporate chains, and more complex navigation through the regulatory system. Many small-scale schemes are sidelined, as they stand outside what the local state perceives as inward investment. As one small-scale bar owner remarked: 'It seems almost that if some corporate big guy comes in they're going to give him whatever he wants because it's inward investment' (Bar owner 4, Newcastle upon Tyne). Of more concern, it also allows the local state to literally divest responsibility for how central nightlife areas should develop and who they are for.

Behind the carnivalesque mask of mainstream and increasingly gentrified nightlife, then, there is a reinforcement rather than liberation of social roles. Mainstream nightlife often represents instrumental and choreographed rather than liminal serendipitous space. Creswell (1996) discusses the relationship between place and ideology, and specifically that certain places contain particular meanings and expectations of behaviour. Mainstream society has its own set of rather overt taken-for-granted norms and sense of limits, reinforced and circulated by an ever growing media and advertising industry. In day-to-day urban life there are few presentations of alternative possibilities and little questioning of the legitimacy of the dominant social order. In this way, mainstream nightlife is a 'normative landscape' in which particular actions and behaviours have become pre-inscribed, tolerated and accepted, while others are not. This is a geography of common sense which renders unacceptable the other, the different, the dirty. One has to look a certain way (through designer clothes), be expected to pay certain prices (for designer beers) and accept certain codes and regulations (from door staff).

While older nightlife places like community pubs and taverns were based around narratives in which people were regarded as citizens/residents (Gofton, 1983), mainstream gentrified nightlife spaces are 'products' in which people are merely (stratified) consumers. Bounding, regulating and controlling are distinct features of mainstream nightlife and bouncers are the ultimate choreographers, by closely regulating access and ensuring an homology between consumers and venues. In essence, mainstream nightlife spaces are the ultimate 'gated (entertainment) community'. Yet within the mainstream there are a variety of regulatory approaches, differentiating mass venues from specialist ones, and 'style and selection' from more 'strong-arm' tactics.

The regulatory character of the mainstream is enforced directly by door policies but also through a complex set of pressures from peer groups, the media and consumer cultures. As one woman (Jackie, 23 years old, Leeds) commented to us about one new city-centre bar in the UK:

> Everybody inside was all tarted up and I was wearing trainers and looking a bit scruffy ... It's just the way they make you feel once you're in there. The people who sort of sneer at you, and you think, I don't feel comfortable here because I can't relax when everyone else is sort of preening and trying to pull, and essentially, I'm not.

The mainstream, then, is imbued with certain codes which (problematically, we would add) link style with behaviour. As one owner of a large city-centre nightclub told us:

> We like people to come down and make an effort and we do turn people away quite a lot. Our sort of feeling is, you know, this is a club not a pub, so don't come down dressed for the pub, basically. We like people to make an effort to look smart. Well, not necessarily smart, but just to look trendy rather than that they've just thrown on any old t-shirt and any old scabby pair of trainers and a baseball cap. We want people to come down thinking, 'I'm on a night out, I'm looking good.' That's the kind of feeling we want. Also helps to attract a slightly better class of clientele. You find that people that have actually made an effort are usually a little bit more social and civilised than people that just come down dressed in whatever.
>
> (Nightclub owner 2, Leeds)

However, many established sports and casual brands have been tainted by their association with mainstream laddish nightlife culture, and as a result have come to signify an 'anti-style'. As one chain bar owner commented: 'People who wear Rockport and the like cause trouble ... they're scallies' (Finnigan, 2001: 8). Many venues, then, have outlawed some mainstream branded styles on the door in their quest to further distinguish themselves from the mass consumer experience. In the process, they are promoting exclusive, upgraded nightlife environments, tailored more to cash-rich groups (Chatterton and Hollands, 2001).

This upgrading of nightlife has generally been welcomed by the police and local government, and is part of an attempt, as one senior police spokesperson told us, to help 'design out' problems of excessive drinking and violence. Whether this approach has been successful or not is another matter (Hobbs *et al.*, 2000; Newburn and Shiner, 2001). However, there is also evidence that such an approach is increasingly welcomed by city councils and city management teams (Reeve, 1996). As one local authority representative commented to us:

> I think the slight emphasis on that is trying to move away from the loutish party image to a more up-market one ... but also looking to the future, the medium to longer term, it is more beneficial to have, what shall we say, a better class of clientele but probably a more mature clientele who are perhaps not so irresponsible as some of the younger ones.
>
> (Local authority representative 2, Newcastle)

Many more style-conscious venues are keen to ensure entry is restricted to elite groups and have 'unconscious dress codes', enforced by 'style pickers' and backed up by bouncers. In other places, a new orthodoxy of 'anti-fashion' style codes are enforced, based upon discreet but still expensive brands and looking, as one person told us, 'stylishly scruffy'. Additionally, managers of busy central bars are increasingly under pressure from shareholders and the corporate centre to maximise returns, which often equates to compromising on strict door policies, and hence safety. As one door security member in the UK told us: 'The manager is under tremendous pressure and he has to sometimes override door staff to make sure he's got a packed house so he can get all his money in his till' (Door security personnel 3, Bristol). Despite the fact that style codes in the mainstream appear to be continually changing, these new leisure spaces clearly remain zones of permitted, regulated and legitimised pleasure (Shields, 1992: 8).

Experiencing the gentrifying mainstream

One of the difficulties of articulating mainstream nightlife is that while it is bounded by certain material structures, like patterns of corporate ownership, branding and particular forms of regulation, it is also 'experienced' and hence has a more subjective dimension. In many cases, this subjectivity is formulated not just directly, in terms of what people consume, but relationally, in terms of what 'others' consume and do on a night out. As such, while we need to see mainstream nightlife as 'fluid', it is equally important not to collapse all youth groups into a 'postmodern maelstrom' where cultural boundaries are blurred, or ignore the impact of social structures and consumption divisions. Explanations of the construction of nightlife identities are connected to understanding 'other' youth groupings, not just as 'imagined others' but as 'real' social categories. 'We create groups with words', as Bourdieu (1977: 101) argues. Class and gender remain central to understanding the mainstream here, although it is important to move some distance from conventional assumptions about these social divisions (Crompton, 1998), with sexual identity and 'race'/ethnicity also playing a part in creating distinctions within nightlife groupings. In the remainder of this chapter we explore a variety of youth nightlife consumption groupings, by considering the experiences and internal divisions within the social space of the mainstream.

As we outlined in Chapter 4, the social spaces of mainstream nightlife can be unpacked at a number of different levels: actual and relational labour market position, and taste and culture/consumption preferences, both of which are shaped by a host of factors such as gender, current age, ethnicity, geographical location, family background, schooling, etc. Combined, these aspects create preferences for various segmented spaces within the mainstream. Clearly, some people's use of the mainstream cannot be understood just through this schema, as one of the main points about mainstream nightlife is that it is always in flux, always being redefined and struggled over. Many people also consume in fluid, contradictory and unpredictable ways, and there are many overlapping positions and specific subgroups within this general category. In one sense, then, our typology is an ethnographic moment 'frozen in time'.

Our first level of interpretation, which reveals itself time and again, relates to the labour market and concerns young people's positioning within a gradient ranging from stable and well-paid employment to those at the bottom end who are unemployed, or welfare-dependent, with a bunching of young people in the middle in full-time, medium-income but unstable employment (Roberts, 1995). Although there is a range of employment opportunities, then, Beck's (2000) notion of a 'Brazilianization of the west' – a political economy of insecurity in which the majority rely upon unstable, nomadic employment – is increasingly instructive for understanding the changing labour market experiences for many young people. There are a number of other factors to consider here within this labour market gradient. For instance, Savage *et al.*, (1992) point towards leisure and taste divisions within middle-class consumption, and there are distinct leisure differences between specific occupational groupings. For instance, creative industry workers (i.e. those in the fashion, music and nightlife industry itself) are often prolific, stylish and competent consumers of nightlife cultures.

A second level of interpretation concerns taste and cultural preferences in the night-time economy, and ranges from the consumption of mass/themed/commercial environments towards more individualistic and stylised commercial tastes in more gentrified spaces, each of which implies a whole range of clothing, style and social codes. Again, this is a rather fluid gradient, with some groups consuming across the spectrum, while others consume more narrowly. Additionally, one should not automatically assume that all groups consuming mass themed commercial environments are 'cultural dupes', or that they approach nightlife in the same way. Here, gender and sexual orientation also influence taste cultures, particularly in the direction of preferring more up-market and stylish environments (Whittle, 1994). Below we map these occupational and taste cultures on to the differentiated nightlife spaces within the mainstream.

Before unpacking the mass mainstream and discussing some of these different youth consumption groupings, it is worth noting some of its general characteristics. One of the most prominent features is its sheer hedonism and excess or, as one person suggested to us, 'the kind of wanting to go out and get completely mindlessly drunk every single night'(Mike, 18 years old, Newcastle upon Tyne). 'Cheesy' nights out, based around happy hours, cheap drinks, themed environments, thundering commercial chart music, the chance of sexual encounter and the possibly of violence/vandalism, continue to define the mainstream. Heavy drinking, then, is an important symbol. As one of our interviewees stated: 'We're going out with the intention of being absolutely hammered rather than just going out to have a nice time' (Sean, 23 years old, Bristol).

As we outlined in Chapter 3, 'yob culture' has become one of the strong discourses within discussions of the mainstream nightlife (Hobbs *et al.*, 2000). While there has been a concerted attempt to reduce violence in the mainstream through design and upgrading, as our opening example of Port Vell in Barcelona showed, this is not always successful. Problems of violence and vandalism appear endemic to alcohol-led northern European and college-campus binge drinking cultures. In particular, while continental European nightlife is regarded as more cosmopolitan, British nightlife is seen to lack class. One city councillor in the UK commented:

> I mean, I'm all in favour of this continental style but somehow we just haven't got the class that French people have, we just never get it quite right ... you have to drink yourself into oblivion every weekend to prove yourself, which is rather unfortunate, and everything has to be done to excess. You can't just go out for one or two drinks, it's got to be a bender.
>
> (Local authority representative 5, Newcastle)

The commercial mainstream is fundamentally geared towards selling alcohol, and for many people the problems associated with such drinking cultures should be partly acknowledged by those who create them. As one door security staff supervisor told us: 'It's alcohol, it wipes away everybody's inhibitions, takes away all their common sense. And it's all down to alcohol and I think the breweries have a major role to play in the blame' (Door security personnel 1, Bristol). The responsibility of large operators in the night-time economy is a theme we return to later.

The mass mainstream is also characterised by sexual permissiveness, although as Hollands' (1995) work has revealed, images and stereotypes often exceed the reality when it comes to actual encounters. Relationship-seeking, casual sex and flirtation are all, at some point and in some combination, part of nightlife for many young people. One bar owner on a mainstream circuit was candid about the role of his venue in facilitating sexual encounter in the city:

> If it was just the drink that people wanted you would buy a six-pack, sit in the house and get a bottle of whisky. But it is actually, as a company we sell, covertly, we sell sex, whether it is just a chance of a meeting of the opposite sex or whether it is actually introductions that go further.
>
> (Bar owner 4, Newcastle upon Tyne)

This formula, although successful in profit-making terms, creates an atmosphere for sexual voyeurism rather than interaction. As one (male) bar worker commented: 'Who cares about talking when you can look at her?' (Bar worker 1, Newcastle upon Tyne).

Finally, the popularity of the mass mainstream results from predictable environments which offer easy-to-read sensory cues and familiar pathways through consumption choices. This partly accounts for the popularity of the mainstream, which allows consumers to satisfy their needs without much stress (Gottdiener, 2001). Familiarity and risk aversion are key aspects which many people seek from consumer experiences. Hannigan (1998) has described the feel of these themed environments through the term 'riskless risk', in that they offer the dual image of 'safe excitement' and predictability. Work by Hubbard (2002) has made similar points in relation to the growth of multiplex, out-of-town cinemas in the UK. In this sense, such people are wary of experimenting beyond what they know. One person commented to us: 'I think I'm quite wary of going off the main track' (Charles, 22 years old, Bristol). 'Self-exclusion', from environments which consumers may be wary about or have little knowledge of, is also a key element in the maintenance of segregated nightlife spaces. Many people, then, are often unaware of other opportunities outside the growing mainstream. As one of Charles's friends commented, in their pursuit of more variety in Bristol's nightlife: 'The question for me is, well, I wouldn't know what was there even if I did want to go out' (Geoff, 19 years old, Bristol).

Despite maintaining these general features, commercial mainstream nightlife spaces are differentially consumed. As we mentioned in Chapter 2, Urry's (1990) 'collective gaze' of the sociable working classes and the 'romantic gaze' of the more detached middle classes is useful here. This has some use for understanding that while some mass mainstream venues have become recognisably predictable, and many groups themselves contain a degree of uniformity in terms of dress sense, symbols and argot, other groups consume themed environments in more individualistic ways, often through the use of irony and pastiche. The most stylised environments encourage a greater degree of individuality, spontaneity and use of cultural or subcultural capital (Thornton, 1995).

One of the largest general consumers of the mass mainstream is a broad middle of working-class youth. Many young working-class people use these spaces to develop a strong sense of identity and place within them, often drawing on distinctive local cultural resources and habits. One interviewee (Dave, 31 years old, Newcastle upon Tyne) mentions here how the consumption of mainstream nightlife literally identifies him with the city in which he lives:

> It's great to be in one of those groups. It feels really powerful in a horrible sort of way ... you are accepted, you know, it sounds, you know, really spooky, but it's like you own the city, you know ... and it's like you've joined it at last, you've joined the real world, you know.

Here, many young people have found a limited degree of labour market mobility and security through service sector employment. As such, many choose an escapist hedonistic 'weekend' nightlife culture, despite the fact that they may be in part-time employment and could also go out on mid-week nights. A number of specific media texts, images and stereotypes have attempted to define this largely working-class mainstream. In the UK, films like *Letter to Brezhnev*, TV programmes like *Tinseltown* (set in Glasgow) and many of the party-night docu-soaps (such as *Ibiza Uncovered*, *Club Reps Uncut*) as well as the classic 'Essex man' and 'Essex girl', lager lads and ladettes stereotypes,[2] are examples of this cultural form. One recent youth consumer survey from Carat Insight (2001) typecast this group as 'L-Plate Lads' (and we would add ladettes) – single, still living with their parents, into lager, relationships and popular culture (video games, lifestyle magazines, designer brands). On the female front, Thornton (1995), for example, discusses the 'chartpop' culture of the 'Sharons' and 'Techno Tracys', labels which refer to the unhip, uneducated newcomers on the dance scene. While we are cautious about adding to such stereotypes, our research reveals variations along this typology which coalesce around the mass mainstream.

The following journalistic vignette from Leeds illuminates the nightlife styles of this mass mainstream group:

> The queue is full of large groups of boys in luminous Ben Sherman shirts and spiky hair and girls in tiny tops ... There is a stage where those celebrating special occasions get up to show off and which, tonight, is occupied by a girl wearing a plastic sash reading 'The party starts here' and four of her friends in bikini tops and paste tiaras. Their happiness is representative of the mood of the place, and even the boys are dancing unselfconsciously to the mixture of house and disco music. A conga line takes off before they remember themselves and drop it, but the atmosphere is one of general abandon and festivity.
>
> (Brockes, 2000)

Many of the elements of this culture are familiar. Mainstream venues reflect the expectations of many revellers, especially the desire for fun and escapism among young people in the face of economic hardship. For some young revellers, the weekend and the evening is their time and space, outside the rigours and rules of the workplace and the watchful eye of bosses. One student (Clare, 19 years old, Newcastle upon Tyne) observed about this group:

> My friends call them workies, people who work nine to five, five days a week, you know what I mean, and they just want to go out and just forget about everything ... they're passionate about going out so it takes away everything else that they have to put up with.

Some young people, then, regard the 'weekend' as a sacred time for letting go and self-indulgence, which is a reward for the time and effort spent in routine work during the week. Telesales is an increasingly common employment option for many young people. Many people commented to us about the tedious and repetitive nature of such work and hence the increased importance of free time outside the rigid structure of work. As one person who worked in a telesales centre told us: 'It's just nice to go out and relax, get drunk and enjoy yourself, especially when you've, say, got a job on the phones as well. I suppose it's quite stressful, like constantly stressful' (Paul, 29 years old, Leeds).

Going out, performing and dressing-up has become an important part of life for many young people and is clearly a crucial part of mainstream nightlife. As one young reveller told us: 'I like to get dressed up. I mean going out, I like to make it an affair' (Emma, 25 years old, Newcastle upon Tyne). Further, Robert (23 years old, Bristol) suggested to us: 'It's nice to dress up to go somewhere. It makes an evening of an event.' While part of the explanation here clearly comes from young people's desire to say something about themselves through their appearance, whether it be appearing more 'grown up' or up-market or simply wanting to be 'on display', many mass mainstream venues also play a part here through their imposition of 'stylish' dress codes.

Ironically, many of the dress codes that have evolved in the mass mainstream are now clichéd, and reveal significant disjunctions between 'appearance' and 'behaviour', as well as providing important markers which help to distinguish between groups within the mainstream. For example, one typical nightlife group often identified were locals or 'townies', who were easily distinguished by their particular dress styles, demeanour and attitude, as the following quotes testify:

> They're down there in their Ben Sherman's and their black polished shoes looking very smart, but acting like wankers.
>
> (Paul, 23 years old, Leeds)

> I think you could classify them as they do not really go for the music so much as for the chance just to get pissed and maybe have a fight, and they have had a good night then, look for a lass.
>
> (Cathy, 26 years old, Leeds)

Within this general grouping, there exist various subdivisions, based partly on intra-class hierarchies, age, gender and the particularities of locality. Under-age townies, 'your kids, or for want of a better word, your arseholes', as one pub owner (Bar owner 3, Newcastle upon Tyne) colourfully put it, are distinguished from older, more mature and clued-up locals labelled 'trendies', who gain their increased status from more 'respectable' behaviour, diverse musical tastes and the wearing of more stylish labels. Townies are also subdivided according to gender, with lads being vilified more for violent and drunken behaviour and young women for being catty and sexually promiscuous (Hollands, 2000).

There are also strong regional variations in the depiction of townies. In the UK, for example, various slang terms like 'scallies', 'Kevs', 'Trevs', and 'Tracys' describe local nightlife groupings. In northern cities, townies are often subdivided into 'charvers', which connotes a section of locals who inhabit particular social spaces in the city's mainstream nightlife, as well as those who engage in certain types of behaviour. As a group of people reflected:

Simon: They wear Ben Sherman shirts and they go out and get pissed.
Jane; ... have a shag and have a fight ...
Simon: Shag, fight and a kebab, very loud.
Geoff: You feel a bit of sort of agro in the air and you get sort of clientele like that in clubs.
Simon: Testosterone fuelled.

(Focus group 3, Newcastle upon Tyne)

The term 'charver' as it is used here, is almost exclusively masculine and refers to young men on the margins of the economy and at the 'rough' end of the class spectrum. The gender equivalent, 'slapper', refers to an equally coarse, vulgar and 'promiscuous' female townie (see Chatterton and Hollands, 2001). Central elements of mainstream nightlife, then, are intra-class and gender social divisions, which reflect certain styles and forms of behaviour.

Continuing problems with mass mainstream venues, combined with growing numbers of aspiring service-sector workers and young professionals, not to mention the impact of various city council initiatives to market their cities, has led to attempts to gentrify and further upgrade nightlife. Those associated with elements of yuppie culture – in marketing-speak the 'cash-rich, time-poor' – are becoming a growing focus for mainstream nightlife producers. Significant parts of mainstream nightlife are being upgraded, with young people attracted to more mixed-use style and café-bars, which have attempted to create a more up-market feel with polished floors, minimalist and heavily stylised décor and a greater selection of designer drinks. These new style venues are perceived by young adults as indeed more cosmopolitan. As one of our interviewees exclaimed in relation to them, 'There's just a better class of people' (Sarah, 21 years old, Leeds), while another young woman summed up the link between her identity and the type of venues she frequented with the simple comment: 'I'm a cocktail person' (Clare, 25 years old, Newcastle upon Tyne).

Many of these bars are extremely style conscious and have become, for some, too pretentious. Referring to one new bar which had opened in a popular student area, one undergraduate (Emma, 21 years old, Leeds) commented: 'It's pretentious and showy and everyone is more bothered about what everyone else is doing rather than having a good time.' Another person (Ben, 20 years old, Leeds) commented: 'They're all, like, posh blonde birds who go round in puffa jackets and things like that. They've got rich daddies and you can tell, because they usually have a Moshcino bag or something, and they walk around with these silly handbags.'

Gentrification, then, in terms of both displacing older users and upgrading the expectations of others, is a developing trend within mainstream nightlife, and it is closely tied up with the middle classes as driving forward particular forms of leisure and usurping established social spaces (Savage *et al.*, 1992; Butler, 1997; Savage and Butler, 1995). Historically, representations of the middle classes at leisure have been relatively simplistic, focusing on stereotypes about suburban living and/or images about 'yuppies'. Savage *et al.*, (1992) recognise that different middle-class groups draw upon various resources – property for the petit bourgeois, bureaucracy and organisation for managers, and culture and credentialism for professionals. As such, recent evidence suggests that sections of the middle class consume in differentiated rather than homogeneous ways. Research by Savage *et al.*, (1992) on cultural and consumption habits reveals at least three types of middle-class lifestyles – ascetic (public-sector welfare professionals), postmodern (private-sector professionals and specialists) and an undistinctive group (managers/government bureaucrats). Private-sector professionals, he argues, are more into 'California sports', conspicuous consumption, individualisation and eclectic consumption, while the public-sector middle-classes engage in more ascetic and health-conscious leisure pursuits. However, the important point is that all these different consumption practices are still framed within capitalist entertainment provision.

How might some of these general consumption patterns relate to mainstream nightlife and the tendency towards gentrification and stylisation? The growth of private-sector professionals in many urban centres, with their more eclectic lifestyles, has fuelled a range of mainstream consumption practices ranging from branded and themed environments to more up-market, gentrified style bars. Roberts (1999: 201) typifies this upper strata as 'leisure omnivores' who 'have sufficient money to nurture and indulge their wide-ranging tastes' – a kind of postmodern consumer. As one young woman, Claire (19 years old, Newcastle upon Tyne), told us: 'I'm a regular everywhere ... I go to all the clubs so there really isn't anywhere that I wouldn't really not go. We try to go to as many places as we possibly can before we run out of money.' The idea of the 'post-tourist' relates to those who, often with a certain amount of irony and pastiche, do not seek authentic experiences, choosing to revel in the artificial and ephemeral atmosphere of themed entertainment spaces (Urry, 1990).

Some young professionals, however, did display a return to neo-traditionalist attitudes in their desire for 'authentic' pubs which foster close interaction rather than the 'artificial' atmosphere of bars whose layout encourages disengagement. One group of young professionals told us:

Fred: That's something you definitely miss in bars. It's like a conversation of six of you all sat round a table ...

Brian: You're drinking, you're round a table enjoying yourself [in a pub], but when you're in a bar you're all stood, the worst thing is finding yourself all stood at the bar sort of like looking off in different directions. You've got no interest there. You've got no focal point. You can't just sit in a circle.

(Focus group 1, Leeds)

Such groups of hard-working young professionals are often more motivated by socialising and 'catching up with friends' than the latest nightlife trends. In Wynne and O'Connor's (1998) study of middle-class urban-livers in Manchester, they found that this affluent, mobile and largely childless group were not particularly exploratory in terms of cultural practices and represented a large 'open middle' of consumer taste. Much of this lack of experimentation reflects balancing work, family and social commitments for such 'cash-rich, time-poor' groups. As one person (Sally, 28 years old, Bristol) told us:

> And it's nice to sit round with a nice bottle of wine, have a chat or whatever, and then if you want to go out after that, then fine. But, you know, it's a good way to sort of see people, particularly when you're working, you don't see them every day.

More specific occupational groups within these general class factions also reflect differences in nightlife consumption. For instance, private-sector professionals in style or creative occupations (advertising, fashion, media) are more apt to consume the latest trendy and up-market nightlife styles and venues (Wynne, 1990). Butler and Robson (2001) also present a useful understanding of the increasing social and spatial divisions within the middle classes, which are not simply sectoral (i.e. public versus private) but based mainly around lifestyle and cultural differences. In the context of London they suggest four broad types of middle-class lifestyle or consumption settlements: highly mobile empty-nesters with heavily work-dominated lives; high-consuming gentrifiers who are attracted to 'global' places of consumption; hedonistic counter-culturalists who are drawn to marginal areas; and enclavists who huddle together in protective ghettos. Such an understanding is useful for thinking about the internal segregations within middle-class mainstream nightlife cultures.

It is not difficult to see what is attractive about these newly developing gentrified space. Clearly, part of their success has been in catering for the increasing numbers of high-spending young professionals and service employees (Ley, 1996). These more exclusive places act to separate this affluent group out somewhat from the more traditional mainstream, providing an atmosphere for networking, socialising and meeting other social climbers. However, they also provoke reaction. As one interviewee observed:

> Some of them are just really stuck up. Some of the times that I've been in this Quayside bar, when I come out of there I just want to drop a bomb on it, you know … there is a dress code sort of thing … and people in there think they are something different. I get that feeling it's mostly like a Don Johnson thing.
>
> (Dave, 31 years old, Newcastle upon Tyne)

Many groups in the gentrified mainstream, then, value style over content, social posturing over social contact. The following conversation exemplifies a rather fashion-conscious clientele in one dance-oriented nightclub:

Steve: It is out of the way, it is too expensive to get in, the drinks are expensive on top of that, and it is just, I don't know, you have a certain type of person that goes there, that is into their dance music.

Sarah: People who are dressed in all the designer stuff.

Steve: Pretentious people on a Saturday.

Sarah: Yeah, they will not get sweaty, they really should do.

Peter: Your obnoxious twats will go there and stand and they will not move in case they sweat, and it is, like, why do you come clubbing for? To get dressed up? To look good?

Steve: And I see a girl that gets up and dances and sweats. I hate to see these beautiful perfect people with their perfect make-up and their perfect hair.

(Focus group 2, Newcastle upon Tyne)

Such stylised spaces also appeal to the wealthier elements of the student population, young women and gay consumers and an increasingly older, more mature and upwardly mobile section of local working-class populations, who view such places as sites to express their perceived mobility, status and maturity. This latter group in particular could be seen as literally trying to 'escape' from their occupational and class background. Research by Savage *et al.* (1992) reinforces this by suggesting that lower-level technical and service workers were as likely to patronise wine bars as higher-income groups (see Box 5.3) and 'This perhaps indicates one function of wine bars is as a training ground for young people aspiring towards high levels of social mobility' (Savage *et al.*, 1992: 118).

Box 5.3: A night out in the mainstream after a hard day's work; Leeds, 21 June 2001

We met in Cosmos at 9p.m., the newest bar to open in Leeds over the summer of 2001. It was bang in the centre of the city's newly emerging southern quarter, a part of town experiencing a rapid growth of independent bars and record and clothes shops. Cosmos was mainly frequented by a young, stylish crowd, attracted by the novel blend of a noodle bar, an exotic if over-priced cocktail menu, and some of the city's coolest DJs playing music before they went on to play in the clubs. Silvia and Julie turned up late and brought some other friends, Brian and Dave, we hadn't met before. Gary and Paul, despite promises on the telephone, didn't turn up at all. Apparently they had gone straight out from working at the call centre at 5p.m., to play pool with their mates in their work section. This had become a bit of a tradition on a Friday, and they were usually tanked up by 8p.m., so it was of no surprise that they didn't make it. 'We might bump into them later at Bonkers Beach Club,' Julie said, 'but don't expect to get any sense out of them. They'll be well up for it and sharking girls. Gary's a right smooth talker after a few pints.'

They all knew each other from working at the call centre, and all were either still at university or had just finished. To break the ice early in the evening we talked more about working life. Silvia told us: 'It's a right laugh working there. Although we have to sit at the phones and constantly be alert, in the breaks we have a right giggle and we can't wait to go out as a group. We get slaughtered together and go clubbing and mess about and who knows …' Silvia and Julie were dressed stylishly in short black dresses, constantly smoking Marlboro lights with Bacardi Breezers their choice of drink all night. 'I like getting dressed up,' Silvia commented, 'it's the only time I get to really go for it, and not have to wear a crappy old blouse at the call centre.' Brian, explaining that he felt the cocktails were a right rip-off in here, still bought a Havana Club and coke at £3.50 per shot. He told us that he liked coming here as it made him feel 'in with this new cool crowd, which is what Leeds is about these days'. Dave, dressed more casually in worn Levi's and Nike trainers, explained to us that 'the bouncers used to be right wankers. Would only let you in if you had crappy shoes and proper trousers on. Now this new string of bars round here has finally caught up and you can go out looking casual, which is cool, more like London.'

After Cosmos we went to Razzmatazz, a chain pub which opened about a year ago, with the slogan 'All YOU need for a night's entertainment'. As we entered, Brian explained that although it was 'one of those wanky chains where everything is the same and the bar staff are fucking miserable, it's the only place you can go on a Friday night in the centre and get a cheap pint'. It was 'Office Party Night' in this huge 1,000-capacity super-pub, decked out with banks of TV screens and a large stage on which two people were singing a karaoke version of a old Wham song from the 1980s. The throng of people, mostly a post-work crowd in suits, were eagerly consuming cheap bottles of Becks for £1 each. We met four more workers from the call centre in there, all with the same idea of making their slim salaries go further on a Friday night.

Our next stop was the Glass and Trumpet, one of the city's oldest pubs, which had recently been bought out by one of the country's largest pub chains. 'I used to come down here all the time,' Dave said. 'It was a great old man's pub, you know, and then they ripped it all out and put in all this fake shit which is meant to look traditional, right. There's no music or gamblers [betting machines] either. The bastards put the beer up 'n'all. I don't get it.' In this rather tranquil atmosphere Silvia and Julie started to look bored, and we headed back out into the night. Our final destination before clubbing was Bar

Che, a new chic theme bar which had just opened in the centre, complete with revolutionary memorabilia and cocktails such as Castro's Revenge. Silvia was particularly keen to try it out as she knew one of the bar staff in there who, she told everyone, was well sexy. However, the 200-strong queue and the £3 price tag on the drinks deterred us and we headed into the night again.

We ended up at Bonkers on the middle of the High Street next to a group of other large theme bars, Downunder, Flanneries and Bar Samba. Dave told us some of these were his favourites: 'They've got all the footy games on, top tunes on a night and the birds in here are all right. Sometimes we have a right laugh. Just get pissed up and make an arse of ourselves and forget about it all ... work and that. Some of 'em are open late, which is wicked, so we can go on dancing and drinking till at least 1 o'clock. Sometimes I get a bit of hassle from the bouncers for being scruffy, but fuck 'em.'

It was £5 in to Bonkers, £4 if you were lucky enough to be handed a promotional flyer from the young woman in a leopardskin jumpsuit roaming the street outside. The bar staff in Bonkers were all in either bikinis and hot-pants, which Brian said was the best reason for coming here. Cheesy house music was playing; it was too loud to talk so we all headed for the dance floor, and, anyway, by this time too many designer bottled lagers had taken their toll. We had lost Silvia and Dave. Julie said it had been on the cards for ages that they'd get it together as they'd spent loads of time together at work. A whole bunch from the call centre were reunited on the dance floor and were jumping around, laughing and grabbing each other to 1970s and 1980s classics. Paul and Gary were there, looking worse for wear, but with their hands raised to the ceiling, singing along to 'It's raining men, hallelujah'. Gary muttered something incoherent to us, which sounded half like an apology, as they headed off to the bar with two women. About half past three we stumbled out of the club in search of taxis and chips, £40 lighter after a night on the town.

Despite examples of weekend hedonism and various forms of escapism, what can also be observed in the gentrifying mainstream are the links between work and leisure. In this sense, just as Fordism implied a broad socio-economic system which did not stop at the factory gate, so too post-Fordist forms of work have reverberations well beyond the boundaries of the call centre or the office (Sennett, 2001). Contrary to the pattern of escaping work discussed earlier, some companies, particularly those involved in financial services, are encouraging their employees to attend particular venues. Brannigans bar chain in the UK, for example, has set up a working relationship with a number of large corporate firms in the service sector, such as Directline, and offer workers discount cards, drink deals and free transport to the bar. Many bars have emerged exclusively to satisfy the demands of after-work consumers, epitomised by the minimalist chic branded venue of All Bar One, which also promotes its female-friendliness to attract both men and women from the business world.

Many young professionals in the private business sector consuming gentrified nightlife adopt a set of values and aspirations from the corporate world of work which respects the value of the brand, corporate loyalty and a 'work hard, play hard' ethic. Corporate workplaces are eager to foster a sense of sociability both in the workplace and outside it, by linking up with certain nightlife operators. In the series *Slave Nation*, screened on Channel 4 in the UK in 2001, presenter and writer Darcus Howe questioned the value and motivation behind some of these new company identities. In such 'work-friendly' venues, nightlife has become an extension of work, another arena in which business is conducted, networks extended, lifestyles consolidated. Sennett (2001: 4) has argued that new flexible working practices are creating parallels in terms of social relations and spaces in cities: 'now, just as the workplace is affected by a new system of flexible working, so the city, too, risks losing its charm as businesses and architecture become standardised and impersonal'. Gorz (1999) further suggests that the dominance of the 'full-time work' ethic has downgraded the right to work and play in other ways.

In summary, clearly, there are some young people who travel across the boundaries of the mainstream and into more alternative and older residual forms of nightlife (see Chapters 8 and 9). Many, with limited financial resources, seek ways to maximise their participation in mainstream nightlife, through buying no-name brands and taking advantage of 'drinks specials'. In contrast, traditional and 'respectable' working-class groups who pursue steady relationships and save up to buy a house and/or get married (which some marketing agencies (see Carat Insight, 2001) have labelled 'new traditionalists') do not particularly prioritise 'going out'. One nightclub operator in the UK lamented about this tendency:

> At one time all they wanted to do was get their pay, or get their unemployment benefit, and go out and spend it on drink ... they're interested in mortgages and holidays and things like that. Now I think they're a lot more responsible. Their capital is tied up in other things.
>
> (Nightclub manager 3, Newcastle upon Tyne)

Others, such as art college students, young artists and subcultural groups on the fringe of mainstream nightlife, both participate in the mainstream and critique it.

Further, as the population in many western countries begins to 'grey' (get older), mainstream nightlife operators are looking towards older 'post-adolescents' and couples with larger disposable incomes, as well as increasingly courting the 'less rowdy' professional female and gay market. The mainstream, then, is constantly shifting ground, fuelled by competition within and between groups. However, these shifts increasingly point to a sanitisation, purification and upgrading of experience.

Limits of the gentrifying mainstream

Arguing that mainstream nightlife in general is more structured and intentional should not imply that it is an immutable monolith. It is continually produced and reproduced by different groups of young adults pursuing their own version of a night out, and is also defined through various contradictions and disillusionments. Ritzer (2001), for instance, highlights a number of general aspects of mainstream production, such as efficiency, control and technology, which are contributing to the 'disenchantment of the consumer world'. Additionally, many countries (especially continental European ones) and regions have competitive local nightlife markets, which makes it more difficult for brand penetration and corporate domination to succeed. Certain companies with global aspirations have also experienced brand saturation, often due to over-exposure of their products. The Planet Hollywood chain is perhaps the most well-known example of such hype, decline and failure in the mainstream.

Smaller urban centres sometimes also overestimate their cosmopolitanism in an attempt to embrace the corporate world. Teatro, an exclusive private members' club owned by actress Leslie Ash and ex-Leeds footballer Lee Chapman, originally opened in London's Soho, with a second venue opening in Leeds in the north of England. One local fashion magazine suggested that they 'hope to inject a shot of glamour and glitz into Leeds' burgeoning social scene, and provide a much needed haven in which local celebrities and high-profile types can socialise and relax, away from snapping cameras and harassing fans' (*Absolute Leeds*, June 2000). However, one year later Teatro had failed to attract a celebrity crowd and was duly closed. While some blamed the location, others knew better. Leeds may have a chic new centre, but it is still located in England's northern industrial belt, far from London.

Further, while many aspects of these new nightlife spaces appear positive and choice has increased, especially through the decline of male-dominated drinking environments, even flexible brands and niche stylised nightlife environments offer largely standardised, sanitised and non-local consumption experiences. Underneath the façade of cosmopolitanism and diversity, a more corporate uniformity and placelessness is growing in downtown areas through expanding pub and bar branded experiences. One journalist lamented to us:

> I think this last, say, especially five years, you know, your designer labels have come in more ... obviously if you have got these middle-of-the-road mainstream looks, you are going to get smart middle-of-the-road mainstream bars ... Well, eventually, essentially you are still getting the same names and the same products. You know it's global and it's horrific. I hate that.
>
> (Journalist 1, Newcastle upon Tyne)

This dominant mode of nightlife production based around serial reproduction and expansion of chain or theme bars continues to provoke criticism from groups such as the Campaign for Real Ale, civic trust associations and residents' associations. Concerns and disillusionment are also raised by consumers, as the following quotes highlight

Simon: The thing is ... all the bars and clubs are owned by big organisations and a lot have taken over, and that's what needs to change. It needs some individual investors.

Peter: I would not like everywhere owned by the same place because it would just be the same and nothing more. Boring.

Simon: If it is a new pub, you think, ah it is just like a replica.

Julie: It is the same club and the same pub, it has just got different names, the same people ... The bar staff are just clones.

(Focus group 3, Newcastle upon Tyne)

There is some evidence of a 'brand backlash' (Klein, 2000; see also Box 5.4), with battle-weary consumers rejecting the serial monotony of high-street nightlife brands. New tastes on the mainstream can quickly turn sour. Like the neo-traditionalists mentioned earlier, many consumers are eager to recapture more 'authentic' and 'characterful' nightlife away from the world of corporate brands. As one person told us:

> Corporate clubs and corporate pubs have got it so wrong, throwing money at it. It's the bland globalisation thing that someone in London has decided that's what Bristol needs and actually it's not. If you look what's happened with the big pubs, that are aimed at students and middle-class drinkers. I think there is a reaction to that and people are starting to go back to corner pubs and back-street boozers. There's something interesting about them.
>
> (Rick, 26 years old, Bristol)

Box 5.4: Mine's not a Guinness

At least ten bars, mostly Irish, have joined a protest towards Guinness Corporation's involvement in starting new Irish theme bars in Philadelphia in the USA. Most of the pubs have discontinued the sale of Bass and Harp, which Guinness also imports, but four have even taken the popular Guinness Stout off tap, among them the city's oldest bar, McGillin's Olde Ale House. 'This is America. There's a lot of great beers here,' McGillin's owner Chris Mullins said. 'We're the oldest bar (since 1860) in Philadelphia and we did not have Guinness before I bought the bar in 1993. We survived a long time without Guinness and we can make it without them now.' The bar owners are upset with Guinness' commercial development division and its relationship to the Irish Pub Co., which has built hundreds of Irish theme pubs around the world since 1991. The company put together its first pub in the United States in 1996 in Atlanta, Georgia, and scores have followed. Most import the physical components of the pub, sometimes even the bar itself, from Ireland, and cost up to $2 million to open. Some, such as the popular Fado pubs, are part of chains, while others operate independently. Their success has led other entrepreneurs to use firms other than the Irish Pub Co., to import Irish pub parts to the USA and open large pubs (Real Beer, 2000).

Many people felt that one of the main problems with the mainstream, then, was its lack of local embeddedness with local cultures. Moreover, for many, the upgrading of nightlife venues often meant all style and no substance. As one young consumer (Colin, 25 years old, Newcastle upon Tyne) exclaimed: 'People want to belong to that elite crowd, but what people do not realise is that it's actually McDonald's with a marble bar.' While the mainstream is a negotiated and often contested space, its influence and acceptance continues in many cities around the world. We explore in more detail in Chapter 9 some of the groups who have more directly challenged this growing corporate sameness of urban nightlife and have sought to provide some alternatives.

6

SELLING NIGHTLIFE IN STUDENTLAND

This chapter is about studentland – a term which alludes to a bounded social and geographical space which leaves a distinctive imprint on many localities, especially at night. While it is a growing, differentiated and indeed fluid space, our argument is that mainstream corporate nightlife operators are increasingly targeting 'traditional' students as part of their general strategy of attracting 'cash-disposable' groups (like professional women, young urban-livers and gentrified gay cultures). As such, commercial nightlife operators help to construct studenthood and student experiences, as much as students help to create nightlife.

Life in studentland throws up many, often contradictory, images. The inter-war splendour of Oxbridge sits alongside hippies, radicals and beatniks of 1960s campus counter-cultures. Similarly, today tens of thousands of students in Indonesia, Korea and China continue to risk their lives in pro-democracy movements. Their peers across Europe and North America worry instrumentally about future careers while engaging in various night-time socialising rituals, with a minority turning their concerns to political issues including globalisation and the environment. Studentland, then, is a complicated place. It is a set of discursive practices brought together by several groups – students, staff, parents, locals, business people, the police – who in their own ways define, delimit and create it. Groups have different motives for engaging with studentland – for younger students it is learning, fun and an important rite of passage experience; for city bosses universities may mean having a skilled and educated future workforce; and for businesses it is a potentially important source of revenue and profit.

Over the last few decades studentland has become gradually occupied by a more diverse population which has moved slowly away from a purely 'elitist' model, with more blurred student–local distinctions and the segmentation of the student body into various subcategories (Hollands, 1995; Chatterton, 1999). Some have argued that the changing context of higher education has led students to become more instrumental in their outlook and more mainstream politically and culturally (Loeb, 1994). Yet there are specific characteristics and trends which mark out students from those outside the university, such as higher levels of free time, disposable incomes and a learning-oriented approach to life. Additionally, there continue to be durable dispositions and habits, common sites and spaces inhabited by students (Bourdieu and Passeron, 1979).

This chapter attempts to unravel studentland, and in particular student nightlife, by looking at how it is produced, regulated and consumed. Social life and nightlife activity are central aspects of student identity. Finding friends, sexual encounters, pranks and nights out are as important, if not more so, to the university experience as the formal curriculum. Much of university and college life, then, is about balancing education and work with fun, drinking and socialising. However, surprisingly little research has looked into the cultural and nightlives of students (although see Moffatt, 1989; Chatterton, 1999; Hollands, 1995). We begin by providing a context for understanding student life today by looking at both change and continuity. In particular, we highlight the creation of a more instrumental and mainstream outlook, in terms of both work and play, which means that studentland has also become ripe for commodification, theming and branding. The second part of the chapter takes a closer look at some aspects of nightlife in studentland, including the role of students as consumers and spenders, and the growing dominance of middle-ground mainstream tastes. In particular, we examine the trend towards a 'corporate campus' in which studentland is increasingly packaged, sold and commodified, especially through the machinations of large nightlife operators eager to cash in on this important consumer group.

Changing contexts for studentland

Different national contexts have a significant bearing on the character of studentland, and hence affect nightlife activities and styles. Each country has developed its own distinctive system of higher education, university cultures and patterns of student nightlife. The UK and the USA, in particular, have some notable differences worth mentioning here. While there was a rapid increase of student numbers in the UK in the late 1980s and early 1990s, access to studentland is still largely restricted to the wealthiest strata of young people, and real differences remain between the elite 'finishing school' experience of the older universities, with a large upper- and middle-class, adolescent, white cohort of 'traditional' wealthy students, and the 'service station' experience of the newer teaching-based former polytechnics (Ainley, 1994; Chatterton, 1999). Studentland in the USA is a far larger and diverse place with over 14 million students spread over 3,400 institutions, and with participation rates at almost 40 per cent as far back as the 1960s. This is not to say that studentland is open to anyone in the USA. Many poorer social groups are virtually absent in higher education (Males, 1996), and there is a long tradition of racial segregation and black struggle here.

In terms of nightlife, differences also exist between national contexts. As we shall see, students in the USA are often more constrained in their leisure time, due to high numbers in employment and engaging in sport/leisure, not to mention a more home-based and private drinking culture and a higher legal drinking age (see figure 6.1). The UK, on the other hand, is often seen as an oasis of student hedonism where young people travel away from home (the 'Great Teenage Transhumance' (Walker, 1997)) to party for three years, while their more sober continental European counterparts usually remain in the parental home during their university careers and maintain existing social bonds. Underpinning such characteristics are a number of structural differences, especially in terms of geographical mobility within countries (Charles, 2000).

Constructing studentland: continuity and change

Students are one of the 'ideal types' which many of us use to understand the world around us (Schutz, 1972). The dominant perception of this ideal type of student is that they are somehow separate from the rest of the population, suspended from the mundane and at a distance from securing the necessities of life, living, as Bourdieu and Passeron (1979: 29) put it, 'in a special time and space ... flouting the distinction between weekends and weekdays, day and night, work and playtime'. Echoing such views, Brake (1985: 26) described student life as a 'moratorium from wage labour' which places them in a unique and privileged position. Such separateness is often reflected through a range of contradictory images. On the one hand, Evelyn Waugh's *Brideshead Revisited*, and the film *Chariots of Fire* evoke the nostalgia and romanticism of upper-class Oxbridge student life in inter-war England. On the other, there is a bundle of images associated with alternative, radical and hedonistic student cultures captured in films such as *Animal House* and the campus scenes in *Born on the Fourth of July* and *Forrest Gump*. Students have also been associated with 'Generation X' (Coupland, 1992) characterised by a preoccupation with lifestyle and disaffiliation. A cluster of images of students in a British context emerged from the TV comedy *The Young Ones*, which depicted a group of student no-hopers, each stereotyped by a particular characteristic; Rick (middle-class 'lefty'), Vivian (nihilistic punk), Neil (hippie) and Mike (wheeler-dealer mature student).

What is the basis for this student 'ideal type' and associated stereotypes? As we discussed in Chapter 4 in relation to Bourdieu's (1984) concepts of field and habitus, studentland allows participants to adopt fairly durable and common dispositions which are developed and maintained within regulated and segregated residential and entertainment environments such as the library, shared student house, dorm, hall of residence (and, in the USA, the fraternity and sorority)[1] and the pub, bar, club or café (Chatterton, 1999). Student spaces act as sites of 'social centrality' within university life in which the rituals of studenthood are undertaken (Hetherington,

1996: 39). Although students are only brought together temporarily, they opt for strong forms of 'elective sociality' (Maffesoli, 1996) based upon their own rituals of initiation, closure against outsiders and a desire for belonging and association. Moreover, the experience of the (often overwhelming) mass campus means that students are keen to seek manageable forms of identification.

Campus cultures are also particularly strong definers of students' leisure and significant contrasts exist between big city campuses and smaller rural campuses in terms of choices and the range of cultural opportunities. Course cultures, too, mark out particular student groups, often separating the more culturally oriented arts and humanities students from their less cultured science-based peers (Gasperoni, 2000). Finally, participation in sports clubs and societies is an important factor which shapes student identity and leisure. Sports teams, and more generally exercise and health, structure large parts of social life, especially for US students (see Figure 6.1). Life for sports students is often far removed from their more hedonistic liberal-arts classmates, and many stick together when socialising, as one college American Football player recounted:

> We have a lot less time, you know. We have to have all our classes done by two o'clock because we have to practice for the rest of the day. It's also important to stay out of trouble. We'd be in the papers for drinking or fighting.
>
> (Ben, 22 years old. Strong safety for the Badgers American Football team, University of Wisconsin–Madison, USA)

Student life is also about 'coming of age' and age grading (Moffatt, 1989). Lifestyle and going-out cultures between different age groups are particularly pronounced, especially between the more carefree freshmen and sophomores and the more serious and worldly seniors. Most students alter their self-identity and socialising patterns around work and financial pressures in later years of degree courses. As such, they may begin to adopt a new set of aspirations and styles, to prepare themselves for life as young professional graduates in the workplace.

Many of the habits, places and rituals within student life have been weakened by the quantitative and qualitative changes within university populations in most western countries. In particular, the number of women has increased dramatically in higher education and in many contexts is equal to, or exceeds, the proportion of men, and there has also been an overall increase in the numbers of previously underrepresented groups within higher education such as mature students, working-class youth, overseas students, ethnic minorities, people with disabilities, access entrants, recurrent and returning learners and part-timers. In the USA, Horowitz (1987) has discussed how these 'new insiders' sit uncomfortably alongside the more established and affluent WASP (White Anglo-Saxon Protestant) elite on college campuses.

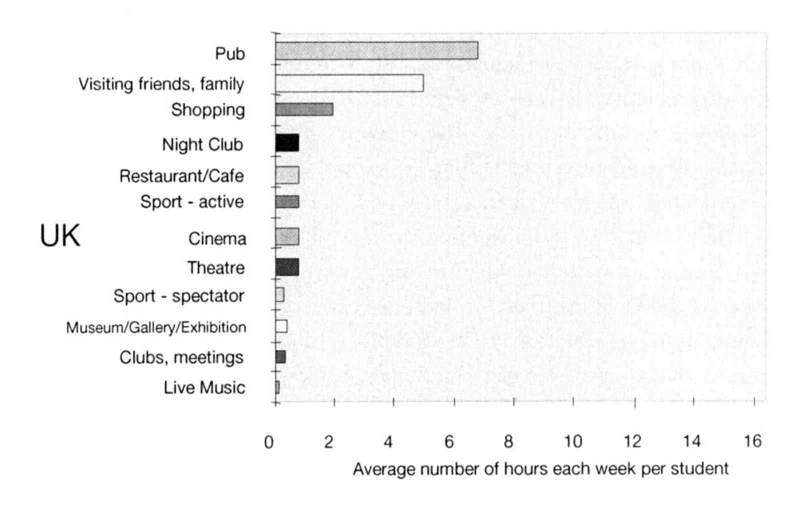

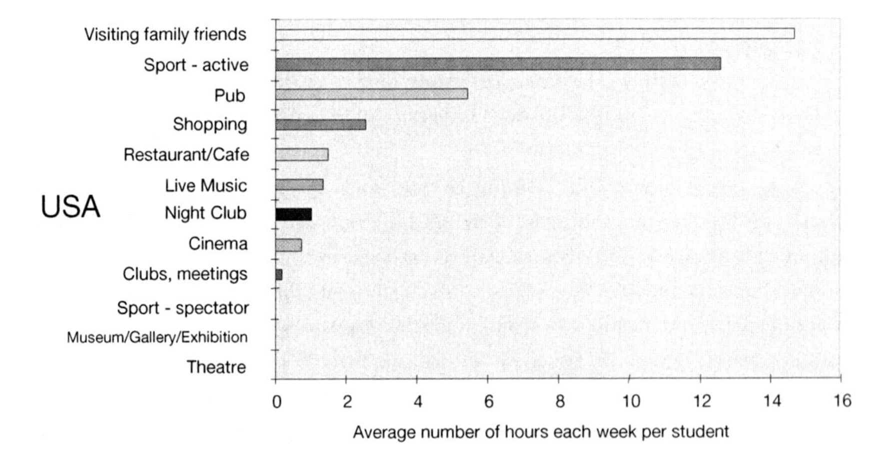

Figure 6.1 Student leisure time in the UK and the USA, away from home
Source: Chatterton, 1998

As such, studentland is now more than just an adolescent transition, especially through the increase of young adult learners, mature students and those taking gap years. In many ways, studenthood is now a broader post-adolescent life stage 'characterized by a prolonged experimentation with life's possibilities' (Johansson and Miegel, 1992: 85–6). More people have joined in the extended 'socialising ritual' of being a student, played out in the consumption-oriented city. This rapid growth of university populations over the last few decades, and extension of what it means to be young, has led to a more generalised college and higher education culture, and hence students have become a focus for, and indeed have taken a more leading role in, youth styles and consumption generally (Roberts and Parsell, 1994).

In reality, then, the student body of today is a more complex and internally divided social group, and rather than representing one rather homogenous lifestyle it consists of several communities such as the 'Sloanes', the 'intellectuals', the 'arty' crowd, the 'jocks' or sports students, the 'lads' (and increasingly the laddettes), the beer boys and the good-time girls, the radicals or the eco-warriors, the skaters, and the entrepreneurs, among others. Such subcommunities may not be so readily observable to the outsider, but hold much sway in the internal discourses of studentland[2]. However, all these student communities still share a common cultural archetype of the 'student' characterised by higher levels of free time, disposable income, socialising, experimentation, youthfulness and a more learning-oriented lifestyle.

One of the most visible student groups around campus are those from wealthy and privileged families, where going to university has long been a rite of passage. Such students have various labels and are known notoriously as the 'Sloanes' in the UK or the 'preppies' in the USA, attired in slacks, loafers and Oxford shirts (Moffatt, 1989). Marked out by their designer labels, blasé attitudes, mobile phones and convertible cars, they are often the source of jibes from fellow students and conflict from local young people. Moreover, they are also a key source of the public's stereotypical views of the 'traditional' student, not to mention forming the archetypal 'student consumer' which much of the business community targets.

Of course, not all students come from this wealthier and more privileged section of society, and the expansion of numbers in higher education to include some of the less well-off population has resulted in issues like debt and a rise in part-time employment. Debt has become a particularly important context for understanding student spending and consumption patterns (McCarthy and Humphrey, 1995). The abolition of the maintenance grant in England and Wales in 1999 has increased levels of genuine student poverty, while the introduction of entrance fees is deterring many working-class youth from enrolling on courses. The National Union of Students in the UK has estimated that students can now expect to owe up to £12,000 upon graduation. In the USA, the combination of readily available credit with loans to cover rising tuition costs means the average student now graduates owing $14,000, rising up to $80,000 for medical students. One student at New York's Niagara University commented that 'It's pretty common for us to sit around discussing why we put ourselves $10,000 in debt to make $25,000 a year.' (Chatzky, 1998). Rising levels of debt not only have an impact on the ability of poorer students to go out, but also on well-being, with those amassing excessive debts on graduation more likely to suffer anxiety and depression compared with other students (Scott *et al.*, 2001). A key part of student life, then, is mixing periods of hedonism and release with periods of stress and money worries.

As such, some students are turning to part-time work to supplement their income and avoid high levels of debt. While this has long been the case in the USA, a recent study in the UK by Barke *et al.* (2000) found that 54 per cent of students had been employed sometime during the academic year, working an average of 12 hours per week. The main area of employment was 'sales assistants or checkout operators', with an average weekly take-home pay of £49. It was also found that in the UK younger students, under 26, now have more money at their disposal than any time since 1988/9, but only because more of it is earned, received as gifts, drawn from savings or borrowed against future earnings (Callender and Kempson, 1996). Rather than spending less, then, students faced with money worries are becoming more selective, instrumental and focused in their use of leisure time.

Today's students are often seen as more career-oriented yet less critically engaged than their predecessors (Bloom, 1987). Due to increasing work loads and money constraints, it may be fair to say that there has been a mainstreaming of attitudes and apathy among students. As one newspaper in the USA put it: 'Tomorrow's leaders are a bunch of uninformed, apolitical, and apathetic college kids worried about little more than their own self-interest' (Brook, 1996: 12). What is also evident is that studentland is increasingly becoming a more 'professionalised' arena through a cadre of student support-service professionals, who manage, and to some extent control, adolescent development, and create 'post *in loco parentis*' structures in the mass university. Such trends are more prevalent in the USA, while UK students retain more autonomy over their extra-curricular lives (Silver and Silver, 1997). Many universities are also significant cultural, entertainment and nightlife providers in their own right, which further extends the formal reach of the university into the extra-curricular lives of students. As the formal curriculum extends outside the classroom and into the private lives of students, both worklife and nightlife in studentland are becoming less a series of possibilities, or serendipitous encounters, and more a programmed and directed experience.

The important point here is that despite increased diversity within the university population and the development of a wider range of identities in studentland, counter-forces are at work which are leading towards the creation of more mainstream tastes, instrumentality and the need for readily packaged educational and socialising experiences, as well as working to construct an 'ideal type' of student consumer. In essence, it is still the traditional white middle-class adolescent, with relatively high levels of free time and disposable income, that is held up as the archetypal student, and this is often the specific grouping which businesses, including nightlife operators, specifically orient themselves towards in their provision of goods and services to the student body. It is to such issues that we now turn.

Understanding nightlife in studentland

Students as consumers

Since the 1960s, it is generally assumed that societies in the west have become more middle class, or at least focused more upon the tastes and needs of this particular class (Savage and Butler, 1995; Ley, 1996; Smith, 1996). The expansion of higher education and white-collar and professional employment has played a key role here. The growth of a well-educated middle class, it is argued, has stimulated an increase in demand for cultural goods and services (DiMaggio, 1991). Such a class of consumers represents the 'more-more principle' in that they are likely to consume more goods and services across more areas (*ibid.*). Moreover, 'there is a tendency on the part of some groups (especially the young, highly educated, sectors of the middle classes) to take on a more active stance towards lifestyle and pursue the stylisation of life' (Featherstone, 1991: 97).

Students are clear examples of this active 'more-more' approach to lifestyle. One of the most significant divides within the consumer tastes of the young occurs at the age of 18, when a proportion of young people go to university and the rest enter into employment, training, unstable employment or indeed unemployment. Research by Roberts and Parsell (1994), however, has shown that the difference is not so much the 'type' of leisure which these two groups participate in, but the 'amount'. As Table 6.1 shows, in the UK, at least, students do not necessarily go on to pursue a different set of leisure options, but simply do more of them compared to their non-student peers.

Due to the sheer size of the student population, not to mention their purchasing power, students increasingly play a dominant role in the reshaping of youth styles and

Table 6.1 Percentage of students participating in activities over previous three months

	Pub	*Nightclubs*	*Cinema*	*Live music*	*Museum/ gallery*	*Theatre*
Henley Centre survey (16–24 years)	82	70	72	16	23	21
Sample of UK students[a]	90	81	65	46	31	28

Source: Chatterton, 1999

Note: [a] This data is based upon a survey of 500 students at two universities in Bristol, UK

cultures generally, and are now a favoured target for the entertainment industry. Student life is part of a larger lifestyle phenomenon which has a pre-student phase (at high school, etc.) and a post-student phase (graduates, post-graduates, young professionals). The expansion of this wider student lifestyle has extended a broad middle ground for youth cultural styles, in which mainstream, middle-class values take centre stage (Miles, 2000; Roberts and Parsell, 1994).

Because of their higher levels of free time and disposable income, students are able to become fuller participants in consumer culture. Unlike their working-class counterparts, students also have legitimate and extended transitional spaces and mechanisms which allow for more personal development and experimentation. During this time, students are offered an array of resources such as consumer goods, easy credit, promotional discounts and employment opportunities, as well as personal and academic support services, which allow them to access consumption opportunities and citizenship rights to a greater extent compared to those outside studentland.

Students, then, play an important economic role in many localities, not just in the housing and labour market but also in terms of spending and consuming. A large part of this impact occurs in the night-time economy. An average UK university city has 50,000 students with an average annual expenditure for each student of roughly £5,000, with one fifth (about £1,000) spent on entertainment (Callender and Kempson, 1996). This equates to a spending power of around £250 million a year, and nearly £50 million spent on entertainment alone. Across the USA, according to the research firm Student Monitor, undergraduates spend more than $21.6 billion per year overall (Marcus, 2001).

Such spending capacities have attracted the attention of many businesses hoping to cash in on this lucrative market. A broad national infrastructure comprised of magazines, books, CDs, web pages and music tours has developed, specifically aimed at a student consumer audience. In the UK, the Virgin Corporation, for example, has launched www.virginstudent.com, offering a range of services for students, including reviews and local guides. Locally, many businesses invest heavily on marketing to attract them, offering student discounts in cinemas, restaurants and, of course, bars, pubs and clubs. Student spending is distinctive due to its seasonal, weekly and daily fluctuations. A range of businesses are keenly aware of, and eager to exploit, these differences in student life outside the customary working day, and in particular students' desire to go out in quiet mid-week periods. Local shops, taxis and fast-food vendors regularly state that, out of term time, business is slack and takings drop. These economic impacts of students are pronounced in smaller college towns and more rural university campuses. Many places have become dominated and even synonymous with their university, and to a certain extent depend, both financially and culturally, on the student body.

In some places, students have also played a key role in the gentrification process and the recovery of downtown populations. This has particularly been the case through the building of university accommodation, where students are new residents for redeveloped urban cores, being suited to reconverted difficult-to-let office space due to their transience, less rigid daily routines, desire to live in the middle of things and high propensity to spend. The publicly funded expansion of universities has also created a stable income stream for many property developers. Many localities, then, 'are busy turning old office blocks or a derelict leisure centre into fashionable mid-town apartment blocks full of undergraduates looking for a good time after a hard day's learning' (Tavener, 2000).

Yet student leisure patterns, while exemplifying the 'more-more' principle, still appear somewhat limited in their horizons. Figure 6.1 gives a flavour of the amount of time spent on various activities out of the home by students in the UK and the USA (see Chatterton, 1998).[3] While there are some similarities, such as an across-the-board popularity for the pub/bar, students in the USA understandably show a greater tendency towards non-alcohol-based leisure, such as sport and visiting friends. What it certainly dispels is the notion that students spend much of their time out of the house on a variety of pursuits, especially high cultural ones such as going to the theatre or visiting galleries.

The leisure tastes of students, then, identifies them as part of the large 'open middle' of consumption who are not particularly high on exploratory cultural practices and may be deterred from certain forms of cultural participation where access is unfamiliar (Chatterton, 1998, 1999; Wynne and O'Connor, 1998: 853). Students' tastes increasingly reflect mainstream and commercial cultural styles, many of which have little resonance with their radical predecessors: 'Mr and Mrs Undergraduate 1996 more closely resemble Saffy, the censorious and infuriatingly sensible daughter in *Absolutely Fabulous*, than Neil, the spaced-out hippy of *The Young Ones*' (Kingston, 1996). Sections of the student body have also become more stylish and brand-conscious, revelling in designer labels and more gentrified forms of consumption:

> Scruffy, baggy. Badly dressed ... Well, you can forget all that: students have moved on and closer to the once unimaginable – absolutely fabulous. Over the past decade designers such as Gucci, Prada and Alberta have been slowly reflected in university campuses.
>
> (*Guardian Education*, 2000)

Students in general are as much, if not more, driven by mainstream values such as pleasure-seeking, hedonism and release through popular and cultural forms such as drink, drugs and nightclubbing, as they are by experimentation or interest in high culture. The majority of students, then, have become rather uncritical and passive in their role as consumers (Chatterton, 1998). Work, time and money pressures may have begun to create a more instrumental outlook among students which pervades both their worklife and their nightlife. In such a context, many nightlife operators have increasingly begun creating nightlife packages, tailor-made around the tastes of brand-aware and fashion-conscious students.

Commercialising the student nightlife infrastructure: studentland up for sale

As Figure 6.1 shows, going to the pub and club is still one of the most significant pursuits in studentland (Thornton, 1995; Malbon, 1999). Many students in the UK choose particular universities on the basis of the nightlife of the locality as much as the academic profile of the institution. Numerous alternative guides have been published to help students choose universities on more than just academic indicators. The *Push Guide* in the UK, for example, offers information on a number of vital statistics such as sex ratio, flunk rate, graduate unemployment, average local rent and average price of a pint of beer, while the *Virgin Alternative Guide* gives universities marks out of twenty for the 'campus/town social scene'. Bristol University, which received the most applications per place in the UK through much of the 1990s, has achieved this not just on its academic performance, but because of the cultural and social vibe of the city, much of which was associated with the 'Bristol sound' of Tricky, Massive Attack and Roni Size. Moreover, in 1996, a survey found that 70 per cent of students at Liverpool's John Moores University said they chose to study in Liverpool because of the presence of Cream nightclub, one of the world's best-known super-clubs (www.clubbed.com).

A key part of nightlife in studentland is a distinctive infrastructure comprised of 'pathways' of venues which weave distinctive time–space patterns through certain areas (Chatterton, 1999). Students colonise, inhabit and modify many places such as pubs, cafés, bars and bookshops within, or adjacent to, student residential areas and university campuses. As outlined earlier, this nightlife infrastructure represents a relatively bounded field (Bourdieu, 1977, 1993), in which students invest and accrue social and cultural capital, which is of equal importance alongside educational capital gained from the formal university calendar. The pub, bar, club and café, then, are spaces which are as central to studentland as the lecture theatre and library.

Students also contribute to nightlife infrastructures themselves through roles they play as producers as well as consumers of nightlife. They are involved in a variety of activities such as small-scale, experimental and underground parties and clubs, DJ-ing and promoting events at local pubs and nightclubs. House parties, in particular, are a source of much creativity and are often used as an experimental space for students involved in producing or organising musical events. Some also remain in their university town after graduation and use their knowledge of the 'scene' to establish or manage venues or to pursue artistic careers. Sid Fox, one of

Leeds' most famous techno club nights, for example, evolved from a group of students running house parties in the city's student areas before moving first to a local community centre and then to residencies at the city's top clubs such as the Warehouse and the Mint Club.

For many young people, being away from home for the first time can initially be somewhat of a hedonistic experience (see Box 6.1). One UK bar manager reflected on what he saw:

> I mean, can you imagine just coming out of high school and going to university, the boys and girls would be like dogs with three cocks. They think, bloody hell, this is fantastic, never been out on their own before, right. Never been out of parental control, and here they are in Sin City.
>
> (Bar manager 3, Newcastle upon Tyne)

Box 6.1: Student drinking and drugs

Drinking, usually to excess, is one of the hallmarks and rituals of student culture. As Silver and Silver (1997: 111) explain:

> There are peer-pressures to drink, and the bar – rather than the political party or the campaign, the concert hall or even the disco – has become for many students the balancing focus for their studies, part-time jobs and tensions. Drinking and getting drunk are for some students a personal and collective response to campus and social pressures, and to some extent an acceptance of traditions associated in the past.

Alcohol and drugs are central to student nightlife and their use is motivated by sex, peer pressure, stress and socialising (Humphrey and McCarthy, 1998; Morgan, 1997). However, their excessive consumption has become a major concern, mainly due to evidence linking it to academic and health problems (Morgan, 1997; MacLeod and Graham-Rowe, 1997). Binge drinking by freshers and recent alcohol-related deaths associated with North American fraternity culture and the practice of 'hazing' have also become issues of concern. Similarly, one report found that 'sensible' drinking levels were exceeded by 61 per cent of men and 48 per cent of women within the student population in Britain (D'Alessio, 1996: 4), while another showed that students at Durham, Edinburgh and Glasgow consumed an average of 36 units of alcohol per week – compared to a recommended weekly intake of 21 for men and 14 for women (Morgan, 1997). Alarm over alcoholism reached new heights at St Catherine's College, Cambridge, when the dean was forced to write to students about excessive intake, following cases of alcohol poisoning and drinking games at formal dinners which required several women students to be carried from their seats, and unacceptable amounts of vomiting around common-rooms (Chrisafis, 2001).

While drinking as part of the undergraduate rite of passage has an enormous international legacy, in the USA the recent history is complicated by the raising of the legal drinking age to 21 in 1987 (Silver and Silver, 1997: 112). This is not to say that under-age students in the States do not drink – alcohol consumption has simply become a more private rather than public affair, framed within house and fraternity parties. Concern over alcohol-related deaths and an increase in alcohol-related violence on US campuses in recent years have prompted colleges to crack down on under-age and irresponsible drinking through educational programming, campus policies (including the creation of 'dry campuses') and interdiction by law enforcement personnel (Engs and Hanson, 1994). For example, in 1996 across US college campuses, alcohol arrests rose by 10 per cent and drug arrests rose by 5 per cent, with Michigan State University leading the league table for alcohol-related arrests, while UC–Berkeley had the greatest number of drug arrests (Honon, 1998). Under-age drinking among students in the USA is dealt with harshly, carrying a $148 fine for a first offence and a $640 for the fourth. At UW–Madison in Wisconsin, the Madison Police Department even embarked upon an undercover operation code-named 'Operation Sting' to infiltrate illegal student drinking parties. In spite of these risks, most students are happy to procure false identity cards for under-age drinking. Moreover, many students in the USA use the spring break as an opportunity to head out of the USA and south to Costa Rica and Mexican resorts such as Cancun to party and drink beer, due to more relaxed licensing laws and cheaper alcohol. Numerous package companies have emerged to service the party needs of college kids during spring break.

Concern also stems from the effects of mixing heavy drinking with illegal drugs, a practice which is common among this young, hedonistic group. Drugs are now a common element of youth consumption and are readily available in studentland. Cannabis, in particular, is a cultural norm for many sections of the student population. Despite moral panics, often fuelled by high-profile Ecstasy-related deaths, drugs continue to play an important role in student culture.

Learning about the perks and pitfalls of nightlife in studentland is mediated through certain initiation rituals such as Fresher's Week and the Freshers' Fair, and a host of student publications, print and web media and radio, as well as through peers. These are arenas where the 'rules of studenthood' are learnt and embodied. Such spaces, while not liminal or carnivalesque in the classic sense of an inversion of social roles, do temporarily stand outside the normal working day.

One of the distinguishing features of nightlife in studentland, however, is its rhythms, routines and rituals. One fresher (Toby, 18 years old, Newcastle upon Tyne) commented on the weekly student nightlife routine:

> First year kind of seems to be dominated by people flocking to certain places on certain nights, so, for example, Monday night would be the Boat nightclub and Tuesday night is a bit of a rest day, and then Wednesday night is Baja Beach Club and Thursday night is Legends nightclub, and then Friday night, everyone goes to the Students' Union.

The pub crawl is another distinctive ritual and rhythm within student life. In Leeds, the 'Otley Run' along the Otley Road has become infamous among students and has been immortalised in a board game in which all of the pubs have to be visited by closing time. Starting at the Boddington hall of residence, it finishes at the main university campus, taking in fifteen pubs and ending at the infamous 'poly bop'.

Students' Unions used to have a monopoly on the supply of entertainment to students in this infrastructure, following a rather formulaic model of 'cram them in, pile it high, sell it cheap'. However, this situation has changed and new 'merchants of leisure' (Hannigan, 1998) increasingly recognise the student market as a lucrative source of income. Competition for a more discerning student 'spend' is now fierce, especially between campus-owned and private operations. Chains such as Taco Bell, KFC, Starbucks and Pizza Hut now litter university campuses in North America (Klein, 2000), while most large nightlife operators in the UK have developed student-oriented venue brands, usually within easy proximity of universities. These operators update their knowledge of the student community through focus groups and other marketing schemes to find out what they want from a night out. Some have recognised that close contact with students is the best way to develop nightlife attuned to their needs. As one nightclub owner commented: 'If you haven't been to university you don't understand the mentality … by being so interactive with the student community you know your audience and how to attract it' (Nightclub owner 3, Bristol).

There are a number of common perceptions which for many bar operators typify student nightlife. One nightclub manager suggested student nightlife continued to be based around a well-established formula: 'The only thing you can do to get students in is to offer them cheap beer and cheap entry on the door or some type of theme night. They're easily pleased' (Nightclub owner 2, Leeds). However, this model of a cheap and simple night out in many cases is responsible for a downgrading of nightlife. As one bar owner commented: 'With the best will in the world, a venue may start out by offering a sedate environment for the over 25s, but soon be tempted into selling cheap drinks to students to avoid bankruptcy' (Bar owner 2, Newcastle upon Tyne). Student nightlife, then, comes with its own distinctive set of rules and regulations, especially in relation to door policies. As one bar owner commented:

> We don't mind students being a complete prat. Our doormen are trained. I dislike doormen who see students messing around and misinterpret it as a problem. The great thing about students is that if one pukes on another they turn round and laugh. If a student pukes on a townie they end up on the floor with no teeth.
>
> (Bar owner 4, Bristol)

Other nightlife operators, however, are aware that some students have moved away from easy stereotypes of being scruffy and only wanting a cheap night. Many are just as likely to listen to the latest Garage and R&B tunes, drink premium bottled branded lagers and wear Dolce and Gabana. As such, many of the mainstream gentrified premises discussed in Chapter 5 also attract their fair share of students, who sometimes opt for a more 'premium' or 'exclusive' night out.

Plate 6.1: Branding studentland: 'It's A Scream' bar brand owned by Six Continents Retail; Middlesbrough, UK, 2002

While nightlife in studentland, then, is many different things to many different people and is a place of possibility, fun, hedonism and experimentation, most large corporate nightlife operators have a specific policy for both attracting and indeed 'constructing' the student audience. While there is nothing new about recognising the financial gains which can be earned from targeting the student body, through such corporate interventions studentland is increasingly being packaged and commodified on a scale and intensity not seen in the past. The structuring of student socialising, once the preserve of the university, is now being undertaken by a host of cultural, media and entertainment firms.

Brannigans, one of the fastest growing city-centre bar chains in the UK, is particularly eager to attract students. It courts them through its student night, 'The University of Pubbing – a degree in drinking', which does little to hide the company's desire to play on stereotypes of student life to increase its profits. The 'It's A Scream' nightlife brand (see Plate 6.1) owned by Six Continents Retail (formerly Bass Retail) is one of the most vigorously marketed student bar brands in the UK, promoted as:

> ... a concept designed to appeal especially to students and those who prefer the student way of life. Located in towns and cities mainly close to colleges or universities with a large student population, It's A Scream gives students value for money on food and drink as well as a great night out.
>
> (Six Continents Retail, 2002)

Each pub comes fitted with chunky tables, bright walls, menus offering burgers and pint deals, discounts on lager, oversize games such as 'Connect 4' and 'Yenga', a pool table, and juke boxes filled with student classics from the Stone Roses to Abba. However, the relationship with studentland doesn't finish there. As they state: 'We believe firmly that university isn't just about lectures and essays but taking advantage of the opportunities to develop extra-curricular skills and talents.' (*ibid.*). As a result, It's A Scream also hosted the UK 2001 Student Radio Awards and sponsors the national Student Talent Programme, which involves encouraging new work by student journalists, writers and photographers as well as, one would imagine, drinking alcohol. Not all students are accepting of the penetration of the 'It's A Scream' brand into studentland. As one student, Chris from Oxford, commented:

> Bass are scum, because they take over previously great pubs and destroy them utterly. When I was at college in Cardiff, a short stagger from our front door was Clancy's, a bona fide Irish pub selling gorgeous Guinness at less than two quid. Came back from summer break to find it had been Scream-i-fied; splattered in yellow and purple with South Park and the

> Simpsons scrawled over the walls, a huge yank up in prices, and (according to the student paper) a policy of serving people under 25 first to drive out the old fellas who used to sit in there and smoke the place out. Tradition and variety just annihilated forever in pursuit of the student pound.
>
> (Chris, 2001)

Student unions have responded to this new competitive arena and many have developed large campus entertainment facilities with a wider public role. The Octagon at Sheffield and the Academy in Manchester, for example, both within their respective universities, function as a main live gig venues across the city. In many ways, student unions are one of the few remaining places left which offer experimental nightlife, as they can subsidise a more specialist product based around music genres such as Goth, rock, indie, drum and bass, electronica and R&B, from profits made from commercial house nights during the weekend. However, student unions are also increasingly becoming commodified and are changing in response to a more mainstream demand from students. In the UK, for example, a youth satellite TV channel called Translucis, backed by drinks manufacturer Diagio, has signed up several campus bars to help them tap into the lucrative student audience. Advertisers signed up on the channel include Bacardi, Budweiser, Sol and Virgin (*The Publican*, 17 September 2001: 4). Further, the norm for campus music tours are tie-in promotions and sponsorship from a range of fashion and media outlets. The Big Break Music Tour featured across UK university campuses in 2001 was held in association with TopMan Clothing Company and *Dazed and Confused* magazine, with limited edition French Connection 'Fcuk alcopops' on sale during the night.

Through such corporate activity, the bricolage, traditions and rites of passage of studenthood are being appropriated, sanitised and sold back to students by corporate nightlife operators eager to cash in on this lucrative market. The commercial world has taken many of the 'authentic' moments of student life and transmitted them to wider mass audiences, diluting their former meanings. The student experience is now for sale. Many ex-students are employed in marketing, promotions and advertising within entertainment, nightlife and retail firms to pass on detailed and up-to-date knowledge about student tastes and cultures. These intermediaries who have been 'cool hunted' by large corporations (Klein, 2000) are a key part in the ongoing insider commodification of studentland.

Within such commercial nightlife spaces, students are sold an identity (fashionable, carefree, young) which they can adopt for a few years and then discard, before moving on when they graduate to adopt other lifestyle personas such as office worker, young professional or creative entrepreneur. As we highlighted in

Chapter 5, student tastes are being shaped within a broader mainstream nightlife linked to other cash-rich groups in the city. Hence, clusters of corporate, branded and themed nightlife operators develop portfolios of venues to socialise and prepare young people as they move from nightlife in 'studentland' to nightlife in 'workland'.

Many modern nightlife venues, then, decrease the experimental and often spontaneous character of student life with heavily scripted spaces regulated through prompts and stimuli, often of banal proportions. As with most other areas of life, this selling of studentland is part of the processes of globalisation and the increasing commodification of lifestyles and cultural forms. Selling it, along with selling other youth cultural styles, is a way of defining and controlling youth cultures (Massey, 1998). Behind the seemingly vast array of consumer choices, nightlife consumption options for students are often highly curtailed and limited. Nightlife in studentland, then, is increasingly a homogenous space, driven by large nightlife producers eager to get a slice of the market and narrow possibilities in the name of profit. Moreover, the expansion of student night spaces is a powerful gentrifying force in many localities, displacing established and lower-order residential and leisure uses, not to mention resulting in instances of community conflict (see Box 6.2).

Resistance and incorporation in studentland

Radicalism and resistance have long been hallmarks of university campuses (Loeb, 1994) which have been a source of both social change and moral panics concerning hedonistic and rebellious youth (Wilson, 1970). A common perception of studentland today is that student radicalism is on the decline. Campus life by the twenty-first century, it seems, has become more apolitical and apathetic. In general, student politics have shifted from wider political causes (such as militarism, government corruption and global poverty) to more specific campus- and student-oriented issues such as debt and an insecure job market, which have made them more instrumental rather than experimental (Seale, 1972).

In reality, radicalism in studentland has always waxed and waned, and in general there has always been a more apolitical majority and a more radical minority (Loeb, 1994). Alternative and radical currents still remain and come in many different guises. For example, in many campuses across the USA, students have embarked upon widespread campaigns against corporate influence such as Nike and McDonald's (Klein, 2000), while in the wake of the September 11 attacks, Stop the War movements have spread to campuses in the USA and Europe. The housing co-operative movement, seen as a haven for 'veggie', 'lefty' and more politically active students, remains a focus for alternative student cultures, especially on certain US campuses where many are organised through the North

Box 6.2: Conflict in studentland

Images of students as carefree, wealthy adolescents have been the source of many longstanding tensions between students and locals, especially at night. With the expansion of student numbers, some of these tensions have dissipated, while others have grown. One of the current tensions in studentland stems from the mobility and temporality of students and the territoriality of many local young people. As one student Kay (19 years old, Newcastle upon Tyne) commented to us: 'You get people coming up to you and saying, oh you're a student, aren't you, you're not from round here, and it's very intimidating when they single you out because you don't sound the same as them, you don't look the same as them.' One bar worker (Barworker 1, Newcastle upon Tyne) reinforced such a division from the 'other side of the fence':

> It seems that they're [students] doing a lot more kind of irresponsible things because they don't feel they have any responsibility to anyone since they're going home pretty soon anyway. And they're just going to be up here and party round in someone else's back yard, and just do what the fuck they like.

The presence of students in a community can also raise mixed feelings from local residents (Kenyon, 1997). They are often catalysts in the regeneration of local facilities, create a lively, youth-oriented environment, add to safety through their street presence, and are sometimes viewed as less problematic and more polite than local young people. However, they are also the source of a number of tensions concerning parking, waste disposal, noise, deterioration of housing stock, burglary and a decline of community feeling. As one newspaper in the UK put it: 'Neighbours who party all night, streets littered with takeaway containers and trash, neglected, crumbling houses, rat-infested discarded mattresses in back gardens; university students are destroying our neighbourhoods in Birmingham, Leeds, Cardiff' (Collinson, 2001).

In the UK, several post-code areas in the inner city have gained notoriety due to large student populations and associated problems such as transience, crime (Bond, 2001b) and low levels of owner-occupation. Many areas are being overrun by students and the subsequent development of student bars, cafes, pubs and shops. Headingly, a once prosperous and leafy suburb in Leeds in the north of England, has been colonised by large numbers of students, graduates and young professionals due to the location and high amenity value of the area. However, this increase in young transient people is causing a rapid downturn in the quality and maintenance of the local housing stock, with Leeds Civic Trust claiming that the increase in multiple occupancy was reducing housing conditions to those of 'nineteenth century slums' (Bond, 2001a: 15). Local Member of Parliament Harold Best has embarked upon a crusade to solve the issue, stating that the two universities were 'devastating the community that has been its host for generation after generation' (*ibid.*). Large nightlife operators continue to move into the area to exploit the market opened up by the existence of a large youth population, resulting in outrage from many sections of the older, established community. The Whitbread company's largest grossing pub in the country is located there, and, to the dismay of many local residents, Six Continents Retail have recently been granted permission to convert an old banking hall in Headingly into a new student pub.

In the USA, since 1994 the Housing and Urban Development Department (HUDD) has established a number of Community Outreach Partnership Centers (COPCs) to try and build better community–university relations. However, in spite of the best of efforts, the problem can often get out of hand. After an end-of-semester party at the University of Colorado, for example, more than 1,500 students overturned dumpsters, set bonfires and pelted police officers in full riot gear with rocks, bricks and bottles. The riot was largely a result of a year of simmering tensions between police and students over alcohol consumption which resulted in the closing down of Greek-sponsored parties due to under-age drinking, and the loss of charters by a number of fraternities. One member of the student executive commented that 'in an attempt to curb under-age drinking throughout the Boulder community, students have been treated as a nuisance rather than valued members of the community' (Zaret, 2000). Due to the disturbances, the University Hill Action Group was established, one aim of which was to 'educate university students about responsibilities as good neighbours'.

American Students Co-operative Organisation (NASCO). Finally, the Coalition to Defend Affirmative Action By Any Means Necessary (BAMN), founded in July 1995 in Berkeley, California, is a national organisation dedicated to building a new mass civil rights movement to defend affirmative action, integration and the other gains of the civil rights movement of the 1960s, and in particular to oppose the racial re-segregation of higher education (Liberator, 2001).

Pockets of resistance exist in elements of the student movement in Europe as well, where global environmental campaigning remains a strong thread. People and Planet in UK, predecessor of the Third World First Movement, became active in the late 1990s across most UK university campuses and organises campaigns on climate change, fair trade and the arms industry. Similarly, across Spain in November 2001, up to 200,000 students and staff took to the streets to object to the imposition of the 'Ley Organica de Universidad' by Spain's governing Partido Popular, as it was seen as both an erosion of university autonomy and an intrusion of business interests.

Corporate influences on campuses continues to cause outrage from some quarters. In particular, the advent of malls and food courts on university campuses has significant consequences for the student experience. As Klein (2000: 98) comments: 'the more campuses act and look like malls, the more students behave like consumers'. In many ways, the university campus in some European countries is one of the few remaining spaces which has not been penetrated by corporate brands and commercial culture, representing the nearest thing left to a public culture open to a 'dialogue of difference' (Bender, 1988). Remaining radical threads means that there is much to studentland which is still uncommodified, shadowy and unregulated, and future attempts to corporatise the experience may continue to meet resistance from sections of the student population. In the sphere of nightlife, students continue to make their own fun and indulge in hedonism, carnival and transgression.

Some of these experiences encapsulate a do-it-yourself ethic, a step outside the corporate controlled spaces of studentland. Additionally, the right to party and drink alcohol outside formal venues and events can also lead to conflict and resistance. For example, clashes between police and students on college campuses in the USA over the right to drink alcohol in familiar areas and during traditional party weekends have become widespread over the last few years (Sanchez, 1998). In July 1998, for example, a rowdy, largely drunk, crowd of 1,500 people gathered near Penn State University after the bars closed, and set fires, damaged street signs, vandalised cars and smashed three storefront windows after one long weekend. Fourteen police officers were injured during a two and a half hour riot and twenty people were arrested, with damage estimated at $50,000 (Weininger, 2001). Meanwhile, at Michigan State University in Lansing on 1 May, about 2,000 students rioted after administrators announced a ban on alcohol at Munn Field, a popular spot for tailgate parties before football games. Police barred students from entering the field and protesters poured into city streets, rioting and lighting fires. Police in riot gear confronted the protesters, and several people were treated for tear-gas-related injuries (Lively, 1998).

What of the future of nightlife in studentland? Despite the fact that it contains potential alternatives, transgressions and different ways of thinking about the relationship between work, leisure and play, it is still a bounded space, which is slowly being encroached upon by mainstream entertainment and nightlife providers who are increasingly commodifying its rituals and places. Financial and work pressures are also creating students who are more instrumental in their outlook, which enables nightlife operators to create and sell easily consumed and packaged experiences. The 'corporate campus' and the surrounding corporate city are increasingly intertwined. The future of work and play in studentland is not just up for grabs: it is also up for sale.

7

SEXING THE MAINSTREAM

Young women and gay cultures in the night

Two of the most fundamental influences on nightlife over the last thirty years have been shifts in gender relations and the impact of gay and queer cultures. The first half of this chapter discusses young women and their changing experience of nightlife. It has been suggested that they have been quietly leading a 'genderquake' revolution (Wilkinson, 1994a) in this sphere, reflected in 'ladies-only' nights (Hollands, 1995) and greater involvement in rave and dance cultures (Henderson, 1997), not to mention an increased feminisation of nightlife venues (Difford, 2000a). In particular, there has been an obvious targeting of young, single and professional women by the nightlife industry (Lleyelyn-Smith, 2001; Chaudhuri, 2001). The second part of the chapter looks at the rise of gay bars, clubs, dance cultures and 'villages', and in particular how they have been pulled them towards the commercial mainstream (Wallis, 1993). Most accounts of the origins of 'acid house', for example, credit the role gay clubs in the USA played in terms of inspiring the music, dress styles and general argot of rave culture (Redhead, 1993). At the same time, nightlife spaces for gay men and lesbians are also increasingly being consumed by heterosexual youth populations (Aitkenhead and Sheffield, 2001), and increasingly packaged and commodified (Chasin, 2001).

While the general 'sexualising' of urban nightlife is a positive trend, it is also important to remain aware of a range of continuing obstacles and problems raised in relation to young women and gay men and lesbian revellers. Despite some of the changes alluded to above, dominant mainstream forms of nightlife remain highly masculinised in terms of the male domination of space and the policing of compulsory heterosexuality. Young women's nightlife experiences often continue to be structured by less financial resources, leisure time and actual involvement in the 'production' of nightlife, continuing assumptions about their sexual availability (by young men), and fear of attack and harassment. At the same time, the 'equality' that many young women have gained in the nightlife sphere has often been on male terms, and contains their own negative consequences like increased levels of drunkenness, violence and drug consumption. Additionally, not all young women have equally benefited from the spoils of a feminised nightlife culture, with the gentrification of the mainstream being as discriminating against the female poor as the male poor.

Similar caveats also extend to understanding nightlife spaces for young gay men and lesbians. For example, the domination of downtown areas by heterosexual space has historically meant that what little gay nightlife provision existed often found itself on the 'urban margins' or 'fringe' (Whittle, 1994). Gay men and lesbians generally only really experience dominant mainstream nightlife spaces as 'invisible gays'. While this isolation has sometimes resulted in the creation and development of 'gay enclaves' or villages, these spaces also have their downside, including commercial incorporation. Those enclaves that have become stylish are partially a victim of their own success, being increasingly infiltrated by straight women, gay people's straight family members, and students (Aitkenhead and Sheffield, 2001). Finally, it is important not to typify and stereotype gay nightlife as a monolith (Buckland, 2002) and to accept that within it there are significant generational and gender (between gay men and lesbians) differences and subject positions, not to mention huge variations in the provision of spaces and premises for gay men and lesbians in cities around the world. The general argument in this chapter about 'sexing the city', then, is that while there are clear transformations in terms of the feminisation of nightlife and a greater acceptance and influence of premises and nightlife styles aimed at gay men and lesbians, these changes are also bound up with the ongoing corporatisation and gentrification of nightlife.

Young women and nightlife

The impact young women have had on transforming the character and atmosphere of urban nightlife is huge. This is particularly the case when one considers the limited role women played in the traditional pub (Hey, 1986) and historically in the city in general (Wilson, 1991). Much has changed even in the last decade, when Lees (1993: 81) wrote: 'the pub is a male environment where girls may go with their boyfriend but do not feel confident to go on their own or even in a group of girls', following McRobbie and Garber's (1976) earlier lament over the way young women were once trapped within the confines of a 'bedroom culture'. While Wilson's (1991: 7) contrary comment that 'The city offers women freedom' is somewhat overstated, numerous commentators have noted the powerful influence young female consumers are having on the transformation of nightlife premises and cultures of cities (Barnard, 1999; Wilkinson, 1994a; Hollands, 1995; Andersson, 2002: 264).

Yet, while one of the issues is explaining the roots of this change and assessing the new opportunities it provides in terms of enjoyment not to mention identity and power, the debate about gender and nightlife also needs to consider some of the economic differences between young women. For example, the feminisation of nightlife is also tied up with the dominant trend towards gentrification in the mainstream, a shift that disqualifies poorer women from participating fully. Within nightlife, one also needs to recognise that there are a range of 'new' femininities developing, such as aspiring young professionals versus laddettes as only two examples (Laurie *et al.*, 1999; McDowell, 1999). In order to more fully understand the various dimensions of this gender shift, one needs to look at the changing economic, educational and domestic (marriage) position of young women, not to mention their roles as consumers of entertainment and their bridging of the gender 'divided' city (McKenzie, 1989).

Women in most developed economies now constitute a considerable proportion of the workforce (ranging from 40–50 per cent, see Brush, 1999: 172). While some of these gains have come in 'routinised' export processing jobs in areas like textiles, electronic components and garments, there has also been an increase in women in 'professionalised' public-service employment (Moghadam, 1999: 136). While male economic activity rates continue to fall in countries like the UK, women's participation rates continue to rise (Wilkinson, 1994a: 10), and it has been estimated that around 80 per cent of new jobs created in the EU since the 1960s have been filled by women (Brush, 1999: 172). In the UK and USA respectively, women who run small businesses form one quarter and one third of the total respectively (Wilkinson, 1994a: 32–3).

The important point here is that there has been an increase in economic activity among young professional females (see Walby, 1997). As young women begin to take advantage of increased educational opportunities and gain qualifications, they account for a higher percentage of professional jobs and gain higher disposable incomes with which to consume. For example, in the UK girls now out-perform boys in terms of high school qualifications, with 54 per cent and 25 per cent achieving five or more GCSE grades (A to C), and two or more A levels respectively, while in comparison boys managed only 44 per cent and 21 per cent (Kelso, 2000). They also outnumber young men in higher education (HEFCE, 2001), and are less likely to be unemployed following graduation (Wilkinson, 1994a: 24). While still behind men, women make up 38 per cent of all professional jobs, with significant gains being made in medicine, law and accountancy (Wilkinson, 1994a: 11). While a gendered wage gap stubbornly persists, it appears to narrow somewhat for the younger generation. In the USA, the average hourly pay rate for a young male high school graduate fell 28 per cent from 1973 to 1995, while for females it fell only 19 per cent. Young women especially are gaining on their male counterparts in the labour market (Brush, 1999: 166).

Historically, women's involvement in leisure has been constrained not only by their lack of fiscal resources but by domestic/family commitments. USA figures show that while hours of domestic work have decreased in the last twenty years, women still did twice as much as men (Brush, 1999: 176), and British Social Attitudes Surveys continually demonstrate that women continue to do the bulk of cooking, cleaning, washing and ironing. Again, however, there are generational differences. Griffin's (1985: 37) work revealed that 45 per cent of young women did any domestic labour as opposed to 100 per cent for their mothers. Wilkinson's (1994a: 20) research shows evidence of young women prioritising careers and jobs over and above having children and raising a family, reflected in the fact that the mean age for getting married in the UK for women is now 28 (30 for men) (Kelso, 2000),

All of these factors – increased economic activity and professionalisation of at least a section of the female job market, greater educational success, a decline in domestic duties and delays in getting married and starting a family – have had an impact on young women's capacity for leisure and consumption, including their involvement in nightlife activity. As such, numerous commentators have noted that young women in particular have begun to traverse the 'divided city' from the more private spaces of home (bedroom culture), community and neighbourhood, to occupy the public spaces of the street, the bar and the club (Barnard, 1999; McDowell, 1999; Hollands, 1995). For example, McRobbie's (1993) research discusses the increased involvement of young women in many contemporary youth cultures, and Henderson (1997: 69) notes the significant presence of young women at raves. Hollands' (1995) research in Newcastle upon Tyne also comments on equal numbers of young women on nights out in the city and supports the notions that their desire and commitment to nightlife cultures rivals that of young men. The research found that young women went on nights out to the city centre more frequently, and that a higher percentage of their income was spent on nights out, when compared to their male counterparts. Furthermore, women reported that they would feel worse than men if they were somehow restricted from going out for a period of time, and more said they would continue to go out as much as men, even when married (Hollands, 1995: 23–4).

Young women have also been at the forefront of transforming nightlife cultures away from a purely sexual and marriage marketplace (Griffin, 1985), the so-called 'cattle market' phenomenon (de Vries, 2001), through a greater emphasis on the 'socialising' function of nights out. Hollands' (1995: 42) research notes the importance of 'socialising with friends' as one of the primary reasons given for going out, and young women are at the forefront of the creation of these 'mini communities' (87 per cent of local women gave this as one of the main reasons why they went out, as opposed to 40 per cent of local men). Represented most strongly by what have come to be known as 'ladies-only' nights out, there has been a move away from traditional 'best friendships' patterns (Griffin, 1985) and the desire to find a potential husband on an evening out among young women, to increased group loyalty and sisterhood. For instance, consider the quote from a young woman (Jan, 31 years old, Newcastle upon Tyne) interviewed by Hollands (1995: 87):

> It's a shame really 'cos we used to have such good nights out when we used to all go out in a foursome with two other girlfriends. That was great, such

> good fun. Because we just laughed all the time, we weren't interested in men, because, like, I was married, me friend was married, me other friend was married, and the last one, she was quite a plain girl who was quite fat, and she wasn't, she used to pretend not to be interested in men, because she couldn't get them, basically, you know. So we went out, we weren't interested in men, we weren't relying on men for conversation.

While the effect of this growing female solidarity on young men is often threatening, with some anecdotal evidence of young women being assaulted for their lack of interest (de Vries, 2001), it is also indicative of a deepening of female friendships (Griffin, 2000).

It has also been suggested that young women have helped fuel a more general shift towards androgyny and equality on the dance floor, particularly in what has conventionally come to be called 'rave' or dance culture (Malbon, 1999). Henderson's (1997) research shows the attraction many young women felt towards the intense social interaction here, heightened by the music and, of course, drugs (particularly Ecstasy). Pini's (1997) work also supports the idea that rave culture worked to 'unlock' the heterosexual coupling associated with the dance floor, with its emphasis on the relationship between the individual dancer, the music and the collective group. Many young women spoke to Henderson about the 'erotic' nature of the dance experience and the move away from outdated notions of the club as a place to meet men. As one interviewee said: 'We do fancy blokes at raves and enjoy flirting with them … but it's like going back to when you were younger, you don't want to get them into bed, you're just friendly' (Henderson, 1997: 71). Similarly, Pini (1997: 122) quotes one young women who states: 'I never really focus on men when I'm dancing. In fact, I like looking at women much more,' with another saying: 'I think my feelings – if they're sexual – apply to everyone, women and men alike. I do like looking at other women dance, and sometimes this gets very strong, almost like an attraction.'

The most sustained political challenge, however, has come from young females occupying alternative nightlife spaces. The emergence of the 'riot grrl' phenomenon in the USA in early 1990s, and its subsequent development in the UK, is a good example of how young women, through the formation of their own music youth cultures, have challenged the status quo of male commercial nightlife (Gottlieb and Wald, 1994; Kearney, 1998). Coming out of the post-punk movement, the riot grrl philosophy was both anti-capitalist and anti-patriarchal. Described as a 'self-styled subgenus of 14–25-year-old feminists' (Van Poznak, 1993), and 'girl-punk revolutionaries' (Matthewman, 1993), riots grrls were involved in creating music through bands such as Huggy Bear (UK), Bikini Kill, Bratmobile, Babes in Toyland (USA) and communicating through zines, impromptu meetings and the internet. Bikini Kill, a riot grrl band of the early 1990s, had a manifesto which suggested 'envisioning and creating alternatives to the bullshit Christian capitalist way of doing things' (quoted in Gottlieb and Wald, 1994: 262).

Diluted by both their popularity and extensive media defamation, the riot grrl phenomenon appeared to fade as quickly as it arose. However, it might be suggested that it has left a lasting legacy of 'invisible' change among at least a section of young women. Blackman's (1998) UK ethnography of a resistant female youth culture, which he called 'new wave girls', is an example of the effect of such a cultural shift. His study reveals a group of confident young women challenging conventional femininity through clothing styles, non-conformist attitudes (towards both boys and school) and heightened female solidarity. Similarly, the Hell Raising Anarchist Girlies (HAGS), a Brighton based collective of anarcha-feminists, were established in 2000 and have been organising pro-women events, publications, squats and conferences.

While Blackman's (1998) study is set largely within the school, it also looks at how these young women challenged the male status quo through their use of public spaces such as the street, parks, cafés and pubs. With respect to the latter space, and in contrast to assertions made by Lees (1993) in her early research about conventional girls seeing the pub as a largely negative male environment, Blackman (1998: 216) asserts that the 'new wave girls' identified the pub as a positive place, and they often went together as a group, unaccompanied by their boyfriends. Similarly, as Chapter 9 discusses, alternative nightlife spaces reveal more equal roles for young women, including involvement in organising, promoting and running such venues. Ironically, such young women often face double discrimination when they step outside alternative nightlife provision, as they are often challenging both traditional images of women and expectant male behaviours within mainstream nightlife.

All this should not imply somehow that either new wave girls are typical, mainstream nightlife has been fundamentally challenged, or that 'rave influenced' nightlife spaces have become a post-feminist panacea. Indeed, Henderson (1997) warns that the commercialisation of dance culture back into the mainstream has signalled the return of both alcohol and increased sexism. It is also crucial to recognise that while young women have become more important consumers of urban nightlife, they have made far fewer inroads into its actual production. In fact, beyond their historical role as 'landladies' (Hunt and Satterlee, 1987) and bar staff, our interviews revealed few young women as brand and regional managers for breweries or pub companies, bar managers, promoters, DJs or musicians. The entire music industry is still very male dominated, even though some research suggests that it is changing slowly (Raphael, 1995; Whiteley, 1997). While young women DJs are growing in number, they are still a minority on the club circuit, especially among A-list DJs (exceptions include Mrs Woods (React), and Rachael Auburn (Truelove) and Kemistry and Storm (Metalheadz)). In a telling quote, Rachael

Auburn, hints at one of the dilemmas of being a female DJ: 'Obviously, you have to deliver the goods, but I do feel that you're not taken quite seriously being a lass. I still kind of get the vibe that you're basically there for attraction value to a degree' (P and Swales, 1997: 10).

Additionally, despite an increased presence of young women in cities, gender inequalities in the occupation of public space continue (Laurie *et al.*, 1999; McKenzie, 1989). As Lees (1993: 69) suggests, 'girls' appearance on the street is always constrained by their subordination', and this is learned early on in life. Seabrook and Green's (2000) work shows that young girls' fear of public spaces is closely linked with the behaviour of men, and Griffin (1985: 65–9) provides ethnographic evidence that young women limit their leisure activities and access to space for similar reasons. Fear of attack and harassment is particularly acute at night. A Canadian survey conducted in Ottawa–Carleton showed that females were six times as likely as men to feel unsafe in their neighbourhood after dark and over 80 per cent limited their activities in some way because of this fear (i.e. didn't go out alone, didn't stay out late, avoided certain places). Additionally, over one third had suffered some form of verbal harassment at night over the previous five years (Andrews *et al.*, 1994). And finally, Hollands' (1995: 65) work in Newcastle upon Tyne, England, revealed that young women were more aware of dangerous areas in the city, and 60 per cent were either physically or verbally harassed on a night out in the city.

As Skeggs (1999) argues, 'respectability' is a crucial component for how women come to occupy public space. The idea that 'nice girls don't go there' is revealed in the following conversation about going out to the city centre at night from Hollands (1990: 133):

Julie: Depends on what part you go to …
Interviewer: Would you go on your own?
Julie: But I wouldn't go on me own. A girl, it's not, it's funny if a girl goes on her own. It's not a thing to do really – not at night.

Others sought protection in large female groups or gangs, and worked hard not to get separated on a night out. However, these safety mechanisms are at least partly compromised by women being more dependent on public transport to get home.

Aside from 'self-policing' and being in a large protective group, a number of young women were consciously choosing nightlife spaces that they perceived were more 'women friendly' or at least safer. Sometimes this involved going to a gay club, or at least a female-friendly bar. Here, two women managers of a Leeds club, who put on gay-friendly nights, talk about the importance of providing such an atmosphere:

Bar manager 5: When so many women come to the club it is because they know that they're not going to get hassled.

Bar manager 6: Because we've always promoted it to be a safe place.

Bar manager 5: As a member you have the right to remove anyone who you feel is detracting from the club, if someone is aggressive to me as a member or a straight guy gets a bit drunk there and he won't take no for an answer. It's not that he's trying his luck 'cos everyone does that, it's that he's bothering me and he'd be out like that. We are, that is another female-orientated thing, we are quite protective about girls getting harassed in there. We don't like it at all and if we see it, they get a warning and if they're physical with it, I've twisted arms behind backs myself and taken them towards a bouncer if I've seen them getting out of control.

Despite evidence that some young women have pushed the boundaries of gendered nightscapes, many are increasingly taking on behaviours characteristic of young men in this sphere. As Wilkinson (1994a: 7) argues, 'Women's increasing outer-directedness is leading them to display what we used to see as male characteristics.' In these cases, rather than challenging male domination of mainstream nightlife spaces by creating alternative female cultures, young women appear to be simply competing on men's terms through a crude 'equality' paradigm (i.e. proving they can be as 'bad' as young men).

One example of this is young women's changing views about sex and traditional courtship on nights out. Wilkinson (1994a: 20) notes that only one quarter of young women believe that they need a stable relationship in order to be fulfilled, and UK statistics show that this generation are far more interested in sex than ever before, with 10 per cent of 25–34 year olds having ten or more partners, compared to only 4 per cent of their mothers' generation (Johnson *et al.*, 1994). Similarly, in the USA the National Health and Social Life Survey (1994) showed that young women aged 20–30 were six times more likely to have had multiple sexual partners by the time they turned 18 than their predecessors born thirty years earlier (Fillon, 1996: 39). While one might interpret such findings as evidence of a new-found sexual confidence, *Sex and the City* style, other commentators have referred to aspects of this phenomenon on a night out as a female 'sexual safari' (Campbell, 1996), with young women adapting a similar 'predatory' sexual attitude to men. Although a minority pattern, there was some evidence of this approach found within our research, as the following quote testifies: 'One time we had a competition to see how many men we could pull in one night, and we were all fierce to get out there. Honestly, we were like man-eaters' (Yvonne, 26 years old, Newcastle upon Tyne). The downside of this so-called 'sexuality equality' is that often young women in general can be differentially stereotyped as 'slags' (Hollands, 2000), and hence seen as being 'sexually available' to men. Much of this boisterous sexual behaviour is symbolic, which ironically occurs during the infamous 'hen night' (the celebration of a young woman's impending marriage). Chatterton and Hollands (2001: 76) also found evidence that beneath the sexual bravado lurked more traditional beliefs about romance and meeting the 'right man.'

Similar arguments hold for young women's increase in alcohol consumption and use of illicit drugs. The Institute of Alcohol Studies (2002: 4) cites factors such as more women working and more female professionals, pubs and bars becoming more women-friendly and drinks targeted at women, as the main reasons behind increased alcohol consumption. Table 7.1 highlights that young women aged 16–24 years old are the heaviest drinkers, and this has particularly increased over the 1990s. One third now exceed the weekly limit and nearly 10 per cent drink over 35 units (Institute of Alcohol Studies, 2002) with this rise fuelled largely by young professional women (*ibid.*). A survey in the USA also revealed that 7 per cent of 12–17-year-old girls reported binge drinking even at this early age (Substance Abuse and Mental Health Services Administration, 1999). Additionally, surveys across Europe and Australia show increases in illicit drug use among young women over the last decade (IAS, 2002; Parker *et al.*, 1998; Australian Institute of Health and Welfare 1999).

Finally, there have been general studies of young women's increasing involvement in crimes of violence in numerous western countries (McGovern, 1995; Chaudhuri, 1994; Morse, 1995; Siddeall, 1993). Perpetrating and/or being a victim of a crime of violence is, of course, strongly linked to alcohol, and hence being present in nightlife spaces. Hollands' (1995: 63) research in Newcastle upon Tyne surprisingly found the percentage of young women reporting that they had ever been involved in a violent incident was actually higher than for men.

Yet two of the biggest threats to a progressive equalisation of nightlife provision for women are ironically tied to some of the gains they have made over the years. First, the emergence of more 'feminised' nightlife spaces directed at attracting young women (Difford, 2000a) has gone hand in hand with a creeping corporate gentrification of bars, pubs, cafés and clubs (Chatterton and Hollands, 2001). Bob Cartwright, Communications Director of Bass Leisure Group which runs the All

Table 7.1 Weekly alcohol consumption for women 1992–2000, UK

	1992		*2000*	
	16–24 yrs	*All*	*16–24 yrs*	*All*
Weekly alcohol consumption (no. of drinks)	7.3	5.4	12.6	7.1
Percentage drinking more than fourteen units[a]	17	11	33	17
Percentage drinking more than thirty-five units	4	2	9	3

Source: Institute of Alcohol Studies (2002) *Women and Alcohol Fact Sheet*.

Note: [a] 14 units per week is the recommended limit

Bar One chain, said: 'We created bars with big open windows, large wine displays and extensive menus, and we put newspapers by the bar for women to read while they are waiting for their friends' (quoted in Chaudhuri, 2001: 53). In essence, such moves to cater for more women customers have subtly been about extending provision to only certain types of female consumers – professional middle-class women with higher disposable incomes. Up-market bars and expensive clubs both attract a certain type of young female while disqualifying others because of dress codes, style policies or expense (see Plate 7.1). Poorer working-class women and those with children can be excluded here (MacDonald and Marsh, 2002), and those who do attempt to consume the gentrified and feminised mainstream are often labelled with derision as 'fat slags' and 'scratters' (see Chapter 8). Additionally, there appears to be little evidence to suggest that women in gentrified mainstream nightlife spaces are any less harassed or exposed to various forms of sexism (Hollands, 1995; Chatterton and Hollands, 2001).

The other main nightlife development of concern, which is linked to the gentrification of nightlife, is the recent rise of corporately owned table- or lap-dancing clubs (see Box 7.1). Curiously, one of the arguments in favour of such spaces is that with supposed 'gender equality' in the nightlife sphere, such places are no longer an affront to women. In fact, some venues suggest that women are among their clientele, while others point to female attendance at male strip events such as the

Plate 7.1 A group of young women enjoying the newly gentrified waterfront nightlife; Newcastle upon Tyne, UK, 2002
Source North News & Pictures

Chippendales to argue the case that equality of provision is a reality. However, it is the corporate infiltration into this sphere of nightlife entertainment which provides justification for arguments about the 'respectability' of premises and their increasingly gentrified clients. As a spokesperson for Surrey Free Inns Group, owners of For Your Eyes Only, table dancing club argues: 'We are a highly reputable business, appealing to the corporate market' (*Herald and Post*, 14 November 2001: 14).

Box 7.1: Lap-dancing goes up-market: Spearmint Rhino

One of the world's largest lap-dancing chains, with thirty-one clubs in the USA, six in the UK (with ambitious plans to expand to twenty or more) and even one in Moscow, Spearmint Rhino has 3,500 employees and a turnover of $60 million (Doward, 2001). Its Las Vegas branch is reputed to have made more than $1 million in a single month (Aitkenhead and Sheffield, 2001), with its Tottenham Court Road branch (see Plate 7.2) turning over more than £300,000 per week before Christmas (Morris, 2002). John Gray, the 44-year-old owner of the company, is estimated to have built up a fortune of £38 million, and is now concentrating the company's expansion plans in the UK (Morris, 2002). The company describes lap-dancing as 'theatrical entertainment' (Aitkenhead and Sheffield, 2001) and in the USA bills itself as the 'largest, most elegant gentlemen's club in the world' (Morris, 2002). The company points towards its up-market décor and five-star food menus as an indication of its quality and respectability.

In defence of his business, Gray claims dancers earn between £5,000 and £25,000 monthly, have private health insurance, there is a 'no-touching rule' on stage and the company prohibits any extra-curricular activity with clients (Ellis, 2001). However, earnings figures have been hotly disputed, with some dancers saying that their takings can fluctuate between £50 and £400 a night (Morris, 2002), and the no-touching rule does not appear to apply to sessions offered to individual clients in private booths (*ibid.*). Allegations that some dancers were soliciting in and around the club (Brigett, 2001) were dismissed by the company, who claimed that on at least one occasion the woman in question was a 'plant' (Ellis, 2001). More seriously Adelle Hamilton, an 18-year-old dancer with the Slough branch, was found murdered at home, after last being seen getting into a car outside the club (Demetriou, 2001).

Spearmint Rhino has faced numerous protests and forms of opposition. Community groups in Camden failed to prevent the opening of the Tottenham Court Road branch, despite concerns about safety, traffic congestion, noise and rowdy behaviour (Salman, 2000). The same branch was under investigation by Camden police with respect to allegations of prostitution nine months after opening, but despite this had its licence renewed for another six months (Brigett, 2001). The opening of the club's latest branch in the prosperous conference destination and market town of Harrogate in the Yorkshire Dales has also created some cause for concern among residents and councillors (Morris, 2002).

Plate 7.2 Spearmint Rhino Gentlemen's Club, Tottenham Court Road, London, UK, 2002

Half a decade ago there were very few lap-dancing clubs in the UK and those that existed were small-scale and seedy venues. The UK industry is said to be worth £300 million, and there are more than 300 venues, set to increase to 1,000 over next five years (Doward, 2001). The sector is still in its infancy here compared to the USA, where the sector is worth $5 billion, with 700 table-dancing venues in Manhattan alone (Aitkenhead and Sheffield, 2001). Prominent companies like Coca-Cola, NatWest and Merrill Lynch have allegedly had lap-dancing venues host their Christmas parties in London, and numerous celebrities have tacitly lent their support through their attendance at such nightlife spaces (Ellis, 2001; Aitkenhead and Sheffield, 2001). Ronnie Scott's Jazz Club in Birmingham was recently turned into the city's ninth lap-dancing establishment (Morris, 2002: 13) in a belief that it would be more profitable. City leaders have become worried that this proliferation of lap-dancing venues, especially in the Broad Street and New Street areas, is giving Birmingham a sleazy image which could damage its bid for European Capital of Culture in 2008, and is planning new legislation to restrict their growth (Guthrie, 2002: 14). Piers Adams, owner of the London nightclub Rock, has also admitted to applying for a licence (Aitkenhead and Sheffield, 2001).

In short, the growth of lap-dancing venues points to the normalisation of the sex industry within urban nightlife on the basis that it is a stylish, up-market and respectable form of entertainment, which can enhance the urban economy and benefit young women by providing employment. Table-dancing is portrayed as 'theatrical entertainment', while other operators euphemistically refer to it as an 'art form' and 'the glamour industry' (Aitkenhead and Sheffield, 2001). Profitability is ensured by corporate expense accounts and a core audience is drawn from the business community during the week and stag parties at the weekend (Doward, 2001). Trade appears to be flourishing, and many existing operators are expanding and others are switching to this area due to higher rates of return. Dancers usually pay the company a fee to dance (around £50–90 a night), and in some of the best establishments are reputed to earn in the region of £500 in an evening, although such figures are hard to substantiate. Some premises have booths at the back for 'private dances', which is where most dancers earn the bulk of their money. Although companies are keen to stress they are providing dance entertainment only, there have been stories and incidences which reveal that sexual transactions do take place both on and off the premises (Aitkenhead and Sheffield, 2001).

While some city councils have been reticent about the issueing of licences to lap-dancing clubs and have sought to ensure that adequate regulations are put in place, in the main their hands are tied. For example, a loophole in the law allows any establishment with a public entertainment licence to convert into a lap-dancing club, as long as it complies with the conditions of its original licence (Morris, 2002: 12–13). Some councils, like Hammersmith and Fulham, have sought to mandate regulations like the 'three-foot exclusion zone' (between patrons and dancers), as well as banning total nudity at tables (i.e. dancers must wear G-strings) (*Evening Standard*, 3 November 1998: 6), while Nottingham County Council has a clause in its entertainment licence that prohibits stripping and table-dancing nudity altogether (*Evening Standard*, 16 March 2001: 8).

Protests were sparked in Newcastle upon Tyne over the opening of the city's first lap-dancing club (Olden, 2001). A liquor and public entertainment licence was granted to For Your Eyes Only (FYEO), a corporate lap-dancing club owned by the Surrey Free Inns (SFI) group, in the city in 2001. SFI also own well-known bar brands like the Slug and Lettuce, Bar Med and Litten Tree chains, and currently has five FYEOs around the UK. It is reputed that its London Royal Park branch makes £1 million annually (Olden, 2001). Earlier in 2001, the proposed site of the lap-dancing club was occupied to raise awareness of issues including the safety of women in the city and the lack of affordable nightlife. The protesting group put on various alternative music events, a women's day and a café to demonstrate what the building could constructively be used for. Despite opposition, the magistrates, backed by the city council, granted a licence for the club.

On the club's VIP opening night, a protest march of 200 people was organised, drawn from student unions, a rape crisis centre and other community, women's and local groups. The march culminated at the club and various members of the group took photos of men going in and out of the club, which were then posted on a 'name and shame' website on the internet. The main arguments of the protestors focused upon issues of sexism and the exploitation of dancers (including issues around prostitution), but also around the impact such venues have on how young women in the vicinity are likely to be viewed and treated. Protesters argued that the venue sent out the wrong message about young women in the city, as well as drawing attention to the dancers. As one spokesperson said:

> There's a young women's project 300 metres away and this is going to bring loads of sleazy blokes into the area. We're also concerned about the women who'll work here. They'll have to pay £50 a night to work, then charge £10 a dance.
>
> (Olden, 2001)

Box 7.2: Hooters: 'Delightfully tacky, yet unrefined' (the company motto)

Possessing a 'casual beach theme', this North American bar-restaurant is known specifically for its scantily clad Hooters Girls. Popular among university students, sports fans and – in downtown areas at least – businessmen, it is primarily a male-group-dominated environment, although it is claimed that around 25 per cent of their customers are women. Describing them as 'all-American cheerleaders', the company believes the Hooters Girl is 'as socially acceptable as a Dallas Cowboy cheerleader, *Sports Illustrated* swimsuit model, or Radio City Rockette'. The company employs approximately 12,000 Hooters Girls, who also make promotional and charitable appearances in their respective communities as part of their job. Hooters also insist that women occupy management positions all the way from Assistant Manager to Vice President of Training and Human Resources, and they claim that they have a longstanding non-sexual-harassment policy for waitresses which works.

Critics argue that the whole 'Hooters Girl' ethos is inherently sexist and exploitative to women in general. For example, women and community groups protested against the conversion of a pizza restaurant near Penn State University into a Hooters, and in Ahwatukee Falls, USA, local residents opposed the siting of one of their premises near two schools, arguing that the restaurant was not compatible with the 'family-oriented neighbourhood'. While the chain acknowledges that many consider 'Hooters' a slang term for a part of the female anatomy, they argue that their logo also contains an owl (whose eyes fill in the two Os in the company name), thereby creating ambiguity over its meaning. Curiously, the company's only brush with equal opportunities legislation came from a somewhat unexpected direction, when they were accused of carrying out hiring practices that discriminated against men (the case is now in abeyance). They also point out that they are good corporate citizens through their involvement in raising money for local charities via their Hooter Community Endowment Fund.[1]

To sum up, there are changes afoot in gender relations within the nightlife industry. Some of these transformations are positive, especially an increase in young women's participation as consumers of nightlife, enhanced through the growth of female-friendly venues and the decline of male-dominated ale-houses and saloons over the past few decades. Such changes have sought to challenge existing gendered relations and barriers. However, in spite of a gradual feminisation of nightlife in terms of style and provision, some of these gains have been made in 'male' terms. For example, many young women have problematically adopted heavy drinking and involvement in violence, and little has been done to maintain access for less well-off sections of the female population, especially those more home-bound through child-rearing (see Chapter 8). The increasingly corporate nightlife city, then, is far from developing spaces which promote sexual equality and question established gender roles.

Gay cultures and nightlife

Gay and queer cultures have increasingly played an important role in the development of nightlife. While most 'world cities' like London and New York (Buckland,

2002; Chauncey, 1994) have a long history of gay activity, others like San Francisco in the USA (Castells, 1982) and Manchester in the UK (Hetherington, 1996; Walker, 1993; Whittle, 1994), have become internationally renowned for their gay politics and cultural provision. Guidebooks to many European and North American tourist destinations invariably contain a section on gay services and spaces, particularly gay nightlife, as a marker of cosmopolitanism, diversity and tolerance. (See Plate 7.3.)

The development of nightlife for gay men and lesbians not only has important consequences for the creation of gay identity and pride, but also has a major impact on heterosexual cultures of 'going out'. For example, youth dance cultures throughout the 1980s and 1990s have been influenced by gay fashion styles and musical anthems, not to mention taking up aspects such as androgyny, pleasure-seeking and questioning the dominance of 'compulsory heterosexuality' (Henderson, 1997) or 'heteronormativity' (Buckland, 2002).

At the same time, gay cultures have been beset by a range of problems, contradictions and internal/external pressures. Labels such as that the gay community are rather clumsy, and terms such as 'gay', 'lesbian', 'homosexual' and 'queer' are ambiguous descriptive categories invoking a multitude of political, social and cultural affiliations. The acceptance of homosexuality and its cultural expressions has been a rather uneven story. The rise of AIDS-related 'moral panics' of the 1980s in both the

Plate 7.3 Sitges is a popular holiday destination for gay men from all over Europe; Parrot's Pub, Plaça de l'Industria, Sitges, Spain, during the February Carnival, 2002

USA and the UK, and the aggressive re-zoning and regulation of 'adult establishments' (including gay nightlife premises) by the Giuliani administration in New York in the mid 1990s (Buckland, 2002) has tempered the growing acceptance of gay cultures within nightlife. The ongoing problem of gay-bashing and harassment is also reflected in figures suggesting that 33 per cent of men and 25 per cent of women in the UK had experienced violent attacks because of their sexuality (Mason and Palmer, 1996). In addition, nightlife spaces for gay men and lesbians continue to be marginalised or ghettoised in many cities, and even where such enclaves have been deemed 'successful', there have been problems of the usurpation of such spaces by heterosexuals and a general trend of upgrading and sanitisation (Chasin, 2001). Finally, gay nightlife spaces continue to be dominated by gay men at the expense of the lesbian population (Valentine, 1993), and influenced by changing notions of masculinity (Nardi, 2000) and gender politics (Edwards, 1994).

There is now a significant body of gay, queer and feminist studies, which helps to locate the issue of sexuality and sexual identity within wider economic and social contexts (Richardson, 2001; Plummer, 1994; de Laurentis, 1991; Weeks, 1989). Similarly, there is a developing body of work which has begun to map out a more spatial dimension to the question of gay identity and look at the ways 'in which space, place and sexuality are implicated in the constitution of each other and society as a whole' (Knopp, 1992: 651). Finally, there is a range of more specific studies looking at the role nightlife in particular plays in the development of gay identities, and the impact that gay cultures and spaces have on the night-time economy (Buckland, 2002; Whittle, 1994; Hollands, 1995; Chatterton and Hollands, 2001; Taylor, 2001).

It is clear that gay cultures cannot be understood outside general social theories concerned with gender, capitalism and power (Knopp, 1992; Edwards, 1994). For example, with regard to patriarchal capitalism, heterosexuality helps to maintain existing gender divisions and male domination, not to mention reproducing traditional notions of the nuclear family which aid capital accumulation. The historical separation of men's public (work, politics, the street) and women's private spaces (home and the community) in the industrial city provided a spatial dimension which buttressed the existing patriarchal order (McKenzie, 1989), allowing little room for ambiguity and, indeed, 'otherness' to arise (Knopp, 1992). As Buckland (2002: 3) states, denied access to many state, church, media and private institutions, many 'people who identify as queer are made worldless, forced to create maps and spaces for themselves'.

However, economic restructuring, the rise of social movements and the search for identity through consumption in the contemporary period have all meant a gradual loosening of traditional sexual categories, allowing a range of gay identities to emerge and flourish. This process has been far from even. As Edwards (1994: 108) argues: 'The paradox of the 1970s was that gay and lesbian liberation did not produce the gender-free communitarian world it envisioned, but faced an unprecedented growth of gay capitalism and a new masculinity.' Adams (1987) also links the commodification of gay lifestyles with an increased masculinisation of gay men's culture. A crucial divide, then, concerns the dominance gay men have enjoyed economically, politically, culturally and spatially over gay women, for instance (Peake, 1993; Edwards, 1994). It has been suggested that lesbians historically have tended to operate more around interpersonal networks, rather than through occupying particular territories or premises in cities (Valentine, 1993), replicating women's general position of urban subordination (although see Podmore, 2001).

Despite these differences, one of the key shifts has been not only the emergence of a more public gay identity, but a transformation in the types of spaces occupied. This has not been a completely linear progression, with, for example, interventions like the re-zoning and policing of many gay spaces in the city of New York in the mid 1990s (Buckland, 2002), nor has it been achieved equally (i.e. between gay men and lesbians). Edwards (1994), however, notes a historic move from informal, non-institutional and functional gay spaces (parks, public conveniences, cruising in cars), to more formal and institutional contexts such as gay bars, nightclubs and bath-houses/saunas, while Buckland (2002: 7) talks about 'third' spaces of recreation opening up for gay consumers (work and home being the first and second spaces).

The importance of nightlife spaces for expressing alternative sexual identities, then, should not be underestimated. For example, the Stonewall Riots in the USA, often held up as a crucial turning point in the fight for gay rights and identity, were triggered by a police raid on the Stonewall bar in the West Village, New York City. Buckland (2002) argues that club culture in the USA has been an important arena for the construction of what she calls the 'queer life-world'. A crucial piece of legislation in England was the 1967 Sexual Offences Act, which first made it possible to open a bar catering to gay men or lesbians without creating an offence (Whittle, 1994: 32). In the UK context, Hindle (1994: 11) argues, pubs and clubs 'are probably the most important single feature of a gay community'. While one might contest the adequacy of the term 'community' here, and recognise that such spaces may not be equally important to all gay men and lesbians, it is clear that bars and clubs remain an important space for such groups, particularly in relation to sexual activity (Mutchler, 2000). Of course, such contexts and localities vary tremendously (Altman, 1997), ranging from a single gay venue in some small towns to gay tourist destinations (see Box 7.3) and entire gay villages and enclaves in major cities, encompassing not just bars and clubs but a whole range of employment opportunities and services. Research in the USA and the UK suggests that a minimum population of 50,000 is needed to support one gay bar (Hindle, 1994).

Box 7.3: Sitges: gay nightlife tourism by the sea

Sitges is one of Europe's top gay tourist destinations. Located about 35 km from Barcelona, Sitges has a longstanding gay appeal, being renowned as a bohemian hang-out and artists' colony since 1900. Christened 'the gayest place on earth' by the Spanish, it has become a very popular gay tourist destination for Europeans and a summer hang-out for the gay population of Barcelona. Gay men come to this area in great numbers, especially since the mid 1990s when an increasing number of gay businesses were established. Calle San Buenaventura is the main gay area of the town, with the renowned Calle del Pecado (Street of Sin) packed with bars. During the February Carnival, 300,000 people flock to the town to see a series of parades and floats packed with an abundance of glamorous drag queens. Various tourist websites note the town's distinct lack of a visible lesbian scene, citing the domination of masculine Mediterranean culture as the reason. The lesbian scene is less tied to the bar scene and, while gay men are more than catered for, there are no visible female-only bars.

While much stigmatisation and prejudice continues to exist (see Buckland, 2002), it has been suggested that cultures and communities for gay men and lesbians have increasingly become more accepted and seen as an important part of urban cosmopolitanism (Aitkenhead and Sheffield, 2001; Simpson, 1999; Hetherington, 1996). The specific role nightlife premises have had on this transformation is crucial. As one young gay man (Steve, 25 years old, Leeds) told us about his city:

> I just think society's attitudes are changing. It's a sign of the times, I think, because gay pubs and gay clubs have become so much more established now and so much more open. I think gay people have been accepted into the community a lot easier now.

In contrast stands a description of the same city ten years ago, by Jim (28 years old, Leeds), a transvestite and a regular on the gay scene:

> But there used to be a point when you wouldn't come out on a Saturday night because there were gangs going round with baseball bats, chains, and they'd literally just pick a violent fight with anybody and now you can go out safely on a Friday, Saturday, Sunday night, any night you want. It's completely different.

Despite changes, this should not imply that homosexuality and gay spaces have been accepted by all sections of society, or even by the young, or that incidents of violence against the gay community do not continue to occur. In fact, one of the crucial issues concerning gay nightlife premises and/or enclaves in the UK, at least, concerns their marginal location (Hollands, 1995). Similarly, while many areas for gay men and lesbians in North America have been gentrified, most originally sprang up in marginalised, run-down areas of the inner city (Bouthillette, 1994: 65). As Hindle (1994: 12) states, they are 'often in what otherwise might be down-at-heels inner-city areas' and, once established, 'inertia has ensured that gay venues stay there'.

City councils have often exploited the potential contained within these marginal areas and have promoted the idea of gay villages. The Old Market in Bristol is one such area which became the focus for the city council's aspirations to create a gay village for the city in 2000, while also finding a way to breathe new life into an area desperate for regeneration after years of neglect. The area had become populated with several sex shops and lap-dancing clubs and is home to the city's largest gay nightclub. However, some gay men felt uncomfortable about being designated out of the city centre into the seedy margins. As one commented:

> It's a rough area of town. And although the venues are so close together it's on a major road ... , so anyone who is openly gay, openly effeminate, anyone dressed in drag, is open to abuse, or to violence or to threats, intimidation. So to try and make that into a gay area is just ridiculous. There's a lot of drug problems in Old Market. Lot of heroin users in that area. So it's not really a safe part of town.
>
> (Craig, 21 years old, Bristol)

While the city council has been actively supporting businesses among the gay community moving into the area, there needs to be an awareness that the emergence of a gay village is a complex, slow and organic process. City council intervention can often have the opposite effect. Marketing can raise hope values and property values, and thus squeeze out established entrepreneurs who could develop the scene more organically. A similar process has occurred in Newcastle upon Tyne, where the city council became vocal about its wishes to see a gay scene thrive, in part to compete with cities such as Manchester and Leeds as a gay consumption destination. However, as one gay venue owner commented:

> The unfortunate thing is that Newcastle City Council decided to make their mouths go about the fact that we want our own little gay village; they had watched *Queer as Folk* (a TV programme) because they had been on a bus trip to Manchester. Basically it has been one of the worst thing that

> could have ever happened because everybody who has got property now does not want to sell it. What might have cost you £40,000 last year will now cost you £1 million. Nobody will sell anything round here now, which is crippling the development of the gay scene.
>
> (Bar owner 6, Newcastle upon Tyne)

The location of gay nightlife in run-down areas of cities continues to have a range of implications, ranging from gay spaces being seen as 'seedy' through to higher incidences of violence, theft, and the impact of poverty and drug abuse. Designating areas as gay villages can also lead to them becoming a focus of violence, as reports of gangs of young heterosexual men targeting the gay community in Manchester in the late 1990s attest (Prestage, 1997). As such, a crime and disorder audit in the city in 1998 showed that the Central ward (including the gay village) was a 'crime hotspot' with the city's highest levels of reported assault.

This should not imply, for example, that gay spaces and enclaves are less safe than heterosexual spaces. In fact, in a study of the so-called 'Pink Triangle', the main gay area in Newcastle upon Tyne, Taylor (2001: 23) found that an overwhelming 83 per cent of his sample identified 'safety from the threat of homophobia' as an important characteristic of the area. Generally, gay men and lesbians felt much more confident expressing same-sex desires (i.e. kissing, holding hands) in this space than they did in straight areas of the city's nightlife (Taylor, 2001). Outside gay premises and enclaves, homosexuals, and even those individuals perceived to be gay, often cannot express their sexual identity differently for fear of ostracism and potential violence (Mason and Palmer, 1996).

While Newcastle is something of an anomaly because of its peripheral location and masculinist history (Lewis, 1994), a group of transvestites from a more cosmopolitan English city, Leeds, also mentioned their reluctance to walk though mainstream/straight areas in the city on a night out:

Sarah: ... not many people will walk here though. They prefer to get a taxi and pay two or three quid.
Giles: Yes, the first time we came to Speed Queen [a night club] we came out and had to walk for a taxi all the way round.
Jim: My God. Yes.
Derek: And the comments were like, oh, flying.
Jim: You get wolf whistles. They come up and you get wolf whistles.
Derek: They're usually derogatory wolf whistles.
Giles: I mean, now we just ignore them.
Jim: We did have some and you've ignored them, but I look more like a woman when I'm dressed up than some of the women.

(Focus group 4, Leeds)

Moreover, the often-cited case of a 33-year-old man in Canada, Alain Brosseau, who was dropped to his death from Alexandria Bridge in Ottawa by a group of youths who 'perceived' him to be gay, demonstrates the frequently ingrained nature of intolerance to sexual differences. This led to the formation of a Bias Crime Unit in the city's police force in 1993, the first unit in Canada formed exclusively to combat prejudice-motivated crimes (Lynch, 1996). Bailey's (1999: 262) work on the city of Philadelphia also noted the formation of a police–gay community relations committee in the 1980s, to discuss issues surrounding gay–police relations. However, Messerschmidt (1993: 182–3) suggests that police agencies actively move to regulate 'normative' sexual behaviour, through the criminalisation of gay activity in premises like steam-baths. He also cites research which shows that approximately 10 per cent of reported anti-gay violence involves physical assault from police officers (p. 183).

Yet the largest, and perhaps most invisible and contradictory, 'threats' to gay space and nightlife come from without and involve the usurpation of gay premises by young straights and the emergence of 'gay-friendly' premises. With regard to the first phenomenon, even by the early 1990s, five of the twenty-three premises in Manchester's gay village were described as 'mixed' (Hindle, 1994: 12), and this number has since increased (Aitkenhead, 2001). By the late 1990s, the gay-led Mardi Gras festival (Prestage, 1997) in Manchester was seen as becoming too straight as the heterosexual population rushed to join in the fun. More recent reports suggest a virtual take-over of premises in Canal Street, the city's main gay area, by gay sons' mums and relatives, not to mention straight females (Aitkenhead, 2001), following the airing of the popular English ITV hit *Bob and Rose* (where a gay man falls for a straight woman). Young single female interest in the gay village bars is apparently motivated by their view of them as stylish and safe premises, not to mention being full of good-looking (gay) men.

The issue of 'gay only' space appears to be particularly important to the lesbian population. Taylor's (2001: 30) survey revealed that lesbians ranked safety from the threat of homophobia as more important to them than gay men. Valentine (1993) argues that lesbians suffer double discrimination, first for being gay and second for being female. As such, there was some suggestion in our research that lesbians were more protective of gay space, as the following quote implies:

> Well, it is, we were shocked, honestly. We went round, we had been out all night and we went round on a Sunday [to a gay venue] just to get really pissed and have a good laugh and there was lesbians shouting abuse at some of our straight friends who were with us. 'You're straight so get out – what are you doing here?' And I think that is just as bad really, it is just, really it is.
>
> (Dave, 28 years old, Newcastle upon Tyne)

One of the paradoxes here is that as gay premises upgrade, they also open themselves up to straight women and men. A manager of a gay premises in Leeds mentions how a 'successful' conversion using trendy décor influenced the mix of clientele:

> Back then, the scene that came in were predominantly male, with a small selection of gay women coming in. Since we've had the re-fit we get such a wide range of people in here, we got a lot of students and a lot of straight people coming now.
>
> (Bar manager 4, Leeds)

While there are clearly potential progressive elements in this latter example, notably the inclusion of lesbian women, the mainstreaming of venues has some serious implications about the usurpation of gay space by the heterosexual population.

Another threat that also comes from without paradoxically derives from what are conventionally called 'gay-friendly' bars and clubs. One bar owner interviewed described the philosophy behind such places rather naively as: 'Gay people, straight people, white, black, everybody is welcome. We want people to have a good time. Like the world, this place is mixed' (Bar manager 6, Bristol). One of the problems for gay youth moving outside 'safe' spaces is that once they are spatially scattered or enter heterosexual territory, even if 'gay-friendly', 'they are not gay, because they are invisible' (Castells, 1982: 138). And yet the development of such nightlife premises not only begins to blur the binary distinction between gay and straight venues: it can also work to challenge entrenched categories of sexuality. This is particularly the case in relation to young gay men's interaction with, and changing views about, straight people involved in 'rave'-inspired dance cultures. As Lee (25 years old, Leeds) said:

> I mean, I will admit when I was younger I used to be frightened of going to straight places. The people that I have met, the lads that I have met, straight lads, big butch skinhead lads, who will come up to me and hug and kiss me and they will freely give me a new light on straight people and ... now I will go anywhere in the country straight, gay, whatever. I would rather go to straight places actually.

In this scenario, a greater tolerance and acceptance of sexual orientation within this particular dance culture not only opens up opportunities for the expression of gay identity, it also affords a similar rethinking of stereotypes of straight male attitudes. The next quote, from John (29 years old, Leeds), provides a further example of how a 'mixed' but obviously gay-friendly venue allows a young gay man to reject gay male stereotypes that clubbing always has to be related to 'cruising' and looking for sex:

> I think the best club I used to go to was the Paradise Factory in Manchester every month, and it was the best club because it was mixed and there was no attitude. Everyone was just there for a good time and a good laugh, and nobody was copping off with each other – it was not that type of club.

Finally, as we have already hinted, gay-friendly premises are also indicative of a wider shift of preference away from older seedier gay venues towards those perceived as trendy and modern (Simpson, M., 1999: 211–13). As one club promoter exclaimed about a number of these types of premises: 'They're not pushing themselves as gay bars, as such, but because it's the new and trendy area, gay men love to go there' (Club promoter 2, Leeds).

These points are tied up with a broader generational transformation in gay attitudes and cultures (see Nardi, 2000, for a historical discussion) and reflect what we have outlined as the shift towards corporate and gentrified commercial mainstream nightlife. Attitudes, places and people have all begun to change in interesting yet contradictory ways. In particular, the following conversation highlights tensions between older and newer gay communities and venues:

Derek: Well, we usually go to up to [name of pub] first because it's a gay pub that's been around for years ...

Jim: It's for the older people, the older crew.

Derek: The older cruisers.

Jim: I was going to say that. The old fogies that go around in forties clothes.

Derek: Which is weird for a gay pub.

Giles: Yes, it's like an old man's pub, isn't it?

Derek: They haven't changed it. It was altered in 1992, extended, and it was oak beams and things like that, and the new owners just never changed it. It just stayed like that. It's very weird, and the clientele are just stuffed like the pub is, I think, in a time capsule.

Giles: It seems strange because all the other big gay pubs are all very glamorous and quite cutting edge.

Derek: Yes, nice and loud.

(Focus group 5, Newcastle upon Tyne)

More serious than the derogatory language used here, Buckland (2002) in her work on gay clubs in the USA actually recalls a lesbian friend of hers being physically beaten up, to the words, 'old against the young'. Many gay men and lesbians also stressed to us the importance of this generational divide in terms of the link between age and style of premises, and here an owner of a recently up-graded gay pub talks about the different clientele:

> [It] used to be an old person's drinking/watering hole, I mean really old leathery queens, bit seedy, but just a general older crowd, pensioners-type age, gay people, you know. And that's shifted now and you've got very much a younger market, and one way we can tell that is that draft sales have gone down, bottle sales have gone up and wine has gone through the roof. You know, wine sales have gone through, you know, thirty times, thirty-fold, you know. A lot more people are sat down now with a, you know, a glass of wine on an afternoon, and reading one of the gay papers, because they feel so relaxed, and it's a proper pub; when they walk in it's like, wow.
>
> (Bar owner 6, Newcastle upon Tyne)

Whittle (1994: 38), in his work on gay space, critically refers to this phenomenon as the 'beautiful people syndrome', to describe the impact that the younger generation of gay men and lesbians were having on gentrifying their own nightlife. For him, the gay village in Manchester is a 'marketplace in which queer people are now seen as cultural consumers' (p. 37), and 'it is the time and age of beautiful young people' (p. 38). Whittle (1994) asks what there is for those left behind, through age or lack of money. He also questions what becomes of gay politics here, lost beneath the celebration of cultural consumerism.

Gay nightlife, then, is at the heart of the upgrading, stylisation and gentrification of downtown areas, much of which is fuelled by increasing corporate interest in the gay market. One manager of a tenanted gay pub in the UK hinted at how corporate interest in the 'pink pound' had changed over the years:

> You know, all they [the breweries operating tenanted pubs] are in it for is the money. Now back then they were quite happy – you know, have a tenant in there, he was going to make them money and sell their beer, but that is as much support as you would get. Now you are in the new millennium where you have got big operators like Bass, you have got your divisions who deal with gay premises. Scottish & Newcastle have got their own brand of gay pub. You know, all these big breweries are identifying the value of a pink pound.
>
> (Bar manager 6, Bristol)

Back in the early 1990s, the Allied Domecq Group was one of the first large brewery companies to cash in on the gay market in the UK through the venue Jo Jo's in Birmingham (Simpson, 1999). Scottish & Newcastle Retail Pub Company have also embraced the idea of gay venues. In Edinburgh, for example, after failed experiments with both Irish and sports bar themes, the company decided to revert one venue back to the Laughing Duck, one of the city's most well-known and hedonistic gay bars, and since has seen profits soar. The S&N Bar 38 brand has also become synonymous with gay lifestyles, mainly because of the location of these bars in Soho and Manchester's Canal Street. Moreover, gay and lesbian lifestyle brand Queercompany continues to develop branded venues such as hotels, spas, bars and restaurants as part of a planned property portfolio on sites around London and Brighton (Leisure and Hospitality Business, 2001). Larger, more corporate businesses have also invested in Manchester's Canal Street, mainly because of the profits to be made (Prestage, 1997). A similar trend towards the corporatisation of gay bars in Leeds has also been noted in the local press: 'Although a boost to gay Leeds social life, bars like Metz and Velvet are merely exports of Manchester brands, with Queens Court keeping it corporate by being part of Bass Leisure' (*The Leeds Guide*, September 2000: 5). Corporate penetration and infiltration into the gay nightlife market, then, reflects some of the general trends we have highlighted in the commercial mainstream, namely a tendency towards upgraded premises, yet a growing homogenisation of provision. As one gay consumer (Jez, 22 years old, Bristol) perceptively argued:

> Unfortunately a lot of the venues are trying to compete for the same customers. Instead of looking at what one pub is doing and looking for a niche in the market and trying to take on that, everybody seems a little bit frightened of trying something new.

The drive to retain corporate profits and keep shareholders happy with moves to open up trendy gay venues to more mainstream and straight consumers may compromise the gay character of many areas. Gerard Gudguin, of Manchester's Village Charity (an amalgamation of pub/club owners and leading figures of the gay community), stated: 'If we're not careful, the village will be like everywhere else, with hordes of testosterone-driven yobbos wandering around' (Prestage, 1997: 9). In this sense, spaces for gay men and lesbians will become less about a gay community, and more about mainstream culture and commercial profit. Chasin (2001) has outlined how gay and lesbian cultures have been absorbed by mainstream cultures, sanitised and marketed, and how in the process the gay movement's political identity has been whitewashed. Popular gay areas, particularly those in the USA, are especially prone to intense gentrification, such that many non-professional gay men and lesbians can no longer afford to live there. Markowitz (1995) cites such areas as Greenwich Village in New York, the Castro district of San Francisco and DuPont Circle in Washington as no longer affordable to the majority of gay men and lesbians, with many now forced to move to the suburbs. In the UK, similar concerns were expressed, in particular that gay communities were being formulated around commercial corporately backed premises and pleasure-seeking which did little in the way of encouraging a wider community feel and infrastructure based around shops, services, etc.

Geoff: I do not like the gay scene. It is not a gay scene; it is not a gay community. It is anything but a gay community because people do not talk to each other.

Steve: It is not the sort of community that we would want to be involved with ... What I would think of as a community is, people look out for each other, and they do not. They are just out for a shag.

(Focus group 5, Newcastle upon Tyne)

To sum up, it is clear that gay cultures have both impacted upon youth nightlife cultures and encouraged some of the wider processes of corporatisation and gentrification. The emergence of spaces for gay men and lesbians has to be understood against the backdrop of the domination of heterosexual and masculine ideologies, yet they have also provided an important outlet for gay identities, politics and 'safe spaces' to develop (Bailey, 1999; Moreton, 1996; Whittle, 1994). Buckland (2002: 2) argues that queer club culture is 'vital to the cultural life of individuals, groups and lifeworlds' and can be seen as a 'carrier of utopian imagination' (p. 3). A concrete example here of the mixture between leisure, pleasure and politics is Renegades, a club night at the Substation venue on Wardour Street in London for those who are HIV positive (Tuck, 1997).

At the same time, as gender and sexual identities are challenged and broadened through nightlife spaces like 'gay-friendly' bars and clubs, not to mention spaces for gay men and lesbians being infiltrated by both straight culture and corporate provision, a number of contradictory scenarios are emerging. One of these is a partial blurring of sexual types, which opens up the possibility of new sexual cultures and spaces developing among the young. Another contrary pressure is the increased infiltration of safe space for gay men and lesbians by heterosexual youth, and the corporatisation and gentrification of provision which has implications for the future of gay politics and exclusion within the community. The inherent difficulty would appear to be that the gradual opening up of sexual identities goes hand in hand with a tendency towards depoliticisation and an incorporation into the commercial mainstream.

8

RESIDUAL YOUTH NIGHTLIFE

Community, tradition and social exclusion

In the shadows of the bright neon of youthful gentrified nightlife consumption, there exists the 'residue' of near-forgotten groups, community/public spaces and traditional drinking establishments marginalised by new urban brandscapes and the commercial mainstream. Numerous commentators have drawn attention to increased polarisation within the contemporary city (Sassen, 1991; Zukin, 1995; Sibley, 1995) and this is no less true in the night-time economy. Nightscapes of decadence, bright lights and pleasure sit uncomfortably alongside landscapes of despair, poverty and the industrial past in many localities around the world. Inhabiting the less desirable spaces of the street, neighbourhood ghettos, council estates, cardboard 'homeless villages', ale-houses, market taverns and saloon bars, often in downmarket parts of town and clinging on to the vestiges of non-commercialised public space, groups of excluded young adults[1] are a well-recognised feature of many cities. Film images ranging from Richard Linklater's *Slackers* and John Singleton's *Boyz in the Hood* in the USA, to Danny Boyle's *Trainspotting* in Scotland and Mathieu Kassovitz's Paris-based *Le Haine*, are all examples of representations of various disaffected and displaced urban youth groups. These spaces and people, useful originally in the industrial city, are now a residual feature and are surplus to requirements in the newly emerging and redeveloping post-industrial corporate landscape, replete with its themed fantasy world of excessive hedonism and forceful consumer power.

This chapter examines the background to some of these excluded youth groups and residual nightlife spaces, charting their history and decline in the context of the post-industrial city, and explores their subordinate position in the night-time economy. Crucial to the analysis will be an understanding of the historic production of residual spaces like the traditional pub and saloon bar (Kneale, 1999; 2001), and their subsequent demise with the decline of manufacturing and the ushering in of new forms of work and leisure, making previous institutions, traditions, cultures and peoples redundant (Roberts, 1999; Clarke and Critcher, 1985). Also important here are historic and contemporary modes of regulating these traditional nightlife spaces and increasingly 'sanitising' the night-time economy through processes of incorporation, containment and exclusion. In some cases these exclusionary mechanisms are subtle (for example, through what Zukin (1995) refers to as 'pacification by cappuccino'), while in others actions are more direct, such as seeking to 'sweep the streets' clean of unwanted groups altogether.

Yet, despite their marginalisation and attempts to cleanse the newly emerging gentrified night spaces of any 'undesirable' elements (Smith, 1996), some examples of residual youth street cultures manage to subsist, despite having few resources. Unfortunately, the commodification of some youth street cultures, like rap and hip hop, by corporate capital demonstrates how commercialisation and incorporation can succeed where repression and laws fail. We begin by defining what is meant by residual nightlife, and outline some of its main elements before turning to a discussion of the wider context and history of the emergence of these spaces and charting their subsequent decline and subordination. We then examine in more detail a number of marginalised urban youth groups drawn from the UK and North America, who struggle to retain a foothold in the corporate city through creating their own spatial practices. While some of these residual youth groups, activities and spaces initially appear to be somewhat removed from what might conventionally be viewed as 'nightlife', their various attempts to exist within the context of the rapidly gentrifying commercial mainstream demonstrate that these 'two worlds' are, in fact, intimately related (also see Smith, 1996; Toon, 2000).

Understanding residual urban nightlife: industrial decline, economic marginalisation and social polarisation

How can we best understand the position of groups of young people and night-time spaces that are increasingly marginalised and excluded by processes of gentrification and corporatisation? Where did these spaces emerge from and how might they be best contextualised and comprehended? First, what do we mean by the term 'residual' and, more specifically, what does residual nightlife consist of? To begin, the *Oxford Handy Dictionary* (1978: 766) defines residual as 'remainder, what is left or remains' and 'left as residue or residuum'. In a more sociological vein, Williams (1977: 122) utilises the term to mean 'effectively formed in the past, but … still active in the cultural process' and 'certain experiences, meanings and values which cannot be expressed or substantially verified in terms of the dominant culture, are nevertheless lived and practised on the basis of the residue … of some previous social and cultural institution or formation'. Williams (1977) goes on to remind us that although elements of the residual may be alternative or oppositional, generally major aspects have been incorporated or diluted by the dominant culture. For example, the 'traditional' working-class pub has been usurped by the middle classes and the corporate producers of 'artificially themed' traditional pub environments (Everett and Bowler, 1996).

By residual nightlife we are referring to those more traditional urban spaces inhabited by increasingly excluded sections of young people, including community bars, pubs and ale-houses, as well as the street, the neighbourhood ghetto or the council estate. Residual nightlife, then, is characterised by a number of features. First and foremost, these spaces are defined against mainstream commercial development. In contrast to the modern – and indeed postmodern – veneer of gentrified nightlife, residual spaces are seen as outdated, obsolete and anarchic. Unlike the bourgeois and middle-class appropriation of the traditional pub, residual spaces are viewed as 'down-at-heel' dilapidated reminders of a lurid era of poverty, vice and debauchery associated with the lumpenproletariat of Victorian times, now occupied by their twenty-first-century counterpart, the 'urban underclass' (Murray, 1990; 1994).

So, a second characteristic is that these spaces are inhabited, or perceived to be occupied at least, predominately by the urban poor – the unemployed, welfare-dependent and shifty criminal classes. Finally, located either on the fringe of city cores, in poorer suburbs and council estates, or in marginal urban areas characterised as the 'rough' and 'down-market' part of cities (occasionally in traditional market areas and/or near the 'back end' of transportation networks), residual spaces represent the 'other' city of dirt, poverty, dereliction, violence and crime (Illich, 2000; Patel, 2000; see Plate 8.1). Created by industrial capitalism originally to meet the leisure needs of the working classes, these residual nightlife spaces have dramatically declined, are perceived as ripe for redevelopment, and are increasingly regulated by stigmatisation and formal policing. However, they continue to play a vital role in terms of meeting the food, entertainment and leisure needs of many urban groups.

These residual groups and spaces are rooted in the creation of industrial cities, their labouring classes and the rise and fall of the first phase of industrial capitalism. The development of the industrial capitalist city in the UK, North America and parts of Europe in the nineteenth century created not only different classes and separate spaces, but particular spheres of activity represented by the division between work, home and leisure. Characterised by rapid industrialisation ('smokestack' cities) and urbanisation, with many localities increasing their population many times over within half a century, the industrial city became infamous for its separation into rich and poor areas, with the former moving to the urban periphery and the latter experiencing over-crowding, bad housing, squalid and unhygienic conditions, pollution and moral decay in the urban core. In the UK, nineteenth- and early twentieth-century studies by Engels (1968) and Booth (1902) described the social structure of this industrial landscape, while in the USA at the turn of the century, the Chicago School's concentric ring model became urban

Plate 8.1 A few traditional ale-houses remain on the margins of rapidly gentrifying downtown areas; Newcastle upon Tyne, UK, 2001
Source: Mike Duckett

spatial orthodoxy in terms of understanding social differentiation (Park, 1967). Additionally, the industrial city saw the formal separation of work (the factory) from the domestic (household/residential) and leisure spheres.

As such, 'redundant' populations have existed from the very beginnings of industrial urbanised capitalism. For example, Marx and Engels (1981) in *The Communist Manifesto* referred to the 'dangerous classes, the social scum, that passively rotting mass thrown off by the lowest layers of old society'. This was the reserve army of labour, the lumpenproletariat, the residuum left at the bottom of the emerging class hierarchy (Byrne, 1999: 20). While Marx and Engels clearly saw this group as a by-product of capitalist urbanisation, their representation of them was one imbibed with negative ideas about 'danger' and, indeed, political conservatism. The urban poor were also negatively represented in many of the early poverty studies conducted in the UK, such as Rowntree's study of the city of York and Booth and Mayhew's work on nineteenth-century London. Booth's description of some of the urban poor as a 'savage, semi-criminal class of people' (see Booth in Thompson and Yeo, 1973) was replicated in numerous commentaries of that period about the habits of the undeserving poor; of fecklessness, violence and over-indulgence of drink (Mann, 1990). In the USA, rapid urbanisation and the rise of capitalist manufacture in the late nineteenth and early twentieth centuries produced sprawling slums and surplus labouring classes (many of them ethnic groups from European or Afro-American descent), viewed in need of reform and social control (Harring, 1983; Sennett, 1974; Jones, 1998; Sugrue, 1998: 120).

As we outlined in Chapter 3, the continued imposition of Fordism in both work and leisure in the industrial city had numerous effects, including the taming of the dangerous classes through their uptake into employment, and various attempts to civilise and educate the working masses. Mass production and mass consumption meant a predictable link was forged between work and lifestyle. As the late Marxist historian E.P. Thompson reminded us, nineteenth-century working-class struggles were over space, time and leisure, as well as over the nature of work discipline itself (Thompson, 1967). Industrial capitalism represented an increasing encroachment into the 'proletarian public sphere' (Habermas, 1991), rationalising leisure and free time as much as work (Clarke and Critcher, 1985). The street, the neighbourhood, working men's clubs, recreation and sport, entertainment, the pub and saloon bar were all sites of this broader 'conflict of culture' (see Palmer, 1979; Cunningham, 1980). The general struggle over working-class leisure and space is crucial because it is here that residual nightlife spaces were originally produced.

Of particular relevance to our discussion is the historical role of the UK public house and the North American saloon bar, both of which form part of what is left of residual nightlife (Harrison, 1971; Harring, 1983). In the UK, the importance of the public house for the working classes was exemplified by Harrison's (1971) famous quip that it was the working man's 'voluntary association', while Bailey (1978), Walvin (1978) and Jones (1986) reiterate its role as a central hub of working-class life and leisure throughout much of the nineteenth and early twentieth centuries. Mass Observation's study of Bolton in the 1930s suggested that: 'In Worktown more people spend more time in the public houses than they do in any other building except private houses and work places' (Mass Observation, 1970: 17), and Jones' (1986: 79) study of work and leisure in the early part of the twentieth century stated that the pub 'continued to play a central part in the social and cultural life of the working-class community'.

More specifically, early observers of pub life in the UK viewed it as an important space for the development of working-class identity, community and – in some cases – political education. Mass Observation's (1970: 19) analysis in 1930s England saw the traditional pub as an 'unfettered space' and 'participator rather than spectator' in its orientation. As Adler (1991: 391) writes, the 'pub, as a central node in this community of men, served as their centre for conviviality, political discussion and the rituals of drinking through which men affirmed in shared communion a collective identity'. Cunningham (1980) and Walvin (1978) specifically noted that it was the source of numerous libraries and political discussion groups historically, while Jones (1986: 79) mentions that it was a focal point for numerous activities not 'conducive to capitalism'. The working-class pub, then, was historically a site of group, not individual, sociality (Mass Observation, 1970: 78), including social ties of obligation, reciprocity and community. It is worth recalling, at least in the UK, that the traditional 'pub' is defined in the *Oxford English Dictionary* as a 'building with a bar and one or more *public* rooms licensed for the sale and consumption of alcoholic drinks' (emphasis added). In this sense, the pub fulfilled its historic function as a public meeting place and a source of community interaction.

This, of course, should not imply that the traditional UK pub of the past did not reflect wider social inequalities. Most commentators mention that it was primarily a male domain (see Harrison, 1971; Gofton, 1983) although Jones (1978: 78) cites one study of Fulham pubs in the 1930s which suggests that around one third of patrons were in fact women. Historically, class divisions in premises were exaggerated in the separation of the lounge from the vault and tap room (Mass Observation, 1970). However, Figure 8.1, reproduced from the Mass Observation study, shows that such spatial and social divisions were sometimes contained within one venue, contrasting greatly with upmarket mainstream venues today, who cater for only fragments of the gentrified middle classes (see Figure 5.1). In the traditional pub, the lounge was the largest room of the three, aimed at a higher class of consumer with correspondingly more chairs and tables, an enhanced décor and beer a penny a pint more. It was also the only room open to women (and hence couples). The one element the lounge had less of was spittoons for spittle, another class indicator. The vault or tap room usually just had linoleum on the floor and was generally used by workers. Clothing differences were apparent, with caps, scarves and overalls more apparent in the vault and tap room, and bowlers, trilbies, ties and 'good suits' characterising the lounge goers (Mass Observation, 1970: 141).

The crucial issue here, however, concerns attempts to wrest control of this working-class space through a combination of legislation and repression on the one hand and commercialisation and gentrification on the other. For instance, there were many examples of legislation directed specifically at working-class drinking establishments and habits, prompted by the middle classes and the temperance movement; these including restricting licensing hours, and targeting under-age drinking and certain types of behaviour. Ale-houses, as opposed to inns where middle-class patrons often spent the night (i.e. literally a hotel with a bar), had early reputations as a 'place of drunkenness, disorder and squalor' (Mass Observation, 1970: 81). While the temperance movement in the UK was adamant about the link between poverty and alcohol (Booth, for example, argued that around 14 per cent of poverty was due to drink), control here was additionally concerned with types of sociality created in such places (Kneale, 1999; 2001: pp. 44, 45), including the role social drinking played in aiding worker solidarity and political mobilisation.

Also important here were broader social changes and trends which weakened working-class cultures and appeared to democratise leisure and entertainment (Hannigan, 1998). Early attempts to legislate the problem, including restricting licensing hours, were largely unsuccessful in comparison to more insidious forms of middle-class control through commercialisation, displacement and gentrification (Adler, 1991; Everett and Bowler, 1996). Williams' (1977) earlier assertion about residual cultural forms becoming incorporated or made redundant by more dominant interests is useful here. For example, in the UK, Bailey (1978: 171) argued that although it was difficult 'to supplant the public house in the affections and habits of the working man', the drive towards more 'rational' forms of recreation throughout the nineteenth century began to take its toll on this important institution. Jones (1986: 79), among others (see Cunningham, 1980) notes an early drive towards the commercialism and gentrification of pubs – typified as 'public house improvements' – around the first quarter of the twentieth century in some cities, first to stave off state interference in the industry (some pubs in Carlisle actually became state-owned during this period) and second to deflect criticism away from the brewing industry. Mass Observation (1970) also related changes in pub culture to a wider institutional decline encompassing the influence of the church, politics and other working-class public spaces and pastimes. While they argued that, compared to many of these other spheres, the pub retained a relatively strong position as a fundamental maker of community, for the male working class at least, subsequent economic trends have meant that brewers and publicans were increasingly forced to sacrifice conviviality for efficiency and competition.

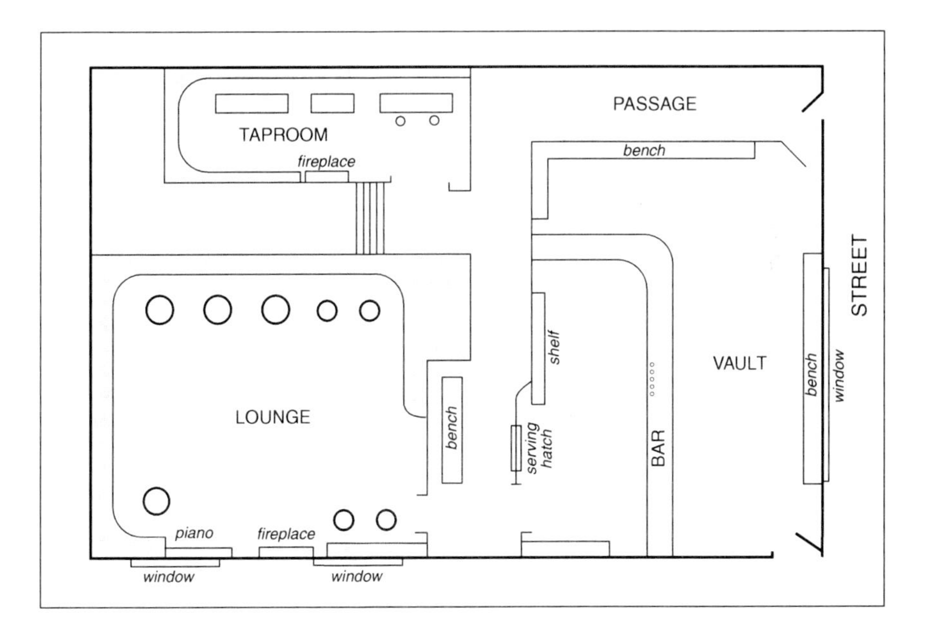

Figure 8.1 A layout of a traditional public house in England
Source: Mass Observation, 1970

With the growing commercialisation of the brewing and pub industry in the UK in the post-war period, there has been a dramatic decline in the community function of the pub for the working classes (Adler, 1991: 391). The changes in the British pub in this period have been described in one text as a shift from folk to a more popular, commercial culture (Everett and Bowler, 1996). Components of this shift include the impact of the Monopolies and Mergers Commission in the UK (Beer Orders Act 1989), demographic changes (going out became more youth-oriented, town-centred and gendered), cultural influences (both from European café culture and a North American McDonaldisation – i.e. the move towards efficiency and standardisation) and changes in tastes (from beer to wine, etc.). We would add here the consolidation of monopoly ownership (with large, acquisitive pub companies moving in), and increased gentrification, and regulation (i.e. the sanitisation of nightlife excluding cash-poor groups).

Similar arguments have been extended to the traditional role of the North American bar and saloon. Harring (1983, chapter 7) historically documents early twentieth-century attempts to regulate workers' recreation, particularly those activities that involved drinking and/or socialising in and around places where liquor was sold or consumed. A coalition of employers, the middle classes and the religiously motivated temperance movement in many US cities pressured the police to enforce licensing hours, impose higher licensing fees and crack down on the sale of liquor to minors. However, due to the fact that many of the saloon owners were petty bourgeois (not to mention that high numbers of them were often city politicians) and that licensing fees were an important component of municipal funds in many localities, the emphasis was placed primarily on negative working-class behaviours like crime, laziness and immorality, rather than being just 'anti-saloon'. In fact, saloon owners were rarely prosecuted for breaking the law and the imposition of higher licence fees in many cities resulted in the differential closure of many smaller working-class premises, while aiding larger and more middle-class premises composed of wine rooms, where more affluent mixed couples were found (Harring, 1983: 159–62). The development of 'saloon bars' – larger and better-decorated premises – signalled that the North American middle classes were beginning to threaten the sanctity of spaces usually reserved for the US labouring classes (Kneale, 1999: 335). Johnson (1986: 527) suggests that arguments over the [American] saloon, were 'a struggle between middle-class notions of privacy and domestic happiness and working-class uses of public space'.

Post-war economic growth and welfare consensus in the political arena in both North America and the UK meant that there was a brief period of economic, political and cultural stability (Bell, 1961). Nearly full employment, growing post-war affluence and mass production and consumption created the impression of a growing democratisation of leisure, despite underlying inequalities (Aronowitz, 1973; Williams, 1961; Hannigan, 1998; Clarke and Critter, 1985). However, the 'consensus' began to break down with industrial conflict, economic recession and the oil crisis in the late 1960s and early 1970s. Free market economics and a declining state, exemplified by Thatcherism in the UK and Reaganomics in the USA, led to a major 'manufacturing shakeout' in the 1980s, resulting in unprecedented unemployment levels, economic decline and social polarisation. For example, unofficial unemployment levels in the UK rose to over 4 million people, while world cities like Paris lost 400,000 unskilled jobs from the mid 1970s to the 1990s (Body-Gendrot, 2000: 182). Economic decline was more severe, however, in those cities with an industrial history. For example, Chicago lost 326,000 (24 per cent) of its manufacturing jobs between 1967 and 1987 (Body-Gendrot, 2000: 153), while Newcastle upon Tyne lost 25 per cent of its manufacturing jobs in the early 1980s (Robinson, 1994). Working-class communities were particularly hard hit, and all western societies experienced social polarisation and increased inequality during this period to some degree.

In the USA by the 1990s, the income of the richest 1 per cent rose by 91 per cent, while the income of the poorest 1 per cent diminished by 21 per cent (Body-Gendrot, 2000: 36). Poverty became very much a black urban problem, encompassing a thirty-year 'white flight' to the suburbs. By 1990 only 25 per cent of North American whites lived in the central city areas, compared with half in the 1960s, while the proportion was 57 per cent for blacks and 52 per cent for Latinos (*ibid.*: 30). Poverty rates in the so-called 'Black Belt' area of Chicago were double that for blacks elsewhere (Wilson, 1996: 14–16), which in turn was substantially higher than for the white population. While many European cities were less polarised, thanks partly to a history of state intervention, inequalities still existed, while England experienced some of the highest rates of income inequality in all of Europe, during the Thatcher years in particular (Hall and Jacques, 1989).

Those left behind in this economic recession and restructuring form the new residual classes, made up primarily of the working classes, ethnic and immigrant groups and unemployed young people. Modern-day residual groups are a product of the dissolution of the 'respectable' working classes as they cope with industrial decline and the shift towards a post-Fordist labour market. One popular, if not problematic, way of thinking about this group is the contemporary idea of the 'underclass' (Murray, 1990, 1994; Darendorf, 1989; Jencks and Peterson, 1990). As MacDonald (1997: 1) explains: 'The idea that Britain and other late capitalist societies are witnessing the rise of an "underclass" of people at the bottom of the social heap, structurally separate and culturally distinct from the traditional patterns of "decent" working-class life, has become increasingly popular over the last ten years.' Numerous critiques of the concept exist (i.e. see Bagguley and Mann, 1992; Morris, 1994), while others have sought to provide more sympathetic labels such as Byrne's (1999) term the 'dispossessed working class', Wilson's (1987) 'ghetto poor', Gans' (1990) 'undercaste' and the UK Labour government's favoured term, 'the socially excluded'.

It should be noted that there are more liberal and social democratic versions of the underclass emphasising economic conditions (Field, 1989; Darendorf, 1989), but the concept has largely been hijacked by the right as a moral and cultural phenomenon, focusing on idleness, crime and immorality (Murray, 1990, 1994). While many on the left completely reject the underclass idea as a 'moral panic' or an 'ideology of the upper classes' (Bagguley and Mann, 1992), MacDonald (1997) persuasively argues that there has been little good empirical work to test out the concept either way. Still others have adopted the more social democratic notion of social exclusion, defined fundamentally as multiple disadvantage (Social Exclusion Unit, 1998). MacDonald and Marsh's (2001) ethnographic work has been perhaps the most helpful and detailed here. While their findings reject the underclass thesis, they are also reticent about various elements of the social exclusion paradigm, preferring to emphasise the complexity and multiplicity of disadvantage, while stressing the continuing importance of economic marginality.

Important to our discussion here is a recognition that such exclusion is not just economic but is also cultural and spatial. Byrne's definition (1999: 2) does include exclusion from 'cultural systems'; this and Bauman's (1998) recent thesis that the poor today are excluded from seductive cultures of consumption rather than just the labour market are both instructive. Bauman (1998) argues: 'The way present-day society shapes up its members is dictated first and foremost by the need to play the role of consumer' (p. 24) and in 'a society of consumers, it is above all the inadequacy of the person as a consumer that leads to social degradation and "internal exile"' (p. 38).

Spaces of consumption, then, are an important component of social exclusion. As Byrne (1999: 100) argues, 'spatial exclusion is the most visible and evident form of exclusion' and the post-industrial city has seen an unprecedented polarisation of space (Lash and Urry, 1994) or what Sibley (1995) has called 'spaces of exclusion'. While there have always been historical attempts to create 'purified communities' (Sennett, 1974) – for example, keeping middle-class communities immune from contagious, dirty working-class ones – various writers have argued that recent attempts to sanitise and purify are exaggerated through place promotion and gentrification (Smith, 1996; Christopherson, 1994). Power is expressed through monopolisation of space by wealthy groups and the exclusion of weaker marginal groups, creating a socio-spatial hierarchy of winners and losers.

Smith (1996), for example, has argued that gentrification is endemic to the capitalist economic system. His idea hangs upon economic arguments about housing, whereby the 'rent gap' created by derelict inner-city properties, leads inevitably to their redevelopment in order to increase return to landlords and property developers, with capital flowing back into the city from the over-priced suburbs. He sees such developments as an attack by the middle classes on the poor, and typifies this scenario as the 'revanchist' city – revenge by the powerful to punish the poor, pushing them further to the margins and worst parts of the city. Ley's (1996) work on the middle-class city is based more on the rise of professional and service classes and their desire to escape suburbia and create a somewhat different lifestyle in inner-city areas. While this explanation is partly dependent on housing, it also involves the development of an appropriate cultural economy including wine bars, gourmet coffee shops, delicatessens and trendy restaurants.

Box 8.1: *La Botellón*

Thousands of young Spaniards have taken to gathering in streets and squares across Spain, drinking ready-mixed potent blends of *calimocho* (wine and cola), spirits and beer costing around □2. The phenomenon of outdoor drinking called the *Botellón* (the big bottle), although nothing new, is now attracting a larger, younger and more rowdy audience. In Madrid, the number of young madrileños gathering for the *Botellón* has particularly grown and has incurred the wrath of neighbouring residents. Many cities have banned open-air drinking and the sale of alcohol after 10 p.m., while councils have also been ordered to compensate neighbours for the noise. Many teenagers suggest that they cannot afford to drink in expensive city-centre bars and discos and so take to the streets instead. This phenomenon is most pronounced in Madrid, where it is claimed by the authorities that 200,000 young madrileños drink in the city's public squares. At the beginning of 2002, the city's right-wing mayor, Jose Maria Alvarez del Manzano, pledged to put an army of police on the street to stamp it out (*Guardian Weekly*, 14–20 February 2002: 13). Legislating against the *Botellón*, then, is eroding youth street cultures. Considering that a recent survey found that 79 per cent of 14–18 year olds in Spain drink alcohol, new youth spaces rather than increased surveillance and coercive legislation may be more realistic solutions.

One can see these processes of exclusion and sanitisation at work on a number of fronts. For instance, Key Note (2001), a market analyst company, suggests that the number of traditional pubs in the UK has fallen from 54,000 in 1990 to 49,500 in 2001, with part of this decline due to the conversion of many premises to other types of fully licensed drinks outlets, including café-bars, brasseries, sports bars and wine bars. Moreover, numerous writers (e.g. MacDonald and Marsh, 2002; Toon, 2000; Blackman, 1997) argue that there is a greater tendency for underprivileged groups to use the street as their domain, hence increased policing and street surveillance here is an explicit attack on the poor. The move towards making city streets 'dry zones' (no alcohol consumption legally enforced through fines), which has long been a feature of many North American cities, is now being adopted across many European cities, and is also an erosion of space for those groups who cannot afford to drink in even the cheapest bars and pubs (see Box 8.1).

To sum up, both residual groups and spaces are the product of the industrial city and its transformation into a post-industrial economic landscape. This has resulted in the marginalisation of groups, not only economically but also in terms of consumption and space, with increased gentrification and attempts to sanitise whole parts of cities. Below, we look at how various residual youth groupings have been affected and have sought to cope with some of these changes.

Marginal youth and residual nightlife spaces

Historically, young people have played a particular role in the emergence of the industrial working classes and urban street cultures. Concerns about the state of the working-class youth labour market, female reproductive behaviour and rowdy urban youth cultures formed part of the early debates about troublesome adolescence (see Griffin, 1993). Pearson (1983: 74–5) mentions varied terminology for the street urchins of the late nineteenth and early twentieth centuries, including 'street arabs', 'ruffians' and 'roughs'. The well-known term 'hooligan' was also used to describe gangs of rowdy youths during the August Bank Holiday celebrations in London in 1898 who were involved in drunk and disorderly behaviour, street robberies and fighting.

However, such boisterous street behaviour was not limited to the workless youth of the day. Employed youth were seen to use their wages to ill effect in Victorian times, with complaints of both sexes 'mingling promiscuously in the cheap theatres and taverns, sometimes in the company of prostitutes, drinking, singing and swearing in a lewd manner' (Pearson, 1983: 166). In Manchester at the beginning of the twentieth century, the so-called 'street scuttlers' became notorious for their 'monkey runs' – a cross between loafing and cruising, in which gangs of working youth would circulate around the city, using the streets as a playground to avoid the police, encounter women and stay in contact with their mates (Haslam, 1999: 48). Harring (1983: 154) mentions that in the USA, at least one important aspect of the anti-saloon movement was to curb under-age drinking. By the turn of the twentieth century, youth disorder was not just a problem of the urban poor, but also of young labourers who used their wages to drink and hang around street corners and amusements, engaging in gambling and leading to worries about creating a loafing, street-corner society (Pearson, 1983: 57, 60; Harring, 193: 147). Much of what came to be seen as traditional working-class pastimes formalised over this period, including music and dance halls, moving pictures, pool halls, the pub, holiday resorts, the football crowd and the 'cycle craze'. Youth, in particular, were seen as problematic participants in many of these activities (Harring, 1983, chapter 8; Pearson, 1983; Cunningham, 1980).

The rise of compulsory primary education in England in the late nineteenth century attempted to deal with problems of youth discipline (Griffin, 1993), and was followed by the development of secondary education (Humphries, 1983), the apprenticeship system and two world wars, all of which had an impact on shaping youth. In the USA at the turn of the twentieth century, various attempts to root out crime and delinquency among young people were made through moral training, responsible parenting and raids on pool halls and saloons (Harring, 1983: 228–32). Such attempts to create a culture of conformity in the inter-war period in the UK were broken after the Second World War, with the early rise of youth cultures in the 1950s and 1960s (Hall and Jefferson, 1976), while the USA began to see the beginnings of gang cultures emerging (Miller, 1958). The creation of many of these working-class groupings were traced to an increasingly differentiated youth labour market, including the beginning of the service economy and the early demise of manual labour. The major crisis of youth unemployment began to bite in the late 1970s and early 1980s, with young people making up 25 per cent of the 'out of work' and 50 per cent of the long-term unemployed in the UK (Hollands, 1990), while in the USA the percentage of unemployed black youth in many urban centres reached over 40 per cent (Kelly, 1997).

So, one of the key 'residual' subgroups to have emerged out of the historical transformations discussed earlier is the young disaffected working class – literally the sons and daughters of the last of the industrial workers and various ethnic and immigrant groups. In fact, MacDonald (1997: 19) suggests that 'the underclass debate is in large part a debate about youth', and numerous commentators have charted the collapse of the youth labour market and investigated the changing family/household situation of the younger generation (Males, 1996; Jones, 1995). However, there has been less work on residual youth that has concerned itself with cultural and spatial exclusion (although see MacDonald and Marsh, 2002; Toon, 2000; Skelton and Valentine, 1998; Stenson and Watts, 1998; Loader, 1996). It is to this issue we now turn through a number of case studies.

Disconnected youth? Social exclusion and residual nightlife spaces in the UK

In light of the dominance of mainstream nightlife, and more specifically the growth of up-market style bars discussed in Chapter 5, possibilities for residual, historic forms of nightlife consumption in downtown centres have rapidly diminished. Consumers of more traditional community-based pubs and ale-houses in the inner city, which hark back to an earlier industrial era, today are described rather unflatteringly as the spaces for the 'underclass' or the 'rabble', often stereotyped as petty criminals, welfare scroungers and hardened drinkers by the police, local state, the media and other more middle-class and respectable consumers (Hollands, 2002). Residual youth groupings here include those left on council estates who are largely excluded from urban centres, the young homeless and transients, as well as sections of less well-off youth who consume what might be called the residual or 'bottom end' of the nightlife market.

As mentioned earlier, there is a long history of concern over 'troublesome' street youth in the UK (Muncie, 1999; Pearson, 1983; Corrigan, 1976). More recently, there has been a spate of 'moral panics' around young people's disruptive use of public spaces on estates, parks, playing fields and street corners (Stenson and Watts, 1998; Loader, 1996; Jeffs and Smith, 1996; MacDonald and Marsh, 2002), to the extent that a contemporary survey of communities argued that 'on all estates, and across all age groups, the biggest single issue identified by respondents was the anti-social behaviour of children and teenagers' (Page, 2000: 37). MacDonald and Marsh's (2002) ethnographic study on the leisure careers of a group of socially excluded youth on a deprived north-east estate is perhaps the most comprehensive research on this topic. Their work confirms the prevalence of a well-recognised pattern of street-based socialising among young people living on poorer estates. As one young person described: '[I would] get up and used to go round and they all used to be sat there, like thirty of them, sat in one little corner, smoking and drinking at nine o'clock in the morning' (quoted in MacDonald and Marsh, 2002: 13).

Crucial to this pattern, besides a sense of belonging to a locale, was a lack of things to do and money to do it with, and a dearth of opportunities to consume other activities in other parts of the city. As Loader (1996: 112), in his similar study of Edinburgh, puts it: 'Denied the purchasing power needed to use, or even get to, other parts of the city (and most importantly the city centre) unemployed youths are for the most part confined to the communities in which they live.' A young man (Ian, 24 years old, Newcastle upon Tyne) from one troubled estate in the north of England, the Meadowell, made this point to us even more starkly when describing his community:

> You can't go anywhere without the police following you, thinking you are up to something, some kind of deal. If you walk around the estate, they follow you. If you go across to [a local shopping centre] or into town they follow you. So you just stay here and do nothing. It's like a prison without walls, that you never leave.

This particular section of 'socially excluded' young people, then, have limited choices, socialising within whatever spaces they can find in what is left of their rapidly declining communities. Local pubs and/or social clubs on the estate (if they have not already closed down) and 'the street' (Toon, 2000) remain their main options. Two excluded groups singled out in particular by MacDonald and Marsh (2002) are young unemployed men and young mothers. Lacking social networks created by employment, some young unemployed men maintained longstanding participation in localised street cultures, often getting involved in crime and/or drugs. As one of their interviewees, Stu, said: 'You're just going back to the same place, the same group of people and it's easy to get back into it' (quoted in MacDonald and Marsh, 2002: 10). Young single mothers, constrained by the demands of child care, financial hardship and trying to maintain a home, also found their leisure of nightlife curtailed:

> Gail: I used to go to the Capital [nightclub] and places like that ... [but] I'm too much in my ways now. I don't miss it [going out]. I'm used to lights out, curtains shut, Joe [her son] on the settee. I sit on the chair and watch the telly and I'm in bed for 10 o'clock. It's my routine now since we had the house.
>
> (MacDonald and Marsh, 2002: 17)

Despite the difficulties and constraints, MacDonald and Marsh's (2002) research also showed that by their later teens, some youths had moved on from street-based socialising to consuming more commercial and alcohol-based leisure in bars, pubs and clubs. A portion frequent those declining traditional city spaces made up of 'rough' market pubs and taverns inhabited by their parental generation, described by many city officials as 'potential sites of regeneration', while the local police often view them as criminal 'dens of inequity' (see Box 8.2). Below, a police inspector with responsibility for one UK city centre, had this to say about such places:

> Very difficult some of them. I mean, they are in poor areas ... They have still got to be policed ... it is a place where if you go through the day you might get offered cheap bloody perfume or something that has been nicked from one of the shops, it is where shoplifters tend to get rid of their gear.
>
> (Police inspector 2, Newcastle upon Tyne)

Box 8.2: A night in the traditional boozer; Bristol, 5 September 2001

The Admiral Nelson has been on the corner of the market for several decades. The pub rarely opens past 7 p.m. these days, largely due the fact that, as Ken its manager said, 'Once the market shuts people go home, they've got no reason to be here and they can't afford it anyway.' We had to agree with him. Once the shops and market closed around 5 p.m., this part of town was now completely dead. Hence, the city council were keen to shed the area of its rather seedy and dangerous image after dark. The Nelson had several rooms, a tap room with simple wooden stools and linoleum on the floor, a function room, a lounge furnished with tatty, torn, red velvet seats and curtains and a pool room with a huge TV. Like most traditional market taverns, it was dimly decorated and had a small but loyal – and mostly male – clientele. Most of the rooms apart from the lounge were not used much, although the pool room was always packed on a lunchtime or Sunday afternoons, when the market traders would stop in for a game of pool or to watch the football. One of the regulars, Frank, told us that a lot of young blokes used to come in here, but now they prefer the centre, 'where the booze is fancier and so are the women'. Geoff, another, continued: 'Our Kev sometimes still comes in here with me and his mum and his mates for a few pints before heading off to the discos, like. Cheapest place in town for them, like, to start off and sink a few cheap pints. Still £1.10 for a pint of ale in here, y'know. They all know Ken the barman 'n'all, from up where we live.' They felt that the city centre was changing and 'so should places like this, to keep up to date'. However, they were less clear on what was going to happen in the future. As Geoff said, 'There's all these new coffee places round here and these posh clothes shops. Pubs like this, I dunno ... What can we do? Maybe they've had their day'.

Nevertheless, for the large pub operator and brewer who owns the Nelson, business is good for the time being. Bulk sales of beer and ale are high and overheads are low. There is little need for sophisticated product lines, costly refits or advertising, and customers are steady and loyal. Furthermore, apart from the odd fight at lunchtime and police suspicion of handling of stolen goods on the premises, the Nelson causes very few problems for the regional manager. This is in contrast to other younger, livelier super-pubs he manages in the city centre, where there are frequent police visits due to under-age drinking, fighting, property damage, noise and litter complaints from new city-centre residents, and police demands for sophisticated surveillance and expensive security firms. Yet clearly the Nelson's days are numbered. A neighbouring pub in the area owned by a rival pub company has recently been revamped into a 'canteen bar' to serve coffee and food during the morning and afternoon to the growing number of office workers and city-centre-livers who are moving into revamped buildings in the area. It also stays open later into the night offering music and live performances. Ken mused, 'A refit'll mean more takings and work for the company. But I reckon a change that big'll be the end for me and this lot.'

Similarly, other young people ventured out and attempted to consume the 'bottom end' of mainstream commercial provision, where they are increasingly unwelcome, as the following quote from a bar employee at one new city-centre bar in Leeds exemplifies:

Interviewer:	Are these locals [kids from an estate] being pushed out of the city centre?
Bar worker 2:	I hope so.

These marginalised post-adolescents are viewed hierarchically as even further down the social scale than aspiring working-class groups (or what we called 'townies' in Chapter 5). For example, note the sheer contempt with which this young consumer (Mark, 22 years old, Leeds) describes a so-called 'scratters' night out:

> Nasty, horrible creatures of society, who crawl out from under their stone on Thursday 'cos it's dole day. They put on the same frock every week 'cos they don't wanna buy a new one until they get too fat. Mainly seen wearing the PVC skirts and boob tubes, which are too tight, sort of sagging and not nice. The over 40s, who still think that they are 18.

It is important to note the clear class and gendered nature of this stereotyped lifestyle of the socially excluded. Typified as welfare scroungers, who prefer to spend their social security cheque on a night's drinking, this group is vilified by the media and so-called 'respectable society' along a number of axes such as promiscuity, crassness and selfishness (see Hollands, 2000, 2002, for an extended discussion).

The most marginal elements of the UK's urban youth population – the homeless – fare even less well in the rapidly gentrifying nightlife (for the USA see Ruddick, 1998). Blackman's (1997) ethnography of youth homelessness shows how their nightlife options and spaces are seriously curtailed. Utilising their physical bodies as a last source of 'personal capital' via drug-taking and prostitution, some of the homeless nonetheless continue to play a marginal role in the night-time economy. 'Destroying a giro', to use Blackman's (1997: 122–3) term, means periodically engaging in a somewhat 'normal' and hedonistic pattern of night-time consumption by literally spending their entire social security money on either drinking in a cheap public house or purchasing large quantities of alcohol or drugs and hanging out in the street. However, even this pattern was contoured by the cost of drinks and going to premises with the cheapest prices, as most face barriers not only of cost but also of dress and style codes. In fact, simply being visibly homeless on the street can sometimes invite derision and physical attack. As Blackman (1997: 121) writes: 'Attacking homeless young people was repeatedly seen as a sport by sections of Brighton's nightclub scene, who would play "kick the beggar" or "piss on the beggar"'. Obviously, groups such as the homeless are unwelcome and will continue to be further maligned and increasingly excluded from city-centre nightlife as gentrification and the 'style revolution' in nightlife continues apace.

Afro-American youth, rap, street culture and the 'hood; segregation, incorporation and cultural power

Living in one of the wealthiest and most powerful nations on earth, the USA, Afro-American youth in a relative sense are one of the most socially excluded and spatially separated generational groups in the world.[2] Yet, paradoxically, economic deprivation of this social group sits alongside their growing global cultural influence, particularly in the realm of music and associated nightlife cultures (Pimlott, 1995). As Lott (1999: 125) argues: 'Despite their status as a group of Maroons in America's urban centers, ironically, black urban youth, through their culture, have had a major impact on mainstream popular culture.' Here we briefly examine the paradox that the majority of urban Afro-American youth continue to be excluded, while a few have made a significant intervention into global nightlife culture primarily through the medium of hip hop and rap (Mitchell, 2001). Clearly there are strongly felt differences here, between those artists who have retained some authenticity and connection with the 'hood (neighbourhood) and those who have 'sold out' and joined the commercial mainstream (Forman, 2002).

Comprising just over 10 per cent of the population of the USA (Oswald 2001: 18), Afro-Americans in general experience significantly higher unemployment and incarceration rates, have poorer health care, education provision and leisure amenities, with many confined to hyper-segregated inner-city neighbourhoods and urban ghettos. With around 30 per cent of blacks in the USA living below the poverty line, according to 'official' statistics, a figure four times that of whites (Cashmore, 1997: 3) and with jobless rates double (Oswald, 2001: 67), it is unsurprising that they are also three to four times as likely to be receiving food stamps, Medicaid and public housing than whites. Afro-Americans are also subject to criminalisation, constituting two-fifths of all arrests in the USA, and 70 per cent of births are conceived out of wedlock in the black population as opposed to 22 per cent for whites (Oswald, 2001: 45). The plight of poor Afro-American youth, and males in particular, is indeed much worse. Unemployment rates in inner cities, particularly those in the so-called 'rustbelt' of former manufacturing cities, conservatively exceeds 40 per cent for black youths (Kelly, 1997: 46–7). And Afro-American youth under 20 years of age are four times as likely to die by murder than white youth, and statistically they are more likely to go to prison than to go to university (32.2 per cent of young black men in the USA are either in prison, on parole or on probation (Cashmore, 1997: 3–4)).

Many black families and youths are concentrated in large cities, confined spatially to decaying inner-city areas and urban ghettos. Robert Taylor Homes (also known as

'The Hole') in South Side Chicago, although now undergoing demolition, was a classic example of such a slum, with 96 per cent unemployment, an average annual income of $5,905 and rife with gang warfare and drug abuse (Vulliamy, 1998). Deindustrialisation, and the flight of both white residents and businesses to the suburbs and hinterland USA throughout the 1980s and 1990s, has taken its toll on many of these formerly 'working cities', leading to decay, neglect and dereliction. Cuts in public expenditure in areas like recreation and the growth of privatised urban spaces have meant that the street is the only place left for many black youths to occupy (Kelly, 1997: 50). While a long tradition of residential segregation existed in many US cities prior to economic downturn (Sugrue, 1998), the current economic situation has made racial polarisation even more apparent.

There is also a long history of black subjugation in the labour market ranging from the transition from slavery to 'free labour' in farm and factory work, to the more recent shift from the decline of manufacturing to the rise of service work (see Jones, 1998). A more recent history has been aptly documented in Sugrue's (1998) study of race and inequality in post-war Detroit. This work shows how, from a boomtown in the 1940s and home to some of the highest-paid blue-collar workers in the USA, Detroit has become a city racked with population migration, decay, unrest, high black youth unemployment and racial segregation. The first generation of out-of-work black youth in the city began in the late 1950s/early 1960s, and anger and despair exploded in the Detroit riots of 1967. Black youth had a particularly important role to play here. As Sugrue (1998: 260) argues: 'Growing resentment, fuelled by increasing militancy in the black community, especially among youth, who had suffered the brunt of displacement, fuelled the fires of 1967.'

While some piecemeal measures (i.e. job training programmes) were put into place to try and deal with the employment problem, the riots were a prophetic reminder of what was to come later in cities like Los Angeles in the 1990s. A continuing lack of jobs for young blacks in many US inner cities meant that crime, particularly drugs, prostitution and theft, increasingly became the mainspring of the local economy (Kelly, 1997: 46–9; Jones, 1998: 378). Of course, the hiring practices of the new service economy are not exactly favourable to Afro-American youth. Jones (1998: 378) cites a survey undertaken in the late 1980s of employers' hiring practices in Chicago; this saw 'young black job seekers as too poor, uneducated, and temperamentally ill-suited for the rigors of modern office work'. Young black men in particular, with their increasing association of crime, violence and drugs, were seen as a particularly bad risk by employers, thereby squaring the discrimination circle.

Against this backdrop, rap and hip hop have emerged as potent black cultural forms and one of the world's most widespread youth articulations. Initially, areas suffering chronic economic depression were major sites for rap's consumption and production. In this sense, rap and hip hop are intensely urban in origin and importantly emphasise an intense sense of place and community, epitomised through metaphors of the 'ghetto', the ''hood', 'homies', and shout-outs to various 'posses' (Forman, 2002). Forman (2002: 8) goes on to describe how hip hop and rap, through these various spatial discourses, have been central for young black people in their attempts to construct spaces of their own. Rather than being rootless, then, hip hop and rap cultures transmit a strong sense of place.

However, rather than occupying central visible locations or permanent venues, the street and its associated cultures of tagging, b-boying, break dancing and graffiti art acted as key sites for hip hop (Kelly, 1997). Such ghetto cultures segregated from the surrounding city were largely seen as dangerous and unruly, and hence were frequently the object of surveillance. Mobile sound systems in parks and at street corners were also crucial places for the development of hip hop, as were more private spaces of the car, gym hall, bedroom and black parties. The streets, then, as one of the only spaces open to many poorer black youth, have sometimes been viewed as a potentially resistant force (Pimlott, 1995). Giroux (1996: 31) holds on to the possibility that such 'cultures of the street' potentially represent a new kind of postmodern hybridised cultural performance, resistance and politics. Many of the early meanings of rap and hip hop, especially those associated with 'message rap', reflected such concerns and appealed for greater cultural and political awareness of the plight of black urban America (Forman, 2002: 82). From such informal places, a more legitimate black hip hop club scene grew up in the late 1970s, especially in New York, which more widely disseminated black cultural styles within nightlife cultures (Forman, 2002). Hence, rap and hip hop have spawned an underground network of clubs, bars and radio stations in urban areas across the world.

However, this cultural and spatial articulation has also resulted in quite negative images. A potent combination of youth, race and class has made rap and hip hop the dangerous 'other' of respectable American society. Moreover, the term 'underclass' in America is a racially coded word for black youth, crime and violence, and illegitimacy (Giroux, 1996; Lott, 1999). Of course, cultural representations of Afro-American black youth in the press and in popular culture (such as films, TV, music, etc.) have partly fuelled such stereotypes. Giroux (1996) cites films such as *Juice*, *Menace to Society* and even the more sensitive *Boyz in the Hood* in this regard, not to mention the way media reaction to 'gangsta' rap increasingly represents it as a growing culture of violence, misogyny and nihilism rooted in drugs, urban decay and black masculinity. Rap and hip hop have long been blamed for creating violence, with rap artists themselves becoming central players. A number of widely publicised media examples have spread the perception that violence is endemic in the growing club infrastructure which has emerged to serve rap and hip hop fans. For example, several shootings in nightclubs in the USA have left people dead, with a number of high-profile rap artists implicated – for instance, Sean 'Puffy' Combs was charged with bribery and illegal gun possession in a shootout at Club New York in 1999, and

Eminem was charged with carrying a concealed weapon in a Detroit nightclub (Myers, 2001: 3). Similarly, several clubs in the UK have also been closed or raided because of an escalation in violence and their associations with black music such as jungle and hip hop. Black youth cultural spaces, then, continue to be criminalised and marginalised from central areas, due mainly to their ongoing, rather problematic, associations with violence, guns, drugs and gang-related activity.

The commercialisation of rap and hip hop has also been lamented as it mutated from an underground, street-based musical form which was the preoccupation of an urban minority into an institutionalised, global, multi-billion-dollar industry engaging people from a variety of ethnic and class backgrounds. Successful crossover artists, global media such as MTV, and its cool status as a youth cultural pastime have allowed rap and hip hop to penetrate deep into popular culture. Initial metaphors such as the 'hood have been appropriated and diluted, now often referring to upscale, gentrified enclaves (Forman, 2002: 343). Further, Cashmore (1997), in his study of the black culture industry, suggests that the white power structure inherent in film, music and TV has either effectively appropriated and commodified black culture in the USA, thereby muting its political potential, or has continued to reproduce stereotypical images, hence keeping the racial hierarchy intact. Forman (2002) also points out that, from its origins, middle-class black entrepreneurs and their white peers were responsible for the production and recording of hip hop, while Kelly (1997) suggests that rap should not be romanticised as the authentic voice of black ghetto youth, arguing that it does little to challenge capitalism or raise black incomes (except for the few success stories). In a somewhat different vein,Cashmore (1997: 171) argues:

> Rap transformed racism into fashion: something that blacks wore to impress and whites liked to glare at without actually doing anything ... What started as a radically different, and in many ways dangerous, music was appropriated, domesticated and ultimately rendered harmless.

Rap music in this scenario becomes a 'culture of compensation', substituting for educational advancement and literacy (Gilroy, 1993).

Lott (1999), in distinguishing between hard-core and commercial rap, suggests the former does have the potential for resistance, particularly in the way it plays out knowledge of surviving on the street and dealing with social exclusion rather than believing in social mobility into a white man's middle-class world. At the same time, he suggests that there is no such thing as 'authentic' resistant black culture (also see Kelly, 1997), which is somehow separate from the wider white mainstream culture which surrounds it. Black music forms, like rap, for him represent a 'recoding' of dominant white values about individualism and hierarchy (survival of the fittest), misogyny and violence which lie at the heart of mainstream American culture.

This is a crucial point in relation to how rap and hip hop have been taken up and rearticulated with other cultural forms and traditions elsewhere (Mitchell, 2001). As Mitchell (2001: 1) argues, 'Hip hop and rap cannot be viewed simply as an expression of African American culture; it has become a vehicle for global youth affiliation and a tool for reworking local identity all over the world.' Of particular interest is its repercussions in the UK (see Hesmondhalgh and Melville (2001); also Swedenburg's (2001) discussion of its articulation with Islam in the UK in the same collection). Hesmondhalgh and Melville (2001) suggest that the appropriation of US rap and hip hop in the UK needs to be situated within the context of a pre-existing Caribbean sound system culture and the evolution of British club culture. The UK has a long history of black 'blues party' culture imported from the Caribbean and a strong tradition of reggae-influenced sounds, not to mention Asian influences (exemplified by the Islamic inspired hip hop group, Fun-Da-Mental, among others), which have been fused with rap and hip hop imported from the USA. The continued mutation of rave and club cultures in the UK also took elements of hip hop and mixed them with techno, reggae and ragga to create jungle and drum and bass, while its fusion with various forms of psychedelic styles produced 'trip hop' associated particularly with the 'Bristol sound' of Tricky, Massive Attack, Roni Size and Portishead (Hesmondhalgh and Melville, 2001: 98–106). The impact of US rap and hip hop in the UK, then, has been far-reaching, albeit through the transformation and combination of the musical form through national and local music traditions and experiences (Back, 1996).

Finally, hooks (1994) argues that rap needs to be located in relation to wider structures of power – which for her is white supremacist, capitalist patriarchy. Young black men don't live in a cultural vacuum, hence one shouldn't be surprised to find that elements of their culture are a specific embodiment of wider norms: 'The sexist, misogynist, patriarchal ways of thinking and behaving that are glorified in gangsta rap are a reflection of the prevailing values created and sustained by white supremacist capitalist patriarchy' (hooks, 1994: 116). She argues that gangsta rap is utilised by the industry to accumulate capital and stir up controversy to sell records (largely to white young male audiences), while at the same time acting to demonise black youth culture as steeped in violence and misogyny. hooks is careful to make clear that this doesn't mean black male youth are not, or should not be somehow accountable. In fact, in the summer of 2001, a rap and hip hop summit encompassing black leaders, rappers and record companies met to consider the problems and the future possibilities of the genre under the title 'Taking Back Responsibility' (Younge, 2001; Sturges, 2001). hooks is suggesting that we need to widen our view of the 'problem' of certain forms of male black youth culture, and not just replicate 'moral panics' around these cultures. Interestingly, rap and hip hop are among the few marginalised youth night cultures to reach and invade public consciousness. The key issues are what impact these youth cultural forms have on the ongoing politics

of the street, how they are taken up in other cultural contexts (Mitchell, 2001) and what space they have within contemporary nightlife, as well as questions about their incorporation and commodification.

Cleaning windows and cleansing the streets? Squeegee kids and the Ontario Safe Streets Act

In the late 1990s, numerous Canadian cities were gripped by a new 'moral panic' on their downtown streets – unemployed youngsters armed with squeegees, a sponge with a handle attached for cleaning car windows (Tyyska, 2001). Accused of 'tarnishing the image of the city' (see O'Grady *et al.*, 1998 for a critique) and branded as 'thugs' and criminals by prominent politicians (Spears, 1999), the squeegee kids quickly achieved the status of Canadian 'folk devils'. Furthermore, the streets at night were allegedly filled with these unwelcome visitors, who congregated in parks, street corners and alleyways, allegedly wasting their money on drink and drugs. Rather than welcoming their crude attempts to survive in the harsh neo-liberal Canadian economic climate by actually 'making work', many motorists apparently abhorred these young people invading their private automobile aura with their hand outstretched expecting some remuneration. The Toronto BMW brigade clearly were more concerned about having their paint scratched or having to wind down their window to part with some change, and hence complained long and loud about the phenomenon, particularly to the popular press. Similarly, no one seemed to be asking questions about why such a group was confined to hanging about the streets at night, rather than being able to access downtown facilities, because of their appearance and status.

As Howard Becker (1966) has long argued, there are few wholly deviant acts in society – only those that are perceived as deviant and defined as such. One might, for instance, imagine a university or charity fund-raising event which engaged in exactly the same kind of behaviours as the squeegee kids, albeit for some benevolent cause, which would be wholeheartedly and good-naturedly supported by the wider community. No such kindheartedness, however, could possibly be extended to those kids scraping for a living. Such an exchange meant, of course, actually being exposed to the 'problem' of homelessness and poverty – breaking the sanitised nature of separating rich and poor in society to avoid this uncomfortable issue. Additionally, these people were also primarily young – a second strike against them (Tanner, 1996). Technically, they should have been in school or at home, not destitute on the street or drinking in parks, reminding Ontarians that poverty, homelessness and unemployment do indeed exist in one of the so-called richest countries of the world (see Hagan and McCarthy, 1997).

Furthermore, these were not clean-cut University of Toronto graduates collecting money for charity, but were apparently grubby and strange-looking street

punks trying to legitimise a 'free handout'. For example, consider the following opinion of squeegee kids by a Toronto driver, posted on a website:

> You're driving to work on an average day enjoying the sights as you approach a stop light when you are all of the sudden looking face to face with some punk looking through your windshield as he or she attempts to clean it with a squeegee. Clad in black or other grubby attire from head to toe, bearing chains and other metal objects both hanging off of and somewhat attached to their body, they finish their job and stand there waiting for you to hand them some money.
>
> (Censorfreeworld, 1998)

The key issue expressed here appears to be an abhorrence of coming face to face with Canada's young underclass in their full youth cultural attire. As such, there was only one way to tackle such an affront to the country's glorified car culture, gentrified nightlife and 'respectable' (moneyed) citizens, and that was to crack down on such perpetrators and literally sweep them off the street. As such, the Ontario provincial government forwarded a large collection of bills under the 'Safe Communities' umbrella that demonstrated its commitment to being tough on young people, street life and crime. For example, the Safe Streets Act banned squeegee kids from so much as stepping on to the road. The government has also passed legislation requiring permits for raves, and the Attorney General has written numerous letters to the federal government condemning the Young Offenders Act for its leniency, despite Canada having one of the highest rate of youth incarceration in the developed world (Winterdyk, 1996).

The Safe Streets Act makes squeegeeing and aggressive solicitation illegal, and allegedly protects people's ability to use public places without intimidation (excluding, of course, the squeegee kids themselves). In other words, the Act makes it illegal to give 'any reasonable citizen cause for concern', day and night. According to government sources, the Safe Streets Act was introduced in response to concerns from police, local officials, residents and businesses (Ontario Government, 1999a). In addition to making it illegal to approach a motor vehicle for the purpose of offering a service, and empowering police to arrest all but the most supplicant beggars, the Act makes it against the law to ask for money when under the influence of drugs or alcohol, outlaws ticket scalping and other forms of 'aggressive soliciting', and makes it illegal to dispose of condoms or syringes in parks or school yards (ICFI, 2000). Punishment under this legislation can mean up to $500 in fines, and a subsequent offence can lead to up to six months in jail (Ontario Government, 1999b).

Despite a series of protests and appeals from various civil liberty (Canadian Civil Liberties Association) and anti-poverty organisations (Ontario Coalition Against Poverty (OCAP)), the law came into effect in November 1999 and subsequently an

Ontario court judge sentenced thirteen 'squeegee kids' for contravening it. They were charged in Toronto with aggressive solicitation, approaching stopped cars at intersections and cleaning their windshields in hope of payment. Toronto's estimated 25,000 homeless people had previously been targeted by an earlier drive in the summer initiated by the mayor of the city, Mel Lastman, and his Community Action Policing programme, to crack down on squeegee kids and aggressive panhandlers in order to clear them from the streets of one of Canada's wealthiest cities (ICFI, 2000). With the new legislation and the cold Canadian winters backing their campaign, the city can rest assured that its attempts to 'clean up the streets' will be successful. As one of the police inspectors involved in the clean-up admitted: 'The best crime-fighting tool we have is minus-30 in February' (*Globe and Mail*, 26 July 1999: 10).

The Ontario Conservative Party, in power since 1995, has waged a campaign to stigmatise the poor, slashing welfare benefits by more than one fifth. Subsequently, the government cancelled all support for social housing construction and instituted a mandatory 'workfare' programme for welfare recipients, depicting them as cheats, drug abusers and fraudsters. They have also announced plans to force all welfare recipients to undergo mandatory drug testing. Hence, there is a growing correlation between the increase in homelessness and panhandling and the cuts implemented in welfare, unemployment insurance and social housing and the closing of mental health facilities. But phenomena such as the squeegee kids are also rooted in family breakdown and violence and other social problems that are indicative, although in a less direct fashion, of the tensions and alienation produced by a society wracked by social inequality and economic insecurity (ICFI, 2000).

Shut out of most mainstream institutions, including retail stores, bars and coffee shops, street kids in Canada are even being denied access to the only space they still have – the street (Hagan and McCarthy, 1997). Much of the crackdown on public drinking is in effect a direct attack on the only socialising spaces left to groups like the squeegee kids. Drinking and drug-taking in the shadows of public parks, alley-ways and car parks are the only residual nightlife spaces left to those excluded from all other social institutions. Hostels and squats may offer some refuge, but it is the street where most of these young people live out their daily lives. With few alternative programmes available, resistance against attacks on this space, to make them 'safe' for 'reasonable' and 'respectable' citizens, are growing.

A 'Squeegee Work Youth Mobilization' programme to teach them to get jobs repairing bikes was instituted as an alternative, but, as reported, many of the young people are wary of official agencies and 'make work' programmes (*Globe and Mail*, 27 July 1999: A9). More formal resistance has taken place, namely the 1998 'Hands Off Street Youth' march, jointly organised by anarchists and the Ontario Coalition Against Poverty. And in 1999 OCAP organised a few-hundred-strong occupation of Allen Gardens Park, which had been an earlier target of the Community Action Program, where police cleared homeless people from the street. This was followed by a demonstration, in the summer of 2000 organised by OCAP and allies from unions and community groups, which ended in a full-scale police riot (Shantz, 2001). In the interim, the fate of the squeegee kids appears to have been largely sealed. Legislated off the street, robbed of their livelihood and shunned by mainstream nightlife provision, they have been even further pushed back off the streets into the shadows of the increasingly gentrified city.

In conclusion, then, residual nightlife is by its very nature marginal. Either elements of it are commercialised and absorbed by the dominant culture, or it is viewed with disdain as a feature of a past that conjures up uncomfortable images of dereliction, poverty and moral decay. Many young people are similarly stereotyped in a highly negative fashion, through labels such as 'street kids', 'schemies', 'gang bangers', 'druggies' and 'criminals' or under more general terms like 'the underclass' or 'the socially excluded'. In the post-industrial city, residual spaces are viewed as potential sites for renewal and redevelopment, and marginalised groups are seen as easy fodder for 'make work' programmes or as targets for the latest street cleansing legislation. In light of such negative images and subordinate to the drive to gentrify the night-time economy, what resistant potential do marginalised youth cultures, like black and working-class youth, the homeless and the socially excluded in general, have? Clearly, part of their potential actually rests on their location outside the mainstream – reminding people that poverty and hardship remain and that not all youth are middle-class gentrified consumers. They are a thorn in the side of capitalism, which continues to expound 'trickle down' theories of wealth, general affluence and classlessness, despite evidence to the contrary.

Yet part of what is going on here in the regulation of various parts of residual nightlife is an 'out of sight, out of mind' exercise. The idea is that the issue will in effect be solved if you remove the 'problem' physically, whether it be renovating a derelict area of town, redeveloping a run-down bar or pub, forcibly removing the local 'undesirables' from the street (Jeffs and Smith, 1996), or ghettoising whole social groups into a particular neighbourhood or estate. Some of these approaches are very direct and, in effect, highly discriminatory, while others are much more subtle and understated. In the case of the Safe Streets Act in Ontario, one can literally see marginal youth being 'swept off the street'. On the other hand, the containment of Afro-American youth to their neighbourhoods and their engagement in largely masculine and misogynistic cultures may work to more subtly incorporate acts of resistance and quell political dissent.

In Chapter 5, we suggested how the continuing gentrification of mainstream nightlife is leading to higher levels of corporate control, alongside the increased regulation and sanitisation of youth cultural experiences of going out. The social polarisation of youth into 'good' gentrified nightlife consumers and those marginalised to 'the dark side of the street' is a visual expression of this process. While such changes are rapidly unfolding, there is plenty of disillusionment and resistance. The future role for residual nightlife spaces is part of the unfolding 'politics of nightlife'. Residual forms need to connect up to more active and alternative political acts and movements. If not, they will be likely to fade into the distant past, as they are incorporated, renovated and literally swept away by the new urban landscapes of pleasure.

9

'YOU'VE GOTTA FIGHT FOR YOUR RIGHT TO PARTY'

Alternative nightlife on the margins

As we have outlined earlier in the book, even though corporate ordering is a dominant feature of urban nightlife, it still contains many oppositional places and hence remains inherently transgressive and resistant. While some alternative spaces are simply more bohemian versions of mainstream culture, others openly identify themselves as oppositional. It is this nightlife 'on the margins', encapsulating both play and resistance, which we consider in this chapter.

All urban areas contain marginal spaces in which alternative cultures can grow. Such marginal nightlife spaces come in many guises from working-class cultures (Taylor *et al.*, 1996); gay areas (Whittle, 1994; Castells, 1982); ethnic communities (Rose, 1994); community centres and arts venues; the rave and 'underground' dance scene (Redhead, 1993; Collins, 1997; Chan, 1999); DiY and 'doof' cultures, and direct action and squatters' movements (McKay, 1996, 1998; Chatterton, 2002b; Bey, 1991; Wates, 1980; Corr, 1999; St John, 2001a). San Francisco's Mission, Haight Ashbury and Castro districts, Ibiza's and Ayia Napa's island dance scenes, and Barcelona's squats have all at some point featured in the landscapes of the alternative or bohemian. In this chapter we primarily reserve our discussion to forms of nightlife that involve intentional acts of resistance – which define themselves explicitly against the corporate 'other', the commercial, the mainstream. In particular, we look at how the 'independent' mode of nightlife production distinguishes itself from the dominate corporate form, and explore some examples from the rave and squatting scenes. Our argument is that while nightlife alternatives do exist in the margins, and offer models of resistance, the current tendency is that they appear to be squeezed further to the edges of the city by the growth of the corporate nightlife machine.

Understanding alternative nightlife on the margins

Resistant and marginal spaces are part of urban nightlife, despite often being sidelined by dominant narratives of the city. As a starting point, what do we mean by alternative and marginal nightlife? Alternative nightlife is often defined in opposition to and distinct from the 'mainstream', while the margins form the geographical edge of the centre. To begin to unpack alternative nightlife on the margins, we can first think of it through its physical location. For many reasons, such as lack of consumer financial strength or self-preservation, alternative nightlife spaces are often found in 'fringe locations'. This location is a vital part of the self-regulation of the alternative scene, as it ensures that attendance is based on desire and empathy rather than chance. One dedicated traveller to a Goth night (Beth, 20 years old, Bristol) suggested that she felt 'a breed apart because you have to go quite a long way, you have to be quite dedicated'.

Margins are an important part of social and spatial ordering. They have a binary relationship with the centre – without margins there can be no centre, and vice versa. Clearly, dualisms such as centre and margin, and associated traits such as order and disorder, simplify the fluid and conflictual nature of urban life. Indeed, defining the boundaries between the mainstream centre and the alternative margins is an act of power in itself. For example, boundary drawing is often undertaken by property speculators, planners and marketers. Most people also accept and reinforce this spatial logic and choose to stick to their 'own' territory, often for reasons of self-preservation. Others invade, challenge or subvert such geographical imaginations, engaging in a more fluid sense of space in the night-time economy, inhabiting both the mainstream and the margins.

However, alternative nightlife spaces are more than physical spaces, they are also socially constructed, collectively imagined and ideologically defined. The margins evoke a strong 'sense of place', which goes beyond bricks and mortar. Clearly, both the mainstream and the margins are historically and socially contingent spaces, both continually reconfigured according to shifting cultural norms and styles. Today's margins can rapidly become tomorrow's centre, especially as alternative, deviant and counter-cultural fashions are commodified and the underground is pushed overground (see Plate 9.1). Neat divisions such as those between the commercial and the mainstream, the marginal and the resistant, are continually blurring and changing (Hetherington, 1997). As Negri and Hardt (2001) point out, there are now few outside spaces or margins left, so it is imperative to be resistant from within the heart of the capitalist empire.

Defining the margins also implies the use of a particular set of spatial and social strategies by a variety of groups and players. For many participants of the underground, a corporate 'other' is necessary to define their scene against, and in locations where the commercial centre is strong, marginal underground scenes are often thrown into stronger relief and given impetus. Yet even the mere presence of an underground scene can legitimate the introduction of restrictive legislation from the state. Various media also play a key role here. Thornton (1995) outlines the mass media as the main antagonist of the underground, epitomised by coverage of the acid-house, Ecstasy-led rave scene in late 1980s England

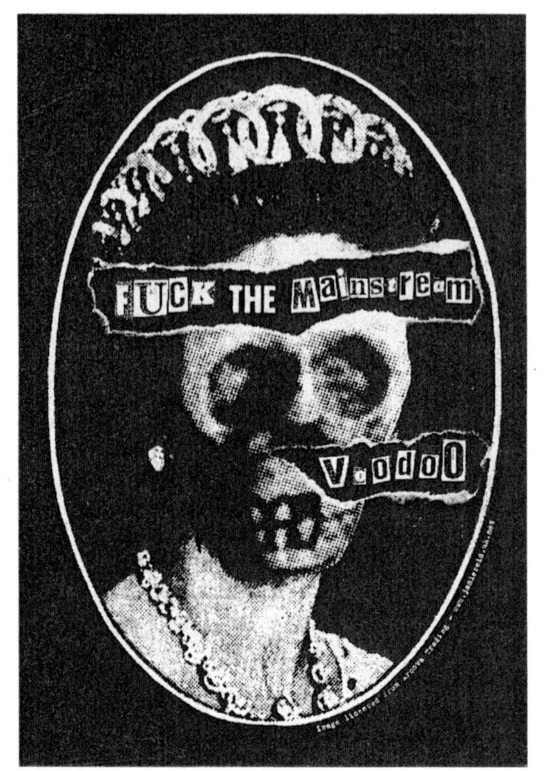

Plate 9.1 Creative dialogue on the under-overground; nightclub flier, Voodoo Nightclub, Liverpool, UK, 2002
Source: Arcova Studio

which vilified young people and induced various moral panics. However, dedicated micro and niche media such as specialist magazines, zines and fliers also play a key self-support role in marginal cultures.

The 'other places' of the margins offer us alternative perspectives or ways of seeing (Hetherington, 1996). Such places play an important role for those who live outside the norms of society, especially through the often romanticised spatial imaginaries of certain groups (Shields, 1991). Many alternative places (Stonehenge, Glastonbury, Woodstock) adopt a mythical status for individuals and subcultures and, as sites of 'social centrality', play a key role in identity-building (*ibid.*). However, such place images can easily be overturned. Longstanding, affordable and sacred nightlife spaces on the margins – a cherished bar, an old restaurant, a corner kiosk, a seedy club, a squat – have been bulldozed to make way for the expansion of corporate culture, literally turning such places into dust through the imperatives of growth and profit (Solnit and Schwartzenberg, 2000).

An alternative viewpoint among participants is a key part of marginal spaces. The margins are places where people on the fringe of society can find a space to articulate themselves (Hetherington, K., 1998). They are often distinguished by shocking, 'out of place' bodily appearances, often displaying an anti-aesthetic, setting them apart from the respectable, fashionable mainstream. Moral panics abound from such appearances, centring around the deviant activities of anarchists, squatters, artists, musicians, drug-users and punks, and other 'non-productive' groups not willing or able to join mainstream consumer and working society. The fringe, then, contains the shadowy and threatening 'other' of urban nightlife. They are part of the unspoken social norms of the segregated city. While many people naturally gravitate to such places, as they feel enfranchised by particular musical or dress codes (or, indeed, lack of them), others, through their own internalised 'sense of limits' (Bourdieu, 1984), exclude themselves by making judgements that such places do not contain 'people like me'. Denizens of alternative nightlife spaces, then, have their own sets of dispositions and practices, or a 'habitus', which differentiates them from the mainstream (Thornton, 1995).

The margins are thus often seedbeds for counter-hegemonic or resistant ideologies (Hetherington, 1997) which involve a commitment to an identity, community or 'cause' as well as an attachment to particular places. Such ideologies, however, come in many different forms. Pile and Keith (1997) have stressed that resistance occurs in many more situations than just those where there is a dichotomy between 'enemy' and 'victim' strung across (pre)defined structured power relations. In this sense, they suggest there needs to be a reconsideration of the relationship between domination and resistance, and that both are 'contingent, ambiguous and awkwardly situated' (*ibid.*: 3). In particular, too literal readings of the margins may regard it as always in strict opposition with mainstream culture, when many aspects of the underground creatively appropriate, plunder, co-exist, and/or simply ignore it (Slater, 1996). However, the whole idea of what counts as 'resistance' needs to be understood alongside the notion of intentionality (Creswell, 1996). In the context of nightlife, this throws up distinctions between actively oppositional cultural forms and more individualised variations in the mainstream, such as an independently run, chic style bar.

On one level, then, for many people resistance evokes images of overt and often heroic acts of defiance with the intention of struggling against identifiable power structures such as capitalism, patriarchy, commercialism and globalisation. Some groups embrace marginality and see it as a backdrop to being provocatively different or confrontational. The growth of radical environmental protests is a great impetus for such resistant lives, with many people, not interested in aligning to mainstream political parties and tired of waiting for change, simply 'Doing it Themselves' (Brass and Koziell, 1997). In the night-time economy, some groups have followed such trends (see Chatterton, 2002b), developing explicit and overt political agendas to resist: for example, staging illegal warehouse and forest parties, and squatting.

However, resistance is often more mundane and pragmatic. Many people simply transgress and disrupt in their own way during their everyday encounters with mainstream commercial culture. Resistance also entails a host of more localised 'backstage' actions, through the use of micro-tactics at the everyday level (Thrift, 1997; De Certeau, 1984), or through alternative conceptions of the self and of play

(Malbon, 1999). Of relevance here are the 'ephemeral spatialised tactics of rebellion' which Flusty (2000) describes in relation to skate kids, street buskers and peddlers evading the watchful eye of downtown security guards. Skateboarding, in particular, is an implicit critique of the limited design and uses to which we put our contemporary cities (Borden, 2001: 173). In some cases, the criminalisation of skateboarding has led to a radicalisation of street skaters into 'pavement commandos' who fight for the right to skate unharassed (Flusty, 2000). Such acts of 'transgression', then, question appropriate codes of behaviours attached to place, and hence the margins often seem 'out of place' in relation to the 'taken-for-granted' invisible boundaries within social life (Creswell, 1996). While such acts are often not overt statements of opposition, they are a semi-permanent feature of the urban landscape which sometimes win small concessions or remind us of the arbitrary, and often banal, norms and rules which entwine urban space (Flusty, 2000). A context such as the city at night is replete with such acts of transgression, often mixed with pleasure-seeking.

On the margins, people are often actively engaged 'auteurs' in the making and remaking of the urban. These are Lefebvre's (1991) spaces of representation, which are directly lived, imagined and reinvented. The sheer scale, in terms of both physical form and interaction, makes cities sites of tremendous possibility. As Castells (1982) pointed out, cities are unfinished products and are the result of an endless historical struggle over the definition of meaning by a variety of opposing social actors. The history of cities, and thus of their nightlife, can be read as an interaction of the dominant commercial urban project, with alternative meanings and visions. Alternative nightlife is very much part of a desire to make an 'other' city, a free city, a wild city, a people's city. Many groups, from squatter movements to arts collectives, are busy exploring strategies for building a range of other 'possible urban worlds' (INURA, 1998). The margins, then, are an appeal for autonomy and self-management outside the influence of big business, government and formal planning. Many individuals and groups have envisioned alternatives based upon new relationships between space and society, challenging prevailing cultural values and political institutions, and exploring new meanings. Urban nightlife is not a mere product of planners, architects and bureaucrats (Borden *et al.*, 2001), and different/marginal groups continue to be involved in its creation, both materially and symbolically.

Beyond conscious acts of resistance, the alternative is also about tolerance, diversity, acceptance or echoing the punk ethic, just being who you are (Muggleton, 2000). It is a space for expressive identities which are less bound by the rigours of fashion and social protocols (Hetherington, 1998). The fringe is a chaotic, unstable, fluid space, an unbounded 'space-between' (Hetherington, 1997: 27) which brings together a collection of unusual things, and contains unsettling juxtapositions and alternative modes of social orderings. Many, often conflictual, identities pass through such spaces, and so while it has its regulars and 'insiders', it is also open to 'outsiders'. Berman's (1986) distinction between 'open-minded' and 'closed or absent-minded' spaces is useful here, with the former having relevance to the spaces of the margins. The concepts of the liminal and the carnivalesque also have relevance to the margins, in that a reversal of normal social roles and a transgression of normative behaviour become possible, and in some cases established practices are replaced with new ones. Such places provide openings for more affective, if looser-bound, neo-tribal forms of identification and lifestyle (Shields, 1991; Maffesoli, 1996), or what Hetherington (1998) has called 'expressive identities' and 'emotional communities'.

While alternative places are often 'melting pots' for a range of marginal groups, they are also characterised by a desire, however fleeting, for affectual solidarity and togetherness (Turner, 1982). Many groups on the margins have come together through disillusionment and frustration with mainstream culture, and represent attempts to recreate a sense of belonging, sociation (or 'bund', according to Hetherington, 1996) and 'authenticity'. Hetherington (1998) comments that notions of belonging among 'outsiders' has led to the development of distinct 'structures of feeling' for many alternative groups. The so-called 'death of society' during Thatcherism and Reaganism, along with new legislation in the UK such as the 1994 Criminal Justice Act, has strengthened the need for such ties of solidarity and togetherness on the margins. Faced with insecure employment and housing, many young people have inverted the selfish individualism of modern consumer culture to create their own, more self-sufficient 'mini-communities'. Solidaristic rituals and means of communication, such as zines, print and web media, gatherings and collectives, are constantly evolving on the fringe to offset the alienating excesses of mainstream culture. Many commentators used the emergence of 'tribal' youth cultures over the 1990s to point towards temporary and highly emotionally bound groups, from sound system collectives, squatters, road protest and peace camps to GM-crop trashers and Reclaim The Streets collectives. While labels such as 'tribal' often have more to do with media hype than sociological rigour (although see Maffesoli, 1996; Bennett, 2000), they do point to groups who, in their own ways, have attempted to create what McKay (1996) has called 'cultures of resistance'.

Finally, nightlife on the margins is also driven much less by commercial styles and music and fads of mainstream culture, blending a variety of uses such as live music, socialising, performing arts, drinking and eating. Marginal nightlife spaces take their inspiration from subculturalists and specific communities such as gay and ethnic minority groups, or alternative musical styles such as hip hop, Goth and post-punk (Bennett, 2000). In this sense, there are important experiential differences between the alternative and mainstream. As one person told us (Joe, 23 years old, Newcastle upon Tyne): 'I think there is no sense of, say, fashion consciousness. It's almost like anti-fashion – people are either making music or listening to music, like non-radio music.' The atmosphere and clientele, which is often described through labels such as 'dark', 'seedy' and 'funky', is vital to the margins.

Clearly, it is important not to over-romanticise such places. While the margins may display strong social and emotional bonds and a feeling of togetherness, they also have their own barriers to entry, stylistic codes, regulations and forms of exclusion. As mentioned in Chapter 4, what Thornton (1995) calls 'subcultural capital' also acts as the lynchpin of an alternative hierarchy which allows those 'in the know' to repel and distance themselves from the 'unhip' mainstream. Moreover, many people on the fringe, because they do not feel part of mainstream society, often consciously build defensive and exclusionary spaces between themselves and the outside world.

The independent mode of production: 'lifestyle over profit'

The margins are based upon a series of beliefs among producers and consumers which set them apart from larger, corporate nightlife operators and include: a greater emphasis on 'use' rather than 'exchange' value; a more prominent role for individual entrepreneurs and collectively run spaces; less emphasis on profit; a more fluid relationship between consumers and producers; and more self-regulatory forms of policing. Such beliefs are described through any number of monikers: 'ethically', 'socially' or 'ecologically aware', 'anti-consumerist', 'pro-lifestyle', 'pro-community' or 'pro-people', rather than 'pro-profit'.

Many alternative places have developed an atmosphere based upon 'a cult of the individual', in which one entrepreneur has catalysed and inspired people and a deep affiliation between the venue and a group has developed. As one DJ from Newcastle upon Tyne commented to us: 'You can't underestimate the power of a person. That person's kind of influence on every aspect of what you experience in that place, and it sounds very corny but it is giving a soul to a place' (DJ 2, Newcastle upon Tyne). Thus, unlike large publicly quoted companies who are ultimately answerable to a board of directors and have a 'fiduciary duty' to shareholders and the stock market, smaller operators are more answerable to themselves and their customers.

Indeed, many independent operators establish bars in order to provide somewhere for themselves, and their consociates (Muggleton, 2000), to socialise. They often feel that they pursue a different philosophy compared to larger commercial operators, in which the venue is part of their lifestyle rather than a vehicle for maximising profits. As one independent bar operator from the north of England eulogised:

> We are one of the few places in the city centre that's offering anything ... that isn't Bar Oz [an Australian theme bar]. And we're not driven by fads. We've got our own rules. And it feels, at times, like running a different country. It feels like you're running an independent nation on the weekend. And everyone's having a good time and it's great. And that's the

> buzz of it for us. I like what I do and I wouldn't sell it for all the money in the world really. I'm far more interested that my children will grow up seeing me doing something that mattered than I am to take a big dollar off someone. Lifestyle over profit every time ... you have to give the people who come here something ... Almost like religion. And you feel, like, 'Ah, that's better, now let's get on with the struggle of life', kind of thing. It's a very romantic notion as opposed to the big-business things that people do.
>
> (Bar owner 4, Newcastle upon Tyne)

Creating an alternative vibe has indeed become one of the aims of some large corporations, eager to compete with their rivals in the independent sector. Corporate operators often track alternative venues hoping to predict new trends in nightlife, often employing 'coolhunters' (Klein, 2000) to extract the essence of 'subcultural capital' (Thornton, 1995) and replicate the 'organic' elements of what makes a great venue. However, the limitations to this process of replication are all too evident, as one music promoter located in Bristol commented to us: 'If you're a big pub and design your pub on market research you are reacting to taste, you are not being creative' (DJ 2, Bristol). Another small-scale promoter from the UK suggested that this 'edge' was not something that could be simply manufactured:

> It is just about having that street sort of edge to what you do and you cannot really define it. People get hundreds of thousands for doing that in London to give brands that kind of thing, and that is because it is, you cannot just write a paper on it and give it to some big company and say do this in your bar and you are going to be cool, man. You cannot do it. It is just intangible.
>
> (DJ 3, Leeds)

What the above comment suggests is that a good venue vibe does not stem from market research or focus groups, but flows from a deeper connection between the owners, staff and customers.

An important feature of the independent mode of production is the challenge it offers to the traditional producer/consumer divide, a split which is often much more explicit in market-based forms of leisure and consumption. On the margins, participation is more about 'active production' than 'passive consumption' and hence there is a more fluid boundary between producers and consumers through the exchange of music, ideas, business deals and networks of trust and reciprocity. DJs, musicians and record label owners have close associations with certain independent nightlife spaces to create an 'authentic' outlet for their work and a meeting place for like-minded people. In turn, clubbers, journalists, reviewers and wanna-be artists visit the venue, hoping to make connections or listen to the music they can't hear elsewhere.

Marginal spaces in this sense are a source of creative innovation, revelling in a desire for novelty, conflict and dialogue. They often form the basis of more localised nightlife production–consumption clusters (Santiago-Lucerna, 1998), characterised by an independent and publicly funded ethos which provides a counterweight to corporate-led nightlife. Hesmondhalgh (1998) has pointed to the independent dance music industry as an example of a more democratic form of popular culture due to the affordability of technologies, which has led to a decentralisation of production epitomised through the 'bedroom DJ' and the lack of a 'star system' which valorises certain big-name artists. The post-rave era has also seen the growth of a network of independent dance clubs, bars, record and clothes shops.

At one alternative bar, the bar manager described the vision of the owners:

> They want it to be a hive of productivity, if you like. They want people to come in here and write music, and produce music and play it where possible. They wanted to come down ... and set up something that would be used by people to produce music and art work, and for it to be a place where you can come socially as well.
>
> (Bar manager 4, Bristol)

For another arts promoter DJ on the underground in the north of England, creating a fluid boundary between consumers and producers entailed allowing people the freedom to express themselves, which stimulated not only creativity but also critical thinking: 'Our aim is freedom of expression. It is to allow people to express themselves ... If you curb creativity you are curbing society's ability to grow ... to deal with big problems' (DJ 2, Newcastle upon Tyne).

Independent and alternative production, then, has a fundamentally different set of parameters compared to the corporate world. In contrast to the mainstream, it aspires to cultural creativity, more democratic and inclusive public spaces and the development of a closer link between consumers and producers. However, as we have seen in Chapter 2, the economic balance between these two worldviews has shifted, especially as large operators are able to gain cost advantages through rational techniques of production and 'synergies' between products (Ritzer, 2001). While central areas at night are teeming with places of corporate consumption, there are now far fewer places for creative interaction between producers and consumers.

Regulating the margins

As argued in Chapter 3, the 'normative landscape' (Creswell, 1996) of mainstream nightlife is imbued with certain styles and behaviours which are expected and accepted, enforced through door policies and technologies such as CCTV. In contrast, the margins offer the possibility of a less regulated, more fluid space defined by an absence, or defiance, of appropriate codes. However, rather than a place of absolute freedom or disorder, marginal spaces and practices have their own modes of social ordering, rules and relations of power (Hetherington, 1997). Alternative nightlife spaces strive for greater freedom and autonomy, not through less regulation but by striving to create more emancipatory forms of social order.

Codes and rules exist within alternative nightlife for a number of reasons. Many normative rules, such as health, licensing and fire regulations, are imposed by the local and national state from which few, apart from the 'unlawful', can escape. Yet, within this legal context, many groups on the fringe establish their own codes of practice (Plate 9.2). The saying that 'there is no such thing as a free party' alludes to the fact that much organisation, regulation and self-policing is needed to hold a successful event. Lighting, sound systems, transport, water, door security, look-outs and even police liaison, all are ingredients in regulating free parties and raves. Many of these rules are self-generated and communally agreed to demonstrate and envision less hierarchical ways of social and spatial ordering.

Among the most visibly different aspects of the margins are its entrance requirements, which are often less formally monitored and based around self-selection and subcultural knowledge. One young woman (Cait, 27 years old, Leeds) reflected on her favourite alternative bar: 'No entrance regulations, I guess, crucially. Like no dress code, pretty friendly, non-pretentious kind of atmosphere. And there's, like, a bit of a kind of a social worker, caring profession, public sector sort of feel to it.' Additionally, some regulars of one bar on the alternative scene in Newcastle upon Tyne, opened by a collective of local actors and musical promoters, commented:

Andy: This is a great pub, loads of different stuff, the arts, playing here, absolutely fantastic Geordie productions. Some of them have got a history, they have not got Bacardi Breezers stacked in front of your face.

Sally: And you have got interesting people in there.

Andy: They are not in the centre of town, they are not full of drunken arseholes ... The hen nights ...

Sally: They play decent music, not just whatever is in the charts. It is quite a safe environment to look different, no skin off my nose, but there are folks, it is a safe place to be down here.

(Focus group 6, Newcastle upon Tyne)

Underpinning such self-regulation is the geographical location of many alternative venues on the edge of central areas – a liminal twilight zone which offers flexibility and anonymity.

Places on the margins, then, strive for self-selection. One manager of a bar/arts venue commented: 'The most powerful form of advertising is word of mouth' in order to attract 'like-minded people' (Bar manager 5, Bristol). For

A FEW POINTS

The free party scene is radically different to the rave scene - It is Smaller, more intimate and far less intrusive. It is both non-profit based and more diverse than the rave scene of the late eighties.

Free party systems have been an important part of Brighton underground culture for almost twenty years - much local talent has emerged from this scene and it is of benefit to the community and the reputation of Brighton and Hove.

It is damaging to society as a whole to criminalise large sections of the community for what they consider to be a relatively harmless leisure pursuit.

Local government and police should try to accommodate the free party scene by talking to party organizers and finding ways of licensing events so that party goers can hold parties, local residents are not disturbed and the safety of people goers can be ensured.

Society changes - and the law and the manner in which it is upheld should reflect These changes. We feel that we are being penalized through knee jurk legislation that was developed to deal with problem that no longer exists. Legislation and attitudes should change to reflects the current situation in order to allow party organizers to lawfully hold smaller gatherings in agreed locations with the consent of both government and police.

Considerable more trouble is caused on an average Friday night in any town or city; free parties are on the whole relatively free of violence and crime. Free party organiz-ers and goers are well educated and aware of health and safety issues that are encountered in the organization and running of parties.

Plate 9.2 Flier given out at a free party; Brighton, UK, 2001

another independent bar owner, attracting such people through personal contacts was one way of avoiding the tension and possible conflict associated with larger downtown bars:

> Go to any of the smaller places, there's a more localised relationship between who is running the place or managing the place, or whatever, and its audiences, clientele. You know, the bigger the place the more of a problem there is. I mean, if you've got a small bar that you know is going to be run individually, you get rid of the social problems associated with the anonymity of the larger places which are just about consumption.
>
> (Bar owner 3, Leeds)

Many underground activities, such as raves, squats and free warehouse and forest parties, involve transient appropriations of material space which transcend established regulations and laws. Hakim Bey's (1991) notion of the temporary autonomous zone highlights the political potential of liberating spaces from their everyday meanings. Within these fluid spaces and among the debris of empty rooms and temporary decorations, young people are active participants in the social and spatial construction of their surroundings. Unlike many of the formalised spaces of urban nightlife, then, free parties and raves are transitory events which are brought together through loose networks of communication. Rather than relying on conventional channels of communication, such events are brought together rapidly and silently through the internet, chat sites, e-mail, mobile phones and word of mouth to sidestep the police and local state bureaucracy. The physical location of raves and free parties and the virtual space of the internet combine to create a 'space of flows' and a range of possibilities to bypass corporate media and entertainment and develop the underground (Gibson, 1999).

The parameters and regulations for staging an event such as a free party are often temporary and fast moving. Once a venue is found and word of mouth is used to attract an audience, the party can commence. As one free party organiser (Simon, 30 years old, Bristol) in the UK put it: 'Personally, I just spot a warehouse, go and eye it up. Two hours before we're ready to do a party someone goes and opens it. Then we drive through the doors, shut it behind, set up, a thousand people turn up. Phone lines, word of mouth.'

Many people are attracted to the free party scene because of the ability to be 'free' and 'party' in the way they like. As an organiser from the 'ResisTrance' free party collective in the south-west of England commented:

> It's free. They're free parties. Obviously, you know, there's a sense of togetherness. But it's very, very different in the free parties. It's very anarchic, chaotic, messy, hazy. You don't have to worry about security

> ... This is all about deregulation, you don't have to ask anybody for permission to do anything. There's a vibe there you won't get in a club.

However, a whole range of restrictive legislation has emerged to crack down on the unregulated nature of alternative nightlife on the fringe. Here, nightlife is often cast as deviant or abject – something 'impure which threatens the purified body of western culture' (Sibley, 1998: 95). Hence, it becomes subject to a whole set of restrictive laws which casts participants further as 'outsiders'. In this sense, many groups on the margins are criminalised and seen as inherently transgressive due to their fluid rather than sedentary use of space (*ibid.*). From such fears of the margins, various mechanisms of social control, such as the Criminal Justice Act (1994) in the UK, have emerged to curb nomadic and spontaneous groups such as New Age travellers and ravers. The media has also fuelled moral panics, often stemming from drug use (McKay, 1996), while other marginal groups, such as gays and ethnic minorities, have been subject to homophobic and racist policing.

Tensions also arise as the margins and the mainstream overlap, collide and antagonise, which brings different ideologies and styles into sharp relief. Many people who have constructed an outwardly alternative identity, for example, can provoke aggressive responses from the 'normal' world around them. As one young woman (Sally, 21 years old, Newcastle upon Tyne) commented:

> I was with a lass with a skin head and we went downtown and blokes would just stop you, would touch you for a start. Stand at the bar, someone runs their hand down your back and goes for your arse – now, I mean, that is not safe – 'Are you a fucking dyke then?' because she has got a skin head.

During the course of a night out, then, such collisions periodically occur between alternative and mainstreams groups as they criss-cross the city to their respective venues.

Resisting corporate nightlife

As we have suggested, resistance to corporate nightlife comes in many guises and there are no ideologically pure spaces of the margins from which to resist. All cultural and nightlife activity is in some way either mediated through, or defined against, commercial culture. Many cultural forms, although linked to the commercial economy, are self-conscious and critical of their relationship with it. The Body Shop treads such a line, while corporate brands such as Sony, Warner, Gap clothing, Starbucks or Bacardi drinks have been 'outed' from the ranks of image-saturated consumer society as corporate 'baddies'. Popular culture, in particular, is a site of both hegemonic and counter-hegemonic ideology and action (Best, 1998). In this sense, while popular culture is clearly enmeshed with commercial culture, it also has a long tradition of thriving in the gaps, rejecting mainstream corporate culture and reflecting the urban margins (Haslam, 1999). From jazz and rock and roll to rap, punk, acid house and break beats, pop culture has borrowed, learnt from and identified with those on the margins.

Many groups and individuals have drawn their own battle lines and have developed tactics for resistance and self-preservation. For instance, one young person we spoke to pointed to the process of standardisation within nightlife: 'You know these super-clubs, you just go, like I say it's just like going into McDonald's. You're like a sheep. You go in for a product, get it, and leave' (Mike, 24 years old, Newcastle upon Tyne). As the following message on a chat board shows, sceptical consumers often expose another reality lying behind the hype of the city boosters and gentrifiers – that behind the glitzy veneer, there is not much going on:

> What the hell is going on? How many more trendy bars and clubs do we need? If Leeds is to live up to the cosmopolitan label it seems to have given itself, then there needs to be a whole lot more variety. Leeds has no arena, no big stadium gigs and is a city of mediocrity ... Don't believe the hype, they're just trying to tempt everyone to Leeds to rent out their extortionately priced flats, drink their over-priced watered-down lager, wear their over-priced clothes – and for what? So you can call yourself cool!
>
> (Geoff, 2001)

Community-based provision continues to play a role in providing diverse alternative nightlife, mainly as public subsidies allow them a degree of flexibility outside the sole focus on profit. However, the growth of standardised nightlife products, especially those geared towards alcohol sales, were seen to have a detrimental effect on local creative and musical cultures and the viability of alternative spaces. As one musician in Bristol commented: 'That's what I worry about, the more it becomes about just drinking and about just money, the more the city's creative cultures will suffer' (DJ 2, Bristol). As larger theme bars come to dominate urban areas, local kids have to try and look harder to connect with creative musical scenes and smaller, more experimental clubs. Another music promoter, from the Rabble Alliance Collective in the north-east of England, reflected that for too long the nightlife market has been driven by established interests rather than being responsive to what people want. In this sense, he said: 'There's a difference between producing what people want and dictating what people get.'

Meanwhile, in Brighton, the anarcho-publication *Schnews*, offering a weekly round-up of radical news and direct action campaigning, outlines the issue graphically:

> The problem is that as we sip our well-earned pint we could be lining the pockets of some corporate nasty we've never even heard of ... the days of your local, friendly, community boozer are definitely numbered. Pubs are being priced out of the range of ordinary working-class people. Soon only the rich or the trendy will be able to afford to drink in a public house. The global capitalist pub of the future will have all the character of the foyer of your local Sainsbury's.
>
> (*Schnews*, 2001b)

What promoted such comments from *Schnews* was the forced closure of several traditional and well-loved pubs in Brighton. After tenants stood up to one of the UK's fastest expanding multi-site pub owners over the issue of tie-ins on beer purchases, rents were hiked to force a change in ownership. The tenants, among 800 nationwide, became embroiled with the pub company in the European Court of Human Rights, in a battle over the legality of tie-ins (Tomlinson, 2002). Regulars from the pubs protested, many helping to organise a nearby temporary autonomous space to highlight the issue of corporate greed, while campaigns have been launched to expose the domination of Brighton pubs by two other large companies (*Evening Anus*, 2000).

While downtown areas are increasingly dominated by such profit-focused nightlife operators, many places on the margins have become well known as seedbeds for radical activity and alternative nightlife. A short walk from Bristol's centre reveals a simmering, subversive nightlife culture, a heady mix of party fused with the politics of the underground. A stroll along Stokes Croft, the city's alternative club destination, and across to Easton in the poorer east side reveals flyposters advertising the underground. Environmental and political posters declare 'Bike It!', 'Vote for Nobody in the coming election', 'World Car Free Day – critical mass carnival', while others advertise nights of party, politics and protest. Tactical Frivolity has recently held a benefit gig for the European tour of the 'Carnival Caravan against Capitalism' at Easton Community Centre, while the 5th of May Group featured a 'resis-dance' event in support of the prisoner Satpal Ram, in collaboration with the local squat-cum-official resource centre, the Kebele, and its radical monthly publication, *Bristle*. In the city's predominantly black inner-city area of St Paul's, the Malcolm X Centre, long-time promoter of black arts and culture, recently organised the 'Resistance Conference', offering a chance to 'discuss and participate in issues you don't hear about on the news'.

So, many groups continue to mobilise around such concerns and disenchantment. Drawing upon their own energies and resources, creative communities of designers, promoters, musicians, fanzine makers, pirate radio producers, DJs, filmmakers, writers, squatters, punks and artists have inhabited the margins, signalling their opposition to corporate activity. Such groups have made more formal attempts to politicise a night out and raise various issues relating to corporate control as well as presenting alternative conceptions of how places might be differently organised. Direct action groups, for example, using techniques of civil disobedience and non-violent action, take the battle on to the streets and into the private space of firms. Earth First!, the network for ecological direct action, for example, is informed by radical notions of urban consumption which question core assumptions about lifestyles in the west (Purvis, 1996; Wall, 2000), and to these ends groups across Europe and the USA have undertaken occupations and sit-ins at shops, offices and workplaces. The Stop the City movement of the 1980s, 'Reclaim the Street' street parties of the 1990s, and the ceaseless flow of protests at international trade and political meetings in, for example Gothenburg, Seattle, Prague, Washington and Genoa, are all wedded to an ideology of direct resistance to corporate power. In San Francisco, the Biotic Baking Brigade took to pie-ing politicians and business leaders, including the city's one-time infamous hard-nosed mayor Willie Brown, for their role in putting profit before people within the urban fabric. The consumption-based, growth-oriented and increasingly corporate-dominated city, then, is a focus for both social and ecological resistance as well as consumerism and play. Below, we look at two specific examples of such organised resistance – the free party scene and squatting.

Staying underground: raving, clubbing and the free party scene

The underground free party scene has long challenged the monopoly of commercial leisure providers, and it often creatively re-uses the debris of the urban environment. In fact, going out at night has changed fundamentally over the last few

Box 9.1: A night out on the margins; Newcastle upon Tyne, 3 October 2001

We met John, Mary and Gaz at the Ferry Arms for a few pints first before going to the party. Mick and Steve were playing some hard house and techno tunes and then going on to play at the warehouse, so no one was going to leave until they did. About 11.30 p.m. we all staggered in convoy down to the river, following the directions of the bar staff, who were eager to close up and follow. Down at the river, we met some friends of John's who were helping out:

Jay: Where's the party?
Pete: It's on the other side of the river. There's a geezer meeting people on the bridge once an hour to take people over.
Jay: What's the place like?

Pete: It's an knackered old pub, man. Wicked. Derelict, but we're only using the downstairs 'cos the upstairs is fucked. We're gunna serve booze from the bar and we've got DJs playing behind there as well.
Jay: Who's playing?
Pete: Techno stuff, man. Bit of gabba as well. Got some live percussion from those blokes who play at the arts centre 'n'all.
Jay: What time's it finish?
Pete: All night, reckon. When the clubs kick out it'll fill up. Security's sorted. The police have been, but we've got a Section 6 notice up so they can't do us.[1]
Jay: See you over there.
Pete: Nice one.

The Cushy Mallet pub had been derelict for several years after the docks had closed. Whoever had broken in and taken the metal grids off the windows had cleaned it up and got the electricity on and fixed the water in the toilets. A bloke on the door, grinning forcefully, repeated to everyone, 'Pound for the sound'. The large bar in the centre of the main lounge was now back in operation after what looked like years of disrepair. Several people were busy rigging up sound equipment and sorting out beer, cigarettes and food to sell at the bar. We bought some bottled organic ale from them, brought in from a local brewer, although the drinks of choice that night were cans of Stella Artois and Kronenbourg from the local drinks store.
We circulated aimlessly round the pub, catching the eye and chatting to people they had met at other similar parties. When the music came on, there was a loud cheer. Banging techno filled the small room, and the majority started to sway to its uplifting rhythms. Mani and his mates arrived about 1 a.m. with some more speakers and started selling Ecstasy pills to those who wanted them. 'Three quid each, like,' he repeated as he drew his large bag of pharmaceuticals out of his coat pocket. By 5 a.m. the party was in full swing, now full of lucky post-clubbers who had heard about it in town. A constant stream of people were coming and going, several DJs had their turn on the decks, a group of people had started drumming along to the music using some crates as an impromptu stage, and the bar staff chatted and negotiated with skint punters who pleaded for late-night discounts. Mad Leslie was dancing on the bar. Gaz told us: 'She always goes for it. Ten pills and she's up for days. Crazy, never stops.' The police drove past a few times that night, but did nothing to stop the party. They knew it was too contained and out of the way, and that it wasn't the type of crowd to cause much trouble. And anyway, it would be more hassle than it was worth to kick everyone out.

decades, mainly due to the advent of DJ-based and prerecorded electronic dance music. Such shifts gained notoriety through the emergence of the 'one-nation' rave cultures in the UK in the late 1980s and early 1990s. Characterised by smiley faces, Ecstasy drug cultures, acid-house music and all-night free parties usually held in disused warehouses, forests and other marginal spaces, 'rave' has grown from a largely underground culture based around places such as London, Detroit, Chicago and Berlin, to become a truly global phenomenon (see Gibson 1999; Saunders 1993, 1995; Collins 1997; Garratt 1998; Bussmann, 1998). The history of rave and dance music is complex and their origins defy easy categorisation (Redhead, 1993). Nevertheless, nightlife changed dramatically after the rave scene and the 1988 'summer of love', when thousands of kids were first united under the banner of PLUR (Peace, Love, Unity, Respect) and DJs played a mixture of hip hop, disco, acid-house and techno music in the rave island of Ibiza.

The Do-it-Yourself (DiY) free party and festival music scenes have played a key role in providing entertainment and music based on the needs, desires and energies of participants, and have politicised a wider community about the dominant influence of the state and private capital play in structuring 'free time' (Rietveld, 1998). In the late 1980s and early 1990s, pioneering free party crews such as Desert Storm, the United Sound Systems, Mutoid Waste Company, Schoom and Psychic TV played a crucial part in developing such cultures of resistance. Spiral Tribe have been one of the most influential hardcore techno sound systems, holding marathon free parties in derelict buildings and warehouses across Europe since the 1990s. Subsequently, they have been central in establishing Network 23, an alternative non-profit information channel. Part internet networking tool, part cosmological doctrine, Network 23 was established around the time of the UK Criminal Justice Act to support and promote the underground scene and bypass the corporate music and entertainment system across North America and Europe.

The free party scene, not surprisingly, has led to the introduction of legislation designed specifically to curb its growth. This was particularly evident in the UK, where the unregulated rave scene was seen as a threat to the 'respectable' hegemonic project of Thatcherism (Hill, 2002: 89). In particular, the Police Party Unit was established in 1989 to co-ordinate national action against the free party scene while the Entertainment (Increased Penalties) Act of 1990 raised fines for organising unlicensed parties from £2,000 to £20,000 plus six months imprisonment (Hill, 2002: 98). The summer of 1992 was a watershed for the scene and its various communities. A free party attracting 40,000 people, organised in the small village of Castlemorton in Worcestershire in the UK by Spiral Tribe, fuelled Middle England's terror of hippy travellers and drug-fuelled ravers and sowed the seeds of the Criminal Justice and Public Order Act (CJA) of 1994. The CJA had specific anti-rave clauses and highlighted the desire of the government to legislate out of existence a particular youth cultural form. In its attempts to do so, the now infamous Part V, Section 63 of the

CJA specifically mentions 'a gathering on land in the open air of 100 or more persons at which amplified music is played during the night' where the music includes 'sounds wholly or predominantly characterised by the emission of a succession of repetitive beats' (Home Office, 1994).

In spite of such attempts to legislate against rave and dance cultures, they have continued to evolve and have had profound effects on many aspects of nightlife. Many young people today acknowledge the vast differences between traditional 'nightclubs' characterised by 'cheesy' commercial music, sporadic violence, excessive drinking and established gender roles, and 'clubbing', based more around musical appreciation, artistic competence of the DJs and the uniqueness of the environment. Although club crowds are often regarded as diverse, they also offer a feeling that you are among 'familiar strangers', your own tribe. As one music promoter commented: 'You know, a club is a club, if you follow what I'm saying; it's like it's a feeling of belonging' (DJ 3, Newcastle upon Tyne).

Drug culture, and especially Ecstasy, has played a crucial role in the experiential aspects of rave and club cultures, taking many participants to a different state of consciousness (Saunders, 1993; Joseph, 2000; Plant, 1999). Ecstasy provided the magic ingredient for the dramatic explosion of dance culture which changed the face of clubbing, immortalised through the 'MADchester' phenomenon of the late 1980s (Haslam, 1999). Although drugs have long played a role within many nightlife cultures, through the growth of dance music they have become pervasive within youth cultures generally (Parker *et al.*, 1998), as well as in mainstream pub and nightclub culture. One of the legacies bequeathed by dance culture and this normalisation of drug use is a current distinction within nightlife between 'beer monsters' and 'E-heads' (Moore, 2000).

Further, rave cultures have to a certain extent challenged, or at least made more transparent, pre-inscribed sexual roles (Henderson 1997; also see Chapter 7). Scanty clothing for boys and girls, ravers sucking dummies and blowing whistles, primary colours, psychedelic doodles, images taken from familiar advertisements, phrases and tunes lifted from children's TV programmes like *The Magic Roundabout* and *Sesame Street* are hallmarks of Ecstasy-fuelled dance cultures (McRobbie, 1994; Malbon, 1999). The club provides a facilitating environment, where temporary and playful regression or infantilisation to various stages of childhood and pre-sexual innocence are possible.

Over the last ten years, the dance music, rave and club scenes have changed dramatically. Legislation aimed specifically at curtailing free parties in the UK, such as the CJA, forced people back into licensed events and corporate-backed nightclubs. Dance music has also diversified and fractured along the lines of class, gender, ethnicity and musical styles, producing a myriad of styles such as trance, garage, drum and bass, hard core, techno, gabba – ad infinitum. Subtle nuances within dance cultures have come to delineate style and music tribes –

the hardcore posse, the trendy trance tribe, the eurotechno ambient crowd and the hippy nouveau tribe (Sharkey, 1993). The unity and the 'we' of the rave generation have given way to schisms and sectarianism (Reynolds, 1998b). While such changes are often taken to represent a growing sophistication and exactitude among contemporary clubbers, they also signify more 'commercial' and 'client-based' pressures.

Once standing on the deviant margins of youth culture, then, dance and rave cultures have been partially legitimised by wider society – commercialised, sanitised and normalised as a global cultural pastime. Music, books, websites, TV programmes, magazines and fashions and wider creative infrastructures linking publishing, books, dedicated club magazines, and record and clothing shops with bars, clubs and galleries have all emerged to feed growing commercial dance cultures. With advances in technologies and the internet, the rave and club scene has also become a virtual commercial place as much as a 'material' space (Gibson, 1999). Over the 1990s, corporate clubbing and super-clubs have emerged and transformed dance cultures into major commercial entities. Club cultures, then, have taken on a truly world-wide meaning and now stretch across the globe from Ibiza and Rio to Tokyo and Bangkok. DJs with global brand recognition command huge paychecks for playing a two-hour music set. The whole meaning of the underground has become weakened as it is reformulated, branded and sold back to new generations of young people. Lifestylers rather than ideological purists abound. Weekend part-time ravers and urban young professionals, who make temporary incursions into the rave scene and retreat back to other 'corporate lives', have been denigrated for diluting the counter-cultural origins of the underground scene. The underground has fused with the overground, resulting in confused class categories such as 'hippy-yuppies', 'ethno-Sloanes' and 'ferals in suits' (St John, 2001a).

In response to such encroachments, some alternative forms of nightlife have emerged which continue to challenge the naked commercialisation of club cultures and, as a result, distinctions have been redrawn between underground and mainstream dance clubs (Collins, 1997; Thornton, 1995). Darker, more hard-core music scenes have developed to distance themselves from creeping commercialism. Such simple dichotomies, however, conceal the many gradations within contemporary club cultures. For example, mainstream nightclubbing can encompass both the traditional 'meat market' venues, which play more commercial, radio-based music and also large super-clubs hosting world-famous DJs, while the 'underground' encompasses both small independent commercial venues and also more illicit squat and warehouse parties.

Nevertheless, many clubbers, especially those who experienced the early spirit of rave culture, view large mainstream clubs with suspicion, being aware not just of their differences in music, but also of attitudes, style and creativity. Some continue to be concerned about the effects of commercial clubbing in terms of limiting the range of night-time consumption opportunities, especially for younger people entering the scene. Josh (35 years old, Bristol), a music promoter, commented:

> Like commercial chain clubs, they go and re-fit their formula like McDonald's and then they get the same national circuit DJs to come and play the same fodder. It's not very innovative, it's very consumer-orientated. I guess if you're raving for the first time, which most of the people are, you know, they are getting served up what they want. Bright lights, loud music, drugs. The kids like that just buy it in. 'Let's pay this guy £2,000 to play here because he plays on the radio and the kids will like him.' It's corporate, they're just serving up a product.

In contrast, for many people involved in the free party scene it is not just about being free to dance, but about being a 'free individual' able to pursue an alternative lifestyle, usually outside the world of 9–5 work. As one free party organiser suggested to us:

> We are living a different life basically than the kids that are going to Mecca. Their Mecca's the 'working week' and going out for their explosion of 6 hours on a Saturday night. And it just seems that most people that turn up to free parties are living some kind of alternative life and want to meet, want to congregate. Don't necessarily want to do it in pubs with alcohol. Don't necessarily want to do it in very dressy, you know, dress-coded, you must be like this or you can't come in. It's just too regulated, and free individuals are looking for a certain vibe. For an entertainment licence you have to pay X pounds per head for those who might possibly attend. It doubles your ticket price. It's massive; entertainment licences don't come easy. I think really the crux of it comes down to freedom issues and whose laws you agree with and obey.
>
> (Geoff, 27 years old, Bristol)

The rave, festival, free party and traveller scene has no doubt helped to develop a critique of the commercialisation of contemporary nightlife. It has liberated a generation and given expression to new musical, cultural and artistic forms. However, as mentioned, it has been subject to repressive legislation, epitomised by the Criminal Justice Act in the UK. Police tactics have also become more sophisticated to deal with and manage the growth of free parties and the rave scene. However, the underground has found new ways to exist, avoid police repression and distance itself from the mainstream. In particular, as laws tighten their grip on free party cultures, they have moved to places where legislation is less coercive, such as the Mediterranean and Eastern Europe, California and Goa. Members of pioneering free party collective

Spiral Tribe have set up communities in southern France, while rave and New Age travelling communities from northern Europe have been part of the 'Spain Drain' by heading south and establishing teepee villages in abandoned settlements in the Spanish Pyrenees and the mountain towns of the Andalusian Sierra Nevada.

Direct actioners, techno-tribes, environmentalists and indigenous groups continue to expand beyond their place and forge global alliances. Australia's underground post-rave 'doof' scene, for example, is marked by a particular brand of inter techno-tribal gatherings, often called 'corroborees', which fuse dance music cultures and direct action with a religious ambience, with strong connections to tribal and indigenous aboriginal rights movements and the natural landscape (St John, 2001a). In South Australia, the EarthDream Festival has brought together eco, rainbow and techno-tribes at the fringe of international youth culture (St John, 2001b). EarthDream, inaugurated in 2000, was a caravan of travelling sound systems, musicians, engineers and performance artists mobilised against copper–uranium mining in the Australian outback on aboriginal land. The year after, over a hundred cities across forty countries participated in the EarthDance festival. Meanwhile, at Black Rock Desert in Nevada, USA, over 20,000 people attend the annual Burning Man Festival to form Black Rock City – a settlement which describes itself as commerce-free, an experiment in temporary community and an exercise in radical self-sufficiency. As the organisers explain: 'You're here to create. Since nobody at Burning Man is a spectator, you're here to build your own new world' (Burning Man, 2001). No events are planned, and participants are encouraged to create their own communities, art installations, stages and performances.

Restrictive legislation has also strengthened the resolve and has politicised many people in the free party community. The direct action, DiY collective Justice? from Brighton in the UK, for example, was born from a desire to raise awareness about the CJA, and has developed into a wider network, including the publication *Schnews*, promoting campaigns on social and environmental justice, especially through squatting. Reclaim The Streets, exporting world-wide their brand of party and politics and carnivalesque street parties which have brought downtown traffic to a halt, have also inspired a wider critique of urbanism, corporatism and capitalism. Street parties use samba bands, sound systems, theatre and live musicians which attempt, temporarily, to transform the sterility of the automobile-choked streets into a place of play, possibility and subversion. However, it is to the squatting movement, which has attempted to envision more permanent alternatives, that we now turn.

Squatting the city: DiY nightlife

> Squatting creates space for much needed community projects. Squatting means taking control instead of being pushed around by bureaucrats and property owners. Squatting is still legal, necessary and free.
>
> (Advisory Service for Squatters, 1996)

While squats often evoke images of Third World squatter settlements and land struggles of indigenous groups, here we are concerned with squatters largely in the western world who have established, if only temporarily, alternative communities for working, living and playing. Outside the temporary nature of free parties and the commercial de-politicisation of much of contemporary dance culture, squats attempt to perform, embody and envisage radically different social and spatial practices. They not only attempt to take back an equal share of the urban fabric from corporate control, but also establish, or re-establish, small-scale human urban communities (Corr, 1999). Squatting as a form of direct action allows people to reassert their rights to self-determination and gives them a meaningful stake in modern city life.

Many groups squat out of genuine need for housing. Other more middle-class groups have squatted to establish counter-cultural ways of life, forming outposts of a new culture (Osborn, 1980). Such latter groups generally seek to make statements about the nature of property speculation, and contemporary ways of living (work, community, leisure). Squatting can transform forgotten urban spaces – dead zones, derelict areas, voids and wastelands (Doron, 2000) – into places of creation and performance. Often, such places are in a state of suspended animation, waiting for the latest plan for a multiplex, casino, fun-pub, restaurant or entertainment centre to emerge from developers, to be fed to the local cash-strapped state. Monbiot (1998: 182) has pointed out the democratic limitations of this development process: 'If ordinary people don't like a local authority decision to approve a development, there's nothing whatsoever they can do about it.'

In this climate of lack of accountability from the local state and the standardisation of most development schemes, squatting and reclaiming land are sensible and increasingly widespread options. Squats often provide an alternative vision to commercial homogeneity and formal planning. While many urban developments point towards sterility and sameness, squatting values diversity and disruption, and represents a desire for serendipity, unpredictability and openness. It refuses to be caught by the bureaucracy of the urban planning system and the rules which currently stifle and regulate play, leisure and entertainment. Squatting also celebrates the power of the local and is one of the few remaining resources which allow cities to retain their connection with people rather than profit. It illuminates a collective and creative use of urban space which sketches out possibilities for radical social change (Chatterton, 2002b).

While land squatting has a long tradition dating back centuries, more recently many European countries have developed a rich variety of urban squatter settlements.[2] For example, since the late 1960s, the urban fabric of Amsterdam's Centrum has been transformed by anarchists intent on promoting affordable housing and keeping corporate developments at bay. Christiania in Denmark is perhaps the best-known and longest-established European squatter settlement. Dubbed 'Freetown' after it declared itself independent of the Danish state, it has survived as a self-governing community since 1971 in a derelict military barracks, and is now home to 800 adults and 250 children (Corr, 1999). Christiania has developed into an almost self-sufficient eco-town, complete with community factories, cafés, bars and stages for cultural events. With half a million visitors a year, it claims to be Denmark's largest tourist attraction (Marshall, 1999). Barcelona's squatting scene, drawing upon Catalunya's strong anarchist traditions, flourished over the 1990s, as many of the country's youth developed their political ideologies in post-Francoist Spain (see Plate 9.3). The city's squatting movement continues to expand and is organised through the Squatted Social Centres Network (Centros Sociales de Okupas), which was established in 1996 to provide some co-ordination among the activities of twenty-one squatted social centres organising events and activities from crèches to bike maintenance courses, yoga, gigs, cafés and dance classes.

Over the 1990s, many squatted centres emerged in the UK, often in reaction to the Criminal Justice Act, which also became focal points for a new wave of politicisation and non-violent direct action. Several houses were squatted on Claremont Road in East London, for example, in an attempt to block the notorious three and a half mile M11 link road, which cost £350 million and led to the destruction of 350 houses. On the squatted road, houses were painted, art-work sculpture filled the street, bands played and an outdoor living room was set up, complete with furniture.

Plate 9.3 The Montanya squat at the heart of Barcelona's squatting scene; Spain, 2002.

During the final eviction of Claremont Road in December 1994, bailiffs and police faced 500 protesters who used a range of non-violent stalling tactics. A couple of years later, 500 activists from the Land Is Ours movement, inspired by the Diggers movement, squatted land on London's waterfront. Here for a brief time they established a community to highlight the shortcomings of property speculation (Halfacree, 1999; Featherstone 1998).

Many squats initially starting out as housing projects, then begin to incorporate working and leisure elements in an attempt to critique and redefine the boundaries of how we work and play, produce and consume. Many squats have become formalised into semi-permanent resource, advice and arts centres for community and campaign groups, run by local people to offer services which the local state is too underfunded or uninterested to provide. On Stoke Newington High Street in London, a group called Hackney Not for Sale set up a squatters' estate agents in 2001, where residents could view over fifty local authority owned properties that were in danger of being sold or privatised. This 'Estate Agents with a Conscience' acquired 'exempt' minutes from a council meeting which listed over 130 local authority owned properties to be assessed for sale as part of their 'disposals programme', including community centres, adventure playgrounds, allotments, nurseries, shops and houses. One of the properties, squatted and reopened as a community centre, carried the spoof advert:

> Formerly a nursery, but we got rid of the kids and their whining parents. We even fibbed in court, to get rid of the squatters who had reopened it as a community centre of all things. We told the master that we were not going to sell this building but, Ha! – It is now freed up to give lots of potential for profit as a yuppie wine bar or a private health club for city workers. Price: Don't bother enquiring, you couldn't afford it. In fact why don't you just move to another borough so that we can speed up the process of gentrification here in Hackney.
>
> (*Schnews*, 2001a: 1)

Against the rising tide of corporate nightlife, people in numerous places have been inspired to take matters into their own hands and develop their own forms of nightlife. In Leeds, an old church, soon to be sold by its owners to make way for a large student theme bar, was squatted and christened 'A-Spire'. The organisers wanted to create a space where the types of activities which occurred there were not, within limits, prescribed:

> It is a place where people can go during the day or night and socialise away from the buy, buy, buy mentality that is present-day capitalism. Once there you can do pretty much whatever you want to (within

reason). You can sit and chat with a cup of fair-trade tea or coffee, bring your lunch to eat in comfortable surroundings or eat some of our vegan, almost organic food. You can go to or put on a gig, read books from our radical library, organise a meeting, juggle, paint, make a date to come back for a specific gig or workshop. Basically we've opened up the space: if you want to use it, then do so.

(A-Spire event flier, October 2001)

Box 9.2: Exodus from Babylon: the 'Respectocracy'

Exodus is a unique urban phenomenon which does not simply confront but intelligently challenges society's assumptions and values. They offer working, viable solutions to many of society's stated ills: poverty, crime, drugs, unemployment and the breakdown of community. Exodus blend a volatile mixture of rastafarianism, new-age punk and street-smart politics. 'We are not drop-outs but force-outs'.

(Exodus, 1999)

The Exodus Collective, based in Luton, 40 miles north of London, emerged from a desire to put derelict land to use by providing entertainment outside of commercialisation, heavy entrance fees and security. Exodus, much more than a group of illegal rave organisers, describe themselves as a self-help collective. The Collective grew out of the free party and rave scene of the late 1980s, and since 1992 have been holding events which have attracted up to 10,000 people, mainly through word of mouth. Exodus suggest that their 'community dance events' are operated on a fundamentally different principle from a commercial event, as the intention is a community gathering free of exploitation, rather than profit generation (Exodus Collective, 1998). Some of the negative aspects of free parties, such as drug-dealing, mugging and theft, are much less visible because, Exodus believe, their events are based upon 'mutual respect and consent, rather than strict policing' (*ibid.*). The raves are much more than dance events. As one member of the Collective said: 'The raves serve as our contact with the community. People come to the raves to party, but while they are there they grow interested in the Collective and what we are about' (*ibid.*).

At such events, entrance fees are replaced with a community levy which provides an income for various community projects. The Collective have been responsible for setting up Long Meadow Community Free Farm, which they initially squatted and eventually were able to buy. It has grown into a showpiece for sustainable living, working and playing, housing a permaculture centre and urban farm, organic food production and alternative energy systems. They have also squatted a number of other buildings, such as HAZ Manor, given to them for a peppercorn rent due to their work with homeless people. While police, local businesses and the local authority were initially hostile, and the police often violent, many of these projects have now gained legitimacy and legal status. A strategic police operation designed to halt the progress of the dance and squatting collective took place during the 1990s, but in 1995 Bedfordshire County Council voted in favour of holding a public enquiry following evidence of police malpractice during their operations. Exodus have gone on to be key players in the regeneration of inner-city Luton.

The Exodus Collective, originally inspired by the free party scene, have been able to negotiate their way through licensing regulations and establish community-owned entertainment, housing and working projects. One reason put forward for the great lengths to which the authorities went to try and close Exodus down is that Luton has a history as a brewery town, where Whitbread have dominated for many years. During Exodus's events, takings in Luton's pubs and clubs are usually down by around 40 per cent. The Respectocracy has taken on the entertainment monopoly and shown that, with enough effort, the people can win.

Examples of the desire to create liveable, leisure alternatives to corporate monopoly continue apace. In London, the London Activists Resource Centre (LARC) opened in 2002 to act as a permanent home for the radical and direct action community, while the London Social Centres Network (LSCN) provides a forum for people interested in setting up social centres or working in existing ones. Jump Ship Rat (JSR) arts venue (see Plate 9.4) occupies a decaying warehouse on the fringe of Liverpool's city centre. In spite of little help or recognition from the local council and official arts organisations, JSR provides gallery space for local artists, performance space, bar, café and venue where bands, multi-media performances and art shows are held. Finally, a group in Brighton bought and opened a collectively owned resource centre in 2002 called the Cowley Club. One member of the Club was motivated by the following reasons:

> I'm fed up with drinking in pubs and watching my money disappear into the pockets of the mass entertainment industry. Why don't we

> start our own club and venue, outside commercial culture, where we can decide where profits from our pockets go?
>
> (Mike, 24 years old, Brighton)

A rich support network has also grown around squatting. The Advisory Service for Squatters was established in London in the 1970s, along with the London Squatters' Union, and continues to this day providing legal, technical and material support for the squatting community, while San Francisco's Homes Not Jails undertakes a similar role. Websites have also developed to create a sense of a squatting community, such as *squ@t!net*, an internet magazine dedicated to promoting squatting throughout Europe.

Possibilities and limits on the margins

What we have sought to provide in this chapter is a discussion of alternative forms of nightlife on the margins of the night-time economy. In the main, we have highlighted organised groups who have sought to identify an adversary (usually the

Plate 9.4 Jump Ship Rat Arts Venue; Liverpool city centre, UK, 2002

corporate city), as well as those who have provided alternative models of urban living, working and playing. This is not to say that we do not acknowledge many different types of resistance in the night, for example through the micro-tactics of play and subversion found in many forms of dance culture (Malbon, 1999; Pini, 1997). Indeed, one of the strengths of the margins is its temporality, unpredictability and fleetingness, its ability to form and reform in order to evade and bypass laws and regulations. It reminds us of how a fluid mosaic of resistance, made up of countless acts of defiance and self-determination both large and small, can help to produce social change and reaction, and provide alternative ways of life.

Yet the margins also contain their own contradictions and problems. Often, their transient nature means that they are more apt to disrupt rather than overthrow. In spite of the earlier discourses of freedom, acceptance and liberation within raves, free parties and the dance club scene, there is little evidence to suggest that they provide a coherent critique of dominant mainstream culture, or solve problems of inequality and hierarchy. As Rietveld (1998: 266) suggests, you can lose yourself to the beat, but this does not question the foundations of society. Similarly, club cultures have also developed their own barriers and hierarchies (Thornton, 1995). In this sense, dance and music cultures are riven with differences and tensions, with notable differences existing between, for example, the largely white rave crowds and the black and Asian followers of jungle and drum and bass. The E-fuelled generation are also regarded as apolitical, 'nowhere people', merely escaping from society's problems through a loved-up world (Reynolds, 1998a).

In many ways, then, neo-tribal forms of sociality on the margins are unlikely to survive for long periods of time. In this sense, they are 'essentially tragic' (Maffesoli, 1996) and are unlikely to sustain a wider critique. Alternative counter-cultural lifestyles are susceptible to incorporation and commodification and have become well-established fixtures of the high street (Hetherington, 1998). Hesmondhalgh (1998) also questions whether alternative and independent cultural production is possible in the context of increasingly concentrated and internationalised cultural industries, especially when there is more emphasis on big-name, big-money artists, increasing collaboration between underground record labels and the music 'majors', and new strategies by large conglomerates to undertake joint ventures to assimilate the independents.

Finally, as we have argued throughout this book, alternative groups and spaces are being forced further out to the margins due to increasing property costs and the changing priorities of nightlife conglomerates, especially through branding and gentrification. Additionally, alternative groups almost invariably provoke strong reactions from the government, the judiciary and the media, which makes their activities harder to sustain. Increasingly, there are fewer opportunities to radically question, subvert and play with one's identity, social role and understanding of the urban. Squeezing out the margins is reducing society's potential for critique and reflection.

With the growth of corporate culture and its spaces, 'rejecting what's sold to us and questioning what's told to us have never been harder' (Haslam, 1999: 280).

Yet despite these odds, many young people continue to try and build something beyond escapist weekend culture. Challenging corporate nightlife means rethinking established notions of night and day, work and leisure, profit and pleasure, and calling into question the values of property speculators, the entrepreneurial state, and the monopolies which dominate ownership, distribution and production. Many alternative nightlife activities and groups are linking up, especially through the internet, with margins outside their immediate localities to create a more global sense of the underground. Perhaps the current shifting of balance between corporate and alternative nightlife and the pressures on sacred marginal spaces is sowing the seeds of greater discontent. It is to some of these wider political issues and scenarios that we turn in the conclusion of the book.

10

URBAN NIGHTSCAPE VISIONS

Beyond the corporate nightlife machine

In this book, we have tried to untangle urban nightlife, to show how it is made and remade in various guises. Through this process, we have pointed to a number of contradictory trends. Fun, hedonism, socialising, sexual encounter and drunkenness remain long-held motives for a night out, while moral panics about disorder and lawlessness continue to increase. Corporate entities are increasing their grip on nightlife infrastructures, despite the fact that at the same time nightlife is marked by seeds of resistance. Major changes are also under way in terms of deregulation and re-regulation. In the UK, for example, gambling and a wider casino culture is becoming a more prominent feature of urban nightlife (Travis, 2001), and some optimistically suggest that the declassification of cannabis to a category 'C' drug may signal the early beginnings of an 'Amsterdam'-style cannabis café culture. Yet moves towards further restrictive legislation in the night-time economy seem to be the norm across much of the UK, USA, Australia and many parts of continental Europe, where, for example, drinking in public is increasingly curtailed through new government laws directed towards young adults.

Our work is part of a tradition of those who feel an ill wind blowing through the contemporary capitalist city (Harvey, 2000), which seeks to draw attention to the central influence of large non-local corporate entities in the making of urban nightlife. In particular, what we have highlighted throughout the book is the growing dominance of mainstream commercial elements. As we outline, this is to the detriment of more historic, older forms of nightlife which are increasingly seen as redundant and residual as urban areas reinvent themselves as post-industrial, high-value centres of activity. Alternative, independent and resistant night-time activities also find it harder to carve out spaces in expensive downtown areas and are often criminalised or treated with suspicion by the local state and police. In this final chapter, we reiterate some of the inherent problems and contradictions in current nightlife trends and flesh out some possible routes beyond the corporate nightlife machine.

Key arguments of the book

One of the main tenets of this work is the fact that our day-to-day lives are being increasingly compromised, narrowed and subdued by profit-hungry entertainment conglomerates (Gottdiener, 2001). While this is not a completely new trend, the range and intensity of commercial domination has grown for many people. As Monbiot (2000: 4) states: 'Corporations, the contraptions we invented to serve us, are overthrowing us.' This process of growing corporate control has been well documented at the city level (see, for example, Solnit and Schwartzenberg, 2000; Hannigan, 1998; Davis, 1992; Zukin, 1995; Sorkin, 1992). More than ten years ago, Michael Sorkin (1992) spoke of the 'Disneyfication' processes – the development of 'ageographical cities', increasingly devoid of place and hermetically sealed from the surrounding reality of urban life. Meanwhile, Hannigan (1998: 7) described entertainment destinations as 'urbanoid environments' and as 'glittering, protected playgrounds for middle-class consumers'. At the beginning of the twenty-first century, the serial reproduction (Harvey, 1989a, 2000) of downtown areas into US-style theme parks, multiplexes and 'casino culture' seems to be continuing apace.

It is against this backdrop of the corporate and gentrified city that we have explored the growth of branded, stylised nightlife. We have highlighted some of the global players, backed by international financial houses, who feature in the ongoing restructuring and concentration of the night-time economy. We have also sought to move beyond rather seductive arguments surrounding 'flexibility' and postmodern consumption, and have stressed more neo-Fordist interpretations of the nightlife industry. The 'new' nightlife mode of production is characterised by some novel features, but also by a continuation and intensification of many Fordist trends, such as a concentration and conglomeration of ownership and production, a lack of real consumer choice and diversity in spite of increases in design and branding, and continued social and spatial segregation as companies vie to attract a number of 'cash-rich' groups of young adults.

We have also outlined an emerging 'consensus' in the governance of urban nightlife. Although central and local governments remain key actors in the regulation of the night-time economy, this consensus is increasingly biased towards the needs of larger property and land developers and entertainment and nightlife conglomerates (Jessop *et al.*, 1999; Fainstein, 1994) and their target audiences, who are often more cash-rich consumers, wealthy urban-livers and professional service workers. For example, attempts by the local state and police to protect the interests of local residents and reduce problems associated with nightlife, such as disorder, noise and vandalism, are increasingly being superseded by the influence of large companies eager to protect their profits and expand their operations, using significant legal capacities to do so (also see Hobbs *et al.*, 2000). Priorities here include both deregulation, in order to aid entrepreneurialism and capital accumulation (i.e.

extended hours, private–public partnerships and public development monies, etc.), and the 're-regulation' of gentrified nightlife premises, increasingly through an emphasis on style codes and pricing polices. This emerging consensus is understandable given that many large-scale nightlife developments represent significant inward investment decisions and hence a sizeable revenue source for cash-strapped urban governments, while the public activities of 'cash-rich' groups provide vital ammunition for city 'imaging-building' and urban gentrification.

So, while we see the emergence of new styles of governance and deregulation for the wealthy few based around entrepreneurialism and lifestyle promotion, regulation priorities like restriction, moral order and control still remain solidly in place for the many. As such, there is evidence of intensified social and spatial control of nightlife spaces via formal mechanisms such as increased surveillance and door security staff, restrictive bylaws and design of the built environment (Christopherson, 1994), in conjunction with attempts to 'sanitise through style'. In particular, the commodification and theming of nightlife spaces have introduced increasingly restrictive codes which filter and control access and movement. Urban and interior design strategies are now replete with attempts to 'script space' in certain ways and civilise its users, while public urban space is increasingly 'managed' by adopting techniques from privately managed malls (Reeve, 1996). Much of this speaks to more apocalyptic readings of the urban, which have portrayed the post-industrial city as a combat zone in which surveillance and control techniques have eroded democratic urban space (Davis, 1992; Zukin 1995). While a night out is still unpredictable and fluid, the 'security-obsessed urbanism' to which Davis (1992) alludes, where whole areas are removed from the realities of urban life and streets are sanitised of evidence of inequality, poverty and homelessness, has increasing resonance with the experience of nightlife.

What of the consumer experience of contemporary urban nightlife? Many young adults told us stories of how a branded, standardised, sanitised and largely non-local experience increasingly marks out urban nightlife. In this sense it is important to remain critical about rather seductive arguments of 'post-Fordist flexibility' (see Pollert's critique, 1991), which often assume the existence of a new set of differentiated consumption practices (Warde, 1994 also makes this point). The ongoing expansion of nightlife branding actually equates to *fewer* opportunities for non-market cultural space and alternative nightlife venues in downtown areas. What options are left for those who seek a range of differentiated products, serendipity and excitement in the contemporary city at night? We as consumers are offered a narrow selection rather than a real range of alternatives (Clarke and Critcher, 1985). As Monbiot (2000: 16) suggests, we have 'a profusion of minor choices and a dearth of major choices'. Similarly, the ability to select from the latest range of nightlife brands is still dependent on economic resources.

The reality, however, is that commercial, branded nightlife continues to be popular. It is not difficult to see why. At first glance, there appears to be a greater variety of places and hence new resources for more consumer choice and reflexive consumption. Indeed, the physical appearance and design of urban nightlife spaces has changed dramatically over the last twenty years. This has mainly occurred through a decline in the number of male-dominated ale-houses, saloons and working men's clubs, and also lager-fuelled 'cattle market' discos and pubs (Gofton, 1983; 1990) and the growth of upgraded, mixed-use, stylish nightlife venues blending food with alcohol, with more seating and higher-quality interiors, which have opened up the nightlife sector to different social groups (young professional women, gentrified gays, 'traditional' students). Yet despite these changes in style, appearance and opportunities, there are clear limitations here as well. As we have outlined, there is an underlying uniformity of ownership and rationalisation of production within nightlife. Moreover, many aspects of nightlife culture continue to be 'awash on a sea of alcohol' (Hobbs *et al.*, 2000), with heavy circuit drinking, vandalism and violence commonplace. Theme bars, for example, have recently been highlighted as the new 'palaces of drunkenness' (Newburn and Shiner, 2001). There is also little evidence that contemporary nightlife provision is really all that 'female-friendly', with continuing evidence of sexual harassment and assaults (Hollands, 1995). Finally, as we have highlighted in Chapters 6 and 7, the supposed opening up of nightlife to social groups like young women, gays and students has often been limited to wealthier sectors of these populations.

The growing boundaries of corporate-owned entertainment also raise issues about the fate of public urban spaces which offer opportunities for human creativity, dialogue and understanding. What do new forms of urban nightscapes mean for concerns such as access, civility and public culture? Clearly, the current round of nightlife restructuring has moved us some way from the historic predecessor of the pub as a public meeting place. The control of the night-time economy by corporate interests and the reduction of social meaning to superficial images within themed environments raise serious questions about the quality of our daily life (Gottdiener, 2001: 207) and the types of encounters within it. The crowds we encounter here are both selectively chosen and highly regulated (*ibid.*).

Downtown nightlife, then, is being sanitised and cleansed of 'undesirable' elements through continued gentrification of housing and leisure markets and the growth of a central urban professional service class. Smith's (1996) analysis of the revanchist city – a vengeful programme aimed at displacing certain types of activity and people, especially non-consumers, the homeless, the urban poor, punks and skaters – has much relevance here. Non-consumers in the corporate-dominated city are cast as deviants – if you are not buying, why are you here? (Atkinson, 2001). While the night does retain some fluidity, there are clear trends towards the demarcation, sanitisation and privatisation of nightlife spaces, not to mention marginalisation of alternatives. As Smith (1996: 105) comments: 'The more likely scenario is of a sharpened bipolarity of the city in which white middle-class assumptions about civil society retrench as a narrow set of social norms.'

In terms of urban nightlife, we are clearly only at the tip of the gentrifying iceberg. Brain (2000), echoing Bauman (1998), highlights that while those in stable employment (the puritans by day and hedonists by night) are seduced by the delights of pleasurable consumption, there are many who are excluded from such fun. The exclusive nature of recent developments in the night-time economy is simply out-pricing many social groups and reinforcing perceptions that it is not a place for them (Chatterton and Hollands, 2001). There are groups of young people living in outer estates or ghettos who have never felt enfranchised by the bright lights of the urban centre, for reasons of price, access, racism, safety or style. What is clear is that the current wave of restructuring is likely to further disenfranchise such groups, as well as some current traditional users with only minimal resources.

Moreover, younger city-centre nightlife consumers are exposed to little choice and have few opportunities outside a narrowly defined role as a 'consumer'. As a result, rather than finding a wealth of opportunities for alternative or independent cultural styles, most young people simply seek escapism in the less risky world of corporate-packaged nightlife on the mainstream. This perhaps says more about the lack of actual choice downtown, and the difficulties involved in locating and travelling to alternative fringe nightlife spaces. Klein (2000: 130) has argued that:

> [t]his assault on choice is happening structurally, with mergers, buyouts and corporate synergies. It is happening locally, with a handful of superbrands using their huge cash reserves to force out small and independent businesses ... And so we live in a double world: carnival on the surface, consolidation underneath, where it counts.

Although global cultural practices are digested, adapted and resisted by places and certain social groups, there is an increasingly transnational secular ideology (Held *et al.*, 1999) where the brand is king and queen. In the context of nightlife, then, consumption options continue to be curtailed. A further aspect of this non-local branded space is a clear functional separation between the spheres of consumption and production. To quote Klein (2000: 346) again: 'The planet remains sharply divided between producers and consumers, and the enormous profits raked in by the superbrands are premised upon these worlds remaining as separate as possible.' In this sense, larger non-local venues do little to promote or connect with existing cultural practices, and in general they are directed from remote head offices and have scant interest or knowledge of local musical tastes, styles and habits. Corporate entities are given free reign to produce, market and sell nightlife and alcohol, and accrue vast profits. Yet they are largely left free to regulate themselves, taking little responsibility for nightlife problems such as violence, noise and social segregation. Increasingly, then, what we are witnessing is the steady growth of 'cities of indifference' (Fincher and Jacobs, 1998), in which corporate, profit-led and alcohol-dominated nights out do little for developing meaningful and humane public cities.

Ways beyond the corporate nightlife machine

Are there ways beyond present forms of work and play which are more bound to human creativity and expression, rather than just the profit motive? It is worth keeping in mind that entertainment remains strongly linked to the world of capitalist work, and the general orientation of nightlife is towards profit rather than creativity and social inclusion. As long as this is the case, the hallmarks of urban nightlife are likely to be standardisation, exclusion and social and spatial segregation. Current regulatory practices also largely fail to address the core of the problem – the lack of a diverse range of nightlife activities and a predominance of mainstream alcohol-drinking cultures.

However, just as there are possibilities beyond a work-based society (Gorz, 1999), there are alternatives to current entertainment and leisure provision, including nightlife. Beck (2000), for instance, talks of the antithesis of the work society – the creation of a civil society built on willing, community labour which addresses some of the problems of the current capitalist system, such as employment shortages and routinised, demoralising jobs. Similarly, what continues to be under-represented in contemporary urban nightlife, are activities which do not simply reproduce capitalist work and leisure relations (Kane, 2000). Of course, such ideas are nothing new. Several decades ago, Bertrand Russell (1932) highlighted the need to reduce work and increase leisure, weaving the two together to tap into and increase human creativity. The pertinent question which remains to be answered is: how do we transcend current patterns of profit-led, urban nightlife when they are so closely tied to the whims of the capitalist labour market, and consequently to many people's means of survival? Under what conditions can oppositional forms of nightlife emerge which are genuinely democratically controlled rather than constrained by both the state and market? And what is the genuine potential for developing entertainment and nightlife spaces which spring from people rather than relations of profit?

There is, of course, no ready-made blueprint for researchers or activists in terms of challenging corporate orthodoxy – instead, there are only a range of strategies and practices emanating out of the politics of the post-Fordist city (Mayer, 1994). In this book we have highlighted examples of oppositional and alternative youth cultures such as the squatters' movement, DiY culture and the rave scene, as well as from more independent clusters of nightlife producers (Richards, 1997; Bey, 1991;

McKay, 1998; Wates, 1980; Corr, 1999). What is the value of such alternative spaces beyond the corporate nightlife machine? It is often hard to gauge the potential of such transient spaces because many defy definition and occupy constantly shifting ground. While some participants suggest that oppositional cultures need to remain 'underground' and 'on the margins' in order to be protected from co-option or annihilation by those in power, others look for points of contact with the mainstream and wider communities. Whatever their hopes and aspirations, and however temporary some of them are, oppositional and alternative spaces as outlined in Chapter 9 are clearly a vital part of urban life (see Plate 10.1). They function as important sources of creativity and provide opportunities for experimentation and transgression. Beyond the abstract, commodified, rigid spaces of capitalism, these differential spaces contain potentialities and prescriptions for different urban futures. Some commentators have suggested that this 'third space' may be the basis for new types of politics and notions of civil society – places where a more hybrid sense of identity can truly flourish (Pile, 1994; Soja, 1996). According to Sibley (1998), we need to be more open to such 'spaces of difference' and find new ways of appreciating and accounting for marginal activity 'in its own terms', rather than simply as a 'subversive other' for respectable society (see also Sibley, 2001).

Alternative marginal nightlife spaces also represent the importance of the 'use' rather than the 'exchange' value of the city. The margins halt the drift towards passivity – the tendency for market forces and vested interests to encourage us to consume rather than create (Haslam, 1999). Grass-roots independent culture signals a desire to be involved and to produce, not just to consume. Live to work, not merely work to live, as the saying goes. In many ways, nightlife on the margins is a constant reminder of the need to challenge, resist and transgress the 'taken for granted' boundaries of the city. While many forms of resistant nightlife do not and cannot provide concrete answers for issues such as under-employment, global capital flows or unequal gender relations, they do present powerful critiques of the boundaries we live by. As Creswell (1996: 166) suggests, 'Within transgression lies the seeds of new spatial orderings.' As we have shown, squats and free parties subvert established geographies and normative landscapes. They sketch out possible alternative worlds based upon collective ownership, non-hierarchical decision-making and ecological and social awareness. Yet we also have to be aware of imitations here, not just in terms of the failed experiments by groups who attempt to create more equal and inclusive social relations, but also trends towards commodification, commercialisation and gentrification which incorporate and limit alternative experiments. As we outlined in Chapter 9, it is also important to temper some of the rather utopian readings of rave culture and the associated possibilities for radical social change beyond the dance floor.

Resistance continues to come from a variety of quarters, not simply anarchists, environmentalists, punks and urban social movements, but also citizens' groups, local

Squatters take over doomed building

Protest at big business dominating city centre

By Tony Henderson
Environment Editor

PROTESTERS have taken over a building at the centre of a planned £60m leisure complex.

The building in Pilgrim Street, Newcastle, which has been empty since being vacated by the Children's Warehouse recycling body two years ago, is due to be demolished as part of developer J J Gallagher's Electric City leisure scheme which has been backed by city councillors.

The group which has seized the building say the action is to highlight the way they feel many people are being squeezed out of a city centre increasingly dominated by expensive leisure developments like "style bars" and multiplex cinema plans.

"We are witnessing a corporate takeover of the city centre by big companies which can afford to develop buildings and sites," said a member of the group, Newcastle University lecturer Paul Chatterton.

"Big business is dictating what happens in city centres and are concentrating on quick profit schemes such as bars, chain pubs, multiplexes, casinos and what are called family entertainment centres."

Mr Chatterton said that this was leading to a lack of variety in the city centre with no space allocated to smaller groups, interests and activities – and that such a lack of diversity could hinder Newcastle and Gateshead's bid for the Capital of Culture title in 2008.

"There is nowhere for local bands to showcase their music, for resource centres or meeting places, studios or craft workshops despite the fact that there is so much empty space in the city centre," said Mr Chatterton.

"We need places which are affordable and accessible for a diversity of uses. But a lot of people are being priced out of the city centre.

"We are increasingly living in corporate city which means a lot of expensive sameness with more and more people, especially younger people, being excluded and feeling they don't belong."

The group has opened a free cafe in the building, which it has christened Eclectic City, and has also provided a music venue using bicycle-powered generators, library and environmental centre with plans for recycling facilities for local businesses.

The Electric City development will include a family entertainment centre, multiplex, restaurants and bars.

Work has also started on another £60m leisure development, The Gate in Newgate Street, Newcastle, which will also include a multiplex, Sky Bar and other bars and restaurants.

Mr Chatterton said: "The group includes social workers, doctors, students and a whole spectrum of people. It is a commonsense group which cares about the city centre."

He said that there were concerns that rising city centre property prices were also excluding small scale activities and the architecture of the new developments fell short of what was needed for a historic city centre.

Kevan Jones, city council cabinet member for development, said: "Newcastle is a regional capital and it has to provide these sort of leisure facilities and if it means jobs and investment I make no apologies for that."

tony.henderson@ncjmedia.co.uk

Picture: Simon Hobson

Determined: Protesters outside the building in Pilgrim Street, Newcastle.

Plate 10.1 Challenging the corporate city: activists open the squat 'Eclectic City' in Newcastle upon Tyne, UK

Source: *The Journal*, 17 October 2000: 6

communities, individuals and, in some cases, even radical local states. Ironically, while Klein (2000) mentions that young people in their role as avid consumers were seen as saviours of the brand in the early 1990s, a section of them also form the most voracious pockets of resistance. The growing anti-globalisation agenda has in various ways brought attention to the corporate control of urban space, while the 'free party', traveller underground (McKay, 1996) and 'rave' scene (Reynolds, 1998) have provided a constant thorn in the side of corporately packaged nightlife. Examples of resistance also exist at the local political level, such as the selective purchasing agreements signed by at least twenty-two cities in the USA barring the purchase of goods and services from firms which break certain ethical codes (Klein, 2000). France has been particularly resistant to the growing power of multinationals and their brands. The French government's attempts to ensure 51 per cent EU output on national TV broadcast channels are legendary, as is the country's anti-globalisation hero, Jose Bove, who highlighted the deleterious effects of large foreign multinationals such as McDonald's on local social and economic life (Bove *et al.*, 2001). There are also Italy's Slow Food and Slow City movements, which have highlighted the ills of global culture on issues such as food, health and community life (Carrol, 2000). Much opposition exists despite the fact that organising against corporations is difficult and often illegal in the face of supra-national legislation such as the World Trade Organisation's anti-protectionist rules, while international human rights law only applies to states and not to companies.

Nightlife has not been a sector which has received much attention from the anti-globalisation and anti-corporate movement (although see Fawcett, 2002). In many ways this is understandable, as the reality of ownership concentration and foreign control by highly acquisitive international financial houses largely hides behind traditional, place-based brands. Consumers too have little knowledge of, or interaction with, companies such as Nomura or Interbrew, whose explicit logic is market growth and domination. A greater awareness of such corporate operators, and their strategies of branding, market domination and segmentation within contemporary urban nightlife, is a first step towards producing more democratic, creative and diverse nightlife infrastructures in our cities, and we hope that this book has contributed something here. The anti-globalisation movement might do well to widen their sights to include the entertainment industries in their analysis, including nightlife, as ironically this is where they will find young people most vigorously interacting with the corporate machine. As we have highlighted in this book, a desire for change and more effective and widespread resistance towards some of the global brand bullies may also come from within the mainstream, in the form of disillusioned consumers who are rejecting the blandness of corporate branded environments and seeking healthier, fairly priced and more 'authentic' lifestyle experiences.

What future lessons arise from our exploration of the making and remaking of urban nightscapes? We have shown a number of pathways which researchers and students can pursue in their own work; for example: looking beyond the corporate gloss, unpacking economic relations, mergers and corporate strategies across time and space; thinking about the strategies of regulatory bodies such as the police and judiciary, and the motives of the contemporary entrepreneurial state; examining the everyday desires, aspirations and motives of young people. More practically, whatever city we look at across the west, there seems to be an air of inevitability or macroneccessity (Jessop *et al.*, 1999) associated with how nightlife may develop in the future. Greater influence of brands, more corporate domination, more market concentration, increased moral panics about a youth generation out of control – all are hallmarks here.

However, there are a number of different choices and ways forward, each of which have different policy implications for nightlife entrepreneurs, the local state, police and consumers. First, downtown areas can continue to become 'Anywheresville' and embrace the global corporate world, hoping that they can become its 'flavour of the month' bringing in all the big brands such as Starbucks, Hard Rock Café, McDonald's or Gap. This very much appears to be the current trend. Localities can get lost in hype and begin to substitute image for reality, advertising over people. In terms of nightlife, they can continue to go to great lengths to attract major developers and entertainment and nightlife operators. As such, smaller, locally owned nightlife spaces will continue to be squeezed and marginalised, with cities experiencing what Harvey (1989b) refers to as 'serial reproduction', losing their uniqueness and distinctive flavour as they become like other urban places in the west. The problem is that, even if successful in the short term, when a particular city eventually falls out of favour and corporate capital moves elsewhere (Harvey's (1989c) 'spatial fix') there will be little local infrastructure to build on.

Balancing the global, the national and the local is seen as a middle pathway. This would involve the local state working together with all interested parties in the night-time economy, and not allowing sectional interests and the profit motive alone to influence the types of nightlife growth. In such a context, there is a need for the local state to play a stronger role in the development of the night-time economy, especially to strike a balance between commercial and local need and all users of the city, whoever they may be. While many city authorities claim that they are making inroads here, the privatisation and corporatisation of downtown areas will carry on apace as long as they seek even greater returns on property, with only large commercial developers having sufficient resources to put derelict buildings back into use. Moreover, the views of nightlife consumers themselves are rarely heard, and there remain often large 'experiential gaps' between those who consume nightlife and those who govern it.

Alternatively, the local state and other formal regulators could take a step back and concede decision-making, space and finances to local communities and political groups to create grass-roots local nightlife cultures which emphasise diversity, creativity and difference. To encourage such a pathway, governance needs decentralising and mechanisms would need to be established to restrict large operators who are geared towards solely maximising profit and alcohol sales, and to open up many more opportunities for communities and local entrepreneurs. Moreover, it would point to a significant change in cultural values and philosophies based around a more shared sense of public space, encouraging the intermingling of different age groups and mixed night-time activities in which alcohol consumption, on its own, plays a much smaller role. The objectives of this approach would be to stimulate diversity, creativity and more democratic relations between producers and consumers – involving young adults as active contributors to nightlife culture, rather than passive consumers. Which way many cities go is still up for grabs. We have suggested that unhindered urban nightlife appears in many parts of the west to be heading down the corporate route. Much effort, activism and support for alternative nightlife provision will be needed to steer another course. Only then can nightlife cultures be built which counter-balance the seductiveness of corporate glam and the entrepreneurial city, and instead respond to the interests of people rather than just the profits of the corporate nightlife machine.

NOTES

1 Introduction: making urban nightscapes

1 'El Maremagnum no renovara las licencias de los bares nocturnes', *El Periodico*, 11 February 2002: 2–3; 'El huido del Caiprinha se entrega', *El Periodico*, 8 February 2002: 1–3; 'Prisión para los tres detenidos por el crimen de un ecuatoriano en el puerto de Barcelona', *El Pais*, 31 January 2002: 16.

2 For example, young adults in the UK are 20 per cent more likely to visit pubs and ten times more likely to frequent clubs than the general population (Mintel 1998, 2000), hence our assertion that night-time spaces are primarily youth spaces.

2 Producing nightlife: corporatisation, branding and market segmentation in the urban entertainment economy

1 This chapter is a version of an article to be published in the *International Journal for Urban and Regional Research* (see Hollands and Chatterton, 2003).

2 Fordism was largely based around a social consensus involving Keynesian welfare state management within a coherent national space, collective wage bargaining, tight fiscal and macroeconomic policies and the simultaneous creation of uniform mass-consumer goods and uniform mass-consumer markets. National markets for mass-consumer goods such as the motor car, electrical goods, home furnishings and clothing were all fuelled by increasing the purchasing power of the mass of workers.

3 Scott's (2000) definition of the cultural economy is rather broad and focuses on the cultural-product sector, where he provides case studies on the jewellery, furniture, film, recorded music and multi-media industries. Zukin (1995), meanwhile, utilises the term 'symbolic' economy, which includes finance, media and entertainment, although she gives little attention to either popular culture (outside of Disney) or the nightlife scene. Our use of the term 'entertainment economy' borrows more specifically on Hannigan's (1998) usage, as it includes themed restaurants and bars, and nightclubs, as well as casinos, sports stadia, arenas and concert hall/music venues, multiplex cinemas and various types of virtual arcades, rides and theatres. Again, unfortunately he largely ignores looking at bars and nightclubs, a crucial element of the night-time economy (see Chatterton and Hollands, 2002).

4 Traditionally, gentrification relied very much on a physical description of the phenomenon and its effects, rather than a theoretical explanation of why the process

exists or why it has accelerated and declined in particular historical periods (Butler, 1997). Originally the term was coined by Ruth Glass in the early 1960s to describe the invasion and displacement of working-class areas by the middle classes in urban areas in the UK (particularly London). Primarily it referred to this process in terms of housing, but also generally the transformation of neighbourhoods (i.e. schools, shops, leisure provision), and only more recently entertainment and nightlife (see Ley, 1996).

5 See http://assembly.state.ny.us/leg/?cl=5&a=4.
6 See http://192.75.156.68/DBLaws/Statutes/English/90l12_e.htm.

3 Regulating nightlife: profit, fun and (dis)order

1 Part of this chapter appears in the Journal *Entertainment Law* (see Chatterton, 2002a).
2 While the legal age is 21 in most of the USA, twenty-nine states have reduced the legal age for drinking alcohol.
3 See 'Blair to propose 48-hour shutdown for rowdy pubs in summit on lawlessness', *Guardian*, 3 July 2000: 12; 'Police win powers to shut down thug bars', *Observer*, 2 July 2000: 5; 'Colonising the night', *Guardian*, 12 September 2000: 5; and 'Straw to target drink-related crime', *Guardian*, 18 July 2000: 6.

4 Consuming nightlife: youth cultural identities, transitions and lifestyle divisions

1 See Dennis Potter's *Lipstick on Your Collar* for a filmic representation of this transformation in the UK; also see Edwards (2000) and Chaney (1996) for discussions of the rise of 'consumer society'.

5 Pleasure, profit and youth in the corporate playground: branding and gentrification in mainstream nightlife

1 For more information see: http://www.hooters.com/companyinfo/media/; http://www.firkinpubs.com; http://www.daveandbusters.com; http://www. primerestaurants.com/whois.htm; http://www.primefranchise.poweredbyego.com/Student.PDF.
2 The Essex man and girl stereotype refers to the county of Essex near London in the UK, and stems from popular opinion and jokes associated with a group of young, upwardly mobile people who lived there and have since benefited from the economic boom of the 1980s in the south-east. This group spent their newfound money on ostentatious consumer goods and going out. The 'ladettes' phenomenon in the UK refers to those women who, again according to popular images, mimicked and outcompeted men in excessive drinking and overt shows of sexuality during a night out.

6 Selling nightlife in studentland

1 In the USA, the Greek system of fraternities and sororities is a distinctive element of American college life, a particularly pervasive structure reverberating throughout the life of students. Many of the frat and sorority houses are huge and elaborate, occupying thirty-bedroom detached houses employing full-time cooks, cleaners and house-moms, networked computer rooms and private weekend retreats. The Greek system, of course, has received much criticism for holding wild parties, encouraging excessive drinking

and dangerous initiation ceremonies. However, certain rituals in the Greek community, such as hazing, are being outlawed as they involve practices such as 'paddling', public stunts, physical and psychological shocks and the forced use of excessive alcohol, which has in some cases led to the death of rushees. Further, 'the fraternities have traditionally had an image of macho attitudes, drunkenness and intolerance ... and discrimination against black, Jewish and other groups of students' (Silver and Silver, 1997: 52). This closed world of tradition and ritual, then, can be a divisive as well as an integrating element on campus. The Greek community is structured around three councils: the National Panhellenic Conference (NPC) of social sororities for women, the National Interfraternity Council of Fraternities (NIC) for men, and the National Pan-Hellenic Conference (NPHC) of historically African-American fraternities and sororities. Each council contains several national chapters, all of which are represented by Greek letters such as ΔΣ (Delta Theta Sigma) or BΦ (Pi Beta Phi). Joining the Greek system entails life-long membership of a particular chapter and adherence to its principles such as leadership, scholarship and philanthropy.

2 While students are constructed by the media, businesses and the university itself, they also define, generate and circulate self-images through niche and micro-media such as fliers, posters, zines, campus newspapers and websites. Many are acutely aware of their own and others' specific subcultural 'student' identity (Hollands, 1995). Many also actively resist being labelled as a 'typical student', as this is regarded as insulting to their own individual take on constructing the experience, and is reserved for someone less experienced at playing the game.

3 These graphs were constructed from time diaries completed by fifty students at Bristol University in the UK and at UW Madison in the USA.

7 Sexing the mainstream: young women and gay cultures in the night

1 For more information see: http://www.hooters.com/companyinfo/media/; http://www.ahwatukee.com/afn/community/articles/010110a.html.

8 Residual youth nightlife: community, tradition and social exclusion

1 While we recognise that there are a range of other, especially older, social age groups that inhabit residual nightlife spaces, our focus in this chapter is specifically on young adults. Furthermore, we are aware that while some may argue that such residual youth groupings, activities and spaces do not technically constitute 'nightlife' *per se*, we would argue in fact that there is a close relationship between the gentrifying commercial mainstream and attempts to incorporate, accommodate and/or sanitise the city of such groups and residual activity.

2 It is important to note that half of Afro-American households are in effect middle class (Oswald, 2001) and that some of the upward trends regarding crime, poverty and teenage birth rates began to reverse in the 1990s. This has two important implications. First, overall figures in fact downplay the desperate plight of the very poorest black youth and their families. And second, we should not assume that black cultures of the street equate to all black youth. Middle-class black youth cultures are very different from their poorer counterparts and may indeed be more comparable with mainstream white American youth cultures.

9 'You've gotta fight for your right to party': alternative nightlife on the margins

1 Section 6 of the Criminal Law Act 1977 in the UK provides squatters with some minimum legal rights.

2 In the late 1980s, rough estimates suggest that the Netherlands had nearly 45,000 squatted settlements, while London had 31,000 squatters and West Berlin had 5,000 (Corr, 1999).

REFERENCES

Abrams, M. (1959) *The Teenage Consumer*, London: Press Exchange.

Adams, B. (1987) *The Rise of the Gay and Lesbian Movement*, Boston: Twayne Publishers.

Adler, M. (1991) 'From symbolic exchange to commodity consumption: anthropological notes on drinking as a symbolic practice', in S. Barrows and R. Room (eds) *Drinking: Behaviour and Belief in Modern History*, Berkeley: University of California Press, pp. 376–89.

Advisory Service for Squatters (1996) *Squatter's Handbook 10th Edition*, London: ASA.

Aggleton, P. (1987) *Rebels without a Cause*, Lewes: Falmer Press.

Aglietta, M. (1979) *A Theory of Capitalist Regulation*, London: New Left Books.

Ainley, P. (1994) *Degrees of Difference: Higher Education in the 1990s*, London: Lawrence and Wishart.

Aitkenhead, D. (2001) 'Village people', *Guardian*, 24 October: 7.

Aitkenhead, D. and Sheffield, E. (2001) 'G-strings join A list', *Evening Standard*, 11 July.

All About Beer (1997) *Government Probes A–B Distribution Practices*. Online. Available http://www.allaboutbeer.com/news/1997archives.htm (accessed 12 March 2002).

Altman, D. (1997) 'Global gaze/global gays', *GLQ: A Journal of Lesbian and Gay Studies* 3: 417–36.

Amin, A. (1994) *Post-Fordism: A Reader*, London: Blackwell.

Andersson, B. (2002) *Oppna Rum: Om Undomarna, Staden Och Det Offentliga Livet* (Open Space: Youth, the City and Public Life), Goteborg: Goteborg University, Department of Social Work.

Andrews, C., Klodawsky, F. and Lundy, C. (1994) 'Women's safety and the politics of transformation', *Women and Environments* Summer/Fall: 23–6.

Aronowitz, S. (1973) *False Promises: The Shaping of American Working Class Consciousness*, New York: McGraw Hill.

Ashby, E. and Anderson, M. (1970) *The Rise of the Student Estate in Britain*, London: Macmillan.

Atkinson, R. (2001) *Domestication by Cappuccino or a Revenge on Urban Space*, Occasional Paper, Glasgow: University of Glasgow.

Australian Institute of Health and Welfare (1999) *1998 National Drug Strategy Household Survey: First Results*, Canberra: Australian Institute of Health and Welfare.

Back, L. (1996) *New Ethnicities and Urban Cultures: Racisms and Multiculture in Young Lives*, London: UCL Press.

Bagguley, P. and Mann, K. (1992) 'Idle, thieving, bastards: scholarly representations of the "underclass"', *Work, Employment and Society*, 6 (1), March: 113–26.

Bailey, E. (1997) 'Party on! You're only thirtysomething', *Independent*, 27 April: 3.

Bailey, P. (1978) *Leisure and Class in Victorian England: Rational Recreation and the Contest for Control 1830–1885*, London: Routlege & Kegan Paul.

Bailey, R.W. (1999) *Gay Politics, Urban Politics: Identity and Economics in the Urban Setting*, New York: Columbia University Press.

Ball, S., Maguire, M. and Macrae, S. (2000) *Choice, Pathways and Transitions Post-16: New Youth, New Economies in the Global City*, London: Routledge/Falmer.

Banks, M. (ed.) (1991) *Careers and Identities*, Milton Keynes: Open University Press.

Bannister, J. and Fyfe, N. (2001) 'Introduction. Fear and the city', *Urban Studies* 38 (5–6): 807–13.

Bannister, J., Fyfe, N. and Kearns, A. (1998) 'Closed circuit television and the city', in C. Norris, J. Moran and G. Armstrong (eds) *Surveillance, Closed Circuit Television and Social Control*, Aldershot: Ashgate.

Barke, M., Braidford, P., Houston, M., Hunt, A., Lincoln, I., Morphet, C., Stone, I. and Walker, A. (2000) *Students in the Labour Market*, Department for Education and Employment. Research Report 215, London: DfEE.

Barnard, M. (1999) 'Ladettes large it in Liverpool'. *The Times*, 'Weekend', 18 September: 29.

Batchelor, A. (1998) 'Brands as financial assets', in S. Hart and J. Murphy (eds) *Brands: The New Wealth Creators*, Basingstoke: Macmillan.

Bauman, Z. (1988) *Freedom*, Buckingham: Open University Press.

——(1997) *Postmodernity and its Discontents*, Cambridge: Polity Press.

——(1998) *Work, Consumerism and the New Poor*, Buckingham: Open University Press.

Beccaria, F. and Sande, A. (2002) 'Young people's use of alcohol in the rite of passage to adulthood in Italy and Norway', unpublished paper, University of Bodo, Norway.

Beck, U. (1992) *Risk Society: Towards a New Modernity*, London: Sage.

——(2000) *The Brave New World of Work*, Cambridge: Polity.

Becker, H. (1966) *Outsiders: Studies in the Sociology of Deviance*, New York: Free Press.

Bell, D. (1961) *The End of Ideology* New York: Crowell-Collier.

——(1973) *The Coming of Post Industrial society*, London: Heinemann.

Bellas, M.C. (2001) 'Worldview. Beer vs CSDs: routes to concentration', *Beverage World International Magazine*, May/June.

Bender, T. (ed.) (1988) *The University and the City: From Medieval Origins to the Present*, Oxford: Oxford University Press.

Bennett, A. (2000) *Popular Music and Youth Culture: Music, Identity and Place*, Basingstoke: Macmillan.

Berger, B. (1963) 'Adolescence and beyond', *Social Problems* 10 (Spring): 294–308.

Berman, M. (1986) 'Taking it to the streets: conflict and public space', *Dissent* Fall: 476–85.

Best, B. (1998) 'Over-the-counter culture: retheorizing resistance in popular culture', in S. Redhead, D. Wynne and J. O'Connor (eds) *The Clubcultures Reader*, London: Blackwell.

Bey, H. (1991) *TAZ: The Temporary Autonomous Zone, Ontological Anarchy, Poetic Terrorism*, Brooklyn: Autonomedia.

Bianchini, F. (1995) 'Night cultures, night economies', *Planning Practice and Research* 10: 121–6.

Bilefsky, D. (2000) 'Interbrew offer set to raise Euros 3.3bn', *Financial Times*, 9 November: 1.

Blackman, S. (1997) 'Destructing a giro: a critical and ethnographic study of the youth "underclass"', in R. MacDonald (ed.) *Youth, the 'Underclass' and Social Exclusion*, London: Routledge.

——(1998) 'Poxy Cupid! An ethnographic and feminist account of a resistant female youth culture: the New Wave girls', in T. Skelton and G. Valentine (eds) *Cool Places: Geographies of Youth Cultures*, London: Routledge.

Bloom, A. (1987) *The Closing of the American Mind: How Higher Education has Failed Democracy and Impoverished the Souls of Today's Students*, New York: Simon and Schuster.

BLRA (Brewers and Licensed Retailers Association) (1999) *Statistical Handbook*, London: BLRA.

Bocock, R. (1993) *Consumption*, London: Routledge.

Boddy, M., Lovering, J. and Bassett, K. (1986) *Sunbelt City?* Oxford University Press: London.

Body-Gendrot, S. (2000) *The Social Control of Cities: A Comparative Perspective*, Oxford: Blackwell.

Bond, C. (2001a) '"Slums" claim as students swamp area', *Yorkshire Evening Post*, 7 November: 8.

——(2001b) 'Crime levels increase', *Yorkshire Evening Post*, 25 April: 8.

Booth, C. (1902) *Life and Labour of the People in London. First Series: Poverty. First Volume: East, Central and South London*, London: Macmillan.

Borden, I. (2001) *Skateboarding, Space and the City*, Oxford: Berg.

Borden, I., Rendell, J., Kerr, J. with Pivaro, A. (eds) (2001) *The Unknown City: Contesting Architecture and Social Space: A Strangely Familiar Project*, Cambridge: MIT Press.

Bourdieu, P. (1977) *Outline of a Theory of Practice*, Cambridge: Cambridge University Press.

——(1984) *Distinction: A Social Critique of the Judgement of Taste*, London: Routledge.

——(1993) *The Field of Cultural Production*, Cambridge: Polity Press.

Bourdieu, P. and Passeron, J.C. (1979) *The Inheritors: French Students and their Relation to Culture*, London: University of Chicago Press.

Bouthillette, A. (1994) 'Gentrification by gay male communities: a case study of Toronto's Cabbagetown', in S. Whittle (ed.) *The Margins of the City: Gay Men's Urban Lives*, Aldershot: Arena.

Bove, J., Dufour, F. and de Casparis, A. (2001) *The World is Not for Sale*, London: Verso.

Brain, K.J. (2000) *Youth, Alcohol and the Emergence of the Post-Modern Alcohol Order*, Occasional Paper 1, London: Institute of Alcohol Studies.

Brake, M. (1985) *Comparative Youth Culture*, London: Routledge.

Brass, E. and Koziell, S.P. (1997) *Gathering Forced. DiY Culture – Radical Action for Those Tired of Waiting*, London: The Big Issue.

Brigett, D. (2001) 'Strip club faces "brothel" probe', *Evening Standard*, 27 July: 18.

Brockes, E. (2000) 'Yorkie bars', *Guardian*, 4 March: 2.

Brook, T.V. (1996) 'Politics studies bore freshmen, survey says', *Madison State Journal*, 17 January: 2.

Brush, L. (1999) 'Gender, work, who cares?! Production, reproduction, deindustrialization, and business as usual', in M. Marx Ferree, J. Lorber and B. Hess (eds) *Revisioning Gender*, London: Sage.

Buckland, F. (2002) *Impossible Dance: Club Culture and Queer World-Making*, Middletown: Wesleyan University Press.

Burning Man (2001) 'What is Burning Man?' Online. Available http://www.burningman.com/whatisburningman/about_burningman/experience.html (accessed 14 December 2001).

Burrows, R. and Loader, B. (eds) (1994) *Towards a Post-Fordist Welfare State?* Routledge: London.

Bussmann, J. (1998) *Once in a Lifetime – the Crazy Days of Acid House and Afterwards*, London: Virgin Books.

Butler, T. and Robson, J. (2001) *The Middle Classes and the Future of London*, Cities Summary No. 12, Economic and Social Research Council Cities Programme, Liverpool: John Moores University.

Butler, T. (1997) *Gentrification and the Middle Classes*, Aldershot: Ashgate.

Byrne, D. (1999) *Social Exclusion*, Buckingham: Open University Press.

Byrne, E. (2001) 'Bristol's style revolution', *Venue Magazine*, July: 23.

Callender, C. and Kempson, E. (1996) *Student Finances, Income, Expenditure and Take-up of Student Loans*, London: PSI.

Campbell, B. (1993) *Goliath: Britain's Dangerous Places*, London: Methuen.

——(1996) 'Girls on safari', *Guardian*, 15 July: 4–5.

Campbell, C. (1995) 'The sociology of consumption', in D. Miller (ed.) *Acknowledging Consumption*, London: Routledge.

Campbell, J.R. (1995) *Reclaiming a Lost Heritage: Land-grant and Other Higher Education Initiatives for the Twenty-first Century*, Iowa: Iowa State University Press.

Cantelon, H. and Hollands, R. (1988) (eds) *Leisure, Sport and Working Class Cultures: Theory and History*, Toronto: Garamond.

Carat Insight (2001) *Consumer Connection Study*, London: Carat Media Group.

Carrol, R. (2000) 'Protesters to try and halt rise of fast food giants in Italy', *Guardian*, 17 October: 10.

Cashmore, E. (1997) *The Black Culture Industry*, London: Routledge.

Castells, M. (1982) *The City and the Grassroots. A Cross Cultural Theory of Urban Social Movements*, London: Arnold.

CCCS (1982) *The Empire Strikes Back: Race and Racism in 70s Britain*, London: Hutchinson.

Censorfreeworld (1998) 'Four arguments for the elimination of squeegie kids!' Online. Available http://www.Censorfreeworld.com/Articles/skids.html (accessed 15 April 2002).

Chan, S. (1999) 'Bubbling acid: Sydney's techno underground', in R. White (ed.) *Australian Youth Subcultures: On the Margins and in the Mainstream*, Hobart: Australian Clearing House for Youth Studies.

Chaney, D. (1996) *Lifestyle*, London: Routledge.

Charles, D. (2000) *Universities in Regional Development. Final Report*, Newcastle: CURDS/University of Newcastle.

Chasin, A. (2001) *The Gay and Lesbian Market Goes to the Market*, London: Palgrave.

Chatterton, P. (1998) 'The university and the community: an exploration of the cultural impacts of universities and students on the community' unpublished thesis, Bristol: University of Bristol.

——(1999) 'University students and city centres – the formation of exclusive geographies. The case of Bristol, UK', *Geoforum*, 30: 117–33.

——(2000) 'Will the real creative city please stand up', *City* 4 (3): 390–7.

——(2002a) 'Governing nightlife. Profit, fun and (dis)order in the contemporary city', *Entertainment Law* 1 (2) (Summer): 23–49.

——(2002b) '"Squatting is still legal, necessary and free." A brief intervention in the corporate city', *Antipode* 34 (1): 1–7.

Chatterton, P. and Hollands, R. (2001) *Changing Our Toon: Youth, Nightlife and Urban Change in Newcastle*, Newcastle: University of Newcastle.

——(2002) 'Theorising urban playscapes: producing, regulating and consuming youthful nightlife city spaces', *Urban Studies* 39 (1): 95–116.

Chatzky, J.S. (1998) 'So young – and so deep in debt', *USA Weekend*, 6–8 March: 2.

Chaudhuri, A. (1994) 'Deadlier than the male', *Guardian*, 13 October: 14.

——(2001) 'Too much of a good thing?' *Cosmopolitan* magazine: 52–7.

Chauncey, G. (1994) *Gay New York: Gender, Urban Culture and the Making of the Gay Male World 1890–1940*, New York: Basic Books.

Chris (2001) 'Pub/Oxford point', Nightshift discussion list. Online posting. Available http://nightshift.oxfordmusic.net/wwwboardb/messages/780.html (accessed 24 September 2001).

Chrisafis, A. (2001) 'Behaving rather badly', *Guardian*, 6 November: 12.

Christopherson, S. (1994) 'The fortress city: privatised spaces, consumer citizenship', in A. Amin (ed.) (1994) *Post-Fordism: A Reader*, London: Blackwell.

Clark, A. (2001) 'Thin end of the dynastic wedge', *Guardian*, G2 supplement, 25 July 2001: 2.

Clarke, B. and Bradford, M.G. (1998) 'Public and private consumption and the city', *Urban Studies* 35 (5–6): 858–88.

Clarke, D. (1997) 'Consumption and the city, modern and postmodern', *International Journal of Urban and Regional Research* 21 (2): 18–37.

Clarke, J. and Critcher, C. (1985) *The Devil Makes Work: Leisure in Capitalist Britain*, Basingstoke: Macmillan.

Cochrane, A. (ed.) (1987) *Developing Local Economic Strategies*, Milton Keynes: Open University Press.

Coffield, F. and Gofton, L. (1994) *Drugs and Young People*, London: Institute for Public Policy Research.

Cohen, P. (1997) *Rethinking the Youth Question*, London: Macmillan.

Cohen, P. and Ainley, P. (2000) 'In the Country of the Blind? Youth studies and cultural studies in Britain', *Journal of Youth Studies* 3 (1): 79–95.

Collins, D.C. and Kearns., R.A. (2001) 'Under curfew and under siege? Legal geographies of young people', *Geoforum*, 32 (3): 389–404.

Collins, M. (1997) *Altered State: The Story of Ecstasy Culture and Acid House*, London: Serpent's Tail.

Collinson, P. (2001) 'Buy-to-let misery for students' neighbours', *Guardian*, 5 May: 7.

Comedia and Demos (1997) *The Richness of Cities: Urban Policy in a New Landscape*, London: Comedia and Demos.

Confederation of Belgian Breweries (2001) 'Profile of the Belgian brewery industry', press release, Brussels, 27 June.

Connell, J. (1999) 'Beyond Manila: walls, malls, and private spaces', *Environment and Planning A*, 31 (3) (March): 417–40.

Corr, A. (1999) *No Trespassing. Squatting, Rent Strikes, and Land Struggles Worldwide*, San Francisco: South End Press.

Corrigan, P. (1976) 'Doing nothing', in S. Hall and T. Jefferson (eds) *Resistance through Rituals:Youth Subcultures in Post-War Britain*, London: Hutchinson.

Coupland, D. (1992) *Generation X*, London: Abacus.

Craine, S. (1997) 'The "Black Magic Roundabout": cyclical transitions, social exclusion and alternative careers', in R. Macdonald (ed.) *Youth, the 'Underclass' and Social Exclusion*, London: Routledge.

Creswell, T. (1996) *In Place, Out of Place. Geography, Ideology and Transgression*, Minneapolis: University of Minnesota Press.

Crompton, R. (1996) 'Consumption and class analysis', in S. Edgell, K. Hetherington and A. Warde (eds) *Consumption Matters: The Production and Experience of Consumption*, Oxford: Blackwell.

——(1998) *Class and Stratification: An Introduction to Current Debates*, Cambridge: Polity Press, 2nd edition.

Cunningham, H. (1980) *Leisure in the Industrial Revolution 1780–1880*, New York: St Martin's Press.

D'Alessio, V. (1996) 'Carry on campus', *The Times Higher Education Supplement*, 15 November 1996: 4.

Darendorf, R. (1989) 'The future of the underclass: a European perspective', *Northern Economic Review*, 18 (Autumn): 7–15.

Davis, M. (1992) *City of Quartz: Excavating the Future in Los Angeles*, London: Vintage.

Davis, S. (1999) 'Media conglomerates build the entertainment city', *European Journal of Communication*, 14 (4): 12–35.

De Certeau, M. (1984) *The Practice of Everyday Life*, San Francisco: University of California Press.

De Laurentis, T. (1991) 'Queer theory: lesbian and gay sexualities. An introduction', *Differences* 3 (2) (Summer): iii–xviii.

De Rituerto, R. (2002) 'El gobierno se fija en el modelo de EE UU para sancionar a los menores que beban alcohol', *El Pais*, 13 February: 13–14.

De Vries, C. (2001) 'Cocktails, cleavage, catfights ... my cattle-market night out', *Mail on Sunday*, 2 December: 58–9.

Demetriou, D. (2001) 'Man quizzed on lapdance murder', *Evening Standard*, 2 August: 3.

Difford, S. (2000a) 'Bar boom!' *Livewire Magazine* (Great North Eastern Railways), June/July: 28–30.

——(2000b) 'The third Industrial Revolution ... it's all @ Leeds', *Livewire Magazine* (Great North Eastern Railways), October: 4.

DiMaggio, P. (1991) 'Social structure, institutions, and cultural goods: the case of the United States', in P. Bourdieu and J.S. Coleman (eds) *Social Theory for a Changing Society*, Oxford: Westview Press.

Ditton, J. (1998) 'Public support for town centre CCTV schemes. Myth or reality?' in C. Norris, J. Moran and G. Armstrong (eds) *Surveillance, Closed Circuit Television and Social Control*, Aldershot: Ashgate.

Dorn, N. (1983) *Alcohol,Youth and the State*, London: Croom Helm.

Doron, G. M. (2000) 'The dead zones and their architectural transgression', *City* 4 (2): 247–63.

Doward, J, (1999) 'Tiger Tiger plans to burn bright in city nightlife' *Observer*, 16 May: 1.

——(2001) 'Putting it on the table', *Observer*, 11 February: 7.

Du Chernatony, L. and Malcom, M. (1998) *Creating Powerful Brands in Consumer, Service and Industrial Markets*, Oxford: Butterworth Heinemann.

Du Gay, P. (1997) *Doing Cultural Studies: The Story of the Sony Walkman*, London: Sage.

Duke. C. (1992) *The Learning University. Towards a New Paradigm?*, Buckinghamshire: Open University Press.

Edwards, T. (1994) *Erotics and Politics: Gale Male Sexuality, Masculinity and Feminism*, London: Routledge.

——(2000) *Contradictions of Consumption: Concepts, Practices and Politics in Consumer Society*, Buckingham: Open University Press.

Ellis, J. (2001) '60 second interview: John Gray', *Metro*, 16 November: 14.

Emmons, N. (2000) 'No myth: Greece set for first themer', *Amusement Business*, 13 November.

Engels, F. (1968) *The Condition of the Working Classes in England 1844*, London: Allen and Unwin.

Engs, R.C. (1991) 'Resurgence of a new "clean living" movement in the United States', *Journal of School Health* 61 (4): 155–89.

——(1995) 'Do traditional Western European practices have origins in antiquity?' *Addiction Research* 2 (3): 227–39.

Engs, R.C. and Hanson, D.J. (1994) 'Boozing and brawling on campus: a national study of violent problems associated with drinking over the past decade', *Journal of Criminal Justice* 22 (2): 171–80.

Epstein, J. (ed.) (1998) *Youth Culture: Identity in a Post Modern World*, Oxford: Blackwell.

Esping-Andersen, G. (1990) *The Three Worlds of Welfare Capitalism*, Princeton: Princeton University Press.

European Commission (2001) *Common Organisation of the Market in Hops NAT/100. Opinion of the Economic and Social Committee on the Proposal for a Council Regulation amending Regulation (EEC) No. 1696/71 on the common organisation of the market in hops* (COM(2000) 834 final – 2000/0330 CNS) Brussels, 25 April.

Evans, G. (2002) *Cultural Planning. An Urban Renaissance? An International Perspective on Planning for the Arts, Culture and Entertainment*, London: Routledge.

Evening Anus (2000) 'Who killed Kenny?' Issue 251, 17 March: 2.

Everett, J.C. and Bowler, I.R. (1996) 'Bitter-sweet conversions: changing times for the British pub', *Journal of Popular Culture* 30 (2): 101–22.

Exodus (1999) 'Movement of Jah people'. Online. Available http://www.exodus.sos.freeuk.com (accessed 8 December 2001).

Exodus Collective (1998) 'Exodus movement of Jah people', in INURA (ed.) *Possible Urban Worlds. Urban Strategies at the End of the 20th Century*, Basel: Birkhauser Verlag.

Fainstein, S.S. (1994) *The City Builders: Property, Politics and Planning in London and New York*, Oxford: Blackwell.

Fawcett, M. (2002) 'A toast to globalisation!' *Ethical Consumer Magazine* February/March: 6–11.

Featherstone, D. (1998) 'The Pure Genius land occupation: re-imagining the inhuman city', in INURA (ed.) *Possible Urban Worlds. Urban Strategies at the End of the 20th Century*, Basel: Birkhauser Verlag.

Featherstone, M. (1991) *Consumer Culture and Postmodernism*, London: Sage.

Featherstone, M. and Lash, S. (eds) (1999) *Spaces of Culture: City–Nation–World*, London: Sage.

Field, F. (1989) *Losing Out: The Emergence of Britain's Underclass*, Oxford: Blackwell.

Fillon, K. (1996) 'This is the sexual revolution', *Saturday Night*, February: 37–41.

Fincher, R. and Jacobs, J. (1998) *Cities of Difference*, London: Guildford Press.

Fine, B. (1995) 'From political economy to consumption', in D. Miller (ed.) *Acknowledging Consumption*, London: Routledge.

Fine, B. and Leopold, E. (1993) *The World of Consumption*, London: Routledge.

Finnigan, K. (2001) 'Not tonight mate', *Guardian*, 24 August: 8.

Flusty, S. (1994) *Building Paranoia: The Proliferation of Interdictory Space and the Erosion of Spatial Justice*, Los Angeles: Los Angeles Forum for Architecture and Urban Design.

——(2000) 'Thrashing downtown: play as resistance to the spatial and representational regulation of Los Angeles', *Cities* 17 (2): 149–58.

Forman, M. (2002) *The 'Hood Comes First. Race, Space, and Place in Rap and Hip Hop*, Middletown: Wesleyan University Press.

Foucault, M. (1989) *The Order of Things*, London: Routledge.

Fowler, M. (1995) *The First Teenagers*, London: Woburn Press.

Frow, J. (1995) *Cultural Studies and Cultural Value*, Oxford: Clarendon Press.

Furlong, A. and Cartmel, F. (1997) *Young People and Social Change*, Buckingham: Open University Press.

Fyfe, N. and Bannister, J. (1998) '"The eyes upon the street." Closed circuit television surveillance and the city', in N. Fyfe (ed.) *Images of the Street: Planning, Identity, and Control in Public Space*, London: Routledge.

Gaines, D. (1991) *Teenage Wasteland: Suburbia's Deadend Kids*, New York: Pantheon.

Gans, H. (1990) 'Deconstructing the underclass: the term's danger as a planning concept', *Journal of American Planning Association*, 56 (3): 23–45.

Garratt, S. (1998) *Adventures in Wonderland – A Decade of Club Culture*, London: Headline.

Garreau, J. (1992) *Edge City: Life on the New Frontier*, New York: Anchor Books.

Gasperoni, G. (2000) 'University study, value orientations, cultural consumption', *Rassegna italiana di sociologia*, XLI (1) (January–March): 109–29.

Gayle, V. (1998) 'Structural and cultural approaches to youth: structuration theory and bridging the gap', *Youth and Policy* 61: 59–72.

Geddes, M. (2000) 'Tackling social exclusion in the European Union? The limits to the new orthodoxy of local partnership', *International Journal of Urban and Regional Research* 24 (2): 782–801.

General Household Survey (1999) *Living in Britain 1998*, London: HMSO.

Geoff (2001) 'What's happening to Leeds.' Online posting. Available BBC Leeds website: http://bbc.co.uk/leeds (accessed 12 September 2001).

Gershuny, J. and Fisher, K. (2000) 'Leisure', in A.H. Halsey and J. Webb (eds) *Twentieth-Century British Social Trends*, Basingstoke: Macmillan.

Gibson, C. (1999) 'Subversive sites: rave culture, spatial politics and the internet in Sydney, Australia', *Area* 31 (1): 19–33.

Gillespie, G. (1994) 'The X factor', *Venue* 327: 13–7.

Gilroy, P. (1993) *The Black Atlantic*, London: Verso.

Giroux, H. (1996) *Fugitive Cultures: Race, Violence and Youth*, London: Routledge.

Glenday, D. (1996) 'Mean streets and hard time: youth unemployment and crime in Canada', in G. O'Bireck (ed.) *Not a Kid Anymore: Canadian Youth, Crime and Subcultures*, Toronto: Nelson.

Gofton, L. (1983) 'Real ale and real men', *New Society*, 17 November: 271–3.

——(1990) 'On the town: drink and the "new lawlessness"', *Youth and Policy* 29: 33–9.

Goldscheider, F.K. and Goldscheider, C. (1999) *The Changing Transition to Adulthood: Leaving and Returning Home*, London: Sage.

Goodman, E.J., Bamford, J. and Saynor, P. (1989) *Small Firms and Industrial Districts in Italy*, London: Routledge.

Gorz, A. (1999) *Reclaiming Work. Beyond the Wage-Based Society*, Cambridge: Polity.

Goss, J. (1993) 'The "magic of the mall": an analysis of form, function, and meaning in the contemporary retail built environment', *Annals of the Association of American Geographers* 83 (1): 18–47.

Gottdiener, M. (2001) *The Theming of America. American Dreams, Media Fantasies and Themed Environments*, Boulder, CO: Westview Press, 2nd edition.

Gottdiener, M., Collins, C. and Dickens, D. (1999) *Las Vegas: The Social Production of an All-American City*, Blackwell: Oxford.

Gottlieb, J. and Wald, G. (1994) 'Smells like teen spirits: riot grrls, revolution and women in independent rock', in A. Ross and T. Rose (eds) *Microphone Fiends: Youth Music and Youth Culture*, London: Routledge.

Griese, R. (1999) 'Anger of today's youth in your face reality', *Toronto Star*, 29 July: 9.

Griffin, C. (1985) *Typical Girls?* London: Routledge.

——(1993) *Representations of Youth: The Study of Youth and Adolescence in Britain and America*, Cambridge: Polity.

——(2000) 'Absences that matter: Constructions of sexuality in young women's friendship groups', *Feminism and Psychology*, 10 (2): 227–45.

Guardian (2001) 'Brewery's half measures down at the pub', *Guardian*, 18 January: 12.

Guardian Education (2000) Untitled, *Guardian*, 27 June: 11.

Guthrie, J. (2002) 'Council views lap dancing as alien culture', *Financial Times*, 22 February: 14.

Habermas, J. (1991) *The Structural Transformation of the Public Sphere: An Inquiry into a Category of Bourgeois Society*, trans. T. Burger and F. Laurence, Cambridge, MA: MIT Press.

Hagan, J. and McCarthy, B. (1997) *Mean Streets: Youth Crime and Homelessness*, Cambridge: Cambridge University Press.

Hagendijk, R.P. (1984) 'Changes in Dutch student culture', *Netherlands Journal of Sociology* 20: 59–75.

Halfacree, K. (1999) '"Anarchy doesn't work unless you think about it": intellectual interpretation and DiY culture', *Area* 31 (3): 209–20.

Hall, C. (2001) 'Warning as women turn to drink', *Daily Telegraph*, 11 December: 4.

Hall, P. (1999) *Cities in Civilization: Culture, Innovation, and Urban Order*, Phoenix Giant: London.

——(2000) 'Creative cities and economic development', *Urban Studies* 37 (4): 639–49.

Hall, S. and Jacques, M. (1989) *The Politics of Thatcherism*, London: Lawrence and Wishart.

Hall, S. and Jefferson, T. (eds) (1976) *Resistance through Rituals: Youth Subcultures in Post-War Britain*, London: Hutchinson.

Hall, S., Critcher, C., Jefferson, A., Clarke, J. and Roberts, B. (1978) *Policing the Crisis: Mugging, the State, and Law and Order*, London: Macmillan.

Hammersley, M. (1983) *Ethnography*, London: Routledge.

Handy, C. (1989) *The Age of Unreason*, London: Hutchinson Business.

Hannigan, J. (1998) *Fantasy City. Pleasure and Profit in the Postmodern Metropolis*, Routledge, London.

Harring, S. (1983) *Policing a Class Society: The Experience of American Cities 1865–1915*, New Jersey: Rutgers University Press.

Harrison, B. (1971) *Drink and Victorians: The Temperance Question in England 1815–1872*, London: Faber and Faber.

Hart, S. (1998) 'The future for brands', in S. Hart and J. Murphy (eds) *Brands: The New Wealth Creators*, Basingstoke: Macmillan.

Harvey, D. (1989a) *The Urban Experience*, Oxford: Blackwell.

——(1989b) 'From managerialism to entrepreneurialism: The transformation of urban governance in late capitalism', *Geografiska Annaler* 71B (1): 3–17.

——(1989c) *The Condition of Post-Modernity*, Oxford: Blackwell.

——(1994) 'Flexible accumulation through urbanization: reflections on postmodernism in the American city', in A. Amin (ed.) *Post-Fordism: A Reader*, London: Blackwell.

——(2000) *Spaces of Hope*, Edinburgh: Edinburgh University Press.

Haselgrove, S. (ed.) (1994) *The Student Experience*, Buckingham: SRHE and Open University Press.

Haslam, D. (1999) *Manchester, England. The Story of the Pop Cult City*, London: Fourth Estate.

Hasse, R. and Leiulfsrud, H. (2001) 'From disorganised capitalism to transnational fine tuning? Recent trends in wage development, industrial relations and "work" as a sociological category', *British Journal of Sociology* 53 (1): 107–26.

Haughton, G. and Whitney, D. (1994) *Reinventing a Region: Restructuring in West Yorkshire*, Aldershot; Avebury.

Heath, T. and Stickland, R. (1997) 'The twenty-four hour city concept', in T. Oc and S. Tiesdell (eds) *Safer City Centres: Reviving the Public Realm*, Chapman: London, pp. 170–83.

Hebdige, D. (1979) *Subculture: The Meaning of Style*, London: Methuen.

HEFCE (Higher Education Funding Council for England) (2001) *Supply and Demand in Higher Education Consultation, 01/62*, Chelmsford: HEFCE.

Held, D., McGrew, A., Goldblatt, D. and Perraton, J. (1999) *Global Transformation*, Polity: Cambridge.

Henderson, S. (1997) *Ecstasy: Case Unsolved*, London: Pandora.

Hendry, L. (1993) *Young People's Leisure and Lifestyles*, London: Routledge.

Herbert, S. (1999) 'Policing contested space. On patrol at Smiley and Hauser', in N.R. Fyfe (ed.) *Images of the Street: Planning, Identity, and Control in Public Space*, London: Routledge.

Hertz, N. (2000) *The Silent Takeover*, London: Heinemann.

Hesmondhalgh, D. (1998) 'The British dance music industry: a case of independent cultural production', *British Journal of Sociology* 49 (2): 134–51.

Hesmondhalgh, D. and Melville, C. (2001) 'Urban breakbeat culture: repercussions of hip hop in the United Kingdom', in T. Mitchell (ed.) *Global House: Rap and Hip Hop outside the USA*, Connecticut: Wesleyan University Press.

Hetherington, K. (1996) 'Identity formation, space and social centrality', *Theory, Culture and Society* 13 (4): 33–52.

——(1997) *The Badlands of Modernity. Heterotopia and Social Ordering*, London: Routledge.

——(1998) *Expressions of Identity. Space, Performance, Politics*, London: Sage.

Hetherington, P. (1996) 'Painting the town pink', *Guardian*, 27 October: 7.

Hey, V. (1986) *Patriarchy and Pub Culture*, London: Tavistock.

Hill, A. (2002) 'Acid house and Thatcherism: noise, the mob and the English countryside', *British Journal of Sociology* 53 (1): 89–105.

Hindle, P. (1994) 'Gay communities and gay space in the city', in S. Whittle (ed.) *The Margins of the City: Gay Men's Urban Lives*, Aldershot: Arena.

Hobbs, D., Lister, S., Hadfield, P. and Hall, S. (2000) 'Receiving shadows: governance, liminality in the night-time economy', *British Journal of Sociology* 51 (4): 701–17.

Hollands, R. (1990) *The Long Transition: Class, Culture and Youth Training*, London: Macmillan.

——(1995) *Friday Night, Saturday Night: Youth Identification in the Post-Industrial City*, Newcastle: University of Newcastle upon Tyne, Department of Sociology.

——(1997) 'From shipyards to nightclubs: restructuring young adults' employment, household and consumption identities in the north-east of England', *Berkeley Journal of Sociology* 41: 41–66.

——(1998) 'Crap jobs, govvy schemes and trainspotting: reviewing the youth, work and idleness debate', in J. Wheelock and J. Vail (eds) *Work and Idleness: The Political Economy of Full Employment*, Massachusetts: Kluwer.

——(2000) 'Lager louts, tarts and hooligans: the criminalisation of young adults in a study of Newcastle nightlife', in V. Jupp, P. Davies and P. Francis (eds) *Doing Criminological Research*, London: Sage.

——(2002) 'Division in the dark: youth cultures, transitions and segmented consumption spaces in the night-time economy', *Journal of Youth Studies* 5 (2): 153–73.

Hollands, R. and Chatterton, P. (2003, forthcoming) 'Producing nightlife in the new urban entertainment economy: corporatisation, branding and market segmentation', *International Journal of Urban and Regional Research*, March.

Home Office (1994) *The Criminal Justice and Public Order Act*, London: The Stationery Office.

——(2000) *Time for Reform: Proposals for the Modernisation of Our Licensing Laws*, CM4696. TSO, London: The Stationery Office.

Homel, R. and Clark, J. (1994) 'The prediction and prevention of violence in pubs and clubs', in R. Clarke (ed.) *Crime Prevention Studies*, vol. 3, New York: Criminal Justice Press.

Honon, W.H. (1998) 'Campus alcohol and drug arrests rose in '96, survey says', *New York Times*, 3 May: 3.

hooks, b. (1994) *Outlaw Culture: Resisting Representations*, London: Routledge.

Horowitz, H.L. (1987) *Campus Life: Undergraduate Cultures from the End of the Eighteenth Century to the Present*, New York: Knopf.

Howe, N. and Strauss, W. (1993) *13th Gen: Abort, Retry, Ignore, Fail?* New York: Vintage.

Hubbard, P. (2002) 'Screen-shifting: consumption, "riskless risks" and the changing geographies of cinema', *Environment and Planning A* 34 (7): 1239–58.

Hudson, R. and Williams, A. (1994) *Divided Britain*, Chichester: John Wiley, 2nd edition.

Humphrey, R. and McCarthy, P. (1997) 'High debt and poor housing. A taxing life for contemporary students', *Youth and Policy* 56: 55–64.

——(1998) 'Stress and the contemporary student', *Higher Education Quarterly* 52 (2): 221–42.

Humphries, S. (1983) *Hooligans or Rebels?* Oxford: Blackwell.

Hunt, G. and Satterlee, S. (1987) 'Darts, drink and the pub: the culture of female drinking', *Sociological Review* 35: 575–601.

Hutton, W. (1995) *The State We're In*, London: Vintage.

ICFI (2000) 'Ontario law aims to drive poor off the streets'. Online. Available http://www.wsws.org/articles/2000/feb2000/ont-f07.shtml (accessed 7 February 2000).

Illich, I. (2000) 'The dirt of cities', in M. Miles, I. Borden and T. Hall (eds) *The City Cultures Reader*, London: Routledge.

Imrie, R., Thomas, H. and Marshall, T. (1995) 'Business organisations, local dependence and the politics of urban renewal in Britain', *Urban Studies* 32 (1): 31–47.

Institute of Alcohol Studies (1999) *Alcohol and Young People*, IAS Fact Sheet, London: IAS.

——(2002) *Women and Alcohol*, IAS Fact Sheet, 7 March, London: IAS.

Institute for Brewing Studies (2001) *Craft-Brewing Industry Fact Sheet*, Minneapolis: Association of Brewers.

INURA (1998) *Possible Urban Worlds. Urban Strategies at the End of the 20th Century*, Basel: Birkhauser Verlag.

Irwin, S. (1995) *Rights of Passage*, London: UCL Press.

JD Wetherspoon (2002) 'About our pubs'. Online. Available http://www.jdwetherspoon.co.uk/pages/body01.html (accessed 24 May 2002).

Jeffs, T. and Smith, M. (1996) 'Getting the dirtbags off the street: curfews and other solutions to juvenile crime', *Youth and Policy* 53: 1–14.

Jencks, C. and Peterson, P. (eds) (1990) *The Urban Underclass*, Washington: The Brookings Institution.

Jessop, B. (1997) 'The entrepreneurial city: reimagining localities, redesigning economic governance or restructuring capital', in N. Jewson and S. McGregor (eds) *Transforming Cities*, Routledge: London.

Jessop, B., Kastendeik, H., Nielson, K. and Pedersen, O. (1991) *The Politics of Flexibility: Restructuring State and Industry in Britain, Germany, and Scandinavia*, Aldershot: Avebury.

Jessop, B., Peck, J. and Tickell, A. (1999) 'Retooling the machine: economic crisis, state restructuring, and urban politics', in A. Jonas and D. Wilson (eds) *The Urban Growth Machine. Critical Perspectives Two Decades Later*, New York: State University of New York Press.

Johansson, T. and Miegel, F. (1992) *Do the Right Thing: Life Style and Identity in Contemporary Youth Culture*, Lund: Almqvist.

Johnson, A., Wadsworth, J., Wellings, K. and Field, J., with Bradshaw, S. (1994) *Sexual Behaviour in Britain*, Harmondsworth: Penguin.

Johnson, L., MacDonald, R., Mason, P., Ridley, L. and Webster, C. (2000) *Snakes and Ladders: Young People, Transitions and Social Exclusion*, Bristol: Policy Press.

Johnson, P. (1986) 'Drinking, temperance and the construction of identity in nineteenth-century America', *Social Science Information* 25: 321–50.

Jones, G. (1995) *Leaving Home*, Buckingham: Open University Press.

Jones, J. (1998) *American Work: Four Centuries of Black and White Labor*, London: W.W. Norton.

Jones, S. (1986) *Workers at Play: A Social and Economic History of Leisure 1918–1939*, London: Routlege & Kegan Paul.

Joseph, M. (2000) *Agenda: Ecstasy*, London: Carlton Books.

Julier, G. (2000) *The Culture of Design*, London: Sage.

Justice Clerks Society (1999) *Good Practice Guide*, London: Justice Clerks Society.

Kane, P. (2000) 'The play ethic: why believe in work, when it doesn't believe in you?' *Observer*, 'Life' magazine, 22 October.

Karp. D., Stone, P. and Yoels, W. (199) *Being Urban: A Sociology of City Life*, London: Praeger.

Kearney, M.C. (1998) 'Don't need you: rethinking identity politics and separatism from a grrl perspective', in J. Epstein (ed.) *Youth Culture: Identity in a Post Modern World*, Oxford: Blackwell.

Kellner, D. (1994) *Media Culture*, New York: Routledge.

Kelly, R.D.G. (1997) *Yo' Mama's Disfunktional: Fighting the Culture Wars in America*, Boston: Beacon Press.

Kelso, P. (2000) 'UK youth: the good, the bad and the sickly', *Guardian*, 29 June: 7.

Kenyon, L. (1997) 'Studenthood in the 1990s: is there life beyond the lecture theatre?' *Youth and Policy* 56: 29–42.

Key Note (2001) *UK Public Houses Market Development*, London: Market and Business Development (July).

Kingston, P. (1996) 'Absolutely sensible', *Guardian Education*, 2 April: 2.

Klein, N. (2000) *No Logo*, Flamingo: London.

Kneale, J. (1999) 'A problem of supervision: moral geographies of the nineteenth-century British public house', *Journal of Historical Geographies* 25(3): 333–48.

——(2001) 'The place of drink: temperance and the public 1856–1914', *Social and Cultural Geography* 2 (1): 43–59.

Knopp, L. (1992) 'Sexuality and the spatial dynamic of capitalism', *Environment and Planning D: Society and Space*, 10: 651–69.

Kumar, K. (1995) *From Post-Industrial to Post-Modern Society*, Oxford: Blackwell.

Landry, C. (2000) *The Creative City. A Toolkit for Urban Innovators*, Earthscan: London.

Lange, J.F. and Voas, R.B. (2000) 'Youth escaping limits on drinking: binging in Mexico', *Addiction*, 95 (4): 521–8.

Lash, S. (1989) *The Sociology of Postmodernism*, London: Routledge.

Lash, S and Urry, J. (1987) *The End of Organised Capitalism*, Cambridge: Polity.

——(1994) *Economies of Signs and Spaces*, London: Sage.

Laurie, N., Dwyer, C., Holloway, S. and Smith, F. (1999) *Geographies of New Femininities*, Harlow: Longman.

Lees, L. (1999) 'Urban renaissance and the street. Spaces of control and contestation', in N.R. Fyfe (ed.) *Images of the Street: Planning, Identity, and Control in Public Space*, London: Routledge.

Lees, S. (1993) *Sugar and Spice: Sexuality and Adolescent Girls*, London: Penguin.

Lefebvre, H. (1991) *The Production of Space*, Oxford: Blackwell.

Leggett, J. (1968) *Class, Race and Labor: Working Class Consciousness in Detroit*, New York: Oxford University Press.

Leisure and Hospitality Business (2000) 'Last orders for traditional pub chains?' *Leisure and Hospitality Business*, 2 November.

——(2001) 'Queer company seeking sites for branded venues', *Leisure and Hospitality Business*, 29 November.

Levitas, R. (1998) *The Inclusive Society: Social Exclusion and New Labour*, London: Macmillan.

Lewis, M. (1994) 'A sociological pub crawl around gay Newcastle', in S. Whittle (ed.) *The Margins of the City: Gay Men's Urban Lives*, Aldershot: Arena.

Ley, D. (1996) *The New Middle Class and the Remaking of the Central City*, New York: Oxford University Press.

Liberator (2001) *Journal for the Emerging New Civil Rights Movement* 5 (1) (September). Published by the Coalition to Defend Affirmative Action By Any Means Necessary (BAMN).

Lincoln, R. and Mustchin, M. (2000) *Clubs and Violence. A Follow-up Evaluation of the Surfers Paradise Safety Action Plan*, Queensland: Centre for Applied Psychology and Criminology, Bond University, Queensland, Australia.

Lister. S., Hobbs. D., Hall. S. and Winlow. S. (2000) 'Violence in the night-time economy; bouncers: the reporting, recording and prosecution of assaults', *Policing and Society* 10 (4): 383–402.

Lively, K. (1998) 'At Michigan State, a protest escalates into a night of fires, tear gas, and arrests', *Chronicle of Higher Education*, 15 May: 3.

Lleyelyn-Smith, J. (2001) 'The girls pile in at six, by eight they are plastered', *Sunday Telegraph*, 24 June: 16.

Loader, I. (1996) *Youth, Policing and Democracy*, London: Macmillan.

Loeb, P. (1994) *Generation at the Crossroads*, New Brunswick: Rutgers University Press.

Longhurst, B. and Savage, M. (1996) 'Social class, consumption and the influence of Bourdieu: some critical issues', in S. Edgell, K. Hetherington and A. Warde (eds) *Consumption Matters: The Production and Experience of Consumption*, Oxford: Blackwell.

Lott, T.L. (1999) *The Invention of Race: Black Culture and the Politics of Representation*, Oxford: Blackwell.

Lovatt, A. (1995) 'The ecstasy of urban regeneration: regulation of the night-time economy in the post-Fordist city', in J. O'Connor and D. Wynne (eds) *From the Margins to the Centre*, London: Arena.

Low Pay Commission (1998) *The National Minimum Wage. First Report of the Low Pay Commission*, TSO (Cm 3976), London: Department of Trade and Industry.

Lynch, S. (1996) 'The hate squad', *Ottawa Citizen*, 12 March: C1.

McCarthy, P. and Humphrey, R. (1995) 'Debt: the reality of student life', *Higher Education Quarterly* 49 (1): 78–86.

MacDonald, R. (ed.) (1997) *Youth, the 'Underclass' and Social Exclusion*, London: Routledge.

Macdonald, R. and Coffield, F. (1991) *Risky Business: Youth and the Enterprise Culture*, London: Falmer.

MacDonald, R. and Marsh, J. (2001) 'Disconnected youth?' *Journal of Youth Studies* 4 (4): 373–91.

——(2002) 'Street corner society: young people, social exclusion and leisure careers', a paper presented at the ESRC *Youth Citizenship and Social Change* Conference, Brighton, March.

MacDonald, R., Banks, S. and Hollands, R. (1993) 'Youth and policy in the 90s', *Youth And Policy* 40 (Spring): 2–12.

MacDonald, R., Mason, P., Shildrick, T., Webster, C. and Ridley, L. (2001) 'Snakes and ladders: in defence of studies of youth transitions', *Sociological Research On-Line* 5 (4).

McDowell, L. (1999) *Gender, Identity and Place: Understanding Feminist Geographies*, Cambridge: Polity Press.

McGovern, C. (1995) 'You've come a long way baby', *Alberta Report*, 31 July 31: 24–7.

McKay, G. (1996) *Senseless Acts of Beauty: Cultures of Resistance since the Sixties*, London: Verso.

——(1998) *DiY Culture. Party and Protest in 90s Britain*, London: Verso.

McKenzie, S. (1989) 'Women in the city', in R. Peet and N. Thrift (eds) *New Models in Geography*, London: Unwin Hyman.

MacLeod, D. and Graham-Rowe, D. (1997) 'Degrees of abuse', *Guardian* Higher Education Supplement, 2 September 1997: i.

McRobbie, A. (1993) 'Shut up and dance: youth culture and changing modes of femininity', *Cultural Studies* 7 (3) (October).

——(1994) *Postmodernism and Popular Culture*, London: Routledge.

——(2000) *Feminism and Youth Culture: From* Jackie *to* Just Seventeen, London: Palgrave, 2nd edition.

McRobbie, A. and Garber, J. (1976) 'Girls and subcultures – an exploration', in S. Hall and T. Jefferson (eds) (1976) *Resistance through Rituals: Youth Subcultures in Post-War Britain*, London: Hutchinson.

McRobbie, A. and Nava, M. (1984) *Gender and Generation*, London: Macmillan.

Maffesoli, M. (1996) *The Time of the Tribes: The Decline of Individualism in Mass Society*, London: Sage.

Malbon, B. (1999) *Clubbing: Dancing, Ecstasy and Vitality*, London: Routledge.

Males, M. (1996) *The Scapegoat Generation: America's War on Adolescents*, Monroe, ME: Common Courage Press.

Mann, K. (1990) *The Making of an English Underclass*, Milton Keynes: Open University Press.

Marcus, D. (2001) 'Students start early, and charge often', *US News*, 19 March: 1.

Marcuse, H. (1964) *One-Dimensional Man*, London: Abacus.

Markowitz, L. (1995) 'Out in the burbs: can gay and lesbian culture thrive outside cities?' *Utne Reader* July–August: 22–4.

Marshall, S. (1999) *Christiania. The Planning Factory No. 10*, London: UCL, Bartlett School of Planning.

Martinez, R. (2002) 'La música pop(ular) i la producció cultural de l'espaisocial juvenil', in C. Feixa, J.R. Saura and J. de Castro (eds) *Música Ideologies: Mentre la meva guitarra parla suaument ...*, Barcelona: SGJ-UdL.

Marx, K. and Engels, F. (1981) *The Communist Manifesto*, Harmondsworth: Penguin.

Mason, A. and Palmer, A. (1996) *Queer Bashing: A National Survey of Hate Crime against Lesbians and Gays*, London: Stonewall.

Mason, C.M. and McNally, K.N. (1997) 'Market change, distribution and new firm formation and growth: the case of real ale breweries in the United Kingdom', *Environment and Planning A* 29: 405–17.

Mass Observation (1970) *The Pub and the People: A Worktown Study*, Welwyn Garden City: Seven Dials Press.

Massey, D. (1998) 'The spatial construction of youth cultures', in T. Skelton and G. Valentine (eds) *Cool Places. Geographies of Youth Cultures*, London: Routledge.

Massey, D. and Allen, J. (1988) *Uneven Redevelopment: Cities and Regions in Transition*, London: Hodder and Stoughton.

Matthewman, H. (1993) 'Rock against men is music to the riot grrls' ears', *Independent on Sunday*, 14 March.

May, T. (1993) *Social Research*, Buckingham: Open University Press.

Mayer, M. (1994) 'Post-Fordist city politics', in A. Amin (ed.) *Post-Fordism: A Reader*, London: Blackwell.

Messerschmidt, J. (1993) *Masculinity and Crime: Critique and Reconceptualisation of Theory*, New York: Rowman and Littlefield.

Miles, S. (2000) *Youth Lifestyles in a Changing World*, Buckingham: Open University Press.

Miller, C. (2000) *We want beer! Prohibition and the will to Imbibe*. Online. Available http://www.allaboutbeer.com/news/states (accessed 10 February 2002).

Miller, W.B. (1958) 'Lower-class culture as a generating milieu of gang delinquency', *Journal of Social Issues* 14 (3): 5–19.

Miller, W.L., Dickson, M. and Stoker, G. (2000) *Models of Local Governance. Public Opinion and Political Theory in Britain*, Basingstoke: Palgrave.

Mintel (1998) *Nightclubs and Discotheques*, London: Leisure Intelligence (September).

——(2000) *Pre-Family Leisure Trends*, London: Leisure Intelligence (January).

Mirrlees-Black, C., Budd, T., Partridge, S. and Mayhew, P. (1998) *The 1998 British Crime Survey, England and Wales*, London: Home Office.

Mitchell, T. (2001) *Global House: Rap and Hip Hop Outside the USA*, Connecticut: Wesleyan University Press.

Mizen, P. (1995) *The State, Young People, and Youth Training: In and Against the Training State*, London: Mansell.

Moffatt, M. (1989) *Coming of Age in New Jersey. College and American Culture*, New Brunswick: Rutgers University Press.

Moghadam, V. (1999) 'Gender and the global economy', in M. Marx Ferree, J. Lorber and B. Hess (eds) *Revisioning Gender*, London: Sage.

Mohan, J. (1994) *American Universities in Their Local Communities: Service Learning, School Partnerships and Community Development*, New York: Harkness Fellowship Final Report to the Commonwealth Fund.

——(1996) 'Re-connecting the academy? Community involvement in American and British universities', in J. Elliott (ed.) *Communities and Their Universities: The Challenge to Lifelong Learning*, London: Lawrence and Wishart, pp. 93–107.

Moi, T. (1991) 'Appropriating Bourdieu: feminist theory and Pierre Bourdieu's sociology of culture', *New Literary History* 22: 1017–44.

Monbiot, G. (1998) 'Reclaim the fields and the country lanes! The Land is Ours campaign', in G. McKay (ed.) *DiY Culture. Party and Protest in Nineties Britain*, London: Verso.

——(2000) *Captive State*, London: Macmillan.

Moore, K. (2000) 'E-heads and beer monsters: researching young people's drug consumption in dance club settings', paper presented at BSA Youth Study Group Conference, *Researching Youth: Issues, Controversies and Dilemmas*, University of Surrey, Guildford, 11 and 12 July.

Moreton, C. (1996) 'Pink tide may turn the Tories out of Brighton', *Independent on Sunday*, 27 October: 10.

Morgan, S. (1997) 'Cheap drinks, heavy costs: students and alcohol', *Youth and Policy* 56: 42–55.

Moron, L., Skeggs, B., Tyrer, P. and Corteen, K. (2001) 'Property, boundary, exclusion: making sense of hetero-violence in safer spaces', *Social and Cultural Geography* 2: 4.

Morris, L. (1994) *Dangerous Classes*, London: Routledge.

Morris, S. (1998) *Clubs, Drugs and Doormen*, Paper 86, London: Police Research Group.

——(2002) 'Dance club craze puts town in lap of gods', *Guardian*, 16 March: 12–13.

Morrow, V. and Richards, M. (1996) *Transitions to Adulthood: A Family Matter?* York: York Publishing Services.

Morse, M. (1995) 'Violent femmes', *Utne Reader* March–April: 22–4.

Muggleton, D. (1998) 'The post-subculturalist', in S. Redhead, J. O'Connor and D. Wynne (eds) *The Clubcultures Reader*, Oxford: Blackwell.

——(2000) *Inside Subculture. The Postmodern Meaning of Style*, Oxford: Berg.

Muncie, J. (1999) *Youth and Crime*, London: Sage.

Murphy, J. (1998) 'What is branding?' in S. Hart and J. Murphy (eds) *Brands: The New Wealth Creators*, Basingstoke: Macmillan.

Murray, C. (1990) *The Emerging British Underclass*, London: Institute for Economic Affairs.

——(1994) *Underclass: The Crisis Deepens*, London: Institute for Economic Affairs.

Mutchler, M. (2000) 'Seeking sexual lives: gay youth and masculinity tensions', in P. Nardi (ed.) *Gay Masculinities*, London: Sage.

Myers, R. (2001) 'Partying too hard. Clubs struggle to meld safety and nightlife', *Christian Science Monitor*, 23 February: 3.

Nardi, P. (2000) '"Anything for a sis Mary": an introduction to *Gay Masculinities*", in P. Nardi (ed.) *Gay Masculinities*, London: Sage.

Nava, M. (1994) *Changing Cultures: Youth, Feminism and Consumerism*, London: Sage.

Negri, A. and Hardt, M. (2001) *Empire*, Cambridge: Harvard University Press.

Newburn, T. (2001) 'The commodification of policing: security networks in the late modern city', *Urban Studies* 38 (5–6): 829–48.

Newburn, T. and Shiner, M. (2001) *Teenage Kicks? Young People and Alcohol: A Review of the Literature*, York: Joseph Rowntree Foundation.

Newey, G. (2001) 'Why Bristol is the most popular university', *Guardian*, 9 January: 21.

Nicol, M. (1999) 'Prescott learns secret of lap-dancing', *Evening Standard*, 4 January.

Norris, C. and Armstrong, G. (1998) 'Introduction: power and vision', in C. Norris, J. Moran and G. Armstrong (eds) *Surveillance, Closed Circuit Television and Social Control*, Aldershot: Ashgate.

O'Connor, J. and Wynne, D. (1995) *From the Margins to the Centre*, London: Arena.

O'Grady, B., Bright, R. and Cohen, E. (1998) 'Sub-employment and street youths: an analysis of squeegee cleaning on homeless youth', *Security Journal* 11: 315–23.

O'Toole, M. (1996) *Regulation Theory and the British State*, Aldershot: Avebury.

Office for National Statistics (2000) *Social Focus on Young People*, London: The Stationery Office.

Olden, M. (2001) 'Squatters take over club to stop table dancing', *Big Issue*, 30 April–6 May: 3.

Ontario Government (1999a) 'Anti-squeegee law, tax cuts among top legislative priorities', news release. Online. Available http://www.premier.gov.on.ca/english/issues/safe.htm (20 September).

——(1999b) *Safer Streets Act*, Ottawa: Ontario Government.

Organisation for Economic Co-operation and Development (1999) *The Response of Higher Education Institutions to Regional Needs*, Programme on Institutional Management in Higher Education, Paris: OECD.

Osborn, T. (1980) 'Outpost of a new culture', in N. Wates (ed.) *Squatting: The Real Story*, London: Bay Leaf Books.

Osgerby, B. (1998) *Youth in Britain since 1945*, Oxford: Blackwell.

Oswald, G. (2001) *Race and Ethnic Relations in Today's America*, Aldershot: Ashgate.

Oxford Handy Dictionary (1978) Oxford: Oxford University Press.

P, E. and Swales, E. (1997) 'Mixing gender', *Paint It Red*, March: 10–11.

Page, D. (2000) *Communities in the Balance*, York: Joseph Rowntree Foundation.

Pain, R. (2001) 'Gender, race, age and fear in the city', *Urban Studies* 38 (5–6): 899–913.

Pakulski, J. and Waters, M. (1996) *The Death of Class*, London: Sage.

Palladino, G. (1995) *Teenagers: An American History*, New York: Basic Books.

Palmer, B. (1979) *A Culture in Conflict: Skilled Workers and Industrial Capitalism in Hamilton Ontario 1860–1914*, Montreal: McGill-Queens Press.

——(2000) *Cultures of Darkness: Night Travels in the Histories of Transgression*, New York: Monthly Review Press.

Park, R. (1967) *The City*, Chicago: University of Chicago Press.

Parker, H., Aldridge, J. and Measham, F. (1998) *Illegal Leisure*, London: Routledge.

Parson, T. (1942) 'Age and sex in the social structure of the United States', *American Sociological Review* 7: 604–16.

Patel, I. (2000) 'Urban violence. An overview', in M. Miles, I. Borden and T. Hall (eds) *The City Cultures Reader*, London: Routledge.

Peake, L. (1993) '"Race" and sexuality: challenging the patriarchal structuring of urban social space', *Environment and Planning D: Society and Space* 11 (4): 415–32.

Pearson, G. (1983) *Hooligan: A History of Respectable Fears*, Basingstoke: Macmillan.

Pearson, K. (1979) *Surfing Subcultures of Australia and New Zealand*, St Lucia: University of Queensland Press.

Pile, S. (1994) 'Masculinism, the use of dualistic epistemologies and third spaces', *Antipode* 26 (3): 255–77.

Pile, S. and Keith, M. (eds) (1997) *Geographies of Resistance*, London: Routledge.

Pimlott, H. (1995) 'Rap, race and resistance', *New Times*, 8 July: 9.

Pini, M. (1997) 'Cyborgs, nomads and the raving feminine', in H. Thomas (ed.) *Dance in the City*, London: Macmillan.

Piore, M.J. and Sabel, C.F. (1984) *The Second Industrial Divide*, New York: Basic Books.

Plant, S. (1999) *Writing on Drugs*, London: Faber and Faber.

Plummer, K. (1994) *Telling Sexual Stories: Power, Change, and Social Worlds*, London: Routledge.

Podmore, J.A. (2001) 'Lesbians in the crowd: gender, sexuality and visibility along Montreal's Boul. St Laurent', *Gender, Place and Culture* 8 (4) (December): 333–56.

Pollert, A. (ed.) (1991) *Farewell to Flexibility?* Oxford: Blackwell.

Pountain, D. and Robins, D. (2000) *Cool Rules. Anatomy of an Attitude*, New York: Reaktion Books.

Pratt, A. (1997) 'The cultural production system: a case study of employment change in Britain, 1984–91', *Environment and Planning A* 29: 1953–74.

Prestage, M. (1997) 'Thugs and drugs move in on the gay village', *Independent on Sunday*, 7 December: 9.

Pryce, K. (1979) *Endless Pressure: A Study of West Indian Life-Styles in Bristol*, Harmondsworth: Penguin.

Purvis, J. (1996) 'The city as a site of ethical consumption and resistance', in J. O'Connor and D. Wynne (eds) *From the Margins to the Centre*, London: Arena.

Raphael, A. (1995) 'Only rock and role', *Guardian*, 'Weekend', 15 April: 32–7.

Rawlings, P. (2002) *Policing. A Short History*, Uffculme: Willan Publishing.

Real Beer (2000) 'Stout protest. Philadelphia bar owners oppose Guinness role in opening pubs' (17 March). Available http://www.realbeer.com (accessed 20 February 2002).

Redhead, S. (ed.) (1993) *Rave Off: Politics and Deviance in Contemporary Youth Culture*, Aldershot: Avebury.

——(1997) *Subculture to Clubcultures*, Oxford: Blackwell.

Redhead, S., Wynne, D. and O'Connor, J. (eds) (1998) *The Clubcultures Reader*, London: Blackwell.

Reeve, A. (1996) 'The private realm of the managed town centre', *Urban Design International* 1 (1): 61–80.

Release (1997) *Release Drugs and Dance Survey: An Insight into the Culture*, London: Release.

Reynolds, S. (1998a) 'Rave culture: living the dream or living death?' in S. Redhead, D. Wynne and J. O'Connor (eds) *The Clubcultures Reader*, London: Blackwell.

——(1998b) *Energy Flash: A Journey through Rave Music and Dance Culture*, London: Picador.

Richards, V. (ed.) (1997) *Why Work? Arguments for the Leisure Society*, London: Freedom Press.

Richardson, D. (2001) *Theorising Heterosexuality*, Basingstoke: Palgrave.

Rietveld, H. (1998) 'Repetitive beats: free parties and the politics of contemporary DiY dance culture in Britain', in G. McKay (ed.) *DiY Culture. Party and Protest in Nineties Britain*, London: Verso.

Ritchie, B (199) *Good Company. The Story of Scottish & Newcastle*, London: James and James.

Ritzer, G. (1993) *The McDonaldisation of Society*, Thousand Oaks, CA: Pine Forge Press.

——(2001) *Enchanting a Disenchanted World*, London: Sage.

Roberts, K. (1983) *Youth and Leisure*, London: Allen and Unwin.

——(1995) *Youth and Employment in Modern Britain*, Oxford: Oxford University Press.

——(1997) 'Same activities, different meanings: British youth cultures in the 1990s', *Leisure Studies* 16: 1–15.

——(1999) *Leisure in Contemporary Society*, Wallingford: CABI Publications.

Roberts, K. and Parsell, G. (1992) 'Entering the labour market in Britain: the survival of traditional opportunity structures', *Sociological Review* 30: 727–53.

——(1994) 'Youth cultures in Britain: the middle-class takeover', *Leisure Studies* 13: 33–48.

Robinson, F. (1994) 'Something old, something new? The great north in the 1990s', in P. Garrahan and P. Stewart (eds) *Urban Change and Renewal: The Paradox of Place*, Aldershot: Avebury, pp. 9–20.

Rock Nightclub (2001) Press releases. Online. Available http://www.rockclubs.co.uk/ (accessed 28 May 2002).

Rojek, C. (1995) *Decentring Leisure. Rethinking Leisure Theory*, London: Sage.

Rose, T. (1994) 'A style nobody can deal with: politics, style and the post-industrial city in hip-hop', in A. Ross and T. Rose (eds) *Microphone Fiends: Youth Music and Youth Culture*, New York: Routledge.

Ruddick, S. (1996) *Young and Homeless in Hollywood: Mapping Social Identities*, London: Routledge.

——(1998) 'Modernism and resistance: how "homelessness" youth sub-cultures make a difference', in T. Skelton and G. Valentine (eds) *Cool Places. Geographies of Youth Cultures*, London: Routledge.

Russell, B. (1932) *In Praise of Idleness*, Massachusetts: Green Party.

St John, G. (ed.) (2001a) *FreeNRG. Notes from the Edge of the Dancefloor*, Altona: Common Ground.

——(2001b) 'Earthdreaming', *Arena Magazine* 53: 41–4.

Salman, S. (2000) 'Driven up pole by club', *Evening Standard*, 20 October: 6.

Sanchez, R. (1998) 'Campus alcohol crackdowns bring mix of rage and relief', *Washington Post*, 10 May: A3.

Santiago-Lucerna, J. (1998) 'Frances Farmer will have her revenge on Seattle: pan-capitalism and alternative rock', in J. Epstein (ed.) *Youth Culture: Identity in a Postmodern World*, Oxford: Blackwell.

Sassen, S. (1991) *The Global City: New York, London, Tokyo*, Princeton, NJ: Princeton University Press.

Saunders, N. (1993) *E is for Ecstasy*, London: Nicholas Saunders.

——(1995) *Ecstasy and the Dance Culture*, London: Nicholas Saunders.

Savage, M., Barlow, J., Dickens, A. and Fielding, T. (1992) *Property, Bureaucracy and Culture: Middle-Class Formation in Britain*, London: Routledge.

Savage, M. and Butler, T. (1995) *Social Change and the Middle Classes*, London: UCL Press.

Schnews (2000) 'Space is the place: direct action and geography', *Antipode* 32 (2): 111–14.

——(2001a) 'National roundup', *Schnews Newsletter*, Brighton: Justice? (14 June).

——(2001b) 'Profitbeering!' *Schnews Newsletter*, 296 (9) Brighton: Justice? (March).

Schutz, A. (1972) *The Phenomenology of the Social World*, London: Heinemann Educational.

Scott, A. (2000) *The Cultural Economy of Cities*, London: Sage.

Scott, A.J., Lewis, A. and Lea, S.E.G. (eds) (2001) *Student Debt: The Causes and Consequences of Undergraduate Borrowing in the UK*, London: Marston Book Services.

Scottish & Newcastle (2001) 'Demanding consumers'. Online. Available http://www.scottish-newcastle.com/retail/retail-index.html (accessed 12 May 2002).

Seabrook, J. (1988) *The Leisure Society*, Oxford: Blackwell.

Seabrook, T. and Green, E. (2000) 'Streetwise or safe: young women negotiating time and space', *Cultural Change and Urban Contexts Conference*, Manchester: Manchester Metropolitan University, Manchester and Salford (September).

Seale, J. (1972) *The Campus War*, Harmondsworth: Pelican.

Sennett, R. (1974) *Families against the City: Middle-Class Homes of Industrial Chicago 1872–1890*, New York: Vintage.

——(1977) *The Fall of Public Man*, Cambridge: Cambridge University Press.

——(1998) *The Corrosion of Character*, Norton: New York.

——(2001) 'A flexible city of strangers', *Le Monde Diplomatique*, February: 2–4.

Shantz, J. (2001) 'Attacks on the poor and homeless continue, resistance builds. Class war in Ontario.' Online. Available http://parsons.iww.org/industrial-worker/2001–09/ontario.html (accessed 12 April 2002).

Sharkey, A. (1993) 'New tribes of England', *Guardian*, 'Weekend', 11 December: 39–44.

Sherman, B. (1986) *Working at Leisure*, London: Methuen.

Shields, R. (1991) *Places on the Margin: Alternative Geographies of Modernity*, London: Routledge.

——(1992) 'Spaces for the subject of consumption', in R. Shields (ed.) *Lifestyle Shopping: The Subject of Consumption*, London: Routledge.

Sibley, D. (1995) *Geographies of Exclusion*, London: Routledge.

——(1998) 'Problematizing exclusion: reflections on space, difference and knowledge', *International Planning Studies*, 3 (1): 93–100.

——(2001) 'The binary city', *Urban Studies* 38 (2): 239–50.

Siddell, R. (1993) 'Girl gangs', *Community Care*, 19–25 January: 16.

Silver, H. and Silver, P. (1997) *Students: Changing Roles, Changing Lives*, Buckingham: Society for Research into Higher Education.

Simpson, D. (1999) 'A short break in Leeds: the inhabitants of the formerly industrial city work hard at being stylish and having fun', *Independent*, 21 March.

Simpson, M. (1999) *It's a Queer World: Deviant Adventures in Pop Culture*, London: Harrington Park Press.

Six Continents Retail (2002) 'What we do.' Online. Available http://sixcontinents.com/aboutus/whatwedo_leisure.htm (accessed 15 January 2002).

Skeggs, B. (1997) *Formations of Class and Gender: Becoming Respectable*, London, Sage.

——(1999) 'Matter out of place: visibility and sexualities in leisure spaces', *Leisure Studies*, 18: 213–32.

——(2000) 'The appearance of class: challenges in gay space', in S. Munt (ed.) *Cultural Studies and the Working Class: Subject to Change*, London: Cassells.

Skelton, T. and Valentine, G. (1998) (eds) *Cool Places – Geographies of Youth Cultures*, London: Routledge.

Slater, D. (1997) *Consumer Culture and Modernity*, Oxford: Polity Press.

Slater, H. (1996) 'Neutral and commercial ... just like everybody', *CTHEORY*, 6 March, Review no. 42. Online. Available http:// www.ctheory.com/r42-neutral.html (accessed 12 January 2002).

Smith, N. (1996) *The New Urban Frontier. Gentrification and the Revanchist City*, London: Routledge.

Smithers, R. (1999) 'Graduate glass ceiling stays put', *Guardian* 17 April: 19.

Snow, D. (1999) 'Skateboarders, streets and style', in R. White (ed.) *Australian Youth Subcultures: On the Margins and in the Mainstream*, Hobart: Australian Clearinghouse for Youth Studies.

Sobel, M. (1981) *Lifestyle and Social Structure: Concepts, Definitions, Analyses*, New York: Academic Press.

Social Exclusion Unit (1998) *Bringing Britain Together: A National Strategy for Neighbourhood Renewal*, CM 4045, London: Social Exclusion Unit.

Society of Independent Brewers (1999) *A Submission to Her Majesty's Customs and Excise on the Subject of Progressive Beer Duty and Market Access*, London: SIB.

Soja, E. (1996) *Thirdspace: Journeys to Los Angeles and Other Real-and-Imagined Places*, Cambridge: Blackwell.

——(2000) *Postmetropolis: Critical Studies of Cities and Regions*, Malden: Blackwell.

Solnit, R and Schwartzenberg, S. (2000) *Hollow City. The Siege of San Francisco and the Crisis of Urban America*, Verso: London.

Sorkin. M. (ed.) (1992) *Variations on a Theme Park: The New American City and the End of Public Space*, New York: Noonday Press.

Spears, J. (1999) 'Mayor's programme called attack on poor', *Toronto Star*, 4 August: 12.

Stedman-Jones, G. (1978) 'Class expression versus social control? A critique of recent trends in the social history of leisure', *History Workshop* 4 (Spring): 162–70.

Stenson, K. and Watts, P. (1998) 'The street: "It's a bit dodgy around there": safety, danger, ethnicity and young people's use of public space', in T. Skelton and G. Valentine (eds) *Cool Places: Geographies of Youth Cultures*, London: Routledge.

Stewart, F. (1992) 'The adolescent as consumer', in J. Coleman and C. Warren-Anderson (eds) *Youth Policy in the 1990s: the way forward*, London: Routledge, pp. 203–26.

Stobart, P. (ed.) (1994) *Brand Power*, Basingstoke: Macmillan.

Storper, M. and Christopherson, S. (1987) 'Flexible specialisation and regional industrial agglomerations. The case of the US motion-picture industry', *Annals of the Association of American Geographers*, 77: 260–82.

Sturges, F. (2001) 'The great hip-hop crisis: it's just too big for its booty', *Independent on Sunday*, 24 June: 9.

Substance Abuse and Mental Health Services Administration (1999) *Summary of Findings from the 1998 National Household Survey on Drug Abuse*, Rockville, MD: US Department of Health and Human Services.

Sugrue, T.J. (1998) *The Origins of the Urban Crisis: Race and Inequality in Post-War Detroit*, Princeton: Princeton University Press.

Swedenburg, T. (2001) 'Islamic hip hop vs Islamophobia: Aki Nawaz, Natacha Atlas, Akhenaton', in T. Mitchell (ed.) *Global House: Rap and Hip Hop outside the USA*, Connecticut: Wesleyan University Press.

Swingewood, A. (1977) *The Myth of Mass Culture*, London: Macmillan.

Tanner, J. (1996) *Teenage Troubles: Youth and Deviance in Canada*, Toronto: Nelson Canada.

Tanska, T. (1997) 'Going undergrad,' *Venue Magazine*, 401: 22–4.

Taverner, I. (2000) 'The city that never sleeps', *Western Daily Press*, 5 December: 12.

Taylor, S. (2001) 'Gays in space', undergraduate dissertation, Department of Sociology and Social Policy, University of Newcastle upon Tyne.

Taylor, I., Evans, K. and Fraser, P. (1996) *A Tale of Two Cities: Global Change, Local Feeling and Everyday Life in the North of England: a Study in Manchester and Sheffield*, London: Routledge.

Tennant, A. (1994) 'Creating brand power', in P. Stobart (ed.) (1994) *Brand Power*, Basingstoke: Macmillan.

The Publican newspaper (2000) *Pub Industry Handbook 2000*, Surrey: Quantum Publishing.

Thomas, C.J. and Bromley, D.F. (2000) 'City centre revitalisation: problems of fragmentation and fear in the evening and night-time city', *Urban Studies.* 37 (8): 1403–29.

Thompson, E.P. (1967) 'Time, work discipline and industrial capitalism', *Past and Present*, 38 (December): 56–97.

Thompson, E.P. and Yeo, T. (1973) *The Unknown Mayhew: Selections from the Morning Chronicle 1849–1850*, Harmondsworth: Penguin.

Thornton, S. (1995) *Club Cultures – Music, Media and Subcultural Capital*, Cambridge: Polity Press.

Thrift, N. (1997) 'The still point. Resistance, expressive embodiment and the dance', in S. Pile and M. Keith (eds) (1997) *Geographies of Resistance*, London: Routledge.

Tomaney, J. and Ward, N. (eds.) (2001) *A Region in Transition. The North-East at the Millennium*, London: Ashgate.

Tomlinson, H. (2002) 'Publicans urged to take Inntrepreneur to the bar', *Independent*, 31 March.

Tomsen, S. (1997) 'A top night. Social protest, masculinity and the culture of drinking violence', *British Journal of Criminology* 37 (1): 90–102.

Toon, I. (2000) '"Finding a place in the street": CCTV surveillance and young people's use of urban public space', in D. Bell and A. Haddour (eds) *City Visions*, Harlow: Pearson Education.

Travis, A. (2001) 'Britain set to embrace "Las Vegas" gambling', *Guardian*, 18 July: 8.

Tuck, A. (1997) 'Club gives a positive welcome to HIV men', *Independent on Sunday*, 7 December: 9.

Turner, V. (1982) *From Ritual to Theatre: The Human Seriousness of Play*, New York: Performing Arts Journal Publication.

Tyyska, V. (2001) *Long and Winding Road: Adolescents and Youth in Canada Today*, Toronto: Canadian Scholars Press.

Urry, J. (1990) *The Tourist Gaze: Leisure and Travel in Contemporary Societies*, London: Sage.

——(1995) *Consuming Places*, London: Routledge.

Valentine, G. (1993) '(Hetero)sexing space: lesbian perceptions and experiences of everyday space', *Environment and Planning D: Society and Space* 11: 395–413.

Valentine, G., Skelton, T. and Chambers, D. (1998) 'Cool places: an introduction to youth and youth cultures', in T. Skelton and G. Valentine (eds) *Cool Places: Geographies of Youth Cultures*, London: Routledge.

Van Poznak, E. (1993) 'Angry young women', *Guardian*, 24 March 24: 3.

Veblen, T. (1994) *The Theory of the Leisure Class*, London: Penguin.

Vulliamy, E. (1998) 'Ghetto-blasting', *Observer Life*, September: 14–21.
Wacquant, L.J.D. and Bourdieu, P. (1992) *An Invitation to Reflexive Sociology*, Cambridge: Polity.
Walby, S. (1997) *Gender Transformation*, London: Routledge.
Walker, D. (1996) 'Tolerance by degree', *Times Higher Education Supplement*, 22 November: 4.
——(1997) 'Why university is a journey too far', *Independent*, 11 September: 15.
Walker, M. (1993) 'It's queer up north, and Manchester's in the pink', *Observer*, 29 August: 12.
Wall, D. (2000) *Earth First!*, London: Routledge.
Wallis, L. (1993) 'Pink pounding', *Guardian*, 25 August: 14–15.
Walvin, J. (1978) *Leisure and Society 1830–1950*, London: Longman.
Warde, A. (1994) 'Consumers, consumption and post-Fordism', in R. Burrow and B. Loader (eds) *Towards a Post-Fordist Welfare State?* London: Routledge.
Warrell, L. (1994) 'Flirting with morality in the law: the booze, the bouncer and adolescence down under', *Anthropological Forum* 7 (1): 31–53.
Wates, N. (1980) *Squatting. The Real Story*, London: Freedom Press.
Weber, M. (1976) *The Protestant Ethic and the Spirit of Capitalism*, London: Routledge.
Wechsler, H., Davenport, A., Dowdell, G., Moeykens, B. and Castillo, S. (1994) 'Health and behavioural consequences on binge drinking in colleges', *Journal of the American Medical Association* 272: 1672–7.
Weeks, J. (1989) *Sex, Politics and Society*, London: Longman.
Weightman, G. (1992) *Bright Lights, Big City. London Entertained 1830–1950*, London: Collins and Brown.
Weininger, A. (2001) 'One night, many consequences. Students find that someone had to be responsible', *The Daily Collegian. Penn State University*: 1.
Weinzierl, R (2000) *Fight the Power!*, Vienna: Passagen.
West Yorkshire Police (2001) Personal correspondence, West Yorkshire Police, Millgarth Division, Leeds.
White, R. (ed.) (1999) *Australian Youth Subcultures: On the Margins and in the Mainstream*, Hobart: Australian Clearinghouse for Youth Studies.
Whiteley, S. (ed.) (1997) *Sexing the Groove: Popular Music and Gender*, London: Routledge.
Whittle, S. (ed.) (1994) *The Margins of the City: Gay Men's Urban Lives*, Aldershot: Avebury.
Wilkinson, H. (1994a) *No Turning Back: Generations and the Genderquake*, London: Demos.
——(1994b) 'Technology is feminising the way we work', *New Times*, 15 October: 6–7.
——(1995) *Generation X and the New Work Ethic*, London: Demos.
Williams, C. (1997) *Consumer Services and Economic Development*, London: Routledge.
Williams, P., Hubbard, P., Clark, D. and Berkeley, N. (2001) 'Consumption, exclusion and emotion: the social geographies of shopping', *Social and Cultural Geography* 2 (2): 202–20.
Williams, R. (1961) *The Long Revolution*, Harmondsworth: Penguin.
——(1977) *Marxism and Literature*, Oxford: Oxford University Press.
Willis, P. (1977) *Learning to Labour: How Working Class Kids Get Working Class Jobs*, Westmead: Saxon House.
——(1978) *Profane Culture*, London: Routledge.
——(1990) *Common Culture*, Milton Keynes: Open University Press.

Wilson, B. (1970) *The Youth Culture and the Universities*, London: Faber.

Wilson, E. (1991) *The Sphinx in the City: Urban Life, the Control of Disorder and Women*, London: Virago.

Wilson, W.J. (1987) *The Truly Disadvantaged: The Inner City, the Underclass and Public Policy*, Chicago: University of Chicago Press.

——(1996) *When Work Disappears: The World of the New Urban Poor*, New York: Alfred A. Knopf.

Winlow, S. (2001) *Badfellas: Crime, Tradition and New Masculinities*, Oxford: Berg.

Winlow, S., Hobbs, D., Lister, S. and Hadfield, P. (2001) 'Get ready to duck: bouncers and the realities of ethnographic research on violent groups', *British Journal of Criminology* 41 (3): 536–48.

Winterdyk, J. (ed.) (1996) *Issues and Perspectives on Young Offenders in Canada*, Toronto: Harcourt Brace.

Wrigley, N. and Lowe, M. (1996) *Retailing, Consumption and Capital: Towards the New Retail Geography*, Harlow: Longman.

Wyn, J. and White, R. (1997) *Rethinking Youth*, London: Sage.

Wynne, D. (1990) 'Leisure, lifestyle and the construction of social position', *Leisure Studies* 9: 21–34.

Wynne, D. and O'Connor, J. (1998) 'Consumption and the postmodern city', *Urban Studies* 35 (5–6): 841–64.

Yeandle, S., Booth, C. and Darke, J. (1996) *Changing Places: Women's Lives in the City*, London: Paul Chapman.

Young, M. (1995) 'Getting legless, falling down pissy-arsed drunk: policing men's leisure', *Journal of Gender Studies* 4 (1): 47–62.

Younge, G. (2001) 'Rhyme and reason', *Guardian*, 'G2', 14 June: 1–3.

Zaret, E. (2000) 'Students riot on the Hill', *The Daily Camera*, 15 September: 2.

Zukin, S. (1988) *Loft Living: Culture and Capital in Urban Change*, London: Radius.

——(1991) *Landscapes of Power: From Detroit to Disney World*, Berkeley: University of California Press.

——(1992) 'Postmodern urban landscapes: mapping culture and power', in S. Lash and J. Friedman (eds) *Modernity and Identity*, Oxford: Blackwell.

——(1995) *The Cultures of Cities*, Oxford: Blackwell.

INDEX

Note: page numbers in italics refer to illustrations

CU00406719

BARKING & DAGENHAM
000 326 584

The London Region
an annotated geographical bibliography

The London Region

an annotated geographical bibliography

by Philippa Dolphin, Eric Grant and Edward Lewis

Foreword by Professor Peter Hall

Mansell Publishing Limited

ISBN 0 7201 1598 1

Mansell Publishing Limited, 35–37 William Road,
London NW1 3ER

First published 1981

Distributed in the United States and Canada by The H. W. Wilson
Company, 950 University Avenue, Bronx, New York 10452.

British Library Cataloguing in Publication Data

Dolphin, Philippa
The London region.
1. London – Description – Bibliography
I. Title II. Grant, Eric III. Lewis, Edward
016.914 21 Z 202 4.L8

ISBN 0–7201–1598–1

Typeset in the U.K. by Computacomp (UK) Ltd, Fort William, Scotland
Printed and Bound by Mansell (Bookbinders) Ltd., Witham, Essex

Contents

Foreword

'London, thou art the flour of Cities all', wrote William Dunbar (1460?–1513?) in an age before urban research properly became an industry. Some will remember that quotation at the beginning of William Robson's monumental work *The Government and Misgovernment of London*, published in 1939 at a time when few academics wrote or cared about London. Since then, however, London has indeed become the flower for serious research into urban structures, urban evolution, urban problems and urban policies. Few other cities in the world can have been so intensively studied. If – as some say – London today is a patient with a serious disease, it is hardly due to lack of attention from the academic physicians at the bedside.

But this attention is in truth very recent. The four centuries from Dunbar to Robson produced very little work; more surprisingly, perhaps, was the small output between 1939 to the end of the 1950s. Since then, the growth has been almost exponential. This of course reflects partly the growth of academic activity in general, but more particularly the development of urban studies as an important field in its own right. If one event triggered this development in Britain it was surely the foundation by William Robson of the Greater London Group at the London School of Economics to initiate and collate the research for the Royal Commission on Local Government in Greater London, between 1957 and 1960. One major outcome of that work, historians would probably agree, was the creation – against overwhelming political odds – of the Greater London Council for which Robson had argued ever since writing his book. The other, embodied even in the structure of the new GLC, was a commitment to research as the basis of policy formation. Two decades later the experts might differ on the performance of the GLC; they might even doubt the contribution that research has made to the quality of decision-making. But what they could hardly doubt is the quantity, the range, and above all the quality of the research itself.

If anyone still questioned that, this bibliography should disabuse him. The 1909 items – which might or might not symbolize the date when London first became subject to statutory town planning – range from the geology beneath London to the air above it, but concentrate especially on the economic and social life of the capital and on the problems of planning that arise. Academics inside or outside of adademia, being the kind of people they are, will immediately rush to the bibliography to find the critical item that the compilers have overlooked. I fear that they may be disappointed. The more recondite will note that foreign sources – such as Kemmann's monumental account of the Victorian railway system, or Pasquet's study of Edwardian industry, or even more recently Chaline's scholarly overview of the London agglomeration – are missing; but this, it seems, is a deliberate and perhaps justifiable policy.

Necessarily, too, the bibliography cannot pretend to catch the millions of references – sometimes densely concentrated – that occur in works not specifically devoted to the subject of London. Thus Spate's contribution to Darby's original *Historical Geography of England* (1936) is duly noted, but not the discussions of London's growth that occur in his new study over thirty years later – simply because no one chapter focuses on the capital. This is inevitable, at any rate until some computerization of knowledge makes it possible to automatically scan all literature for content analysis. Meanwhile, the compilers of this bibliography are to be congratulated on a splendid scholarly accomplishment. They have put not only all serious students of London, but also all serious students of cities anywhere, forever in their debt.

Peter Hall
Institute of Urban and Regional Development
University of California, Berkeley

Introduction

This bibliography aims to fill a major gap in the literature on London. It is intended to provide a basic reference work to help teachers, researchers, librarians and students find the main body of literature on major aspects of the London region. Because of its wide-ranging nature, the bibliography should be useful to those interested in geography, environmental studies, conservation, industry, office location, transport, population, housing, planning, archaeology, history and topography.

The need for an annotated bibliography on the London region has long been apparent. Previous attempts at compiling bibliographies on London have suffered from such drawbacks as only covering official publications or one aspect of London only, or by being too long and poorly organized. The most important previous bibliographies are listed in section 1.1. In the first category, the bibliography by DAWE, D. (1972) only covers certain aspects of the City of London, and that by PERHAM, L. (1972) only those works published by the Greater London Council (GLC) from 1965 to 1971. The more voluminous bibliographies include the Members' Library Catalogue (LONDON COUNTY COUNCIL, 1939) and the unpublished bibliography on Middlesex (MIDDLESEX LOCAL HISTORY COUNCIL, 1959). The sheer size of these works (the latter has over 5,000 entries) reduces their usefulness, and the lack of annotations means that the intending reader is deprived of any knowledge of the contents of the books and articles whose titles interest him and much time may be wasted in fruitless searching. A more manageable bibliography on London is by WILKINS, P. M. (1964) in CLAYTON, R. (ed.) *The geography of Greater London.* This bibliography is however only partly annotated, is weak on social and economic aspects, and now largely outdated.

The student wanting up-to-date references on London is therefore forced to undertake a time consuming search through indexing journals, library catalogues, individual specialized bibliographies, and the bibliographies in recent texts on the region. This very obvious gap in the bibliographical literature has now been recognized and discussed by the London Group of Geographers who have given strong support to the compilation of a bibliography of the London region.

We have tried to meet this need with the present bibliography. A definition of the London region is not easy and the adminstrative area of London and the functional unit of London have rarely been coincident. London originated as the

Roman city of Londinium, covering about 300 acres. After an apparent decline in the Saxon period, the medieval city began to expand and in AD 1222 the administrative boundary was extended to encompass the area that still survives as the essentially self-governing City of London. The adjacent parts of the counties of Middlesex, Essex, Surrey and Kent became progressively urbanized as they were functionally drawn into London's web, but administrative recognition was slow in coming to terms with the expansion. The Metropolitan Police and the Metropolitan Board of Works, established in 1829 and 1855 respectively, were early attempts at encompassing specific activities within the rapidly developing metropolitan sprawl. The establishment of the London County Council (LCC) in 1888 was a major land-mark in bounding and governing a metropolitan area, but almost from the beginning the area under its jurisdiction was too small. Rapid growth in the first 60 years of the twentieth century resulted in a situation whereby the majority of Londoners lived outside the LCC limits. On 1 April 1965, following a major reform of local government in London, the Greater London Council was created with responsibility for the broader functions of a much larger area than the old LCC administrative zone. This area, known as Greater London, covers 610 square miles and comprises the historic City of London and 32 metropolitan boroughs. Greater London provides the initial focus for this bibliography and, as far as possible, a representative cross-section of the literature appertaining to this area has been included.

The physical environment is less easily defined. Geologists recognize a structural unit known as the London basin which comprises the area between the Chalk escarpments of the Chiltern Hills and the North Downs. To a large extent this has formed the spatial bounds for the references in section 3, though the more important broader works on south-east England are also included.

The economic and social influences of London are also difficult to delimit. The post-war growth of the population beyond the green belt, the population overspill into the new towns, and the corresponding growth of commuting from the various dormitory towns and suburbs, have meant that the influence of London is felt well beyond the GLC area. This belt, known as the Outer Metropolitan Area or Outer Ring, lies between 30 and 65 km from Piccadilly Circus, and contains such major towns as Guildford, Welwyn Garden City, Chelmsford and Basildon. Greater London plus the Outer Metropolitan Area form a zone with a radius of roughly 65 km from the centre of London, and can be best equated with the London Region. Although it is not an administrative area, it is recognized as a distinct region for planning and statistical purposes, and forms the wider spatial base for the social and economic activities referred to in sections 5–8.

The nine main sections of the bibliography have been chosen to reflect the broadest range of human activities within the spatial areas already defined. Each has a separate introduction giving a brief account of the scope of the section, the development of the subject, and the nature of its literature.

The number of subsections within each section varies according to the size and breakdown of the literature, but each generally has a first subsection containing bibliographies and statistical sources, and a second listing general works on the topic. Much material does not fit neatly into one section or another and there is therefore a certain amount of similarity as, for example, between the planning and transport sections.

The main focus of the bibliography is the GLC area. Material on the Outer Metropolitan Area has been included where it relates to the metropolis. A book simply on Guildford would be rejected but one which related the expansion of Guildford to that of London, discussed commuting flows, or analysed its contribution to the capital's workforce would be included. Similarly, literature on the south-east region or on national affairs would be included only where relevant; for example, the *Strategic plan for the south east* (1970) and the *Policy for the inner cities* (DEPARTMENT OF THE ENVIRONMENT, 1977) are included.

The question of how much local material to include has also been a problem. Highly localized material has generally been omitted but the level of inclusion varies according to the different sections of the bibliography. The enormous volume of local borough planning material and the fact that a recent bibliography of local plans is already in existence, led us to decide to exclude this type of material from section 8. However, London's long tradition of local history writings ensured that a selection of local bibliographical and historical works was included in subsection 4.10, though rigidly limited to the GLC area.

The selection of material for inclusion in the main categories has been a major problem because of the sheer volume of the literature concerning London and the fact that metropolitan affairs are closely bound up with national ones. Certain categories of material have been deliberately excluded: unpublished material, theses, newspaper articles, legislation, guide books, maps and non-print media. Outside these categories, other material has been excluded for a variety of reasons: because it is very short or ephemeral, because it has been superseded (for example by a later edition or by a longer paper by the same author) or because it is too general or too localized. The question of accessibility has also been considered. Every item in the bibliography has been examined and consulted and references which were very difficult to locate in libraries have been excluded. Only references published up to the end of 1979 have been included.

References are arranged according to the *British Standard for bibliographical references* (BS 1629:1976), but with the year of publication immediately after the author's name. Within each subsection items are organized chronologically by publication date and alphabetically by author within the same year. Annotations are generally restricted to two sentences and are descriptive rather than critical.

A large number of sources have been used in the compilation of this work, and they obviously vary in usefulness according to the individual sections. Some of the more fruitful bibliographic sources were *Geo Abstracts*, *Urban Abstracts*, the Guildhall Library catalogue, GLC research bibliographies, the *British National Bibliography*, the various government publications' catalogues and *Urban History Yearbooks*. Researchers seeking further references are recommended to use these sources in addition to those mentioned in the general bibliography (section 1) and those in certain subsections. Libraries in London which contain sizeable collections of material on London are listed in the appendix.

Articles on London have appeared in a formidable variety of journals. There are a few journals dealing specifically with the London area: the *Greater London Intelligence Journal* (and its predecessors), the *London Journal*, the *Transactions of the London and Middlesex Archaeological Society*, *London Naturalist*, *London Archaeologist*, and the now defunct *East London Papers*. Many articles, however, appear in journals aimed at the subject content rather than the area and it has been a major task for the authors to locate and annotate these articles in such journals as the *Quarterly Journal of the Geological Society*, *Weather*, *Journal of the Marine Biology Association*, *Economic History Review*, *Demography*, *Regional Studies*, *Town and Country Planning*, *Surveyor*, *Political Quarterly*, *Geographical Journal*, *Municipal Journal*, *Housing Review*, *Public Health Engineer*, *Atmospheric Environment* and many others.

This is a selective bibliography and the authors are conscious that a great deal has had to be left out. However, it is hoped that the material which is listed provides a balanced selection of the literature which is available. Inevitably, users will find articles that might have been listed, while those familiar with specific subject areas are bound to feel that some of their best known references are omitted. In the end the choice is a personal one, but we would be grateful for information about omissions and any mistakes that have crept into the titles, for use in future editions.

It is impossible to compile a book like this without the help of others. The librarians, archivists and information officers of a large number of libraries and institutions have assisted the authors in many ways and freely given of their time and information. Several academics with specialist knowledge of the London Region commented on drafts of individual sections, suggesting additions and deletions. The following are thanked for their help and support in this respect: Bruce Atkinson, Annabel Coker, Bryan Ellis, John Hall, Chris Hamnett, David Jones, John Martin, Mary Moody, Paul Moxey, Hugh Prince and Angela Taylor. Several busy secretaries cheerfully helped in typing including Marcia Harris, Liz Spellen, Marlene Mascarenhas and especially Joyce Foster who typed the great bulk of sections. Linda Jones and Steve Chilton of the technical staff at Middlesex Polytechnic are thanked for their help in annotating certain references and drawing the maps.

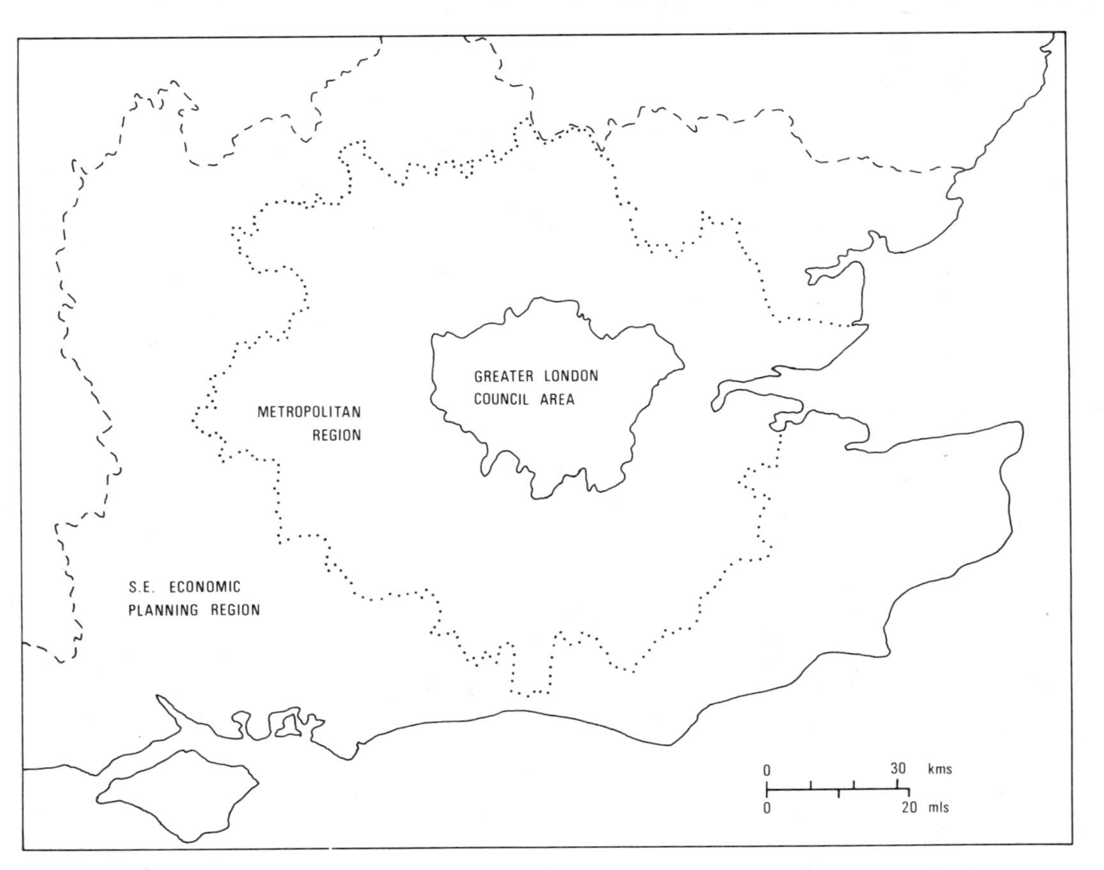

Figure 1: Greater London, the Metropolitan Region and the South-East Economic Planning Region

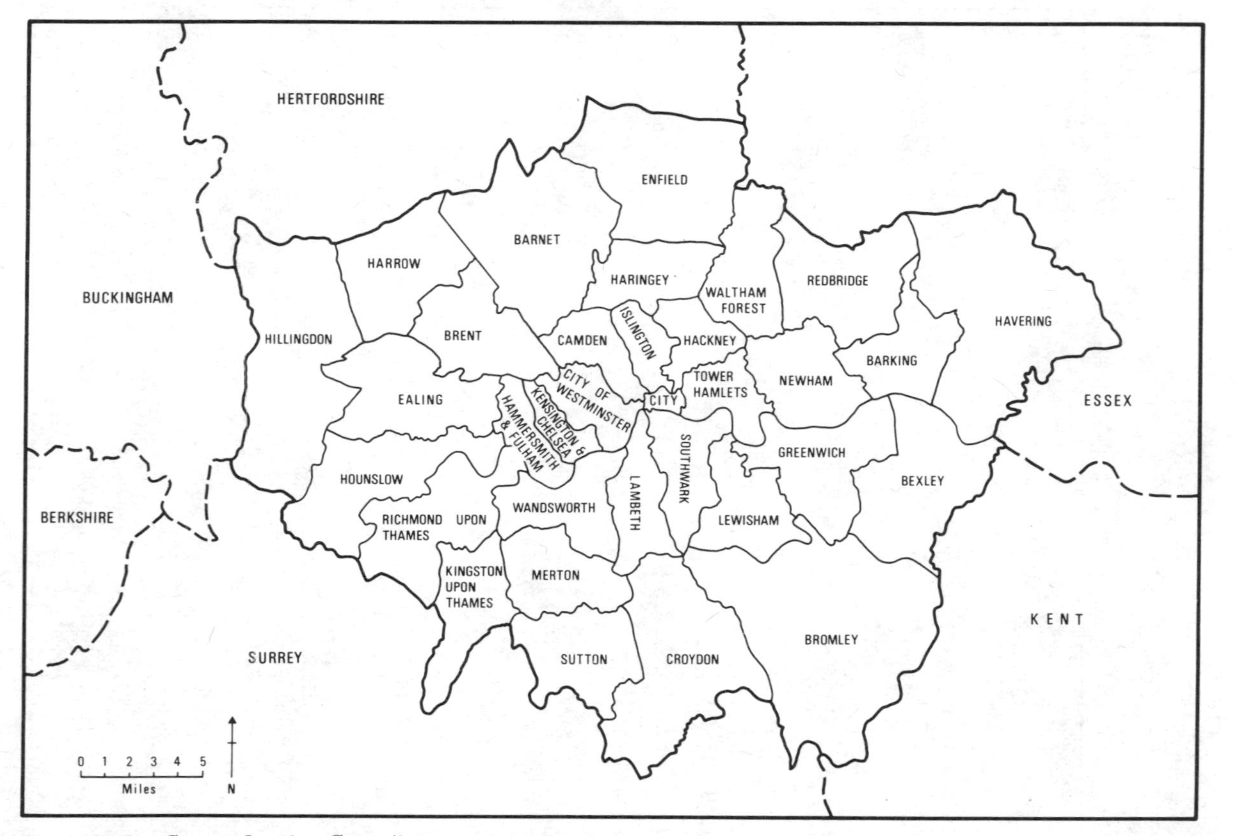

Figure 2: The Greater London Council area and the London boroughs

The Bibliography

Section 1 General bibliographies, statistical sources and atlases

Bibliographical control of information on London and the compilation of statistical data concerning the city have been undertaken primarily by the various local authorities: the London County Council, the Greater London Council, and the individual London boroughs. The London County Council, established in 1888, set up a Members' Library (now the Greater London Council History Library) to provide information to LCC members. The publication of its catalogue in 1939 marks the first major contribution to the organization of literature on London. Since 1958 it has produced a regular accessions list jointly with the London County Record Office (now the Greater London Record Office).

The GLC's Research Library, which was established in 1969, has done more for the collection and organization of information on London than any other body. Its *Planning and Transportation Abstracts* and subsequent *Urban Abstracts* have been, until the launching in 1974 of the ACOMPLIS database, the major tool for the researcher seeking information concerning current developments in the metropolis. Since 1970 over 100 research bibliographies have been produced by the staff of the Research Library. These vary greatly in length and style, and are not always confined to London. A few of the more general research bibliographies are listed in this section. Others, concerned with specific topics such as transport and housing, are listed elsewhere under the relevant sections. In addition, the GLC's *London Topics* series, which have been appearing irregularly since 1974, are extremely useful listings of key documents on particular issues with summaries of their arguments.

Keeping track of the various items published by the London boroughs is a difficult task. The research bibliography compiled by SCOTT, A. J. C. (1973) is a useful index, and *Urban Abstracts* lists many of them. The research bibliography by COCKETT, I. (1978) lists completed and on-going research projects of the London boroughs.

Papers by THOM, W. (1975) and COCKETT, I. (1978) have been included here since they give accounts of computerized information services on London which have become available over the last decade. The GLC Research Library's ACOMPLIS database now holds over 30,000 abstracts on urban documentation. A keyword index derived from ACOMPLIS is available on microfilm.

The Middlesex Local History Council's bibliography of Middlesex (1959) and its supplement (1968) have been included although they are unpublished. It was decided to include them since they represent a unique and very valuable source of references on Middlesex, and typescript copies are available in many libraries in London.

Obviously a large number of abstracting journals and indexes contain material on London. The main ones are *Geo Abstracts*, *British Humanities Index*, *New*

Geographical Literature and Maps, and the *Departments of Environment and Transport Library Bulletin*. Some of these, and many others, are listed in the bibliography by HAMLYN, P. (1978). Only those journals specifically concerned with London have, however, been included in subsection 1.1. The *Daily Intelligence Bulletin* (1971–) which is produced by the GLC Research Library has not been included since it is basically for internal consumption by the GLC and the London boroughs, and is not available on subscription.

The compilation of statistics on London has increased greatly since the nineteenth century. Only the more recent census volumes (1951 onwards) for London are included in subsection 1.2, but the British census has been taken every ten years since 1801 (with the exception of 1941) by the General Register Office, and so demographic statistics are available for the city from this date. In 1891 the London County Council began publication of a regular statistical series *London Statistics* containing a whole range of statistics and LCC services. Another statistical publication *Statistical Abstract for London*, covering ten year periods, was established in 1897. The GLC's *Annual Abstract of Greater London Statistics* (1968–) is now the main source of statistical information concerning the Greater London area. The creation of the 32 new London boroughs and the GLC area in 1965 does, however, mean that there are problems of comparison with statistics collected before this time for the old administrative areas.

The GLC has been responsible for the bulk of the work in collecting and interpreting statistical data on the metropolis. A large number of research memoranda have been produced which provide detailed statistics for the GLC area, interpet census data, and give demographic projections for the city and individual boroughs. More recently, various boroughs have produced brief statistical profiles of the areas they administer, either on a regular basis or as one-off publications. A selection is presented here.

Obviously there are numerous regular national statistical series which include data for London, for example *Regional Statistics*, *Social Trends* and *Population Trends*, all produced by the Central Statistical Office. These series have not, however, been included since they are not specifically concerned with London. The main statistics relevant to London are reproduced in the *Annual Abstract of Greater London Statistics* (1968–) and can also be found in the census volumes for London. Specialized statistics such as purely demographic statistics or transport and economic data can be found together with the other literature of the subject area in the relevant sections of this bibliography. *Guide to official statistics*, (2nd ed. HMSO, 1978) and *Reviews of UK statistical sources* (Pergamon, 1974–) will guide the reader in search of additional or more specialized data.

The only major modern atlas of the London region is the *Atlas of London* (JONES, E. and SINCLAIR, D. J., 1968), which is now somewhat out-of-date. Street atlases of London, of which there are many, have been excluded. Atlases concerned with a particular theme, such as the *Social atlas of London* (SHEPHERD, J. et al., 1974), have been included in the relevant sections of the bibliography.

1.1 Bibliographies

1 LONDON COUNTY COUNCIL (1939) Members' library catalogue: Vol. 1, London history and topography. London County Council, 142p.

The holdings of the London County Council Members' Library which relate to the history and topography of the capital are listed under various subject headings. There are author and subject indexes.

2 *London County Council Library and London County Record Office Accessions List* (1958–1965) London County Council (quarterly). Superseded by *Greater London History Library and Greater London Record Office Accessions List* (1965–).

3 MIDDLESEX LOCAL HISTORY COUNCIL (1959) Bibliography of Middlesex compiled mainly from materials in libraries in the county. 610p. LONDON AND MIDDLESEX ARCHAEOLOGICAL SOCIETY (1968) Supplement. 120p. (both unpublished).

5,645 entries relating to Middlesex as a whole or any of its districts, whether in print, typescript or manuscript, but excluding archives and pictorial records with no text. Entries are listed under the following categories: ecology, meteorology, palaeontology, biology, botany, zoology, history, topography, economic history (including transport), political and military history (including administration), social history (including public utilities and planning), religion, biography, architecture, periodicals, directories and almanacs, verse. Entries 1–1462 are for the county as a whole; entries 1463–5645 for each county district (Acton–Yiewsley). There are subject and personal names indexes.

4 WILKINS, P. M. (1964) Bibliography. In: CLAYTON, R. (ed.) The geography of Greater London: a source book for teacher and student. George Philip, pp. 293–342.

Books and periodical articles are listed with brief annotations under a variety of headings: physical aspects, topographical-historical surveys, growth and history, the face of London, government, economy, demography, geography and planning. There are also sections of general interest and local publications.

5 *Greater London History Library and Greater London Record Office Accessions List* (1965–) Greater London Council (quarterly). Supersedes *London County Council Library and London County Record Office Accessions List* (1958–65).

A select list of recent accessions arranged under subject headings. The library class numbers are given and there are occasional short annotations.

6 *Planning and Transportation Abstracts* (1969–74) Greater London Council, Nos 1–50 (monthly).

Superseded by *Urban Abstracts* (1974–).

Abstracts compiled from journal articles, selected books and pamphlets received in the GLC Research Library. Local government, planning, transport, urban problems, environment, construction, and social services are the main areas covered.

7 DAWE, D. (1972) The City of London: a select book list. Guildhall Library, 38p.

A brief annotated list of books relating to the City of London. References are arranged by subject, such as physical aspects, government and finance, and commerce.

8 PERHAM, L. (1972) Greater London Council publications 1965–1971. Greater London Council, 117p. (GLC Research Bibliography, No. 30)

An annotated bibliography of 665 publications from the GLC and ILEA between 1965 and the end of 1971. Items are arranged in broad subject order and there is an index.

9 DAY, P. (1973) Publications on the environment 1970–1972. Greater London Council, 18p. (GLC Research Bibliography, No. 42)

A select annotated list of 129 British books and reports relevant to London's environment. References are listed under broad headings such as conservation, housing, and recreation.

10 SCOTT, A. J. C. (1973) London borough departmental publications 1965–1972. Greater London Council, 163p. (GLC Research Bibliography, No. 50)

A list of items published by the 32 London boroughs and the Corporation of London between 1965 and the end of 1972. Arrangement is by borough and there is an index but no annotations.

11 *Urban Abstracts* (1974–) Greater London Council, No. 1– (monthly). *Supersedes Planning and Transportation Abstracts* (1969–1974).

A bibliography, with abstracts, compiled from publications received by the GLC Research Library covering literature concerned with metropolitan areas, particularly London. Topics cover: environment and pollution, government, housing, planning and land use, social planning and transport. There is an annual index.

12 THOM, W. (1975) ACOMPLIS: a computerised information service. *Greater London Intelligence Quarterly*, No. 30, pp. 15–17.

A brief account of 'A Computerised London Information Service' (ACOMPLIS) which has been operational at the GLC Research Library since 1974. The service replaces the library catalogue and provides selective dissemination of information and retrospective searches.

13 GOMERSALL, A. (1975) Directory of information resources in the Greater London Council. 4th ed. Greater London Council, 34p. (GLC Research Bibliography, No. 46)

A directory of the 28 library and information units, both inside and outside County Hall, with information about their resources and the services they provide.

14 KING, R. (1976) Catalogue of the London collection. City University Library, 58p.

A listing of the separate collection of books on London, with particular emphasis on the City, which is held at City University Library. Subjects covered include: government, commerce, livery companies, social services, history, architecture, streets, topographical views and maps, natural conditions and the people.

15 COCKETT, I. (1978) London boroughs Research Register. Greater London Council, unpaged. (GLC Research Bibliography, No. 100)

A bibliography of research and survey projects submitted by some London boroughs to the London Research Register in 1978. Completed, on-going and proposed papers are included.

16 COCKETT, I. (1978) The London Research Register. *Greater London Intelligence Journal*, Vol. 41, pp. 24–6.

The London Research Register is a computerized register developed by the GLC's Research Library as a current account and historical record of research and survey projects undertaken by various bodies in its subject areas. Its users, contributors, coverage and operation are described.

17 HAMLYN, P. (1978) Indexing and abstracting services in the Research Library. Greater London Council, 18p. (GLC Research Bibliography, No. 91)

An alphabetical list of 101 indexing and abstracting services held by the Research Library, together with a subject index. Information includes full title, publisher, frequency, dates of holdings, and a brief annotation.

18 WRIGHT, N. H. (1978) Publicly available research memoranda Nos 31–530: a contextual index of titles and authors. Greater London Council, 32p. (GLC Research Memorandum, No. 530)

A computer produced KWIC index covering, in one sequence, the authors and titles of all research memoranda available to the public.

1.2 Statistical sources

19 CORPORATION OF LONDON (1881) Report on the City day-census, 1881. Longman, 140p.

20 CORPORATION OF LONDON (1891) Ten years' growth of the City of London: report of the day-census, 1891, by J. Salmon. Simpkins, Marshall, Hamilton, Kent, 139p.

21 CORPORATION OF LONDON (1911) City of London day-census, 1911. Simpkins, Marshall, Hamilton, Kent, 128p.

Attempts to compensate for the fact that as the official census was taken at night London's daily population was grossly under-represented. The day census enumerated inhabited houses, employees in offices and factories, and persons arriving by foot or vehicle.

22 *London Statistics* (1891–1939) London County Council, Vol. 1, 1890/1–Vol. 41, 1936/8.
Superseded by *London Statistics*, New Series, (1957–68).

23 *Statistical Abstract for London* (1897–1950) London County Council, Vol. 1, 1897–Vol. 31, 1939/48.
Superseded by *London Statistics*, New Series (1957–68).

Provided statistics over ten year periods with little or no explanatory matter. A wide range of population statistics are given, together with details of rates, metropolitan borough expenditure and services provided.

24 GENERAL REGISTER OFFICE (1953) Census, 1951: England and Wales, county report, London. HMSO, 90p.

Statistics for the City of London and the 28 metropolitan boroughs. They include details of populations and acreages, information concerning private households, birth-place and nationality, education, social class and the distribution of the local population according to sex, age and marital condition.

25 GENERAL REGISTER OFFICE (1954) Census 1951: England and Wales, county report, Middlesex. HMSO, 70p.

Statistics for the County of Middlesex (which became part of Greater London in 1965). They cover population numbers and structure, housing, nationality and social class.

26 *London Statistics*, New series (1957–1968) London County Council/ Greater London Council, Vol. 1, 1945/54–Vol. 7, 1964.
Supersedes *London Statistics* (1891–1939); *Statistical Abstract for London* (1897–1950).
Superseded by *Annual Abstract of Greater London Statistics* (1968–).

Statistical tables, with explanatory matter, covering a variety of aspects of the administrative county of London, including vital statistics, services, industry and communications.

27 GENERAL REGISTER OFFICE (1963) Census 1961: England and Wales, county report, London. HMSO, 270p.

Statistics for population generally and by age, sex and marital condition, birth-place and nationality, housing and households. The area covered comprises the City of London and the 28 metropolitan boroughs.

28 GENERAL REGISTER OFFICE (1966) Census 1961: England and Wales, Greater London tables. HMSO, 58p.

Tables cover the same topics included in the London county report of the 1961 Census. The area, however, is larger, since figures have been produced to cover the new Greater London Council area which comprises the City of London and 32 London boroughs, and was established, in 1965, after the 1961 Census.

29 GENERAL REGISTER OFFICE (1967) Sample Census 1966: England and Wales, county report, Greater London. HMSO, 144p.

Data based on a 10 per cent sample and covering population and acreage; sex, age and marital condition; birth-place; private households, dwellings, tenure and household amenities; private cars and garaging; economically active and retired males and their socio-economic groups. Figures are given for the whole GLC area and the constituent boroughs.

30 *Annual Abstract of Greater London Statistics* (1968–) Greater London Council, Vol. 1– (annual).
Supersedes *London Statistics*, New Series (1957–68).

A large range of statistics is presented, concerning population, economic activity, land use, transport and a variety of local authority services. Statistics cover the area served by the GLC, the City Corporation and the 32 London Borough Councils.

31 CITY OF WESTMINSTER Social Services Department (1970–) Profiles of the City of Westminster. City of Westminster, No. 1– (annual).

Statistical profiles covering a variety of social and economic topics are presented for the City, which is divided into four areas.

32 CRAWFORD, K. A. J. and THRASHER, P. A. (1971) Standard statistical sectors for Greater London. Greater London Council, 28p. (GLC Research Report, No. 14)

A brief account of how the GLC's Standardization of Boundaries Working Party defined the standard statistical sectors. These sectors, and divisions used by other organizations, such as the London Electricity Board for administrative purposes, are shown in a series of maps.

33 LONDON BOROUGH OF WANDSWORTH (n.d.) Census 1971. London Borough of Wandsworth, 14p.

Census statistics for the borough are presented down to ward level, accompanied by explanatory text.

34 TROWBRIDGE, B. (1971) 1961 Census data. Greater London Council. 18p. (GLC Research Memorandum, No. 325)

1961 Census tables which are available in the GLC's Department of Planning and Transportation Strategy Branch but not found in published volumes are listed, and their contents defined.

35 GREATER LONDON COUNCIL (1972–) London facts and figures. Greater London Council, No. 1– (irregular).

A quick guide to basic general facts about London which are presented in a series of maps and diagrams.

36 THOMPSON, E. J. (1972) Demographic, social and economic indices for wards in Greater London. Greater London Council, 174p. (GLC Research Report, No. 15)

Largely comprises data describing some of the social, economic and demographic characteristics of the population of London and its housing at the ward level as revealed by the 1966 Sample Census.

37 TROWBRIDGE, B. (1972) 1966 Census – basic summaries for Greater London and the Greater London conurbation. Greater London Council, 36p. (GLC Research Memorandum, No. 369)

Basic summary figures from the 1966 Sample Census are given for Greater London and the conurbation, together with more detailed notes on each table in order to facilitate comparison with future census data.

38 DAVIES, H. and TROWBRIDGE, B. (1973) Reference manual of special data obtained by the GLC from the 1966 Census. Greater London Council, 220p. (GLC Research Memorandum, No. 377)

Details of all special tabulations produced by the GLC using 1966 Census data. Various definitions and notes are given, as are specifications of all the unpublished tables produced by the General Register Office and obtained by the GLC.

39 MORREY, C. R. (1973) 1966 Sample Census: borough totals, reference tables, 2. Greater London Council, 87p. (GLC Research Memorandum, No. 389)

1966 Sample Census data concerned with changes of economic activity, journey to work and migration are considered in relation to differences in base population and variable definitions. Data are collated in an attempt to facilitate comparison.

40 OFFICE OF POPULATION CENSUSES AND SURVEYS (1973) Census 1971: England and Wales, county report, Greater London, Parts 1/3. HMSO.

Part 1 gives general information about the population by age, sex, marital status, country of birth and economic activity. Part 2 gives statistics about households covering size, type of household space, and density of occupation. Part 3 contains further tables about households including information on tenure and amenities, and statistics of both occupied and vacant dwellings. Figures are generally given for both the Greater London area as a whole and the individual London boroughs.

41 DAVIES, H. and MORREY, C. R. (1974) 1971 Census: constituency data for Greater London. Greater London Council, 374p. (GLC Research Memorandum, No. 437)

Unpublished 1971 Census data for enumeration districts in London have been used to prepare census data for the new parliamentary constituencies. There are short introductions and notes on each table.

42 LONDON BOROUGH OF RICHMOND UPON THAMES (1974) The population of Richmond upon Thames. London Borough of Richmond upon Thames, 74p.

Demographic and social statistics are presented from the Small Area Statistics of the 1971 Census.

43 MORREY, C. R. (1974) The 1971 Census: further constituency data for Greater London. Greater London Council, 379p. (GLC Research Memorandum, No. 451)

The census data for enumeration districts have been used to prepare data for the new Greater London parliamentary constituencies. These tables cover mainly employment and migration.

44 LONDON BOROUGH OF HILLINGDON (1974) Borough profile – a booklet of facts and figures. London Borough of Hillingdon, 24p.

Data covering population, housing, land use, transport, economic activity, local authority services and climate. Statistics for Hillingdon are often compared with those for Greater London as a whole.

45 GREATER LONDON COUNCIL (1975) GLC population census bulletin. Greater London Council, unpaged.

A guide to population census data for the Greater London area. Details are given of prepublished, unpublished and published 1971 Census data, ward indices, GLC special tables, Small Area Statistics, together with GLC publications concerned with the Census.

46 COLE, B. (1976) Social and economic indices for London constituencies (1971 Census). Greater London Council, 46p. (GLC Research Memorandum, No. 502)

Social and economic profiles for each Greater London constituency have been provided in the form of indices and rankings from 1971 Census data.

47 MORREY, C. R. (1976) 1971 Census: demographic, social and economic indices for wards in Greater London. Greater London Council, 2 vols (GLC Research Report, No. 20)

Data presented at ward level on some of the social, economic and demographic characteristics of the population of London and its housing, as revealed by the 1971 Census.

48 LONDON BOROUGH OF LEWISHAM (1977) Comparative inner London statistics: 74/5. London Borough of Lewisham, 65p.

Gives statistics for the 12 inner London boroughs covering population, transport, health, the environment and recreation.

49 LONDON BOROUGH OF LEWISHAM (1977) Lewisham: a statistical digest, 1976. London Borough of Lewisham, 120p.

A variety of data concerned with demographic and socio-economic issues and services provided by the borough.

50 LONDON BOROUGH OF SOUTHWARK (1978) Southwark, towards a community plan: facts and figures. 2nd ed. London Borough of Southwark, 83p.

A brief, largely statistical collection of information covering services provided by the borough. Data cover population, housing, the work of the Engineer and Surveyor's Department, environmental and protection services, libraries, social services, staffing, revenue and expenditure.

51 LONDON BOROUGH OF GREENWICH Programme Planning Unit (1979) A profile of Greenwich (statistics). London Borough of Greenwich, 111p.

Statistics cover demographic, social and economic activity in the borough, in addition to the various services provided by the council.

1.3 Atlases

52 JONES, E. (1965) The London Atlas. *Geographical Journal*, Vol. 131, pp. 330–40.

The need for an atlas is discussed together with the appropriate methods of data collection and presentation.

53 CASTLE, D. and FIELDING, J. (1968) The Atlas of London: problems in compilation. *East London Papers*, Vol. 11, pp. 85–93.

The problems of selecting, using and presenting data from the 1961 Census and the 1966 Sample Census are discussed.

54 FIELDING, J. (1968) Statistical data and other reference material held by the Atlas of London. London School of Economics and Political Science, unpaged. (LSE Graduate Geography Department, Discussion Paper, No. 13)

Details of the published and unpublished social and economic data which have been gathered for the production of the *Atlas of London* (JONES, E. and SINCLAIR, D. J., 1968). Descriptions are also given of the form of the data and the indices which have been derived from them for use in the atlas.

55 JONES, E. and SINCLAIR, D. J. (1968) Atlas of the London region. Oxford: Pergamon, 70 sheets.

This atlas confines itself chiefly to social and economic factors, using data collected by the household interview questionnaires which were part of the *London traffic survey* (FREEMAN FOX and PARTNERS, 1964, 1966). Altogether there are 70 maps, each accompanied by text and varying in scale between 1:50,000 and 1:1,000,000.

Section 2 General works

There are innumerable general works on London, ranging from tourist guide books to highly literary interpretations of London life. This section is restricted to general studies of the capital which are of an academic nature, so guide books and unreferenced general accounts are excluded.

Wherever possible, works have been listed in the more specific main sections of the bibliography, so that a general book on historical development will be listed in section 4, and a general book on social problems in section 7. The books remaining in the present section are thus those that encompass more than one main aspect of London, for example BRETT-JAMES, N. G. (1951) *Middlesex*, or edited collections of essays covering several topics such as CENTRE FOR URBAN STUDIES (ed.) (1964) *London: aspects of change*. However, more specific individual contributions within edited general works are separately annotated in the appropriate sections.

56 ORMSBY, H. (1924) London and the Thames: a study of the natural conditions that influenced the birth and growth of a great city. Sifton Praed, 189p.

A traditional geographical analysis stating how the physical geography of the Thames has been the controlling factor in the early history of London by influencing the bridge site and the port until about the end of the sixteenth century.

57 ORMSBY, H. (1930) The London basin. In: OGILVIE, A. G. (ed.) Great Britain: essays in regional geography. 2nd ed. Cambridge: University Press, pp. 42–68.

Traces the influence of the relief, soils and drainage on movement and settlement within the London basin.

58 WILLATTS, E. C. (1933) Changes in land utilisation in the south west of the London basin. *Geographical Journal*, Vol. 82, pp. 514–28.

A detailed historical study of land utilization changes in the period 1840–1932 in ten parishes in the Windsor area using the tithe maps as a data source and comparing the results with the maps produced by the Land Utilisation Survey.

59 CARRIER, E. H. (1937) The inner gate: a regional study of north west Kent. Christophers, 136p.

An examination and description of the physical, historical, social and economic geography of the area astride the former Kentish Watling Street, part of which is now in Greater London.

60 WILLATTS, E. C. (1937) Middlesex and the London region. Geographical Publications, 185p. (The land of Britain: the report of the Land Utilisation Survey of Britain, Part 79)

After a brief description of the geography of the area and the different types of land use which predominate, there follows more detailed description of the various land use regions within the area.

61 BRETT-JAMES, N. G. (1951) Middlesex. Robert Hale, 442p. (County Books Series)

A wide-ranging survey covering the following aspects of Middlesex: geology, physical geography, prehistory, history, famous houses and worthies, encroachment and enclosure, churches, roads, education, sport, aviation, local affairs, requirements, wild life, and brief topographical survey.

62 FRANKLYN, J. (1953) The Cockney: a survey of London life and language. André Deutsch, 332p.

A descriptive survey of the history and language of the Cockney, together with various aspects of his work and customs.

63 ROBBINS, M. (1953) MIddlesex. William Collins, 456p. (New Survey of England)

Part 1 covers general, social and economic history, architecture and literature of Middlesex, and its relationship with London. Part 2 is a gazetteer providing brief histories and topographical descriptions of all the districts in the county.

64 RADCLIFFE, C. (1954) Middlesex. 2nd ed. Evans, 240p.

Contains accounts of the historical development of the county of Middlesex, and of the establishment and services of the Middlesex County Council.

65 WOOLDRIDGE, S. W. and HUTCHINGS, G. E. (1957) London's countryside: geographical field work for students and teachers of geography. Methuen, 223p.

A general description of the geomorphology and historical geography of parts of the Weald and North London is given, followed by detailed instructions for 12 excursions or field days.

66 WISE, M. J. (1962) The London region. In: MITCHELL, J. B. (ed.) Great Britain: geographical essays. Cambridge: University Press, pp. 57–85.

Traces the changing human geography of the London region since the 1920s and the attempts of the planners to cope with the growth of the metropolis during this period.

67 CENTRE FOR URBAN STUDIES (ed.) (1964) London: aspects of change. Macgibbon and Kee, 344p.

Ten contributions cover a variety of aspects of London from the nineteenth century labour market to contemporary structure and planning and different communities within the city.

68 CLAYTON, R. (ed.) (1964) The geography of Greater London: a source book for teacher and student. George Philip, 378p.

Combines a survey of London's geography with a description of field trips and visits, and a review of sources of study material including a fairly extensive bibliography.

69 CLAYTON, K. M. (ed.) (1964) Guide to London excursions (of the) 20th International Geographical Congress, London 1964. London School of Economics and Political Science, 162p.

Descriptions of 42 excursions in and around London to a variety of features and organizations of interest to geographers.

70 COPPOCK, J. T. and PRINCE, H. C. (eds) (1964) Greater London. Faber and Faber, 406p.

Fifteen essays on a variety of aspects of the past and present geography of London and its region.

71 COPPOCK, J. T. (1964) The growth of London and its related problems. In: STEERS, J. A. (ed.) Field studies in the British Isles. Thomas Nelson, pp. 90–105.

A condensed account of the growth of London from earliest times to post-1945 redevelopment. The growth of employment in central London and the journey to work is discussed along with the creation of the green belt and new and expanded towns.

72 STEVENS, D. F. (1964) The central area. In: COPPOCK, J. T. and PRINCE, H. C. (eds) Greater London. Faber and Faber, pp. 167–201.

A descriptive analysis of the growth, development and functions of central London (basically the City, Westminster and parts of the surrounding boroughs). The chief functional groups examined are transport, offices, industry, shops, hotels and entertainment, along with an analysis of the future of the central area.

73 HARRISON, M. (1971) London beneath the pavement. 2nd ed. Peter Davies, 280p.

An account of the history and current state of subterranean London. Archaeology; sewers; water, gas and electricity supply; road and rail tunnels; the underground; and crypts and tombs are amongst the topics covered.

74 STAMP, L. D. and BEAVER, S. H. (1971) London. In: STAMP, L. D. and BEAVER, S. H. The British Isles: a geographic and economic survey. 6th ed. Longman, pp. 663–88.

An account of the growth, population, employment and industrial activities of London and the different districts within the metropolis.

75 STEVENSON, B. (1972) Middlesex. B. T. Batsford, 200p.

A description of the face of Middlesex, arranged under topographical areas, with histories of the more important buildings and associations en route.

76 HAYES, J. and PROCHASKA, A. (1973) London since 1912. HMSO, 36p.

A brief history of London's social, physical and planning development between 1912 and 1972.

77 DONNISON, D. and EVERSLEY, D. E. C. (eds) (1973) London: urban patterns, problems and policies. Heinemann, 452p.

A number of aspects of London's economic and social development are discussed, with particular emphasis on employment, social and ethnic distribution, and housing. Policies designed to influence this development, especially those contained in the *Greater London development plan* (GREATER LONDON COUNCIL, 1969), are reviewed.

78 HALL, J. M. (1976) London: metropolis and region. Oxford: University Press, 48p. (Problem Regions of Europe)

An examination of the economic and social problems of London, and of the various plans which have been produced in an attempt to provide solutions to these problems. These problems are illustrated by two case studies of areas which are undergoing considerable change: docklands and Covent Garden.

79 ALLISON, R. (1978) Greater London. Hodder and Stoughton, 140p.

A textbook covering various important aspects of London: the Thames, the port, water supply, manufacturing industry, transport, in addition to certain sections of the city. There are large numbers of maps and photographs, and exercises for pupils.

80 CLOUT, H. (ed.) (1978) Changing London. University Tutorial Press, 151p.

Fourteen contributions covering a selection of systematic and spatial themes which are of importance in the understanding of the changing and increasingly planned environment of the capital. Issues considered include population, transport, industry, water and meteorology, as well as the problems of areas such as docklands and Thamesmead.

Section 3 The physical environment

The natural environment of the London Region has been seriously affected by the rapid growth of population and building that has taken place in the region over the last 100 years. Natural landscape features have been removed, modified or covered as the brick and concrete surfaces have spread outwards from central London; river courses have been straightened, culverted and controlled; the climate of the region has been altered through urbanization and London's ecological environments have been changed as urban development has removed or transformed the habitats of various plants and animals.

Despite the dominance of human activity in shaping the metropolitan environment, the Region's underlying physical environment remains of considerable importance and interest to London's population. The aim of this section is to present the published literature on the natural environment in the London Region – its geology and geomorphology, weather and climate, hydrology and ecology. It is important to note the problem of separating the literature on the natural environment from the literature concerned with applied environmental issues. This is especially the case with hydrology (3.3) where there are strong links with section 9 (environmental problems: conservation, pollution and control), for it is inevitable that subsections on water supply (9.3), flood hazard (9.4) and water pollution (9.5) contain material relevant to hydrologists. The classification of material has been based upon whether the published work is concerned purely with the workings of the natural environment (section 3) or whether it is related directly to an applied environmental problem (section 9). The reader searching for material in these two sections is advised to refer to the index.

The historical depth of the literature on the geology of the London basin (subsection 3.1) reflects the quickening of interest in the scientific study of the natural environment which occured in the second half of the nineteenth century and the need for a deeper knowledge of London's geological resources. The pioneer research was undertaken by the Geological Survey with publications by WHITAKER, W. (1872, 1875, 1889) on the geology of London and its surrounding areas. Building upon these initial efforts, the Geological Survey produced detailed descriptions of the geology of various parts of the region in memoirs published during the first 30 years of this century (1903–33). The book by WOODWARD, H. B. (1909) summarized the geology on the four special maps of the London district, and this was revised by BROMEHEAD, C. E. N. (1922). The standard geological reference *London and Thames valley* (SHERLOCK, R. L., 1960) was first published in 1935.

Following these early geological publications, the results of research into the Pleistocene period and the recent geomorphological evolution of the region were published. WOOLDRIDGE, S. W. was a major contributor and his book, written with LINTON, D. L., *Structure, surface and drainage in south east England* (1955) is a classic work on the geomorphology of the region. He also contributed with others to the history of landscape development in the region (WOOLDRIDGE, S. W. and GOLDRING, F., 1953), (WOOLDRIDGE, S. W. and HUTCHINGS, G. E., 1956). During the last 20 years Wooldridge's work has been examined in more detail with continued interest in Pleistocene chronology and in land-sea level changes (DUNHAM, K. C. and GRAY, D. A., 1972).

Studies of London's weather and climate (subsection 3.2) were also undertaken in the nineteenth century, while MANLEY, G. (1961, 1963) has analysed the historical records of weather observations taken in the seventeenth and eighteenth centuries. However, the bulk of the specialized literature post-dates the production of longer runs of reliable meteorological data collected by the stations of the Meteorological Office. Thus, most of the references on the climate of the London Region have been published since 1945. MARSHALL, W. A. L. (1953) presents statistics of London weather over the period 1841–1949 and the analysis is taken further by BRAZELL, J. H. (1968) in his study of London weather. It was not until the mid 1960s that CHANDLER, T. J. (1965) produced the standard work on the area, entitled *The climate of London*, and since then the detailed literature of London's urban climate has been added to by numerous researchers, notably ATKINSON, B. (1966–79).

There is no basic reference on the hydrology of the whole of the London Region equivalent to Chandler on climate, but on ecology FITTER, R. S. R. (1945) *London's natural history* was an early basic text, reinforced by his studies of *London's birds* (1949) and his joint publication FITTER, R. S. R. and LOUSLEY, J. E. (1953) on *The natural history of the City*. The reawakening of interest in ecology and nature conservation in the 1960s has increased the range and depth of ecological publications on various species – birds, plants and animals and different habitats in the London Region.

The Geological Survey, the Meteorological Office and the Institute of Hydrology have played a major role in research and publication into the patterns and processes affecting London's physical environment, and university lecturers have also contributed a great deal to our present regional knowledge. An interesting feature has been the part played by local societies in publishing relevant literature. The London Natural History Society, via its journal *London Naturalist*, is the obvious case, but there are other societies such as the Essex Field Club, the Hertfordshire and Middlesex Trust for Nature Conservation, the Ruislip Natural History Society and the Croydon Natural History Society, all publishing articles on the natural history of their locality and on individual species.

For further references on London's physical environment, the reader is recommended to consult *Geo Abstracts* and *Ecological Abstracts*. There are no bibliographies to recommend to the reader, but further specialist references, for example on individual species or specific localities, can be found by examining the indexes of the journals of the learned societies mentioned above and referred to in the subsections below.

3.1 Geomorphology and geology

81 WHITAKER, W. (1872) The geology of the London basin, Part 1. The

Chalk and the Eocene beds of the southern and western tracts. HMSO and Longmans Green, 619p. (Memoirs of the Geological Survey, Vol. 4)

Describes the Chalk in general terms, while the Eocene deposits are discussed in great detail, district by district. Appendices include a bibliography, well sections and fossils. Part 2 was never published, and is superseded by WHITAKER, W. (1874).

82 WHITAKER, W. (1875) Guide to the geology of London and the neighbourhood. HMSO, 72p. (Geological Survey of England and Wales)

Provides a brief sketch of the Cretaceous, Tertiary and drift geology of London and its surrounding area with a general explanation of the geological model of London in the Geological Museum.

83 WHITAKER, W. (1889) The geology of London and of part of the Thames valley. HMSO, Vol. 1, Descriptive geology. 556p. Vol. 2, Appendices. 352p.

These two volumes take the place of WHITAKER, W. (1872). This work describes in great detail the geology of London and its surrounding districts, including not only the Mesozoic and Tertiary formations, but also a detailed account of the superficial deposits, together wtih appendices on borings and well sections.

84 BLAKE, J. H. (1903) The geology of the country around Reading. HMSO, 91p. (Memoirs of the Geological Survey, Explanation of Sheet 268)

Describes the geology of the western part of the London basin, centred upon Reading. An account is given of the Chalk rocks, followed by an examination of the Reading Beds and London Clay, and the superficial deposits of the area.

85 JUKES-BROWNE, A. J. and OSBORNE WHITE, H. J. (1908) The geology of the country around Henley-on-Thames and Wallingford. HMSO, 113p. (Memoirs of the Geological Survey, Explanation of Sheet No. 254)

Describes the geology of an area of the Thames valley and the Chiltern hills in Berkshire and Oxfordshire with a small area of Buckinghamshire. An account is given of the Jurassic rocks followed by five chapters on the Cretaceous rocks, plus chapters on the Eocene rocks and the superficial deposits.

86 SHERLOCK, R. L. and NOBLE, A. H. (1912) On the glacial origin of the Clay-with-Flints of Buckinghamshire and on a former course of the Thames. *Quarterly Journal of the Geological Society*, Vol. 68, pp. 199–212.

The authors have surveyed some 260 square miles in Buckinghamshire, Berkshire, Hertfordshire and Middlesex. They consider the origin of the

Clay-with-Flints and associated gravels as well as the origin of the plateau gravels at a lower level than the Clay-with-Flints. They believe that the plateau gravels are of fluvioglacial origin – connected with the former course of the Thames.

87 DAVIES, G. M. (1914) Geological excursions round London. T. Murby, 156p.

A summary of the geology of south-eastern England with directions for 26 geological excursions in and around London.

88 DEWEY, H. and BROMEHEAD, C. E. N. (1915) The geology of the country around Windsor and Chertsey. HMSO, 123p. (Memoirs of the Geological Survey, Explanation of Sheet 269)

Describes the Cretaceous and Eocene deposits in the area which includes south-west Middlesex (West Drayton and Staines) along with an account of the plateau gravels, river deposits and the economic geology of the area.

89 DEWEY, H. and BROMEHEAD, C. E. N. (1921) The geology of south London. HMSO, 92p. (Memoirs of the Geological Survey, Explanation of Sheet 270)

Describes the geology of an area including portions of Surrey and Kent, a small part of Essex and a substantial part of the GLC area. There are sections on the underground rocks and Cretaceous rocks, followed by the main chapters on the Tertiary rocks, the superficial deposits and river deposits, with a final chapter on the economic geology of the area.

90 SHERLOCK, R. L. (1922) The geology of the country around Aylesbury and Hemel Hempstead. HMSO, 66p. (Memoirs of the Geological Survey, Explanation of Sheet 238)

Describes the geology of the 216 square miles of the counties of Buckinghamshire, Bedfordshire and Hertfordshire from Aylesbury to Redbourn and from Watford to Luton Hoo. There are sections on the Jurassic, Cretaceous, Eocene and Pleistocene and Recent deposits with a focus upon the Chalk. A final section deals with the economic geology of the area.

91 SHERLOCK, R. L. and NOBLE, A. H. (1922) The geology of the country around Beaconsfield. HMSO, 59p. (Memoirs of the Geological Survey, Explanation of Sheet 255)

Describes the geology of the 216 square miles of the Thames basin from Maidenhead to Amersham and from High Wycombe to Uxbridge. There are sections on the underground rocks, but the main chapters deal with the

Chalk and the London Clay, with a final section on the Pleistocene and Recent deposits, and the economic geology of the area.

92 WOODWARD, H. B. (1922) The geology of the London district. 2nd ed. revised by C. E. N. BROMEHEAD. HMSO, 99p. (Memoirs of the Geological Survey)

Provides an explanation of the four special maps representing the geology of the London district and includes chapters on the Chalk, Tertiaries, Recent deposits and economic geology.

93 WELLS, A. K. and WOOLDRIDGE, S. W. (1923) Notes on the geology of Epping Forest. *Proceedings of the Geologists' Association*, Vol. 34, pp. 244–52.

Based upon an excursion to the area, the article provides: (1) a general account of the geology of the forest (largely comprised of Claygate Beds and London Clay); (2) discussion of the age of the high-level gravels in the forest; (3) the morphology of the Lea valley to the west of the forest.

94 SHERLOCK, R. L. and POCOCK, R. W. (1924) The geology of the country around Hertford. HMSO, 66p. (Memoirs of the Geological Survey, Explanation of Sheet 239)

Describes the geology in an area mostly in Hertfordshire with small parts of Bedfordshire, Greater London and Essex, stretching from Luton Hoo in the north west to Waltham Abbey in the south east. The memoir provides an account of the Cretaceous and Eocene rocks, followed by substantive chapters on the Pleistocene and the river deposits. A final section deals with the economic geology of the area.

95 DEWEY, H. et al. (1924) The geology of the country around Dartford. HMSO, 136p. (Memoirs of the Geological Survey, Explanation of Sheet 271)

Describes the geology of the south-east side of the London basin mainly in Kent, but including parts of Greater London and Essex. The bulk of the text concerns the Upper Cretaceous and Eocene rocks together with the Pleistocene deposits and a final chapter on the economic geology of the area.

96 SHERLOCK, R. L. (1924) The superficial deposits of south Buckinghamshire and south Hertfordshire and the old course of the Thames. *Proceedings of the Geologists' Association*, Vol. 35, pp. 1–28.

Describes the superficial deposits in an area from Aylesbury to beyond Ware and from Luton Hoo to Enfield Lock, mapped on the six inch scale. It provides further evidence to follow the former course of the Thames to the sea and gives a connected account of the age and mode of origin of the various superficial deposits of south Buckinghamshire and south Hertfordshire.

97 BROMEHEAD, C. E. N. et al. (1925) The geology of north London. HMSO, 63p. (Memoirs of the Geological Survey of England and Wales, Explanation of Sheet 256)

Describes the geology of the area from east of the river Lea to the river Colne in the west, including part of Hertfordshire in the north. The area is composed of rocks that range from the Upper Chalk to the Bagshot Sands and there are numerous drift deposits, including Boulder Clay and extensive beds of Thames river gravels.

98 BULL, A. J. and WOOLDRIDGE, S. W. (1925) The geomorphology of the Mole gap. *Proceedings of the Geologists' Association*, Vol. 36, pp. 1–10.

The authors show that in mid-Tertiary times the river Mole crossed the Chalk in a shallow valley with the bottom at 400 feet O.D. After the Pliocene transgression the stream rapidly entrenched its wide meanders forming a steep sided, flat bottomed valley at about 200 feet O.D. Below this are wide stretches of a terrace near 160 feet O.D., best preserved in the Micklesham Flat.

99 DINES, H. G. and EDMUNDS, F. H. (1925) The geology of the country around Romford. HMSO, 53p. (Memoirs of the Geological Survey of England and Wales, Explanation of Sheet 257)

Describes the distribution of Eocene rocks and Pleistocene and later deposits in the area around Newham, Barking, Havering, Redbridge, Waltham Forest and part of Essex.

100 WOOLDRIDGE, S. W. (1926) The structural evolution of the London basin. *Proceedings of the Geologists' Association*, Vol. 37, pp. 162–96.

An examination of the relationship between sedimentation and contemporaneous earth movement in the London basin. The author presents results which bear on the date of initiation and the rate of growth of various minor structures as well as the evolution of the main synclinal basin to which the other features are subsidiary.

101 WOOLDRIDGE, S. W. (1927) The Pliocene history of the London basin. *Proceedings of the Geologists' Association*. Vol. 38, pp. 49–132.

Studies the physical history of the London basin during Pliocene times covering an area from the longitutude of Reading to the coasts of Essex and Kent. The paper examines the early Pliocene sea, the Pliocene contribution to the present drainage system and the evolution of the dominant landforms during this time.

102 DINES, H. G. and EDMUNDS, F. H. (1929) The geology of the country around Aldershot and Guildford. HMSO, 182p. (Memoirs of the Geological Survey, Explanation of Sheet 285)

Describes the geology of an area containing the North Downs and Surrey Hills mostly in Surrey, but including a part of Hampshire and a small part of Berkshire. The focus of the early part is upon the Wealden Gault and Greensand rocks and the Chalk. This is followed by accounts of the Eocene rocks and Pleistocene deposits with a final chapter on the economic geology of the area.

103 WOOLDRIDGE, S. W. and SMETHAM, D. J. (1931) The glacial drifts of Essex and Hertfordshire and their bearing upon the agricultural and historical geography of the region. *Geographical Journal*, Vol. 78, pp. 243–69.

The distribution of glacial drifts in the two counties is discussed and mapped and related primarily to soils and the agricultural types. The drift covered area of eastern Hertfordshire and northern Essex is considered to have been more favourable for early settlement.

104 DEWEY, H. (1932) The palaeolithic deposits of the lower Thames valley. *Quarterly Journal of the Geological Society*, Vol. 88, pp. 35–56.

Discusses the stratigraphical succession and dating of human artifacts found in four pits along the southern side of the Thames valley.

105 WOOLDRIDGE, S. W. (1932) The physiographic evolution of the London basin. *Geography*, Vol. 17, pp. 99–116.

The different stages in the development of the London basin from pre-Tertiary times are explained, and their relevance to the historical geography of the region is considered.

106 DINES, H. G. and EDMUNDS, F. H. (1933) The geology of the country around Reigate and Dorking. HMSO, 204p. (Memoirs of the Geological Survey, Explanation of Sheet 286)

Describes the geology of an area of Surrey lying across the North Downs and Surrey Hills separated by the Holmdale valley. There are chapters on the structure, scenery and drainage systems followed by an account of the rocks from the Wealden Series to Pleistocene and Holocene deposits, together with chapters on the economic geology and water supply of the area.

107 BULL, A. J. et al. (1934) The river Mole: its physiography and superficial deposits. *Proceedings of the Geologists' Association*, Vol. 45, pp. 35–69. (Weald Research Committee Report No. 18)

Preliminary study of the physiography and superficial deposits of the Mole basin undertaken by six geologists.

108 KING, W. B. R. and OAKLEY, K. P. (1936) The Pleistocene succession in the lower parts of the Thames valley. *Proceedings of the Prehistoric Society*, Vol. 2, pp. 52–76.

Examines the relationships between the various Pleistocene and Holocene deposits of the lower and middle Thames valley. Seventeen cross-sections are drawn illustrating the evolutionary stages of the valley and its deposits.

109 WOOLDRIDGE, S. W. (1938) The glaciation of the London basin and the evolution of the lower Thames drainage system. *Quarterly Journal of the Geological Society*, Vol. 94, pp. 627–67.

A discussion of the riverine and glacial drift belts of the London basin. Their origin is considered, and possible inter-regional correlations are suggested. A model is produced for the evolution of the Thames that involves two deflections to the south by glacial ice.

110 WOOLDRIDGE, S. W. and LINTON, D. L. (1938) Some episodes in the structural evolution of south-east England considered in relation to the concealed boundary of meso-Europe. *Proceedings of the Geologists' Association*, Vol. 49, pp. 264–91.

Considers the form and structure of the Palaeozoic floor that underlies the later sediments of south-east England, and shows how this partly controlled the later surface deposits.

111 DAVIES, G. M. (1939) Geology of London and south east England. T. Murby, 198p.

This historical geology takes a retrospective approach, commencing with the Recent deposits and working downwards to the Palaeozoic floor. There are also chapters on water supply and building stones.

112 HESTER, S. W. (1941) A contribution to the geology of north west Middlesex. *Proceedings of the Geologists' Association*. Vol. 52, pp. 304–20.

Presents the results of a detailed survey of an area of north-west Middlesex and south Hertfordshire. The paper deals primarily with the Lower Tertiaries. The revised geological map of the area excludes the outlier of London Clay at Ickenham and adds areas of Reading Beds in South Ruislip/Eastcote and North Harrow/Headstone.

113 WARREN, S. H. (1942) The drifts of south west Essex. *Essex Naturalist*, Vol. 27, pp. 155–63; 171–79.

Records certain new geological sections in south-west Essex and places them within the framework of the physical history of the region.

114 HARE, F. K. (1947) The geomorphology of part of the middle Thames. *Proceedings of the Geologists' Association*, Vol. 58, pp. 294–339.

A detailed study of the geomorphology of the drift-covered country north of the Thames around Slough and Beaconsfield. Evidence is presented for the development of the Thames modern drainage system and the origins of the landscape of the surrounding countryside.

115 WOOLDRIDGE, S. W. (1953) Some marginal drainage features of the Chalky Boulder Clay ice sheet in Hertfordshire. *Proceedings of the Geologists' Association*, Vol. 64, pp. 208–31.

Examines the relations of the Chalky Boulder Clay drift along the northern margin of the Vale of St Albans and in the several Chiltern valleys entering it from the north west.

116 WOOLDRIDGE, S. W. and GOLDRING, F. (1953) The Weald. Collins, 276p.

A detailed scientific study of the geography of the Weald – an area occupying the greater part of the counties of Kent, Surrey and Sussex, with a fringe of Hampshire. The book deals with both the natural history and the human development of the region.

117 WOOLDRIDGE, S. W. and LINTON, D. L. (1955) Structure, surface and drainage in south east England. George Philip, 176p.

A detailed history of the geomorphology of the whole of south-east England, with particular emphasis on the drainage systems of the Weald, and the Thames valley.

118 WOOLDRIDGE, S. W. and HENDERSON, H. (1955) Some aspects of the physiography of the eastern part of the London basin. *Transactions of the Institute of British Geographers*, No. 21, pp. 19–31.

Traces the physical history of the successive courses of the river Thames through a sequence of events, from the earliest lineal ancestor of the present river in Miocene times to the current course of the river. This reconstruction of events suggests a diversion of the northerly route on the emergent floor of the Pliocene sea in two stages – the first, the work of an early glaciation and the second, the result of the second and major glaciation.

119 CULLING, E. W. H. (1956) Longitudinal profiles of the Chiltern streams. *Proceedings of the Geologists' Association*, Vol. 67, pp. 314–68.

From an examination of the longitudinal profiles of the main valleys in the Chilterns, a denudation chronology is constructed and then compared with the sequence derived from the morphological studies of the Thames terraces in the type area.

120 SEALY, K. R. and SEALY, C. E. (1956) The terraces of the middle Thames *Proceedings of the Geologists' Association*. Vol. 67, pp. 369–92.

An account of the river terrace succession in the middle Thames area between Bourne End and Reading, tracing upstream the succession established by HARE, F. K. (1947) in the Slough–Beaconsfield country. A final section deals with drainage diversions in the Reading–Henley area.

121 HAYWARD, J. F. (1956) Certain abandoned channels of Pleistocene and Holocene age in the Lea valley and their deposits. *Proceedings of the Geologists' Association*, Vol. 67, pp. 32–63.

Describes temporary sections in the Lea valley (Hertford–Edmonton) as evidence of former channels of the river Lea. Attempts to provide a relative chronology of various deposits of Pleistocene and Holocene date, based chiefly on a study of the commoner molluscs.

122 WOOLDRIDGE, S. W. and HUTCHINGS, G. E. (1956) London's countryside. Methuen, 223p.

Deals with the physiography and aspects of the historical geography of the London basin and the Weald. Twelve walking routes are described for teachers and students.

123 CLAYTON, K. M. (1957) Some aspects of the glacial deposits of Essex. *Proceedings of the Geologists' Association*, Vol. 68, pp. 1–21.

Examines glacial drifts of the area around Chelmsford and the drift geology of Harlow new town. Gives an outline chronology for the river Thames in Essex.

124 WARREN, S. H. (1957) On the early pebble gravels of the Thames basin from the Hertfordshire–Essex border to Clacton-on-sea. *Geological Magazine*, Vol. 94, pp. 40–6.

The erratic rocks of the early Pebble Gravels are similar to those of the Oxford Plateau Drift. Reasons are given for concluding that they belong to the drifts of the proto-Thames and are earlier than the southward diversion of that river by an advancing ice sheet.

125 WOOLDRIDGE, S. W. (1957) Some aspects of the physiography of the Thames valley in relation to the Ice Age and early man. *Proceedings of the Prehistoric Society*, New Series, Vol. 23, pp. 1–19.

The glacial deposits of the area are discussed, and the question of the nature of the local boulder clay at Hornchurch is re-examined.

126 CLAYTON, K. M. and BROWN, J. C. (1958) The glacial deposits around Hertford. *Proceedings of the Geologists' Association*, Vol. 69, pp. 103–19.

A number of sections in the drift deposits around Hertford are described and a local stratigraphical succession is established. The deposits are shown to have accumulated in an extensive pro-glacial lake when the former valley of the river Thames was blocked by an ice sheet. The recognition of this lake confirms the diversion of the Thames from the Finchley depression by the last ice advance to reach as far south as the London basin. The lower Lea is envisaged to have originated during the retreat phase of this ice sheet.

127 BROWN, J. C. (1959) The sub glacial surface in east Hertfordshire and its relation to the valley pattern. *Transactions of the Institute of British Geographers*, No. 26, pp. 37–50.

A detailed local study of the evolution of the drainage pattern of the upper Lea basin.

128 BROWN, E. H. (1960) The building of southern Britain. *Zeitschrift fur Geomorphologie*, Vol. 4, pp. 264–74.

Focuses upon contributions to the development of the relief in southern Britain from the Palaeozoic floor through the post-Hercynian and Tertiary cycles of erosion to the wave cut platforms in Pleistocene times.

129 CLAYTON, K. M. (1960) The landforms of parts of southern Essex. *Transactions of the Institute of British Geographers*, Vol. 28, pp. 55–74.

An explanatory description of the landforms of Essex with special reference to the pattern of glacial deposition and inter-glacial erosion.

130 FRANKS, J. W. (1960) Inter-glacial deposits at Trafalgar Square, London. *New Phytologist*, Vol. 59, pp. 145–52.

Fossiliferous gravels containing organic deposits were discovered during building on the south side of Trafalgar Square. The author describes an investigation of the plant remains in these deposits.

131 WOOLDRIDGE, S. W. (1960) The Pleistocene succession in the London basin. *Proceedings of the Geologists' Association*, Vol. 71, pp. 113–29.

The drift deposits of the London basin are divided into three main groups: (1) the summit deposits of the hill-tops and high plateaux; (2) the plateau deposits and, locally, the valley deposits; (3) the valley or terrace drifts. The origin and history of each type of deposit are discussed in turn.

132 SHERLOCK, R. L. (1960) London and the Thames valley. 3rd ed. HMSO, 62p. (British Regional Geology).

An account of the geology of the counties of Essex, Middlesex, Buckinghamshire, Berkshire and Oxfordshire, and parts of Wiltshire, Hampshire, Surrey and Kent. It includes a brief description of the structure and physical features of the area with a chapter on the economic geology.

133 THOMAS, M. F. (1961) River terraces and drainage development in the Reading area. *Proceedings of the Geologists' Association*, Vol. 72, pp. 415–22.

Examines the terrace succession in the middle and lower reaches of the rivers Kennet and Loddon-Blackwater, and of the area around their junctions with the Thames. Correlations are found with the terrace succession of the main stream, established by F. K. HARE (1947) and K. R. and C. E. SEALY (1957). In the light of the new evidence presented, a reassessment is made of the current theories for mechanisms of drainage diversions, in the Reading area.

134 THOMASSON, A. J. (1961) Some aspects of the drift deposits and geomorphology of south east Hertfordshire. *Proceedings of the Geologists' Association*, Vol. 72, pp. 287–302.

Data on the lithology and mineralogical composition of the drift deposits of south-east Hertfordshire are presented and their chronology and derivation are discussed. Attention is drawn to the distribution of asymmetric valleys in this and neighbouring areas, and their origin is debated.

135 ZEUNER, F. E. (1961) The sequence of terraces of the lower Thames and the radiation chronology. *Annals of the New York Academy of Science*, Vol. 95, pp. 377–80.

Outlines the events affecting the middle and lower courses of the river Thames during the second half of the Pleistocene. Argues that the evidence of river terraces connected with sea level changes, ocean temperatures, and astronomical evidence for variations in solar radiation are all beginning to present a coherent picture of these events.

136 LOVEDAY, J. A. (1962) The plateau deposits of the south Chiltern hills. *Proceedings of the Geologists' Association*, Vol. 73, pp. 83–102.

On the basis of field and laboratory studies of the superficial deposits and associated soils of the southern Chilterns, the author attempts to provide a strict definition of Clay-with-Flints and to clarify its geomorphic relationships to the other stony plateau deposits grouped here as Plateau Drift. The several theories of origin which have been proposed for these deposits are re-examined in the light of the available evidence.

137 BROWN, E. H. (1964) Some aspects of the geomorphology of south east England. In: CLAYTON, K. M. (ed.) A Guide to London excursions. London School of Economics and Political Science, pp. 113–18.

Examines the form of the ground on a line, 16 miles (26 km) north-westwards from central London to the Vale of Aylesbury. A brief description of the underlying rocks is followed by an examination of the development of the Thames valley together with sections on the Finchley Depression, the South Hertfordshire Plateau, the Vale of St Albans, the Chiltern Hills and the Vale of Aylesbury.

138 CLAYTON, K. M. (1964) The glacial geomorphology of southern Essex. In: CLAYTON, K. M. (ed.) A Guide to London excursions. London School of Economics and Political Science, pp. 123–28.

Presents the background to an excursion whose aim is to explain the way the landscape of southern Essex is related to the complex events of the Pleistocene period. A first section records the state of geomorphological knowledge of the glacial evolution of the area, and this is followed by a description of the six main stops in the itinerary.

139 LAWRENCE, G. R. P. (1964) Some pro-glacial features near Finchley and Potters Bar. *Proceedings of the Geologists' Association*, Vol. 75, pp. 15–30.

The detailed examination of two valleys and a more general study of erosion levels of south Hertfordshire suggest that three or more stages are present in the drainage pattern and that these are related to successive courses of a proto-Thames.

140 WOOLDRIDGE, S. W. (1964) The minor structures of the London basin. *Proceedings of the Geologists' Association*, Vol. 34, pp. 175–93.

Attempts to describe the minor folds and faults in the London basin, to correlate and compare them over the whole extent of the basin, and to examine the cause of the folding and the age of the folds.

141 CURRY, D. (1965) The palaeogenic beds of south east England. *Proceedings of the Geologists' Association*, Vol. 76, pp. 151–73.

A general account of the distribution and relationships of the individual beds of the Palaeogenic (early Tertiary) for both the London and Hampshire basins.

142 GALLOIS, R. W. (1965) The Wealden district. 4th ed. HMSO, 101p. (British Regional Geology)

An account of the geology of the Wealden district embracing the major part of the counties of Kent, the whole of Sussex, the southern half of Surrey and a small part of Hampshire. It includes chapters on the older concealed and exposed strata, the geological structure, the Pleistocene and Recent deposits and finally, sections on the development of scenery and the economic geology of the district.

143 HESTER, S. W. (1965) Stratigraphy and palaeogeography of the Woolwich and Reading Beds. *Bulletin of the Geological Survey of Great Britain*, Vol. 23, pp. 117–38.

The divisions and facies of the Woolwich and Reading Beds of southern and eastern England comprising up to 160 feet of sediments are described. Relationships with the underlying Chalk and Thanet Beds and the overlying Blackheath and Oldhaven Beds and London Clay are considered. The origins of these beds and the palaeogeography of the period are discussed.

144 TERRIS, A. P. and BULLERWELL, W. (1965) Investigations into the underground structure of southern England. *Advancement of Science*, Vol. 22, pp. 232–52.

Contains a brief review of the geological history and surface structures of southern England, concluding that the underground structure has hardly changed since the Pliocene. The pre-Upper Cretaceous faulting is noted and details of recent deep bores given. The results of a detailed regional analysis of gravity and aeromagnetic maps of southern England are also presented.

145 HEY, R. W. (1965) Highly quartzose pebble gravels in the London basin. *Proceedings of the Geologists' Association*, Vol. 76, pp. 403–20.

A re-examination of localities in Hertfordshire, where Pebble Gravels were identified in 1896, leads to the conclusion that a new subdivision of these deposits, the Westland Green Gravels can be recognized. The deposits discussed in the paper, perhaps provide a link in the interpretation of chronology in Hertfordshire and East Anglia.

146 EVERARD, C. E. (1966) The physique of east London. *East London Papers*, Vol. 9, pp. 59–83.

Describes the geological structure, the physiographic evolution of the rivers Thames and Lea and the landforms of the area dominated by the confluence of the two rivers. The various sources of water supply from the rivers, gravel terraces and solid rocks (Tertiaries and Chalk) are also described.

147 DOCHERTY, J. (1967) The exhumed sub-Tertiary surface in north-west Kent and its relationship to the Pleistocene erosion stages. *South East Naturalist*, Vol. 70, pp. 19–33.

Reports the results of a programme of geomorphological mapping of the Pleistocene erosion surfaces and their associated deposits in the lower Darent basin. Suggests that exhumation of the large fragments of the sub-Tertiary surface in north-west Kent was accomplished by peneplanation to a local base level above the surface, followed by a predominantly aeolian winnowing of the remainder of the overburden.

148 PITCHER, W. S. et al. (1967) The London region (south of the Thames). 2nd ed. Colchester: Benham, 32p. (Geologists' Association Guide, No. 30B)

A brief description of the geology of the area between the Thames and the North Downs is followed by nine itineraries to areas of geological interest.

149 BLEZARD, R. G. et al. (1967) The London region (north of the Thames). 2nd ed. Colchester: Benham, 34p. (Geologists' Association Guide, No. 30A)

The introduction contains a brief account of the geology of the London region north of the Thames including Middlesex, Buckinghamshire, Hertfordshire and Essex. Seven itineraries to interesting geological areas are then described.

150 WALDER, R. S. (1967) The composition of the Thames gravels near Reading, Berkshire. *Proceedings of the Geologists' Association*, Vol. 78, pp. 107–9.

Suggests that by a study of the differing make up of gravels from the past and present drainage areas of the Thames, the Kennet and the Loddon, it may be possible to recognize the points of confluence within the composite drainage system of the Reading area, corresponding to the times during which the river terraces were laid down. The possibility was tested on the dissected terrace now preserved at 140 feet above the level of the present Thames flood plain (the Lower Winter Hill terrace).

151 JARVIS, R. A. (1968) Soils of the Reading district. Harpenden, Soil Survey of England and Wales, 150p. (Memoir of the Soil Survey of Great Britain, Sheet 268)

Describes the soil in the district at the western end of the London basin centred upon Reading, but including parts of the counties of Berkshire, Oxfordshire, Buckinghamshire, Hertfordshire, Surrey and Hampshire. Soil series are described and a final section deals with agricultural, horticultural and forestry land use.

152 CLAYTON, K. M. (1969) Post war research on the geomorphology of south east England. *Area*, Vol. 1, pp. 9–12.

Argues that further research is needed before a coherent statement can succeed *Structure, surface and drainage in south-east England* (WOOLDRIDGE, S. W. and LINTON, D. L. 1955). Reviews post-war research work and cites the need for more work on the Clay-with-Flints, a re-examination of the denudation chronology and the need to relate the Thames terrace to the Oxford area. Argues that work on sedimentological characteristics and on interpretation of the morphological aspects of dated sites in London is needed.

153 LINTON, D. L. (1969) The formative years in geomorphological research in south east England. *Area*, Vol. 1, pp. 1–8.

Describes the background to geomorphological studies of south-east England established by work up to 1925, when these studies were considered an aid to stratigraphy, and ideas of eustatic changes of sea level were foreign to British geological thought. Between 1925–39 these studies owed much to the Weald Research Committee and especially to Wooldridge. By the late 1930s, it was accepted that Pleistocene chronology must be considered in relation to the history of base level.

154 SPARKS, B. W. et al. (1969) Hoxnian interglacial deposits near Hatfield, Hertfordshire. *Proceedings of the Geologists' Association*, Vol. 80, pp. 243–68.

Biogenic interglacial deposits intercalated between two complexes of glacial deposits near Hatfield and at Stanborough are shown to be of Hoxnian age. The presence of a Hoxnian interglacial deposit within the sequence of glacial deposits in the Vale of St Albans establishes the fact that both Lowestoft and Gipping glaciations are represented there.

155 WOOLDRIDGE, S. W. (1969) The physique of Middlesex. In: The Victoria history of the county of Middlesex. Vol. 1, Constable, pp. 1–10.

A general descriptive account of the solid and drift geology and physical geography of Middlesex. The human response to the environment especially to soils and water supply is discussed.

156 HEY, R. W. et al. (1971) Surface textures of sand grains from the Hertfordshire Pebble Gravels. *Geological Magazine*, Vol. 108, pp. 377–82.

The Pebble Gravels of Hertfordshire are among the earliest Pleistocene deposits of the London basin. Examines the current views on their origin and uses electron microscope studies of the surface textures of sand grains to assess these views. Results suggest that they may well be substantially correct.

157 BAKER, C. A. (1971) A contribution to the glacial stratigraphy of west Essex. *Essex Naturalist*, Vol. 32, pp. 318–30.

The author uses newly available data from the M11 Soil Survey in west Essex to examine the systems of subdivision of the Essex glacial deposits. Data from the 622 boreholes along a 36 mile route from Woodford to Great Chesterford provided a consistent picture of stratigraphical relationships and the glacial deposits are considered under three headings – Chalky Boulder Clay, sub-boulder clay sands and gravels, and intra glacial sediments. Concludes that the data favour the essential unity of the west Essex Chalky Boulder Clay as the product of a single glaciation.

158 D'OLIER, B. (1972) Subsidence and sea-level rise in the Thames estuary. In: DUNHAM, K. C. and GRAY, D. A. (eds) A discussion on problems associated with the subsidence of south eastern England. *Philosophical Transactions of the Royal Society of London*, A, Vol. 272, pp. 121–30.

Traces the development of the outline of the Thames estuary since the late Cretaceous period. The positions of the shorelines during the critical phase 9,600 B.P. to 8,000 B.P., of the last transgression of the sea are shown and an estimation is made of the subsequent average rate of subsidence and/or sea level rise based upon the concept of sedimentary equilibrium.

159 PRENTICE, J. E. (1972) Sedimentation in the inner estuary of the Thames, and its relation to the regional subsidence. In: DUNHAM, K. C. and GRAY, D. A. (eds) A discussion on problems associated with the subsidence of south eastern England. *Philosophical Transactions of the Royal Society of London*, A, Vol. 272, pp. 115–19.

Looks at sedimentation in the Thames estuary from Teddington Weir to Southend. The studies suggest that this inner estuary is a sediment trap in which the limits are set in the upstream direction by the salinity gradient and downstream by the point at which loss of muddy sediment exceeds supply.

160 WEST, R. G. (1972) Relative land-sea level changes in south eastern England during the Pleistocene. In: DUNHAM, K. C. and GRAY, D. A. (eds) A discussion on problems associated with the subsidence of south eastern England. *Philosophical Transactions of the Royal Society of London*, A, Vol. 272, pp. 87–98.

Examines the evidence for relative land-sea level changes in south-east England during the Pleistocene period, covering the last two million years. A stratigraphical treatment of the evidence is used and a summary of land-sea level changes is given in diagrammatic form.

161 BRISTOW, C. R. and FOX, F. C. (1973) The Gipping Till: a reappraisal of East Anglian glacial stratigraphy. *Journal of the Geological Society of London*, Vol. 129, pp. 1–37.

Interprets Pleistocene stratigraphy based upon 6 inch geological surveys in the Chelmsford and Norwich areas with a critical look at the evidence for the existence of the Gipping Till and its associated glaciation. Concludes that widespread Chalky Boulder Clay was formed during only one glacial episode, that the Hoxnian and Ipswichian deposits in East Anglia are closer together in time than has been thought hitherto, and that the main glaciation of East Anglia is the penultimate (Wolstonian).

162 HUTCHINSON, J. N. (1973) The response of London Clay cliffs to differing rates of toe erosion. *Geologia Applicata e Idrogeologia*, Vol. 7, pp. 222–39.

Reports observations on the various types of land sliding found on cliffs of London Clay adjoining the Thames estuary and examines the influence of differing rates of toe erosion on their mode of breakdown. Three cases are distinguished. Concludes that the control of erosion is the key to the conservation of natural slopes.

163 JONES, D. K. C. (1973) The influence of the Calabrian transgression on the drainage evolution of south east England. In: BROWN, E. H. and WATERS, R. S. (eds) Progress in geomorphology: papers in honour of David Linton. Institute of British Geographers, pp. 139–57. (IBG Special Publication, No. 7)

The discordant drainage pattern of south-east England has been explained in terms of superimposition from a high level marine surface created by the early Pleistocene or Calabrian marine transgression. This paper argues that there are grounds for rejecting this hypothesis, especially as morphological and sedimentological evidence for the transgression is largely confined to the London basin while the discordant relationships are located in Sussex and Wessex. It is argued that the main elements of the present drainage network are inherited from the concordant pattern created by the withdrawal of the early Tertiary Sea.

164 KELLAWAY, G. A. et al. (1973) South east England. In: MITCHELL, G. F. et al. (eds) A correlation of Quaternary deposits in the British Isles. (Geological Society of London. Special Report No. 4) pp. 45–53.

A detailed account, consisting of notes and a correlation chart for Quaternary deposits found in south-east England.

165 BURNHAM, C. P. and McCRAE, G. S. (1974) The relationship of soil formation to geology in an area south east of London. *Proceedings of the Geologists' Association*, Vol. 85, pp. 79–90.

The soil types in a small area on the south-eastern outskirts of London are described and their mode of formation related to the underlying geology and other environmental factors. The soils studied are a humus iron podzol on the Blackheath Beds, a groundwater gley and peat, a surface water gley and a topo sequence of soils associated with a chalky dry valley.

166 GRUHN, R. et al. (1974) A contribution to the Pleistocene chronology in south east Essex. *Quaternary Research*, Vol. 4, pp. 53–71.

Identifies five terraces to the Thames parallel to the Thames in Essex and analyses the gravels and brickearth deposits. The latter deposits are seen as loess deposits and the evidence for subsidence is assessed in a summary of Pleistocene events in the area.

167 D'OLIER, B. (1975) Some aspects of late Pleistocene-Holocene drainage of the river Thames in the eastern part of the London basin. *Philosophical Transactions of the Royal Society of London*, A, Vol. 279, pp. 269–78.

Results from an extensive survey of the Thames since 1967, (using reflection seismic techniques and logs and samples from some 140 boreholes) have allowed the reconstruction of the late Pleistocene/early Holocene palaeodrainage pattern of the river Thames. The relationship of this drainage system to the tectonic pattern would indicate that there has been late Pleistocene re-emphasis of the existing dominant structures. Some evidence for the existence of a late Pleistocene ice sheet is presented.

168 SHEPHARD THORN, E. R. (1975) The Quaternary of the Weald – a review. *Proceedings of the Geologists' Association*, Vol. 86, pp. 537–47.

Recent work on the Quaternary deposits and processes in the Weald is reviewed and an attempt is made to fit them into a chronological sequence.

169 BRUNSDEN, D. (1976) Even London has a landscape. *Geographical Magazine*, Vol. 48, pp. 282–9.

Argues that despite urban development, London's natural landscape can still be detected, whilst evidence of the ice sheet and changes in climatic conditions can also be found. The article explains how man dominates landform evolution in the area and examines the main influences on London's physical landscape.

170 CATT, J. A. and HODGSON, J. M. (1976) Soils and geomorphology of the Chalk in south east England. *Earth Surface Processes*, Vol. 1, pp. 181–93.

Looks critically at the history of landscape development in south-east England proposed by Wooldridge and Linton (1955) and suggests that soil mapping and soil profile studies on the Chalk of south-east England do not support this widely accepted view. The evidence for Plio-Pleistocene sea levels is reviewed and alternative explanations for the so-called Calabrian marine platform and cliff are considered. Emphasis is placed on periglacial processes in the later Pleistocene denudation of the Chalk and the protective role played by the cover of disturbed basal Tertiary sediment.

171 SKEMPTON, A. W. and HUTCHINSON, J. N. (eds) (1976) A discussion on valley slopes and cliffs in southern England: morphology mechanics and Quaternary history. *Philosophical Transactions of the Royal Society of London*, A, Vol. 283, pp. 421–635.

Contains six papers with two features in common: they are based on detailed sub-surface investigations of slopes, using boreholes, trial pits or sections and the stages of development of the slopes or more generally of the valleys or escarpments have been correlated as far as possible with Quaternary history. The slopes are all in clay strata (Jurassic, Cretaceous or Eocene) and the sites lie outside the ice limits of the Late (Devensian) glaciation.

172 CLAYTON, K. M. (1977) River terraces. In: SHOTTON, F. W. (ed.) British Quaternary studies: recent advances. Oxford: University Press, pp. 153–67.

The terraces of major English rivers, including the Thames, are described, with particular reference to their age. It is concluded that river terraces cannot be used as a key to the fluctuation of Quaternary sea levels.

173 SHEPHARD THORN, E. R. and WYMER, J. J. (1977) South east England and the Thames Valley. Norwich, International Union for Quaternary Research, 76p.

This guide book was produced for a six day itinerary of excursions undertaken before the INQUA Congress in Birmingham. The Quaternary features and deposits fall into two geological regions – the London basin and the Weald. There are three excursions from Reading – the Kennet valley and Marlborough Downs, the Chiltern Hills and Goring Gap and the middle Thames valley; the other three areas described in the guide are the lower Thames and Medway valleys, East Kent and the raised beaches of West Sussex.

174 GIBBARD, P. L. (1977) Pleistocene history of the Vale of St Albans. *Philosophical Transactions of the Royal Society of London*, B, Vol. 280, pp. 445–83.

Reports on a detailed investigation into the stratigraphy of the Pleistocene sequence in the Vale of St Albans. Results from boreholes exposures and sections are used to link the sequences in East Anglia and the middle Thames.

175 BROWN, E. H. (1978) Land under London. In: CLOUT, H. (ed.) Changing London. University Tutorial Press, pp. 138–48.

Examines the physical site and situation of London, and includes sections on: the geological underpinnings, superficial deposits, valley floors, Thames terraces, the earlier valley of the Thames and hill top gravels. A description of landform regions is followed by a conclusion summarizing the influence of the land upon changing London.

176 GREEN, C. P. and McGREGOR, D. F. M. (1978) Pleistocene gravel trains of the river Thames. *Proceedings of the Geologists' Association*, Vol. 89, pp. 143–56.

The gravel trains of the Thames were examined between Bourne End (Buckinghamshire) and Bricket Wood (Hertfordshire) and compared with gravels at higher and lower levels. Results indicate a variety of catchment changes before, during and after gravel train times, and suggest a reinterpretation of the drainage development of the area examined. No diversion of the Thames from its early course through the Vale of St Albans can be detected before or during gravel train times.

177 McGREGOR, D. F. M. and GREEN, C. P. (1978) Gravels of the river Thames as a guide to Pleistocene catchment changes. *Boreas*, Vol. 7, pp. 197–203.

The results from analyses of Pleistocene gravels from the Vale of St Albans are used to show that there is evidence for major changes in the catchment area of the proto-Thames and its tributaries. Glaciation on one or more occasions in the upper catchment of the Thames is suggested to explain the high frequencies of the travelled materials.

178 GIBBARD, P. L. (1979) Middle Pleistocene drainage in the Thames valley. *Geological Magazine*, Vol. 116, pp. 35–44.

Reports recent findings on the drainage in the Thames valley in the area to the north and west of London during the middle Pleistocene period. The results provide the first evidence of south bank tributary drainage in the Thames valley during the middle Pleistocene and confirm earlier suggestions that the river Thames never flowed via the Finchley Depression.

179 DEVOY, R. J. N. (1979) Flandrian sea level changes and vegetational history of the lower Thames estuary. *Philosophical Transactions of the Royal Society of London*, B, Vol. 285, pp. 355–407.

Reports the results of a biostratigraphic study investigating the interleaved Flandrian biogenic and inorganic deposits of the lower Thames estuary between central London and the Isle of Grain. The vegetational and environmental history showing the relation of biogenic deposits to former sea level has been deduced from pollen diatom and other micro fossil studies, while radio carbon dating has been used to establish an objective chronology. Relative sea level curves for the Thames during Flandrian times are deduced and they correlate well with the form and rate of relative sea level changes shown for north-west Europe.

180 BERRY, F. G. (1979) Late Quaternary scour hollows and related features in central London. *Quarterly Journal of Engineering Geology*, Vol. 12, pp. 9–29.

Describes and analyses 26 depressions represented on a map of shallow buried 'channels' formed in the surface of the London Clay in central London beneath the Lower Floodplain deposits of Ipswichian to Recent age.

3.2 Weather and climate

181 METEOROLOGICAL OFFICE (1904) London fogs: report on the fogs in the London area, 1901–2 and 1902–3. HMSO, 48p.

The report of the inquiry initiated by the Meteorological Council in 1901 into the occurrence and distribution of fog in the London district.

182 BROOKS, C. E. P. and MIRRLEES, S. T. A. (1930) Irregularities in the annual variation of the temperature of London. *Quarterly Journal of the Royal Meteorological Society*, Vol. 56, pp. 375–388.

Examines the assertion that warm and cold spells or crests and troughs on the curve of temperature are not entirely haphazard in their occurrence, but show a tendency to cluster closely around particular days. Data for Kew Observatory is used over the period 1871–1929 and analysis shows that it seems improbable that there exists in the climate, an abiding tendency for any part of the year to be either abnormally warm or abnormally cold for the season.

183 WRIGHT, H. L. (1932) Observations of smoke particles and condensation nuclei at Kew. Meteorological Office, 57p. (Geophysical Memoirs, No. 57)

Examines the nature and distribution of smoke particles and condensation nuclei found in suspension in the lower strata of the atmosphere by using measurements taken at Kew Observatory from 1928.

184 DRUMMOND, A. J. (1943) Cold winters at Kew Observatory, 1783–1942. *Quarterly Journal of the Royal Meteorological Society*, Vol. 69, pp. 17–32; 147–55.

Extends the observations made at Kew Observatory back in time, to obtain a homogeneous series of winter temperature data over a period of 160 years by using observations made at Syon House, Isleworth and at Chiswick. These observations are formed into a series of mean monthly winter temperatures and the results are discussed with particular reference to the winters of 1940, 1941 and 1942.

185 BELASCO, J. E. (1948) Rainless days of London. *Quarterly Journal of the Royal Meteorological Society*, Vol. 74, pp. 339–48.

The incidence of the 12,616 rainless days (less than 0.1 mm of rain) which have occurred at London (Kew) during the 70 years, 1878–1947 are examined. The results are analysed and use is made of the synoptic succession which marks the onset, course and termination of the nine longest rainless spells.

186 DOUGLAS, C. K. M. and STEWART, K. H. (1953) London fog of December, 5–8, 1952. *Meteorological Magazine*, Vol. 82, pp. 67–71.

An account of the thick fog in the London basin over a four day period in December 1952, in which visibility over large areas was below 20 yards for many hours on end and was often below 10 yards. The primary cause of the persistent dense fog was the complete absence of any pressure gradient for an exceptionally long period.

187 MARSHALL, W. A. L. (1953) A century of London weather. HMSO, 134p.

Gives statistics and analyses of the weather in London from 1841–1949. Yearly, monthly and seasonal summaries are provided along with annual figures for temperature, rainfall and sunshine.

188 SHELLARD, H. C. (1959) The frequency of fog in the London area compared with that in rural areas of East Anglia and south east England. *Meteorological Magazine*, Vol. 88, pp. 321–423.

Presents and discusses the ten year frequencies of fog for four stations – Kingsway, Croydon, Mildenhall and West Raynham. The author concludes that fog (less than 1,000 yards visibility) is more frequent in central London than in the suburbs or in rural areas, and it is rather more frequent in outer London (Kew and London Airport) than in the country.

189 CHANDLER, T. J. (1961) The changing form of London's heat island. *Geography*, Vol. 46, pp. 295–307.

An account of a co-operative study of London's climate by schools and colleges using climatological readings from 39 stations and a mobile survey.

Focuses upon the most distinctive feature of London's local climate – the heat island – the mass of warm air within and above the built up area.

190 MANLEY, G. (1961) A preliminary note on early meteorological observations in the London region, 1680–1717, with estimates of the monthly mean temperatures, 1680–1706. *Meteorological Magazine*, Vol. 90, pp. 303–10.

Describes the early meteorological observations and deductions with regard to the meteorology of the period 1680–1720.

191 CHANDLER, T. J. (1962) Diurnal, seasonal and annual changes in the intensity of London's heat-island. *Meteorological Magazine*, Vol. 91, pp. 146–53.

A study of London's heat-island showing the significant variations over diurnal seasonal and annual time scales with recognizable periodic trends occurring in spite of perturbations imposed by irregular changes in weather.

192 CHANDLER, T. J. (1962) Temperature and humidity traverses across London. *Weather*, Vol. 17, pp. 235–41.

Explains the equipment and techniques used in the field traverses of London conducted as part of the investigation into London's urban climate.

193 CHANDLER, T. J. (1962) London's urban climate. *Geographical Journal*, Vol. 128, pp. 279–302.

Examines the form and degree of climate modification exerted by London itself by looking at the nature and intensity of contrasts between the climates of the built up area and the surrounding districts.

194 LOWNDES, C. A. S. (1962) Wet spells at London. *Meteorological Magazine*, Vol. 91, pp. 98–104.

Examines and analyses wet spells in the period 1935–59 in order to provide a background to the problems of forecasting wet spells at London.

195 BRAZELL, J. H. (1963) Severe winters and following summers in London. *Weather*, Vol. 18, pp. 322–4.

Examines the idea that severe winters are usually followed by fine summers by studying severe winters during the period 1763–1963, together with rainfall and mean temperature in the following summers. Suggests that over this 200 year period, the probability of a good or very good summer after a severe winter is only 12 per cent, while the probability of a poor or very poor summer is 64 per cent.

196 KELLY, T. (1963) A study of persistent and semi-persistent thick and dense fog in the London area during the decade 1947–56. *Meteorological Magazine*, Vol. 92, pp. 177–83.

Analyses data from three stations – Heathrow, Croydon and Kingsway – for the period 1947–56. Little difference was observed between the stations with persistent, dense fog, but semi-persistent thick fog occurred most frequently at Heathrow and least frequently at Croydon. Heathrow has twice as many semi-persistent dense fogs as Croydon and Kingsway.

197 LOWNDES, C. A. S. (1963) Cold spells at London. *Meteorological Magazine*, Vol. 92, pp. 163–76.

An examination of cold spells at Kew, 1935–59, to provide a background to the problem of forecasting the ending of cold spells at London.

198 MANLEY, G. (1963) Seventeenth-century London temperatures: some further experiments. *Weather*, Vol. 18, pp. 98–105.

Describes meteorological observations taken in seventeenth century London in an attempt to tie them into a continuous table. An account is given of the methods used and the problems met with in this task. This suggests that the uttermost limit of instrumental observations has been virtually reached in this historical quest.

199 BRAZELL, J. H. (1964) Frequency of dense and thick fog in central London as compared with frequency in outer London. *Meteorological Magazine*, Vol. 93, (1102), pp. 129–35.

An analysis of statistics comparing fog frequencies for central London (Kingsway) with those of outer London (Kew, Heathrow and Croydon) over a period from 1947–62.

200 CHANDLER, T. J. (1964) An accumulated temperature map of the London area. *Meteorological Magazine*, Vol. 93, pp. 242–5.

Examines the geographical distribution of accumulated temperatures across the London area using data produced by the Meteorological Office and the London Climatological Survey. Altitude and urban exposure are the two main factors differentiating accumulated temperatures within Greater London. London's heat island and its effects on regional temperatures are summarized in map form.

201 CHANDLER, T. J. (1964) Climate and the built up area. In: COPPOCK, J. T. and PRINCE, H. C. (eds). Greater London. Faber and Faber, pp. 42–51.

A study of the imprint of London's growth upon the mesoclimates of the conurbation as a whole. The climate in central London and suburban London is examined in separate sections.

202 CHANDLER, T. J. (1964) The climate of London. In: CLAYTON, R. (ed.). The geography of Greater London: a source book for teacher and student. George Philip, pp. 171–82.

Examines the distinctive urban climate of London set against the broad climate of south-east England. There are sections on pollution, visibility, sunshine, winds, temperatures, relative humidity, cloud cover and precipitation.

203 CHANDLER, T. J. (1965) The climate of London. Hutchinson, 292p.

A consideration of the nature of London's climate, including the manner and degree to which this varies in sympathy with changes in the morphology of the city and the contrasts between the built up area and the surrounding rural districts.

204 CRADDOCK, J. M. (1965) Domestic fuel consumption and winter temperatures in London. *Weather*, Vol. 20, pp. 257–8.

Using his own estimated figures, the author provides a case for climatologists to consider the effect of domestic fuel consumption on London's temperature, especially under conditions of nocturnal inversion.

205 ATKINSON, B. W. (1966) Some synoptic aspects of thunder outbreaks over south east England, 1951–60. *Weather*, Vol. 21, pp. 203–9.

Presents an analysis of synoptic scale circulations relevant to a larger study of the distribution of thunder rains over south-east England in the decade 1951–60.

206 LAWRENCE, E. N. (1967) Meteorology and the Great Fire of London, 1666. *Nature*, Vol. 213, pp. 168–9.

Consideration is given to the inherent stability or instability of the air mass as one of the relevant factors in the manner and intensity of the development of the Great Fire of London. Use is made of limited meteorological evidence indicating that prevailing easterly winds and persistent fine weather preceded the fire and that therefore atmospheric instability may not have been extreme.

207 ATKINSON, B. W. (1968) A preliminary examination of the possible effect of London's urban area on the distribution of thunder rainfall, 1951–60. *Transactions of the Institute of British Geographers*, Vol. 44, pp. 97–118.

Outlines the need for a detailed investigation of the possible effect of an urban area on rainfall generally and attempts to evaluate that effect on thunder rainfall in particular, using data for London's urban area, 1951–60.

208 BRAZELL, J. H. (1968) London weather. HMSO, 250p.

Extends *A century of London weather* (MARSHALL, W. A. L., 1953) in that it includes records before and after the period 1841–1949, dealt with in that text. A comparison and summary of weather by years, seasons and months is made and special features such as fogs and cold spells are examined. Appendices include outstanding features of climate until 1841, early records of rainfall at Edmonton and summaries of temperature, rainfall and sunshine for Greenwich and Kew.

209 FREEMAN, M. H. (1968) Visibility statistics for London, Heathrow airport. *Meteorological Magazine*, Vol. 97, pp. 214–8.

Based upon hourly observations, these statistics on visibility are analysed to show changes through time (1949–67) on an annual, monthly and diurnal basis, with the conclusion that visibility has generally improved.

210 ATKINSON, B. W. (1969) A further examination of the urban maximum of thunder rainfall in London, 1951–60. *Transactions of the Institute of British Geographers*, Vol. 48, pp. 97–119.

The maximum of thunder rain over London's urban area is examined in an attempt to isolate storms which contribute most thunder rainfall.

211 CANOVAN, R. A. (1969) The 'cold spell' of September 1952 to April 1953. *Weather*, Vol. 24, pp. 199–200.

A report of some of the noteworthy events which occurred during this cold spell such as blizzards, floods and the disastrous London fog.

212 DAVIS, N. E. (1969) Diurnal variation of thunder at Heathrow airport, London. *Weather*, Vol. 24, pp. 166–72.

Hourly analysis revealed a maximum of thunderstorms in the afternoon, a secondary maximum after midnight and a morning minimum. July was found to have the greatest total of thunderstorm hours and the most night time storms, the latter occurring most frequently in odd years, possibly due to more blocking action at this time.

213 JENKINS, I. (1969) Increase in averages of sunshine in Greater London. *Weather*, Vol. 24, pp. 52–4.

Shows that sunshine duration at the London Weather Centre during the winter months increased by approximately 50 per cent since 1958 – compared with small variations shown in a suburban and a rural site. The author concludes that this is probably associated with a decrease in smoke in the city.

214 LAWRENCE, E. N. (1969) Effects of urbanization on long term changes of winter temperature in the London region. *Meteorological Magazine*, Vol. 98, pp. 1–8.

Compares changes of temperature in central London with other stations in south-east England for 1920–68. The mean values of the daily minimum and maximum temperatures for the winter season (December–February) reveal a decrease in the mean daily minimum temperature in central London when related to the surrounding areas, and an increase in the corresponding mean maximum temperature. It is concluded that these changes may be the result of urban effects and the different responses of urban and rural areas to climatic changes.

215 ATKINSON, B. W. (1970) The reality of the urban effect on precipitation – a case study approach. In: Urban climates; proceedings of the symposium on urban climates and building climatology ... Brussels, 1968. World Meteorological Organisation, pp. 342–60. (WMO Technical Note, No. 108)

Describes an analysis of the thunderstorms which occurred over London on 21 August 1959. Concludes that the storms were triggered by the high urban temperatures and that turbulence and potential condensation and ice nuclei in the urban area played a negligible role in their initiation.

216 SELF, R. W. (1970) Periodicities in London rainfall. *Weather*, Vol. 25, pp. 75–6.

Analyses the periodicity of the fine summer weather by examining the records for annual rainfall which are longer than those for sunshine.

217 ATKINSON, B. W. (1971) The effect of an urban area of the precipitation from a moving thunderstorm. *Journal of Applied Meteorology*, Vol. 10, pp. 47–55.

A case study of precipitation from a thunderstorm cloud was made for 9 September 1955, over London's urban area. Data were provided by dense observation networks and the cloud was tracked by radar. It is concluded that in this particular case the urban effect was real, but generalization from this result may not be valid.

218 JENKINS, I. (1971) Decrease in the frequency of fog in central London. *Meteorological Magazine*, Vol. 100, pp. 317–22.

Using data from the London Weather Centre a decrease in the frequency of fog in central London is shown. It is suggested that these changes are connected with changes in smoke emission in central London, though not specifically with the consequences of the Clean Air Act (1956).

219 KELLY, T. (1971) Thick and dense fog at London/Heathrow Airport and Kingsway/Holborn during the two decades 1950–9, 1960–9. *Meteorological Magazine*, Vol. 100, pp. 257–67.

Extends the time period of a previous study (KELLY, T., 1963) by considering the occurrence of thick and dense fog in the 1950s and 1960s at Heathrow and Kingsway/Holborn.

220 LAWRENCE, E. N. (1971) Urban climate and day of the week. *Atmospheric Environment*, Vol. 5, pp. 935–48.

Surveys previous research on the day of the week variation in certain weather parameters, concentrating upon rainfall amount. The results of analysis within the London area: (1) the weekly rainfall cycle is related to wind speed, sunshine duration and the size of the urban heat island; (2) in midsummer a weekly cycle exists in the difference between mean daily maximum temperatures in central London and the surrounding area; (3) support for the theory that thermal connection may cause the difference between rainfall figures for London and the surrounding area; (4) that wind speed is an important factor to consider in relation to urban heat island and rain island phenomena.

221 LAWRENCE, E. N. (1971) Day of the week variations in the weather. *Weather*, Vol. 26, pp. 386–91.

An analysis of the reality of the weekly weather cycle using data from London. Concludes that all the evidence taken together supports the hypothesis that urban areas may show real weekly meteorological patterns.

222 MOFFITT, B. J. (1972) The effects of urbanization on mean temperatures at Kew Observatory. *Weather*, Vol. 27, pp. 121–9.

Examines the effect of local urban development in explaining temperatures changes measured on that site. Rothampstead (Hertfordshire) and Kew are used as case studies and it is concluded that about 1° C of the temperature change at Kew since the 1880s may be attributed to urbanization and that some account of this must be taken when making assessments of climatic change.

223 SMITH, B. G. W. (1973) Evaporation in the London area from 1698 to 1970. *Meteorological Magazine*, Vol. 102, pp. 281–91.

Describes and illustrates the preparation of available data (limited in early years) of monthly estimates of potential evaporation, representative of Kew for the period 1698–1970.

224 ATKINSON, B. W. (1975) The mechanical effect of an urban area on convective precipitation. Queen Mary College, 27p. (QMC Department of Geography, Occasional Paper No. 3)

Discusses four ways in which precipitation amount may be increased by urban areas: (1) urban heat island; (2) increased concentrations of condensation and ice nuclei within such an area; (3) the higher absolute humidities of an urban area; (4) the mechanical effects of such an area. These four are considered in an analysis of precipitation distribution figures for 1 September 1960.

225 BERNSTEIN, H. (1975) The mysterious disappearance of Edwardian London fog. *London Journal*, Vol. 1, pp. 189–206.

Suggests that the disappearance of urban fog in the 1960s and 1970s was only partly due to the Clean Air Act of 1956, as Londoners had already started utilizing fuels other than coal. Furthermore, the preliminary results of detailed weather comparison suggests that cyclical climatic changes over the London basin may also have had more effect than has hitherto been realized.

226 LEE, D. O. (1975) Rural atmospheric stability and the intensity of London's heat island. *Weather*, Vol. 30, pp. 102–9.

Information on the intensity of the urban heat island is often required in calculating the urban mixing depth in air pollution prediction models. When no urban temperature data are available, it would be useful if an index of urban heat island density could be derived from the relationship between nocturnal urban heat island intensity and lapse rate in the surrounding rural area, both being controlled by similar meteorological conditions. This paper reports a test in the London area to see how far rural lapse rate can be used as an indicator of nocturnal heat island intensity.

227 KEERS, J. F. and WESTCOTT, P. (1976) The Hampstead storm – 14 August, 1975. *Weather*, Vol. 31, pp. 2–10.

An analysis of the thunderstorms with exceptionally heavy rain and hail, which lasted about 2.5 hours and affected a large area of north-west London on 14 August 1975. Meteorological conditions were favourable for the development of large thunderstorms, but the storm developed in south-east England only over north London, centred on Hampstead. The local topography may have influenced the precise location of the storm, and other contributing factors were the extra source of heat from the roads, buildings and industry and a sufficient supply of moisture.

228 ATKINSON, B. W. (1977) Urban effects on precipitation: an investigation of London's influence on the severe storm in August, 1975. Queen Mary College, 31p. (QMC Department of Geography, Occasional paper, No. 8)

The effect of London's urban area on the very severe storm of 14 August 1975 is investigated. The analysis suggests that the urban heat island did have a real effect in the development of severe cumulonimbus clouds over London.

229 LEE, D. O. (1977) Urban influence on wind directions over London. *Weather*, Vol. 32, pp. 162–70.

Examines the modification of air flow within London under different conditions.

230 LYALL, I. T. (1977) The London heat island in June–July 1976. *Weather*, Vol. 32, pp. 296–8.

A study of the magnitude and structure of the heat island of London during the exceptional summer drought months of June and July 1976.

231 THORNES, J. E. (1978) London's changing meteorology. In: CLOUT, H. (ed.). Changing London. University Tutorial Press, pp. 128–37.

Examines London's changing meteorology with special reference to changes in air chemistry and air quality. Includes a section on changing weather and climate followed by a summary conclusion.

232 ATKINSON, B. W. (1979) Urban influences on precipitation in London. In: HOLLIS, G. E. (ed.). Man's impact on the hydrological cycle. Norwich: *Geo Abstracts*, pp. 123–33.

The effects of greater warmth and moisture in London are illustrated by three storms showing that the warmer city air played an important role in the development of cloud growth – consequently influencing precipitation distribution. The mechanical effects of the city on the air flow and precipitation are also illustrated by a study of a storm.

233 LEE, D. O. (1979) Contrasts in warming and cooling rates at an urban and a rural site. *Weather*, Vol. 34, pp. 60–6.

A study of warming and cooling rates for the London area carried out for the summer and winter months of 1975 and 1976 using data from fixed stations.

3.3 Hydrology

234 BUCHAN, S. (1953) Ground-water supplies in the Lea valley. *Lea Valley Growers' Newsletter*, pp. 2–8.

A study of the groundwater situation in the Lea valley between Stanstead Abbotts and Chingford with special reference to the extraction of water by the growers in the area.

235 PREDDY, W. S. (1954) The mixing and movement of water in the estuary of the Thames. *Journal of the Marine Biology Association*, Vol. 33, pp. 645–62.

A method of representing the mixing of water in an estuary is described, and the amount of mixing which occurs in the Thames estuary is calculated. Changes in the salinity predicted from the results are found to agree with the observation.

236 INGLIS, C. C. and ALLEN, F. H. (1957) The regime of the Thames estuary as affected by currents, salinities and river flow. *Proceedings of the Institution of Civil Engineers*, Vol. 7, pp. 827–78.

Describes the Thames estuary and the prevailing tidal conditions, and outlines the nature of the siltation problem. Observations of currents, salinities and concentrations of suspended solids are reported, as well as model experiments designed to amplify these data. The results show the existence of a net landward drift of water near the bed of the estuary which carries mud upriver and it is suggested that the best safeguard against the return of dredged material is to pump the material ashore.

237 BUTLER, R. E. (1961) The buried rivers of London. *London Naturalist*, Vol. 41, pp. 31–41.

Examines 16 buried rivers in the London area from the Wall Brook to the Peckham Rye, tracing their courses through the modern landscape.

238 ANDREWS, F. M. (1962) Some aspects of the hydrology of the Thames basin. *Proceedings of the Institute of Civil Engineers*, Vol. 21, pp. 55–90 (plus discussion in Vol. 24, pp. 247–87)

Describes in brief, the history of the earliest records of the flow of the river Thames, and outlines the hydro-geology of the basin. Three methods of assessing groundwater flow are discussed and direct or storm water run-off in relation to rainfall examined. Reference is made to floods on the Thames and its tributaries and their history and frequencies are referred to, with a final section on dry weather discharges.

239 INESON, J. and DOWNING, R. A. (1963) Changes in the chemistry of groundwater of the chalk passing beneath argillaceous strata. *Bulletin of the Geological Survey of Great Britain*, pp. 176–92.

An account of geological change and variations in groundwater of the chalk under the London basin.

240 PRENTICE, J. E. et al. (1968) Sediment transport in estuarine areas. *Nature*, Vol. 218, pp. 1207–10.

A study of the Thames and Medway river estuaries to investigate the sedimentary processes at work in these estuarine areas. The processes of movement are studied and the form and structure of the sediments examined.

241 GRAY, D. A. and FOSTER, S. S. D. (1972) Urban influences upon ground water conditions in Thames Flood Plain deposits of central London. In: DUNHAM, K. C. and GRAY, D. A. A discussion on problems associated with the subsidence of south eastern England. *Philosophical Transactions of the Royal Society*, Series A, Vol. 272, pp. 245–57.

Shows that current conditions of the groundwater in the riverine deposits of the Flood Plain Terrace of the river Thames in central London is dominated by man-made influences, particularly the underground railway systems and the river walls. Operation of the proposed Thames barrier in a half-tide mode would modify these influences and could lead to additional drainage problems and affect basement structures.

242 WRIGHT, C. E. (1974) Influence of catchment characteristics upon low flows in south east England. *Water Services*, Vol. 78, pp. 227–30.

Study of low flows on the Thames at Teddington and correlation with geology, geomorphology and climatic variables.

243 BUTTERS, K. and VAIRAVAMOORTHY, A. (1977) Hydrological studies on some river catchments in Greater London. *Proceedings of the Institution of Civil Engineers*, Vol. 63, pp. 331–61.

Examines the problems involved in determining the relationship between rainfall and run-off on some sub catchments of the river Thames in Greater London, and the various attempts made by the Department of Public Health Engineering of the GLC to overcome these problems.

244 HALL, M. J. (1977) The effect of urbanization on storm run-off from two catchment areas in north London. In: INTERNATIONAL ASSOCIATION OF HYDROLOGICAL SCIENCES Effects of urbanization on the hydrological regime and on water quality, pp. 144–52. (International Association of Hydrological Sciences, No. 123)

A study of rainfall and river flow records from two adjacent catchment areas in the north London suburbs – the Silk Stream and the Dollis Brook – shows that simple measures of urban development, such as the proportion of impervious area are insufficient to describe the variations in catchment response between ostensibly similar drainage areas. Greater attention should be given to the channel conditions and their modification and the distribution of urban area within the catchment.

3.4 Ecology

245 HUDSON, W. H. (1898) Birds in London. Longmans Green, 339p.

An account of the London bird life of the late nineteenth century. Chapters on the birds – crows, the carrion crow, the daw, rooks, wood pigeons and the small birds – followed by a survey of birds in the parks of London.

246 HAMPSTEAD SCIENTIFIC SOCIETY (1913) Hampstead Heath: its geology and natural history. Fisher Unwin, 328p.

Eleven separate chapters on systematic aspects of the natural history of Hampstead Heath – topography, geology, climate, plant life, bird life, mammals, fishes, reptiles, insect life, molluscs, pond life.

247 BAYES, L. S. (1943) A historical sketch of Epping Forest. *London Naturalist*, Vol. 23, pp. 32–43.

Provides a view of the environmental history of Epping Forest from the end of the last Ice Age to the Second World War.

248 FITTER, R. S. R. (1945) London's natural history. William Collins, 282p.

A history of London (20 miles radius from St Paul's) in terms of the animals and plants it has displaced, changed, moved and removed, introduced, dispersed, conserved, lost or forgotten.

249 LOUSLEY, J. E. (1946) Wild flowers in the City of London. *Geographical Magazine*, Vol. 18, pp. 413–22.

Describes the wild flowers that developed in the City of London on its bomb damaged areas – 160 acres, out of a total of 460 acres occupied by buildings before the bombing.

250 CASTELL, C. P. (1947) Nature conservation in the London area. *London Naturalist*, Vol. 26, pp. 17–41.

Provides an historical introduction to nature conservation in the UK since 1912: (1) a summary of recent work on nature conservation; (2) a report on the London Natural History Society Nature Reserves Committee's recommendations; (3) a brief consideration of the *Greater London Plan* (1944) in relation to these proposals and the problems resulting from their implementation. An appendix lists and maps the Nature Reserves and Conservation Areas in the London division.

251 FITTER, R. S. R. (1949) London's birds. William Collins, 256p.

Analyses the way in which birds in London (the London County Council area) have contrived to adapt themselves to a series of habitats in an urban area – buildings; the ground; shrubs and trees; shallow water; deep water; and the air. The role of man as an enemy and a friend is given special attention in the final two chapters.

252 PETERKEN, J. H. G. (1952) Habitats of the London area. *London Naturalist*, Vol. 32, pp. 2–12.

Summarizes the different kinds of habitat in the London area (20 miles radius from St Paul's Cathedral) from the botanical aspect and indicates some of the localities where they can be studied. Woodlands, heathland, grassland, marshland, bogs and salt marsh are considered. Arable fields, hedgerows, ditches and chalk, gravel and sandpits are also examined.

253 FITTER, R. S. R. and LOUSLEY, J. E. (1953) The natural history of the City. Corporation of London, 36p.

An account of the more prominent and interesting features of the natural history of the City written for visitors. The impact of the Second World War bomb damage in extending the natural habitats is examined. Separate sections are included on wild flowers; birds and beasts; insects and other invertebrate animals.

254 ROSE, F. (1956) Vegetation history and environmental factors in the London area. *London Naturalist*, Vol. 36, pp. 29–40.

Looks at two aspects of the plant ecology of the London area – the historical factors of changing climate and increasing human influence that have modified the flora and vegetation since the last glaciation; the factors of soil, topography and human interference that appear to regulate the present day semi-natural vegetation of the area.

255 LOUSLEY, J. E. (1957) Changes in the flora of the London area since 1858. *London Naturalist*, Vol. 37, pp. 35–49.

Offers a general comparison between the plants available to botanists in 1858 and those which could be seen in 1957.

256 WHEELER, A. C. (1958) The fishes of the London area. *London Naturalist*, Vol. 37, pp. 80–101.

A list of fish in the London area accumulated from a variety of personal records and notes together with details from published sources.

257 BANGERTER, E. B. (1959) The botany of the London area. *London Naturalist*, Vol. 40, pp. 6–16.

This presidential address examines the contribution made by the London Natural History Society to the study and knowledge of the vegetation of the London area. Members have contributed to the historical, topographical, systematic and floristic studies of London's plants.

258 HOMES, R. C., ed. (1964) The birds of the London area, Part 1. 2nd ed. Hart Davis, 332p.

Traces the changes since 1900 in the very varied population of birds in and around London (20 miles radius from St Paul's Cathedral). Looks at the effects of London on bird life with attention to principal habitats on migration and on roosts and fly lines. This new edition has a supplementary chapter covering the major changes between 1952 and 1961 with special emphasis on inner London and migration.

259 BURTON, J. A. (1966) The distribution of weasel, stoat, common shrew, roe deer, water shrew and mole in the London area. *London Naturalist*, Vol. 45, pp. 35–42.

Plots the distribution of six mammals in the London area as an interim study of these selected species.

260 CRAMP, S. and TOMLINS, A. D. (1966) The birds of Inner London, 1951–1965. *British Birds*, Vol. 59, pp. 209–33.

Compares the number of birds attempting to breed or breeding within inner London's 40 square miles. In 1965, an increase of three breeds was found compared with 1950. Of these, 30 were regular breeders: 12 species of which had increased in numbers, five had declined, while the number of other species had remained static. Possible reasons are given for the changes, for example rebuilt sites, toxic chemicals, provision of sanctuaries and the felling of old trees. The decline of the house sparrow remains to be explained, for this breed, with the feral pigeon, makes up 90 per cent of London's bird population.

261 MORRIS, P. (1966) The hedgehog in London. *London Naturalist*, Vol. 45, pp. 43–9.

An account is given of the distribution of the hedgehog (Erinaceus europaeus) from observations recorded in the years 1956 to 1964 inclusive.

262 DONY, J. G. (1967) Flora of Hertfordshire. Hitchin Museum, 112p.

A collection of the records of the wild flora in Hertfordshire. An introduction examines the historical background to the study of Hertfordshire flora and is followed by 109 habitat studies, descriptions of the flora and 56 maps of their distribution in the county.

263 TEAGLE, W. G. (1967) The fox in the London suburbs. *London Naturalist*, Vol. 46, pp. 44–67.

Reports the results of a survey of the distribution of the fox in the London suburbs, examining earlier records from the 1930s to 1958 and recent observations of their present distribution.

264 STOTT, P. A. (1968) The nature and conservation of West Middlesex chalk grassland. *London Naturalist*, Vol. 47, pp. 11–8.

A detailed examination of some of the remaining chalk grassland habitats in West Middlesex with a discussion on the question of conserving these habitats in relation to urban–industrial advance and the collection of the rarer species.

265 HURCOMB, Lord (1969) Protection of wild life in London and its outskirts by public authorities. *Biological Conservation*, Vol. 1, pp. 166–9.

Examines ways in which governmental authorities protect wild life as part of public policy in the area within a radius of 20 miles from St Paul's Cathedral.

266 TEAGLE, W. G. (1969) The badger in the London area. *London Naturalist*, Vol. 48, pp. 48–75.

Summarizes the results of a badger survey that collected information between 1959 and 1964.

267 LAUNDON, J. R. (1970) London's lichens. *London Naturalist*, Vol. 49, pp. 20–69.

Describes and analyses the past and present lichen flora within a 16 km radius from Charing Cross. 165 species have been recorded of which 71 have been seen since 1950. Air pollution by sulphur dioxide is the chief factor affecting the lichen flora; the numbers of lichen species increases at advancing distances from the centre of London. Detailed records are given under the heading 'lichen flora'.

268 WEBSTER, A. D. (1970) London trees. Swarthmore Press, 218p.

An account of trees that succeed in London with a descriptive account of each species and notes on their comparative value and cultivation, with a guide to where the finest London trees are to be seen.

269 BATTEN, L. A. (1972) The past and present bird life of the Brent reservoir and its vicinity. *London Naturalist*, Vol. 50, pp. 8–62.

Presents the results of an enquiry into the changing bird life of an area north-west of London in the vicinity of Brent Reservoir since 1830 where there has been a considerable amount of urbanization. A detailed account of the species in each breeding season since 1957 is given and the main habitats in the study area are examined. A complete list of species recorded in this area is inserted at the end.

270 YATES, E. M. (1972) The management of heathlands for amenity purposes in south east England. *Geographica Polonica*, Vol. 24, pp. 227–40.

Examines the cessation of management of the heathlands of south-east England for agricultural purposes and the problems resulting from their increased use for amenity. These problems are exemplified by consideration of three heaths with differing managerial systems: Blackdown, Sussex; Ashdown Forest, Sussex; and Headley Heath, Surrey.

271 MABEY, R. (1973) The unofficial countryside. William Collins, 157p.

Explores the natural history of the city and suburbia using the outskirts of London and west Middlesex as the main areas for observation.

272 BATTEN, L. A. (1974) Blackbird boom in suburbia. *Wildlife*, Vol. 16, pp. 274–7.

Report of a study carried out at the Brent Reservoir in north-west London in which 6,000 blackbirds were ringed. The pattern of recoveries suggests this to be a highly residential and closed population.

273 BURTON, J. A. (1974) The naturalist in London. Newton Abbot: David and Charles, 176p.

An introduction to Greater London for the naturalist visiting the area for a limited period of time.

274 JERMYN, S. T. (1974) Flora of Essex. Colchester: Essex Naturalists' Trust, 302p.

An introduction to the climate, history and geology of Essex is followed by a systematic list of flora, together with maps of their distribution.

275 LEUTSCHER, A. (1974) Epping Forest: its history and wild life. Newton Abbot: David and Charles, 203p.

Examines Epping Forest in a systematic manner – from geology and prehistory to conservation and the modern management of the forest area.

276 SIMMS, E. (1974) Wildlife in the royal parks. HMSO, 48p.

A short history of the royal parks is followed by a description of the wild life that they contain, from insects and fishes to badgers and deer.

277 COMMITTEE ON BIRD SANCTUARIES IN THE ROYAL PARKS (ENGLAND AND WALES) (1975) Bird life in the Royal Parks, 1974: a report. Department of the Environment, 13p.

Reports by the Bird Sanctuaries Committee have been published either annually or biennially since 1928. It includes a table of observations of birds in the royal parks.

278 KENT, D. H. (1975) The historical flora of Middlesex. Ray Society, 676p.

An account of the wild plants found in the Watsonian vice-county 21 from 1598 to the present time.

279 BURTON, J. A. (1976) Fowls in foul air. *New Scientist*, Vol. 71, pp. 400–1.

It is fairly generally believed that birds are returning to London because the air is now cleaner. This article argues that there is no firm evidence to support this theory and that the overall state of each particular species is the most important factor, with climate and disturbance of their usual habitats as contributory factors.

280 BURTON, J. A. (1976) The decline of the common frog in the London area. *London Naturalist*, Vol. 55, pp. 16–8.

Results of a survey concerned with the decline of *rana temporaria* in the London area. Attributes blame to the destruction and pollution of ponds and the collection of frogs by children and for educational use.

281 HARRISON, C. M. (1976) Heathland management in Surrey, England. *Biological Conservation*, Vol. 10, pp. 211–20.

Heathland in Surrey is managed primarily for purposes of amenity and wildlife conservation. Many of the ecological problems associated with the management of the open heath area stem from frequent, accidental fires and from the absence of grazing. This study discusses the causes and consequences of these problems for heathland managers.

282 HARRISON, J. and GRANT, P. (1976) The Thames transformed: London's river and its waterfowl. André Deutsch, 239p.

Describes how the cleaned Thames in its inner and outer reaches has attracted back many species of waterfowl.

283 LOUSLEY, J. E. (1976) Flora of Surrey. Newton Abbot: David and Charles. 484p.

Examines the flora of the county, following on with chapters dealing with the physical aspects of Surrey – topography, climate and geology. The final section comprises 504 maps of selected species in systematic order.

284 BATTEN, L. A. (1977) Sailing on reservoirs and its effect on water birds. *Biological Conservation*, Vol. 11, pp. 49–58.

Considers the problem of disturbance to water bird populations of an increase in sailing activities on the Brent Reservoir in north-west London. The species affected by sailing activities are identified and their tolerance to disturbance by boats is measured. It is suggested that the impact can be minimized by proper screening of refuges and leaving a large enough part of the lake to the birds.

285 BOORMAN, L. A. and RANWELL, D. S. (1977) Ecology of Maplin Sands and the coastal zones of Suffolk, Essex and north Kent. Institute of Terrestrial Ecology, 56p.

The report of a major survey of plants and animals in Essex and north Kent coastlines and the likely effects of an airport development upon the ecology of the area.

286 MONTIER, D. J., ed. (1977) Atlas of breeding birds of the London area. B.T. Batsford, 288p.

An atlas of London's some 120 breeding birds compiled from data gathered between 1968 and 1972 by some 450 observers under a scheme launched by the British Trust for Ornithology.

287 BAKER, C. A. et al. (1978) Woodland continuity and change in Epping Forest. *Field Studies*, Vol. 14, pp. 645–69.

Reports on the results of pollen analysis and radio carbon dating of species change in Epping Forest over 4,000 years. The dominance of lime woodland was broken in Saxon times and the present beech, birch and oak hornbeam vegetational associations of the forest only developed after the Saxon period.

288 SILVERTON, J. (1978) The history of woodlands in Hornsey. *London Naturalist*, Vol. 57, pp. 11–25.

Historical sources are used to show that the woodlands of Hornsey are likely to, and have been, continuously present since prehistoric times. The history of the woodlands since Roman times demonstrates that their proximity to the City of London has been an important factor in their exploitation and more recently in their preservation. The daily management of Highgate Wood by the Corporation of London is briefly discussed and an explanation offered for the poverty of its present flora.

289 STOREY, W. (1978) Woodland management and nature conservation in London. *Greater London Intelligence Journal*, Vol. 39, pp. 21–7.

Gives an outline of London's woodland managed by the GLC Parks Department and the main characteristics of each wooded open space. Discusses some of the problems associated with tree planting on polluted ground and with the effects of atmospheric pollution on plant life. Concludes by giving a brief account of the wildlife to be found in London's woodland, stressing the importance of maintaining continuity and balance in woodland policy.

290 SCOTT, D. (1979) The nature of Hampstead Heath. High Hill Press, 62p.

Describes the various forms of life on Hampstead Heath – structure and topography, grasses, plants and flowers, trees and shrubs, birds, mammals, insects and butterflies, and pond life. Contains four nature walks for the visitor.

Section 4 Historical patterns of growth and development

This section is basically concerned with how London has developed from a Roman town to a rapidly expanding nineteenth century metropolis. The emphasis is very much on emerging urban patterns and topography and the section does not include all aspects of London's history. Additional historical references will be found in the historical subsections of the main sections 5–8; that is, patterns of economic development in 5.3; transport history in 6.2; historical social patterns in 7.3; and the emergence of local government control in 8.3.

Aspects of London history which have been only lightly touched upon in this section include the detailed administrative history of the City, its institutions and ceremonies. There are a large number of books and articles on these aspects and some account of them will be found in the general and period histories that are listed. Furthermore, there are so many histories of the craft and merchant gilds, livery and other companies of London that only comprehensive histories of these institutions and a few individual studies are listed in subsection 5.3.

London has been well served by historians and their work is summarized by RUBINSTEIN, S. (1968). London is also extremely well off for historical sources, with notable coverage of written documents, maps and other illustrative sources. No attempt has been made to include sources per se, but subsections 4.2 and 4.3 note the main books and articles that do list or otherwise describe these sources.

A selection of the many general works on the history of London, or those covering more than one period, is given in subsection 4.4. The principal histories of London are multi-volume compilations, and whenever possible a period volume is inserted in the appropriate subsection. The *Survey of London* by BESANT, W. (1902–12) is a profusely illustrated ten volume work, famous in the annals of London's historical writings, but the content and highly personal written style are now somewhat dated. A more modern multi-volume series is the *History of London* under the general editorship of F. Sheppard. So far only a few volumes have appeared: RUDÉ, G. (1971); SHEPPARD, F. (1971); BROOKE, C. N. L. and KEIR, G. (1975).

The first period subdivision, prehistoric and Roman, also includes more general works on the archaeology of the London region where the main emphasis within these works is on prehistoric and Roman. There is a huge literature on Roman London spread through a wide variety of journals, with many of the articles essentially recording individual finds of Roman material. Few of these are listed as the gazetteer in MERRIFIELD, R. (1965) *The Roman city of London* lists almost all these up to 1965. A wider range of articles on Roman London published after 1965 is included in this bibliography reflecting the increased archaeological activity in London since the late 1960s.

References to prehistoric finds and sites in local sub-areas of London are included in subsection 4.5 as it is considered inappropriate to assign them to subsection 4.10 (local topography and history). The emphasis in that subsection is very much on the post-Roman villages and towns that became drawn into London, and in many cases there is only the most tenuous link between the later settlement history and the prehistoric and Roman periods.

The items in subsection 4.9 (general topography) are either general works on the topography of London or are topographical works covering more than one sub-area. Several important topographical series are listed in this subsection. The *Victoria history of the counties of England* is a magnificent work of reference containing, for each county, general historical volume(s) and topographical volumes that describe local areas on a parish-by-parish basis. Unfortunately, the Victoria history, though commenced about the turn of the century, is still unfinished and only one volume for the County of London (LCC area) has been produced. The Middlesex volumes are well progressed and volume 6 (covering Finchley, Friern Barnet, and East Barnet) was due to be published in 1980. The *Buildings of England* series contains county surveys of buildings of architectural note, arranged on a parish-by-parish basis within counties. The inventories of monuments in London and Middlesex produced by the Royal Commission on Historical Monuments (England) also contain topographical information arranged by local sub-areas. The *Survey of London*, founded in 1894, aims to record the topographical and architectural history of London, area-by-area. Churches, public buildings, large and smaller houses are illustrated and described in detail, giving an indication of the ownership, history, architecture, and known drawings of each building. The survey is now controlled by the GLC, but originally both the LCC and the London Survey Committee produced separate volumes. There are no general volumes in the *Survey* and the 39 parish volumes (but not the monographs on individual buildings) have been entered in the appropriate local lists in subsection 4.10. The student interested in the topography or history of a specific area should look at the relevant parts of the above topographical series as well as the more specific books and articles listed in the appropriate areas of subsection 4.10.

There is a vast literature on local areas within London and we were presented with a major task in selecting what we believed were the most important or accessible references for the specific localities. In addition, there has been an upsurge in local history writing in the last two decades, much of it by local history societies who often publish ephemeral pamphlets or in short-lived local journals. Questions have been raised over the difficulty in assimilating the flood of local writings (*see* REEDER, D. A. (1977) Keeping up with London's past. In: *Urban history yearbook 1977*. Leicester University Press, pp. 48–54). We apologise to inquirers and to authors if we have overlooked significant articles and papers from this flood of local material.

The criteria adopted for the selection of both old and new material in subsection 4.10 were that the references should be concerned with the settlement history or topography of the area or place in question. This meant omitting local works on genealogy, biography, customs, sports, organizations and individual firms. The more popular histories were by and large omitted as were church descriptions and guides.

Material in subsection 4.10 is arranged by present London borough divisions (including the City of London) rather than by any earlier place-name or subdivision. The present-day boundaries of the London boroughs are shown in Figure 2. The names of the pre-1965 boroughs that were amalgamated to form the new boroughs are given after the London borough name throughout the subsection. Inquirers interested in a specific place should consult the index in the first instance.

The arrangement of subsection 4.10 means that only references to places in Greater London (that is, within the GLC area) are included, and areas outside have been rigidly excluded. This is because local areas beyond London have been less influenced by London in a historical sense and the inclusion of references beyond the Greater London boundary would add enormously to the size of this bibliography.

References to books which cover a wider area, such as GOVER, J. E. B. et al. (1934) *The place-names of Surrey* are included because they refer to areas which were formerly in the surrounding counties but are now part of Greater London.

Several avenues can be explored to find additional material on the history of London. The *Bibliography of Middlesex* (MIDDLESEX LOCAL HISTORY COUNCIL, 1959) contains over 5,000 entries on all aspects of the former county of Middlesex. Although unpublished, copies can be consulted in many public and research libraries in London. The Guildhall Library catalogue is useful for up-to-date additions to the literature on London. The *Urban history yearbook* (annual) also gives details on current research and publications on London. Several historical and archaeological journals specializing on London are published, including *Transactions of the London and Middlesex Archaeological Society*, the London Topographical Society's *London Topographical Record*, and *London Archaeologist*. As previously mentioned, a large number of local historical and archaeological societies produce pamphlets, monographs and local journals. These are best consulted in the appropriate public libraries (*see* appendix).

4.1 Bibliographies

291 GOSS, C. W. F. (1932) The London directories, 1677–1855. Denis Archer, 147p.

An annotated bibliography of London directories, listed under year of publication. A long introductory chapter discusses the origin and development of directories and their use to the historian.

292 The Victoria history of the County of Essex: bibliography. (1959) Oxford University Press for Institute of Historical Research, 352p.

A bibliography in three parts covering (1) the whole county, arranged by topics; (2) biography and family history; (3) individual places and regions. There is an author and miscellaneous index. An unofficial supplement to this bibliography was issued by Dagenham Public Libraries in 1962.

293 A list of works in Guildhall Library relating to the plague in London,

together with the Bills of Mortality 1532(?)–1858. (1965) *Guildhall Miscellany*, Vol. 2, pp. 306–17.

Lists contemporary as well as more recent works on the plague and bills of mortality.

294 A select list of printed works relating to the Great Fire of 1666 and the rebuilding of London, from the collections in the Guildhall Library. (1966) *Guildhall Miscellany*, Vol. 2, pp. 369–76.

Lists contemporary as well as modern works.

295 GROSS, C. (1966) A bibliography of British municipal history, 2nd ed. Leicester: University Press, 461p. (Reprinted photographically from the sheets of the 1897 first edition, with a new preface by G. H. Martin)

308 entries in the section on London, covering such topics as town records, general histories, medieval London, charters, laws, courts, gilds, municipal reform and the LCC. All are nineteenth century or earlier.

296 RUBINSTEIN, S. (1968) Historians of London. Peter Owen, 239p.

Subtitled 'An account of the many surveys, histories, perambulations, maps and engravings made about the City and its environs and of the dedicated Londoners who made them'.

297 GARSIDE, P. L. (1977–9) The development of London: a classified list of theses presented to the Universities of Great Britain and Ireland and the CNAA, 1908–77. Guildhall Studies in London History, Vol. 3, pp. 175–94.

A list of all theses relating to the development of London from 1500 to the present day, arranged by topics.

4.2 Documentary sources and place-names

298 GOVER, J. E. B. et al. (1934) The place-names of Surrey. Cambridge: University Press, 445p. (English Place-Name Society, Vol. 11)

For each parish in Surrey (including those now in Greater London), the derivations of the principal and secondary place-names are given, backed by documentary references. Field and minor names are given separately at the end.

299 REANEY, P. H. (1935) The place-names of Essex. Cambridge: University Press, 698p. (English Place-Name Society, Vol. 12)

Gives the derivation of the principal, secondary and minor place-names for each parish in the county, including those now in Greater London. The distributions of certain place-name elements are shown in maps.

300 GOVER, J. E. B. et al. (1942) The place names of Middlesex, apart from the City of London. Cambridge: University Press, 237p. (English Place-Name Society, Vol. 18)

For each parish, the derivations of the principal and secondary place-names are given, backed by documentary references. Field and minor names are given separately at the end.

301 TATE, W. E. (1948) Enclosure acts and awards, County of Middlesex. *Transactions of the London and Middlesex Archaeological Society*, New Series, Vol. 9, pp. 268–82.

A list of enclosure maps and awards, prefaced with a discussion of field systems and early enclosure movements.

302 SMITH, R. (1949) London local collections. *Journal of Documentation*, Vol. 5, pp. 14–30.

Briefly describes the content and location of documentary and printed local collections in London libraries.

303 JONES, P. E. and SMITH, R. (1951) A guide to the records in the Corporation of London Records Office and the Guildhall Muniment Room. English Universities Press, 203p.

Lists a wide variety of records for the two repositories separately. The CLRO archives are a natural accumulation from the legal, administrative and financial activities of the Corporation. The Guildhall Library Muniments preserve the archives of parishes, wards and gilds and other deposited records illustrating the history of the City.

304 Le HARDY, W. and MERCER, E. D. (1951) Manorial documents in the Middlesex County Record Office. *Transactions of the London and Middlesex Archaeological Society*, New Series, Vol. 10, pp. 252–59.

A parish by parish list. (The Middlesex County Record Office is now part of the GLC Record Office.)

305 EKWALL, E. (1954) Street names of the City of London. Oxford: Clarendon Press, 211p.

Discusses the philology and chronology of street names generally, followed by a gazetteer of names found in medieval documents. The gazetteer is arranged under the following elements: -street; -lane; -row; -alley; -hill.

306 JONES, P. E. (1956) The estates of the Corporation of London: property records as a source for historical, topographical and economic research. *Guildhall Miscellany*, No. 7, pp. 3–16.

Uses examples from documents and manuscript maps to underline the scope for better spatial and topographical analysis. The care of the town wall and encroachments on the town ditch are used as examples.

307 DARLINGTON, I. (1962) Guide to the records in the London County Record Office: Part 1, Records of the predecessors of the London County Council, except the Board of Guardians. London County Council, 63p.

Gives short introductions and lists the main groups of records for such bodies as: Commissioners of Sewers; Metropolitan Buildings Office; Bridge Companies; Metropolitan Asylums Board; School Board for London, etc.

308 LONDON TOPOGRAPHICAL SOCIETY (1962–1967) Survey of building sites in the City of London after the Great Fire of 1666, by Peter Mills and John Oliver. London Topographical Society, 5 vols.

A facsimile copy of the descriptions and small plans of the building sites surveyed by Mills and Oliver for the special summary court set up to agree property boundaries after the Great Fire.

309 SHEPPARD, F. H. W. (1968) Sources and methods used for the Survey of London. In: DYOS, H. J. (ed.) The Study of urban history. Edward Arnold, pp. 131–45.

The Survey of London, founded in 1894, aims to trace the topographical and architectural history of London, area by area. The GLC is now responsible for the Survey and this article outlines the main documentary sources used for tracing the ownership and history of buildings and how a survey is prepared. All the volumes produced are listed with a brief guide to their use.

310 SIMS, J. M. (1970) London and Middlesex published records. London Record Society, 66p.

A handlist to the manuscript sources relating to the history of London and Middlesex which have been published.

311 SMITH, A. I. (1970) Dictionary of City of London street names. Newton Abbot: David and Charles, 219p.

An alphabetical list that traces the origins of and changes to the name of each street in the City.

312 BEBBINGTON, G. (1972) London street names. B. T. Batsford, 367p.

An alphabetical list of street names with explanations of their origins, preceded by a general introduction on street names. The area covered is bounded east by the Tower, west by Earl's Court, north by Highgate and south by Lambeth.

313 MASTERS, B. R. (1975–6) The Corporation of London Records Office: some sources for the historian. *Archives*, Vol. 12, pp. 5–14.

Introduces the following groups of (mainly post-medieval) records which have been largely ignored by historians: Bridge House Estate records; City's cash accounts; loan and tax accounts; seventeenth century estate surveys; building plans; records of the Commissioners of Sewers.

4.3 Cartographic and illustrative sources

314 OGILBY, J. and MORGAN, W. (1677–80) A large and accurate map of the City of London and London surveyed: or an explanation of the large map of London, Lympne Castle, Kent: Harry Margary in association with the Guildhall Library (reprinted 1976 with introductory notes by R. Hyde), 48p.

A reprint (available in book form) of this detailed and accurately surveyed map, at a scale of 100 inches to the mile, which was a cartographic milestone in the mapping of London.

315 Collins' illustrated atlas of London (1854), Reissued 1973 with new introduction by H. J. Dyos. Leicester University Press, 45p. (Victoria Library Series)

The original edition claimed to be the first pocket atlas of London, but Dyos shows that others had the lead. It provides, however, a measure of the extent of London in 1854, though it does not include the suburbs beyond the margins of the capital.

316 Stanford's library map of London and its suburbs (1862). Edward Stanford, 24 sheets.

Published in atlas form at a scale of six inches to the mile, the maps present a beautifully executed and detailed view of London and the surrounding country. The successive editions summarize London's growth.

317 New large scale ordnance atlas of London and suburbs (1886). G. W. Bacon, 72 sheets.

318 Bacon's new large scale atlas of London and suburbs (1910). G. W. Bacon, 80 sheets.

These atlases at four inches to the mile show streets and buildings in fine detail as well as summarizing the state of urban growth in their respective years. Both contain geology maps of the London region at a scale of one inch to the mile.

319 WHEATLEY, H. B. (1903) Notes upon Norden and his map of London, 1593. *London Topographical Record*, Vol. 2, pp. 42–65.

Describes Norden's map making activities and provides a descriptive analysis of named places and buildings on his map of London.

320 MITTON, G. E. (ed.) (1908) Maps of old London. A. and C. Black, 28p.

Nine maps of London are reproduced and discussed in detail. They have been selected to show the growth of the city between the sixteenth century (Wyngaerde's panorama of London) and the eighteenth century (Rocque's map of London 1741–5).

321 SPIERS, W. L. (1908) Morden and Lea's plan of London, 1682. *London Topographical Record*, Vol. 5, pp. 117–35.

Describes Ogilby and Morgan's detailed street plan of London published in 1682. This work is often referred to as 'Morden and Lea's plan of London', but they only produced a later edition of Ogilby and Morgan in 1732.

322 ORDISH, T. F. (1909) Visscher's view of London, 1616. *London Topographical Record*, Vol. 6, pp. 39–64.

Provides background information on Visscher's panoramic view, including a transcript of Camden's description of London, based on the view. (Visscher's view was published by the Topographical Society of London in 1883–5.)

323 WHEATLEY, H. B. (1914) Rocque's plan of London. *London Topographical Record*, Vol. 9, pp. 15–28.

Describes the background and product of Rocque's famous survey of London, 1746.

324 MARTIN, W. (1917) The early maps of London. *Transactions of the London and Middlesex Archaeological Society*, New Series, Vol. 3, pp. 255–86.

Classifies maps and panoramic views, for the period 1550s–1660s, into groups according to the source or draft from which they were drawn.

325 HIND, A. M. (1922) Wenceslaus Hollar and his views of London and Windsor in the seventeenth century. John Lane, The Bodley Head, 92p.

Provides a survey and appreciation of the work of Hollar, followed by a catalogue of his etchings of London views and buildings, many of which are illustrated in the 64 plates.

326 MARTIN, W. (1922) The earliest views of London. *Transactions of the London and Middlesex Archaeological Society*, New Series, Vol. 4, pp. 353–75.

A survey of drawings and bird's eye views of London from the Bayeaux Tapestry to Tudor panoramas. A postscript in volume 5 (1929), pp. 1–4 illustrates views of London on a Roman coin and medallion.

327 HOLMES, M. J. R. (1952) A seventeenth century map of London and the Thames. *London Topographical Record*, Vol. 20, pp. 26–33.

Describes Jonas Moore's map of the Thames, which has marginal views of Woolwich, Greenwich, Erith and Gravesend, as well as an elaborate panorama of London, all apparently by Wenceslaus Hollar (c. 1660). (The map was published by London Topographical Society in 1912.)

328 SCOULOUDI, I. (1953) Panoramic views of London, 1600–1666, with some later adaptations: an annotated list. Corporation of London, 87p.

An analytical list of 110 engraved views of London from the South Bank. The views are arranged into 12 main classes and associated with one key view.

329 BULL, G. B. G. (1958) Elizabethan maps of the lower Lea valley. *Geographical Journal*, Vol. 124, pp. 375–8.

Mainly describes a large strip map of the river Lea and the New Cut between Tottenham and Cheshunt, c. 1594. Competing riverine and riparian land use was probably the reason for the map's compilation, under the direction of Lord Burghley, the Lord Chancellor.

330 HOLMES, M. J. R. (1963) Moorfields in 1559: an engraved copper plate from the earliest map of London. HMSO, 34p.

A description of an important map which shows the city densely built up to the wall, ribbon development along Bishopsgate and open space in Moorfields.

331 BULL, G. B. G. (1964) Maps of Greater London. In: CLAYTON, R. (ed.) The Geography of Greater London. George Philip, pp. 183–201.

Describes and assesses the usefulness of: (1) maps showing the physical setting of London; (2) historical maps (with excerpts); (3) Ordnance Survey maps; (4) recent cartographic developments.

332 MARKS, S. P. (1964) The map of mid-sixteenth century London. London Topographical Society, 29p.

Discusses the relationship of the copper engraved map of London (c. 1553–9), Braun and Hogenberg's map (1572), the so-called Agas map of 1570, and a later pewter engraved map. Although only two parts of the copper engraved map survive, it is shown that it was the original from which the other maps were derived.

333 HURSTFIELD, J. and SKELTON, R. A. (1965) John Norden's view of London, 1600. *London Topographical Record*, Vol. 22, pp. 5–26.

A study of London's importance c. 1600 and a description of Norden's view of London and Westminster.

334 HYDE, R. (1967) Ward maps of the City of London. Map Collectors' Circle, 56p.

An illustrated list of maps of wards in the City, arranged by name and with library locations.

335 PRAGNELL, H. J. (1968) The London panoramas of Robert Barker and Thomas Girtin, circa 1800. London Topographical Society, 27p.

The large panoramas of London (north and south of the Thames) prepared by Barker and Girtin have not survived, but contemporary drawings of parts of the panoramas do survive, and are reproduced in this article.

336 TALLIS, J. (1969) John Tallis's London street views 1838–40 (introduction by Peter Jackson). Nattali and Maurice and the London Topographical Society, 301p.

Reproduces Tallis's magnificent views of street façades that illustrate individual buildings on the streets, along with directories of the trades and businesses carried on.

337 FLEETWOOD-HESKETH, P. (1972) The grand architectural panorama of London, 1849. *London Topographical Record*, Vol. 23, pp. 111–8.

Describes an immense panorama published in 1849 that illustrated Nash's and others' great piece of town planning from Regents Park via Regent Street to Trafalgar Square.

338 GLANVILLE, P. (1972) London in maps. The Connoisseur, 212p.

The evolving topography of London from Roman times to the twentieth century is outlined, and illustrated with nearly 70 contemporary maps.

339 JONES, P. E. (1972) Four fifteenth century London plans. *London Topographical Record*, Vol. 23, pp. 35–59.

Four plans of properties belonging to London Bridge Estates (in Deptford; at Southwark Bar; near St George's Bar; and near Carter Lane) were described and illustrated by J. H. Harvey (1952) in *London Topographical Record*, Vol. 20, pp. 1–8. The present article identifies the sites exactly and shows that fifteenth century plot drawings were more common than supposed.

340 HYDE, R. (1973) Mapping London's landlords: the Ground Plan of London, 1892–1915. *Guildhall Studies in London History*, Vol. 1, pp. 28–34.

The Ground Plan of London was an LCC survey of all freehold land in the county, mapped at 25 inches to the mile. Over 100 individual sheets survive and this article describes their compilation and the uses that can be made of them.

341 WELLSMAN, J. S. (1973) London before the Fire, a grand panorama by C. J. Visscher. Sidgwick and Jackson, 8p.

A large pull-out reproduction of Visscher's panoramic view of London, 1616. On the reverse is Hollar's prospect of London before and after the Fire, an engraving of part of Southwark by Hollar, and a colour facsimile of a view by de Witt.

342 HYDE, R. (1975) Printed maps of Victorian London, 1851–1900. Folkestone: Dawson, 272p.

A general introduction to the different types of map produced during this period, followed by a list of every map focused on the original limits of London and published between 1851 and 1900. Maps are listed chronologically by date of publication and locations are given where the maps may be viewed.

343 HYDE, R. (1976) The act to regulate parochial assessments 1836, and its contribution to the mapping of London. *Guildhall Studies in London History*, Vol. 2, pp. 54–68.

The 1836 Act demanded a more reliable method of calculating rateable values, and local vestries were required to have detailed maps of properties made. These survive erratically for the London parishes, but if in existence they offer detailed maps of the early nineteenth-century capital.

344 HYDE, R. (1976) The Ogilby legacy. *Geographical Magazine*, Vol. 49, pp. 115–8.

Describes and illustrates Ogilby and Morgan's survey of the City of London published in 1677 at a scale of 100 inches to the mile. The first large multi-sheet plan of a British town, it shows London rebuilt after the Great Fire, with new buildings, straightened streets and a re-emerging St Paul's.

345 PHILLIPS, J. F. C. (1976) Shepherd's London: four artists and their views of the metropolis, 1800–1860. Cassell, 116p.

Reproductions, with brief descriptions, of some of the famous drawings of London executed by the Shepherd family in the nineteenth century.

346 HOWGEGO, J. (1978) Printed maps of London circa 1553–1850. 2nd ed. Folkestone: Dawson, 296p.

A general introduction to the different types of printed map produced during this period, followed by a catalogue which includes locations where the maps may be viewed.

4.4 Growth and development: all periods

347 WELCH, C. (1896) Modern history of the City of London: a record of municipal and social progress from 1760 to the present day. Blades, East and Blades, 492p.

A year by year account of happenings in London from 1760 to 1895, with special emphasis on the achievements of the Corporation of London.

348 BESANT, W. (1902–12) Survey of London. A. and C. Black, 10 vols.

A profusely illustrated history of London covering: (1) periods from prehistoric to the nineteenth century (7 vols), and (2) topography (3 vols). The scale of the survey is immense but the content and written style are now somewhat dated. Each volume is separately listed in the appropriate section of this bibliography.

349 BESANT, W. (1908) Early London: prehistoric, Roman, Saxon and Norman. A. and C. Black, 370p. (Survey of London)

A detailed descriptive account of the geology and site of London and political and social history from the earliest inhabitants to the reign of Stephen. The founding of Westminster Abbey on Thorney Island is described, as well as a transcription of Fitzstephen's Chronicle and an account of the streets and people of the Norman city.

350 GOMME, G. L. (1914) London. Williams and Norgate, 381p.

A history of London from pre-Roman origins to the end of the nineteenth century. The author argues that continuity of site and municipal organization is a dominant theme in understanding the history of the city.

351 PAGE, W. (1923) London: its origin and early development. Constable, 300p.

An account of the history of Roman, Saxon and Norman London, almost wholly dependent on documentary sources. The origin of sokes and wards is described along with the development of churches, schools and government.

352 SHARPE, M. (1929) The making of Middlesex: its villages, fields and roads. *Transactions of the London and Middlesex Archaeological Society*, New Series, Vol. 5, pp. 237–55.

Attempts to reconstruct the Roman survey and land divisions of Middlesex and how they survived to influence later topographical patterns.

353 SHARPE, M. (1932) Middlesex in British, Roman and Saxon times. Methuen, 240p.

An account of the history and antiquities of Middlesex from the Iron Age to Domesday Book. The author attempts to reconstruct Roman land divisions in the county and states that they influenced the location of early churches and the size and distribution of Saxon and Domesday virgates.

354 SHARPE, M. (1937) Four eras in the Middlesex area. *Transactions of the London and Middlesex Archaeological Society*, New Series, Vol. 7, pp. 193–208.

Describes continuity and change in land patterns from prehistory to Domesday Book, and attempts to reconstruct the supposed Roman, Saxon and Domesday land divisions of Middlesex.

355 RASMUSSEN, S. E. (1948) London: the unique city. 2nd ed. Jonathan Cape, 440p. (Also abridged edition, Penguin Books (1960) 249p.)

The author, an architect, asks why London differs from continental cities and analyses its growth in terms of early town planning and domestic architecture and the contribution of parks and gardens.

356 ROSE, M. (1951) The East End of London. Cresset Press, 275p.

A history of the progressive suburbanization and industrialization of the former rural and riverside East End hamlets, including Whitechapel, Spitalfields, Wapping, Stepney, Shadwell, Limehouse, Poplar and Hackney.

357 MAYNE, D. (1952) The growth of London. Harrap, 144p.

An introductory account covering periods of growth from prehistory to the Second World War. Makes much use of maps and diagrams.

358 GRADY, A. D. (1959) The lower Lea valley, a barrier in east London. *East London Papers*, Vol. 2, pp. 9–18.

The Lea valley forms a clear dividing line within east and north-east London, and the spread of urbanization was delayed until the latter part of the nineteenth century owing to the extensive marshes along the river. However, the natural routeway afforded by the valley has been used by canal and rail transport and a considerable amount of industry has spread along the valley floor. The Lea valley is also an important water supply and urban drainage zone.

359 BARTON, N. J. (1962) The lost rivers of London. Phoenix House and Leicester University Press, 148p.

Discusses the part played by the Walbrook, Fleet, Tyburn and other streams in the development of London, and the uses made of the rivers for military, domestic, commercial and recreational purposes.

360 ASH, B. (1964) The golden city: London between the fires, 1666–1941. Phoenix House, 214p.

Sees the period from the construction of the renaissance city after the Great Fire as an age of splendour and enrichment until its destruction in the Second World War.

361 SMAILES, A. E. (1964) The site, growth and changing face of London. In: CLAYTON, R. (ed.) The geography of Greater London. George Philip; pp. 1–52.

An account of the physical geography of the London basin and the growth of London from the Roman period to twentieth century redevelopment, supported by 21 maps.

362 The growth of London AD 43–1964 (1964) 20th International Geographical Congress, 102p.

A catalogue of an exhibition of maps, drawings, photographs and artifacts illustrating the history of London, held at the Victoria and Albert Museum 17 July–30 August 1964.

363 BROWN, I. (1965) London: an illustrated history. Studio Vista, 156p.

A profusely illustrated history of London from the Romans to the twentieth century, with an emphasis on social history.

364 HARRISON, M. (1965) London growing: the development of a metropolis. Hutchinson, 224p.

A history of London from prehistory to the twentieth century, emphasizing the role of legislators, architects, and developers in shaping and re-shaping the city.

365 TRENT, C. (1965) Greater London: its growth and development through two thousand years. Phoenix House, 282p.

Described as a synoptic history, this book traces the history of London and the area later to become Greater London, from the Romans to the creation of the GLC. The contrast between town and countryside until the expansion of the suburbs is a major theme.

366 EADES, G. E. (1966) Historic London: the story of a city and its people. Queen Anne Press and City of London Society, 298p.

A straightforward account from prehistory to the twentieth century.

367 FREEMAN, T. W. (1966) The Greater London conurbation. In: FREEMAN, T. W. The conurbations of Great Britain. 2nd ed. Manchester: University Press, pp. 17–71.

Examines two main themes: (1) the growth of the urbanized area, population and industry; (2) the problem of the administrative definition and organization of London up to the creation of the GLC in 1965.

368 HANSON, M. (1967) 2000 years of London: an illustrated survey. Country Life, 232p.

A fully illustrated history of the development of London from the Romans to the Barbican scheme, with particular emphasis on buildings.

369 HAYES, J. (1969) London: a pictorial history. B. T. Batsford, 180p.

An introductory chapter traces the history of London from the Romans to the twentieth century, backed up with 153 illustrations, many from the Museum of London collection. The illustrations include maps, plans, topographic views, and prospects of buildings and streets.

370 HIBBERT, C. (1969) London: the biography of a city. Longman, 290p.

A profusely illustrated history from the Romans to 1968, with emphasis on buildings, growth and social conditions.

371 HOLLAENDER, A. E. J. and KELLAWAY, W. (eds) (1969) Studies in London history, presented to Philip Edmund Jones. Hodder and Stoughton, 509p.

A collection of 19 essays on London with an appreciation of P. E. Jones and a bibliography of his writings. Certain chapters are separately listed in this bibliography.

372 GUTKIND, E. A. (1971) London. In: GUTKIND, E. A. Urban development in Western Europe: the Netherlands and Great Britain. New York: Collier-Macmillan, pp. 451–73. (International History of City Development, No. 6)

A history of the internal development and expansion of London from the Romans to the nineteenth century, illustrated with contemporary maps and drawings.

373 KNOWLES, C. C. and PITT, P. H. (1972) A history of building regulations in London, 1189–1972. Architectural Press, 164p.

Examines the various Acts and regulations controlling buildings in London, from the viewpoint of the district surveyor.

374 BARKER, F. and JACKSON, P. (1974) London: 2000 years of a city and its people. Cassell, 379p.

A history of London from the Romans to the twentieth century, fully illustrated with 1,000 contemporary pictures and drawings. The main topics covered within each chronological period are architectural development and political and social history.

375 FRIEDLANDER, D. (1974) London's urban transition 1851–1951. *Urban Studies*, Vol. 11, pp. 127–41.

Identifies four concentric areas around London and analyses the varying patterns of concentration and dispersal of population in these zones, using census data for 1851–1951. Concludes that nineteenth century London was characterized by population concentration and urbanization while twentieth century London is characterized by population dispersal and suburbanization. However, patterns of concentration and dispersal overlapped in space and time.

376 JENKINS, S. (1975) Landlords to London: the story of a capital and its growth. Constable, 310p.

A survey of the ownership of development land in London, ranging from the great estates of the sixteenth century to the property boom of London after the Second World War.

377 PATTEN, J. (1976) Villages in sururban London. *Geographical Magazine*, Vol. 48, pp. 737–41.

Describes the process of suburbanization in Middlesex in the nineteenth and twentieth centuries, and identifies surviving (though usually modified) village centres.

378 BORER, M. C. (1977) The City of London: a history. Constable, 324p.

A history of the City from the Romans to the twentieth century. Special consideration is given to the evolution of commerce, gilds and financial institutions.

379 BIRD, J. et al. (eds) (1978) Collectanea Londiniensia: studies in London archaeology and history presented to Ralph Merrifield. London and Middlesex Archaeological Society, 422p. (LAMAS, Special Paper, No. 2)

Thirty-four articles on London covering all aspects and periods from prehistory to the destruction of nineteenth century theatres. Many of the articles describe artifacts, but those with a spatial content are separately referenced in this bibliography.

380 GRAY, R. (1978) A history of London. Hutchinson, 352p.

A history of London from the Romans to the GLC, for the general reader.

381 SHEPPARD, F. et al. (1979) The Middlesex and Yorkshire deeds registries and the study of building fluctuations. *London Journal*, Vol. 5, pp. 176–217.

Uses the number of property deeds registered annually as an index for determining building cycles and fluctuations for three periods, 1715–1785, 1785–1857, 1857–1914. During the last period the number of registrations fell (between 1903 and 1914) despite increasing population and movement to the suburbs.

4.5 Prehistoric and Roman and the archaeology of the London area

382 ROYAL COMMISSION ON HISTORICAL MONUMENTS (ENGLAND) (1928) An inventory of the historical monuments in London, Vol. III, Roman London. HMSO, 207p.

The inventory covers: the defences; structures within the walls; structures outside the walls (and within the County of London); and appendices on inscriptions, pottery and coins. Pages 1–67 present a history of Roman London by R. E. M. Wheeler.

383 VULLIAMY, C. E. (1930) The archaeology of Middlesex and London. Methuen, 308p. (The County Archaeologies)

A traditional and now out-dated account arranged by periods from palaeolithic to Saxon.

384 WHEELER, R. E. M. (1930) London in Roman times. Lancaster House, 211p. (London Museum Catalogues, No. 3)

An introductory chapter outlining the history and everyday life of Roman London, followed by a catalogue of Roman finds.

385 SHARPE, M. (1940) Roman rural economy and its effect on Middlesex. *Transactions of the London and Middlesex Archaeological Society*, New Series, Vol. 8, pp. 1–13.

Describes the land holdings, agriculture and villages of Roman Middlesex, and compares them with other areas.

386 HOME, G. C. (1948) Roman London AD 43–457. Eyre and Spottiswoode, 302p.

A chronological history of Roman London along with a description of the defences, public life, religion and commerce.

387 HILL, W. T. (1955) Buried London: Mithras to the middle ages. Phoenix House, 192p.

Well-illustrated record of the post-1945 excavations of London's bomb and redevelopment sites. The chief emphasis is on Roman London, especially the Walbrook Mithraic temple, along with descriptions of the damaged churches and livery halls.

388 COPLEY, G. J. (1958) An archaeology of south-east England: a study in continuity. Phoenix House, 324p.

A period-by-period account with distribution maps and gazetteer of the most important sites.

389 LACAILLE, A. D. (1961) Mesolithic facies in Middlesex and London. *Transactions of the London and Middlesex Archaeological Society*, Vol. 20, pp. 101–50.

A detailed survey of the mesolithic period in Middlesex, with accounts of the known sites and finds.

390 MERRIFIELD, R. (1965) The Roman city of London. Benn, 344p.

A comprehensive survey of the history and topography of Roman London with particular emphasis on the fortifications and the public buildings. There is a gazetteer of sites that have yielded Roman remains up to 1965, plus detailed notes on 140 plates.

391 MARSDEN, P. (1967) The river-side defensive wall of Roman London. *Transactions of the London and Middlesex Archaeological Society*, Vol. 21, pp. 149–56.

Reviews the archaeological evidence for the existence of a continuation of the Roman defensive wall along the Thames. The evidence is inconclusive as the fragments that have been found are so variable that they cannot be regarded as all belonging to the same structure.

392 DERWENT, G. (1968) Roman London. Macdonald, 126p. (Discovering London, No. 1)

A guide to the surviving remains of Roman London (in situ and in museums) along with background information on Roman history and institutions.

393 GRIMES, W. F. (1968) The excavation of Roman and Medieval London. Routledge and Kegan Paul, 261p.

A report on 15 years excavations, including major Roman sites such as Cripplegate Fort, Temple of Mithras and London Wall. Many medieval ecclesiastical sites are described including St Brides, Fleet Street.

394 DAWSON, G. J. (1969/70) Roman London Bridge. *London Archaeologist*, Vol. 1, pp. 114–17; 156–60.

Reviews the evidence and theories for the siting of the Roman bridge and concludes that a best fit solution is upstream of the medieval bridge, rather than downstream as has been traditionally stated.

395 MARSDEN, P. (1969) The Roman pottery industry of London. *Transactions of the London and Middlesex Archaeological Society*, Vol. 22, pp. 39–44.

Summarizes the evidence obtained from rescue excavations in 1677, 1908–9 and 1961 for a first century Roman pottery industry in the area around St Paul's Cathedral and Newgate Street, lying west of the town limits at that date.

396 MERRIFIELD, R. (1969) Roman London. Cassell, 212p. (Cassell's London Series)

Examines the origin of London, communications and road network, status and functions, fortifications, Roman life and religion, and Roman sites within and outside the city. The importance and contribution of archaeology is manifest.

397 SHELDON, H. (1971–2) Excavations at Lefevre Road, Old Ford, E. 3. *Transactions of the London and Middlesex Archaeological Society*, Vol. 23, pp. 42–77.

Excavation showed the line of a major Roman highway (London–Colchester) aligned on Aldgate, and that a settlement existed alongside during the later phases of the Roman period. Further evidence of settlement and burials along this road is published in the same volume, pp. 101–47.

398 WARREN, S. (1971) Neolithic occupation in Putney. *London Archaeologist*, Vol. 1, pp. 276–9.

A prehistoric trackway probably crossed the Thames at Putney, and this article argues that the finds of Neolithic pottery and large numbers of flint implements indicate settled occupation of the area.

399 CASTLE, S. A. (1972) Brockley Hill; the site of Sulloniacae? *London Archaeologist*, Vol. 1, pp. 324–7.

A brief review of the various excavations that have uncovered several Roman pottery kilns and vast amounts of pottery sherds. The question whether Brockley Hill (near Edgware) is the Roman posting station of Sulloniacae is still unanswered.

400 CASTLE, S. A. (1972) Trial excavations in field 410, Brockley Hill. *London Archaeologist*, Vol. 2, pp. 36–9 and pp. 78–83.

Reports on the excavations of additional kilns at Brockley Hill, providing further evidence of pottery production there in the Hadrianic-Antonine period.

401 GUILDHALL MUSEUM (1972) Archaeology in the City of London: an opportunity, by Max Hebditch. Guildhall Museum, 9p.

A report made by Hebditch after becoming director of the Guildhall Museum. The threat of redevelopment and the opportunity this presents for excavation is outlined.

402 MARSDEN, P. (1972) Mapping the birth of Londinium. *Geographical Magazine*, Vol. 44, pp. 840–5.

Reports on a survey of the geology and physical geography of London and the topography of the area at the time of initial Roman settlement. The uses made of natural features such as streams, brick-earth and a sandbank are outlined.

403 CHAPMAN, H. and JOHNSON, T. (1973) Excavations at Aldgate and Bush Lane House in the City of London, 1972. *Transactions of the London and Middlesex Archaeological Society*, Vol. 24, pp. 1–73.

The excavation of these two sites provided the first archaeological evidence for early Roman military occupation in London, dating from just after the invasion of AD 43. The line of the London–Colchester road is also shown to have been originally south of its later alignment along Aldgate.

404 MERRIFIELD, R. (1973) A handbook to Roman London. Guildhall Museum, 47p.

Uses the evidence of remains in situ and in the Museum of London to briefly outline the history, buildings, trade, industry, religion and clothing of Roman London.

405 CHRISTOPHERS, V. R. et al. (1974) The Fulham Pottery: a preliminary account. Fulham and Hammersmith Historical Society, Archaeological Section, 26p. (FHHS Occasional Paper, No. 1)

Describes the excavation of the seventeenth to nineteenth century remains of the famous pottery established by John Dwight in 1672–3. The significance of the vast number of stoneware sherds discovered is assessed.

406 MERRIFIELD, R. and SHELDON, H. (1974) Roman London Bridge: a view from both banks. *London Archaeologist*, Vol. 2, pp. 183–91.

Reviews the archaeological evidence and varying theories for the location of the London Bridge. Road alignments, both in the City and Southwark, suggest that the Roman Bridge was on precisely the same site as the later stone-built medieval bridge, rather than upstream or downstream.

407 TATTON-BROWN, T. (1974) Excavations at the Custom House site, City of London, 1973. *Transactions of the London and Middlesex Archaeological Society*, Vol. 25, pp. 117–219.

A detailed excavation report of the uncovering of parts of Roman timber quays, 40 metres north of the present waterfront. No remains were found between late Roman and late thirteenth century suggesting a marine transgression and a Saxon waterfront under Lower Thames Street. A fourteenth century timber quay and seventeenth or eighteenth century river walls lay south of the Roman waterfront.

408 TATTON-BROWN, T. (1974) Roman London: some current problems. *London Archaeologist*, Vol. 2, pp. 194–97.

Reviews some of the outstanding questions about Roman London such as the environmental background, roads, military sites, bridge site, and the southern defensive wall. Much is still unknown and it is suggested that Roman London badly needs a detailed study from the social and economic point of view.

409 MARSDEN, P. (1975) Excavations at a Roman palace site in London, 1961–1972. *Transactions of the London and Middlesex Archaeological Society*, Vol. 26, pp. 1–102.

A full report of the complex remains in Upper Thames Street of what was probably the residence of the Roman governor of Britain, built when London became the capital of Britain. Additional details are given in *Transactions of the London and Middlesex Archaeological Society*, (1978) Vol. 29, pp. 99–103.

410 MERRIFIELD, R. (1975) The Archaeology of London. Heinemann, 96p. (Regional Archaeologies)

An introductory text for students covering the archaeology of the London basin from the Old Stone Age to the seventh century AD. A gazetteer lists sites and museums to visit.

411 MORRIS, J. (1975) London's decline AD 150–250. *London Archaeologist*, Vol. 2, pp. 343–4.

Considers that Sheldon has misrepresented the evidence from pottery and coins and rejects the theory of an economic decline in London in the period AD 150–250 (*see* SHELDON, 1975). Sheldon also provides a short counter-reply.

412 ROBERTSON, B. (1975) Roman Camden. *London Archaeologist*, Vol. 2, pp. 250–5.

Briefly reviews the evidence of Roman sites and finds in the London Borough of Camden. Although there is probably much yet to be uncovered, the main evidence is from finds associated with known or supposed Roman roads in the area.

413 SHELDON, H. (1975) A decline in the London settlement, AD 150–250? *London Archaeologist*, Vol. 2, pp. 278–84.

Argues that evidence from coins, pottery and buildings suggests decline in Roman Southwark in the period AD 150–200 and actual desertion in the first half of the third century. Such an economic decline can be identified in other Roman settlements (urban and rural) in the London region, and the author suggests that military instability, disease or taxation may have been responsible (*see also* MORRIS, 1975).

414 HOBLEY, B. (1976) The archaeological heritage of the City of London. *London Journal*, Vol. 2, pp. 67–84.

Subtitled 'a progress report from the Department of Urban Archaeology, Museum of London', this article reviews the present state of knowledge. Although most work has been conducted on the excavation of Roman London, there are many questions to be considered, while a research programme that embraces sub-Roman, Saxon and Viking London, is a first priority.

415 LAWS, A. (1976) Excavations at Northumberland Wharf, Brentford. *Transactions of the London and Middlesex Archaeological Society*, Vol. 27, pp. 179–205.

Brentford may have originated as a small Roman relay station on the Silchester road, midway between London and Staines, where the road crossed the river Brent. The excavations reported here unexpectedly showed some kind of Roman activity (possibly stock raising) on the west bank of the river Brent.

416 MARSDEN, P. (1976) Two Roman public baths in London. *Transactions of the London and Middlesex Archaeological Society*, Vol. 27, pp. 1–70.

Describes the excavation of and finds from the baths at Huggin Hill (Upper Thames Street) and Cheapside. Both were built in the late first century, extensively modified in the second century and demolished before the end of that century, possibly during the expansion and reorganization of the city. The Cheapside baths had a military appearance and may have served the Cripplegate fort.

417 SCHOFIELD, J. and MILLER, L. (1976) New Fresh Wharf: 1, the Roman waterfront. *London Archaeologist*, Vol. 2, pp. 390–5.

Reports on the excavation of the Roman timber built waterfront wharf near Billingsgate. The method of construction and the dating agreed with the earlier excavated Roman waterfront at the Customs House site further east.

418 SHELDON, H. (1976) Recent developments in the archaeology of Greater London. *Royal Society of Arts Journal*, No. 5240 (74), pp. 411–25.

Examines the impact of increased financial backing for the progress of archaeology in London and discusses the main findings from excavations in the previous five years. Future needs are also assessed.

419 SYKES, M. (1976) The Roman port of London. *Port of London*, Vol. 51, pp. 94–6.

Outlines the history and development of port facilities and trade in Roman London, using mainly archaeological evidence.

420 LONDON AND MIDDLESEX ARCHAEOLOGICAL SOCIETY (1976) The archaeology of the London area: current knowledge and problems.

(1976) *London and Middlesex Archaeological Society*, (LAMAS Special Paper No. 1)

Identifies the knowledge and gaps in London's archaeology and indicates the direction of future research. The six chapters are: Palaeolithic and Mesolithic (Desmond Collins); Bronze Age (John Barrett); Iron Age (Roy Canham); Roman (Ralph Merrifield); Anglo-Saxon and Medieval (John Hurst).

421 DAWSON, G. J. (1977) Roads, bridges and the origin of Roman, London, *Surrey Archaeological Collections*, Vol. 71, pp. 43–56.

Reviews the conflicting archaeological evidence for the site of Roman London bridge and the roads leading to it. The relationship between the roads and the early evolution of a civic centre in Roman London is discussed.

422 HILL, C. (1977) The London riverside wall. *Current Archaeology*, Vol. 5, pp. 308–10.

Reports on excavations conducted in 1974 and 1975 which produced the first archaeological evidence that Roman London possessed a defensive riverside wall. Although the land wall of Roman London was constructed c. AD 200, there was no riverside wall until Saxon pressure from AD 370 goaded the authorities into constructing a riverside wall and adding bastions to the land wall.

423 HOBLEY, B. and SCHOFIELD, J. (1977) Excavations in the City of London: first interim report. *Antiquaries Journal*, Vol. 57, pp. 31–66.

A report on the first results of the excavation of 16 archaeological sites in the waterfront areas and Roman and medieval defences. Evidence of the relatively unknown Saxon occupation has also been strengthened.

424 PHILP, B. J. (1977) The forum of Roman London: excavations of 1968–9. *Britannia*, Vol. 8, pp. 1–64.

The redevelopment of a site at the corner of Gracechurch Street and Fenchurch Street permitted a rescue excavation on the south-east corner of the Roman forum. This article reconstructs the building history of the site, from early structures of the conquest to the construction of the forum and basilica (by far the largest in Britain) as part of a major town centre redevelopment c. AD 100.

425 CANHAM, R. et al. (1978) Excavations at London (Heathrow) Airport, 1969. *Transactions of the London and Middlesex Archaeological Society*, Vol. 29, pp. 1–44.

Excavations in advance of runway extension revealed a number of structures and small finds covering several periods from Bronze Age to Romano-British. A pattern of spasmodic settlement is suggested, with the Iron Age, 550–300 BC, best represented.

426 COLLINS, D. (1978) Early man in west Middlesex: the Yiewsley Palaeolithic sites. HMSO, 57p.

Gravel pits at Yiewsley have yielded one of the largest series of Lower Palaeolithic tools found in Europe, providing evidence of early man c. 200,000 years ago. Tools of middle palaeolithic age (c. 70,000 years ago) have also been recovered.

427 Hampstead Heath: a mesolithic site in Greater London (1978). *Current Archaeology*, Vol. 6, pp. 24–6.

Reports on an excavation in Hampstead Heath which has revealed the first mesolithic site of any size or concentration to be found close to London. Thousands of artifacts and a hearth have been recovered, while a peat deposit at a nearby spring site is being analysed to establish the vegetational history of the area.

428 MARSDEN, P. (1978) The discovery of the civic centre of Roman London. In: BIRD, J. et al. (eds) Collectanea Londiniensia. London and Middlesex Archaeological Society, pp. 89–103.

Summarizes and maps the evidence for the origin, location and enlargement of the forum and basilica in the area of Gracechurch Street.

429 SHELDON, H. and SCHAAF, L. (1978) A survey of Roman sites in Greater London. In: BIRD, J. et al. (eds) Collectanea Londiniensia. London and Middlesex Archaeological Society, pp. 59–88.

Maps all the known sites and finds of Roman remains in Greater London and discusses these under (1) those that lie on the major roads and (2) those beyond the roads in the countryside. Most of the rural sites probably reflect the utilization of productive farmland, as most of the sites are on the sands and gravels rather than the London Clay. Detailed maps and discussion is presented of the following major site areas located on or near to roads: Brockley Hill, Brentford, Staines, Enfield, Old Ford, Crayford and Dartford, Ewell.

430 HILLAM, J. and MORGAN, R. (1979) The dating of the Roman riverside wall at three sites in London. *London Archaeologist*, Vol. 3, pp. 283–8.

Discusses the problems of relative and absolute dating of the oak piles used in constructing the riverside wall, found in excavations at Blackfriars, New Fresh Wharf and the Tower. Dendrochronology and radiocarbon dating suggest construction in the second half of the fourth century, with some timber being up to 20 years old when used.

431 MALONEY, J. (1979) Excavations at Dukes Place: the Roman defences. *London Archaeologist*, Vol. 3, pp. 292–7.

Describes the results of an excavation of part of the Roman Wall near Aldgate. Much constructional evidence was found and it can be suggested that the bastions on the wall were part of a late fourth century comprehensive reorganization and strengthening of the defences.

432 MILNE, G. and C. (1979) The making of the London waterfront. *Current Archaeology*, Vol. 6, pp. 198–204.

Analyses the results of archaeological excavations at several sites along the Thames waterfront. Substantial fragments of Roman timber suggest that wharves extended along much of the waterfront, a pattern repeated in the medieval period with wharfs constructed along a new waterfront 50–100 metres further south.

433 ROSKAMS, S. (1979) The Milk Street excavation. *London Archaeologist*, Vol. 3, pp. 199–205.

Reports on the finds of Roman material at Milk Street, near Cheapside. Six periods of Roman use were identified, from a possible early military structure to a second century building with a fine mosaic, which was subsequently dismantled.

4.6 Pre- and post-conquest medieval

434 LETHABY, W. R. (1902) London before the Conquest. Macmillan, 217p.

Reconstructs the main topographic elements of early medieval London from the best sources then available, as well as rejecting many ill-founded earlier ideas. The main topics covered include the origin of London, Roman London, rivers, fords, roads, bridges, walls, gates, quays, wards, churches, Guildhall and the Royal palace.

435 WHEATLEY, H. B. (1904) The story of London. J. M. Dent, 410p. (The Medieval Towns Series)

A general history and topographic description of the City, derived from primary and secondary sources.

436 SHORE, T. W. (1905) Anglo-Saxon London and Middlesex. *Transactions of the London and Middlesex Archaeological Society*, New Series, Vol. 1, pp. 283–318, 366–91 and 469–505.

Three papers which, though upholding the traditional ideas on the Saxons and other Germanic groups, provide some useful information on Saxon life, trade and estates.

437 BESANT, W. (1906) Medieval London. Vol. 1, Historical and social. A. and C. Black, 419p. (Survey of London)

Covers the political history of medieval London with a chapter for each sovereign's reign, followed by descriptions of the port, trade, streets, buildings, social life, customs, fire, plague and famine.

438 BESANT, W. (1906) Medieval London. Vol. 2, Ecclesiastical, A. and C. Black, 436p. (Survey of London)

Covers the government of London before and after the commune of 1191. Ecclesiastical life is outlined and all the religious houses are described.

439 KINGSFORD, C. L. (1916) Historical notes on medieval London houses. *London Topographical Record*, Vol. 10, pp. 44–144.

An alphabetical and annotated list of the more important houses in medieval London to 1603. The information is almost wholly derived from documentary sources. Supplements to the list are given in *London Topographical Record*, Vol. 11 (1917), pp. 28–81 and Vol. 12 (1920), pp. 1–66.

440 CURTIS, M. (1918) The London Lay Subsidy of 1332. In: UNWIN, G. (ed.) Finance and trade under Edward III. Manchester: University Press, pp. 35–92. (Reissued by Frank Cass, 1962)

The list of taxpayers is reproduced in full and used as a basis for analysing the size, wealth and occupations of the population of London in 1332. The distribution of tax paid (as an index of wealth) shows rich mercantile wards on the river bank, a wealthy shop-keeping area in the centre, and poorer districts on the northern, eastern and western peripheries of the City.

441 HOME, G. C. (1927) Medieval London. Benn, 382p.

A traditional, political history of the period AD 457–1485. Social and religious life and the face of the City are also described.

442 WHEELER, R. E. M. (1927) London and the Vikings. Lancaster House, 55p. (London Museum Catalogues, No. 1)

A brief history of the Viking raids on London, followed by a catalogue of Viking artifacts in the (former) London Museum.

443 MARTIN, W. (1929) The Black Friars in London: a chapter in national history (architectural description by Sidney Toy), *Transactions of the London and Middlesex Archaeological Society*, New Series, Vol. 5, pp. 353–79.

A history of the Black Friars in London with a description of the former great friary near Ludgate Hill, and its subsequent excavation.

444 STENTON, F. M. (1934) Norman London. Historical Association, 39p. (Historical Association Leaflet, No. 93)

A political historical essay, largely derived from primary and secondary sources. William Fitzstephen's description of London (c. 1180) is translated in full by H. E. Butler. A sketch map of London under Henry II by Marjorie Honeybourne shows the wall, gates, roads, streets, churches, markets and wharfs, together with extramural religious settlements.

445 WHEELER, R. E. M. (1935) London and the Saxons. Lancaster House, 200p. (London Museum Catalogues, No. 6)

A brief history of the Saxon period in London, followed by a catalogue of Saxon remains in the (former) London Museum.

446 SHARPE, M. (1937) Post-Roman London. *Transactions of the London and Middlesex Archaeological Society*, New Series, Vol. 7, pp. 353–64.

Describes the political geography of London and its surrounding area (especially Middlesex) in the Saxon period.

447 EKWALL, E. (1956) Studies on the population of medieval London. Stockholm: Almqvist and Wiksell, 334p.

A study of medieval immigration to London using the evidence of toponymous surnames. Migrants appear to have come to London from all parts of England, with the greatest proportion from the East Midlands.

448 CAMPBELL, E. M. J. (1962) Middlesex. In: DARBY, H. C. and CAMPBELL, E. M. J. (eds) The Domesday geography of South-East England. Cambridge: University Press, pp. 97–137.

A historical geographical interpretation of the Domesday Book folios for Middlesex. A considerable contrast in the agriculture, population and woodland is evident between the clayey soil area of northern Middlesex and the loamy soils of the south. (London was not covered by the Domesday survey.)

449 WILLIAMS, G. A. (1963) Medieval London, from commune to Capital. Athlone Press, 377p.

Uses original documentation to trace the emergence of London as a financial and mercantile centre in the period from early self-government in the late twelfth century to the charter of 1327 representing full administrative freedom.

450 HONEYBOURNE, M. B. (1965) The reconstructed map of London under Richard II. *London Topographical Record*, Vol. 22, pp. 29–76.

Describes the compilation of the map using original sources and describes every place marked. (*A map of London under Richard II* was published by London Topographical Society, in 1960.)

451 HONEYBOURNE, M. B. (1966) Norman London. *London and Middlesex Historian*, No. 3, pp. 9–15.

A brief description of the topography and organization of London from 1066 to William Fitzstephen's description c. 1170. London is considered to have been 'populous, well defended and well ordered'.

452 BRECHIN, D. (1968) The Conqueror's London. Macdonald, 125p. (Discovering London, No. 2)

An introductory history of early medieval London along with details on where to view surviving remains in the landscape and in museums.

453 DERWENT, K. (1968) Medieval London. Macdonald, 126p. (Discovering London, No. 3)

Mainly concerned with medieval kings, social life and institutions. There are brief descriptions of Westminster Abbey, Guildhall and the Tower, and a guide to the main museums.

454 ROBERTSON, D. W. (1968) Chaucer's London. New York: John Wiley, 241p. (New Dimensions in History: Historical Cities)

Presents a long description of the topography of the City in the late fourteenth century, followed by an account of City customs, government and historical events, and a discussion of the intellectual and social background of the city Chaucer lived in.

455 HONEYBOURNE, M. B. (1969) The pre-Norman Bridge of London. In: HOLLAENDER, A. E. J. and KELLAWAY, W. Studies in London history. Hodder and Stoughton, pp. 15–39.

Argues that the wooden bridge that crossed the Thames until the first stone London Bridge was built in 1209, was in fact a Saxon and Norman repaired version of the original Roman bridge. Documentary evidence is used to locate the bridge about 100 feet downstream from the stone bridge that replaced it.

456 BAKER, T. (1970) Medieval London. Cassell, 260p. (Cassell's London Series)

This illustrated history examines the re-emergence of London in late Saxon times and the expansion of the medieval city. London's role in government and the church and the problems accruing from narrow streets, refuse and poor water supply are also considered.

457 HASLAM, J. (1972) Medieval streets in London. *London Archaeologist*, Vol. 2, pp. 3–7.

Reports on the archaeological investigations of five medieval lanes between Upper Thames Street and the river. It is suggested that a phase of deliberate planning took place from the thirteenth century onwards, with the reclamation of large areas of the Thames foreshore by the building of a number of imposing stone-built houses, lanes and docks.

458 MYERS, A. R. (1972) London in the age of Chaucer. Norman, Oklahoma: University of Oklahoma Press, 236p.

A wide ranging synopsis of life and institutions in fourteenth century London. Trade, administration and religion contributed to the city's importance, but London was still smaller than many continental towns.

459 BROOKE, C. N. L. and KEIR, G. (1975) London 800–1216: the shaping of a city. Secker and Warburg, 424p. (History of London)

A detailed account of the formative period in London's transformation from a shadow of a town to an important political and commercial capital. The book is in four parts covering: (1) the events, urban background and source materials; (2) the physical shape of the city, based on William Fitzstephen's description of London in the 1170s; (3) social strata in the city and the role of sheriff and mayor, trades and crafts; (4) ecclesiastical history and the role of the church.

460 TATTON-BROWN, T. (1975) Excavations at the Customs House site, City of London, 1973 (Pt. 2). *Transactions of the London and Middlesex Archaeological Society*, Vol. 26, pp. 103–70.

The topography of the south-east corner of the city in the later medieval period is reconstructed and the medieval and later Customs Houses reassessed, together with a description of the medieval finds.

461 BROWN, A. (1976) London and north-west Kent in the later middle ages; the development of a land market. *Archaeologica Cantiana*, Vol. 92, pp. 145–55.

Shows that after c. 1340 London merchants and officials were buying land in north-west Kent as a hedge against the economic fluctuations that affected the city.

462 HILL, D. (1976) London bridge: a reasonable doubt. *Transactions of the London and Middlesex Archaeological Society*, Vol. 27, pp. 303–4.

Argues that the earliest documentary reference to London Bridge (in a tenth century charter) refers to a bridge on a London road in Northamptonshire or Lincolnshire, and not to the Saxon bridge in London.

463 MILLER, L. (1977) New Fresh Wharf: 2, the Saxon and early medieval waterfronts. *London Archaeologist*, Vol. 3, pp. 47–53.

Shows that the timber built Roman waterfront was allowed to decay in Saxon times, though this was followed by a possible Saxon jetty and the erecting of a palisade of defensive posts. Towards the end of the Saxon period a process of land reclamation began which led to a waterfront built out into the river by the twelfth century. No evidence for a pre-Norman bridge was found on the line described by Honeybourne.

464 SCHOFIELD, J. (1977) New Fresh Wharf: 3, the medieval buildings. *London Archaeologist*, Vol. 3, pp. 66–73.

Describes the results of the excavation of 11 buildings near Billingsgate dating from the twelfth century to the Great Fire. The area was developed in association with reclamation of the waterfront up to the twelfth century.

465 McDONNELL, K. (1978) Medieval London suburbs. Phillimore, 196p.

Examines the emergence of East London with especial reference to the manor of Stepney. By the sixteenth century many manufacturing and trading ventures were located in the area, especially along the waterfront.

466 ROSKAMS, S. and SCHOFIELD, J. (1979) The Milk Street excavation: Part 2. *London Archaeologist*, Vol. 3, pp. 227–34.

Reports on the excavation results from an important site near Cheapside. A thick layer of dark earth sealing the Roman levels is interpreted as a cultivated soil of the late Saxon period, suggesting sparse population and farming within the walled area. Tenth and eleventh century wooden structures were superseded by twelfth century stone buildings as the area became progressively urbanized.

4.7 Renaissance London: AD 1500–1800

467 BESANT, W. (1902) London in the eighteenth century. A. and C. Black, 667p. (Survey of London)

Describes the political history of the City, the extent and face of the City, the church, government, trade, manners, customs, social life and crime.

468 BESANT, W. (1903) London in the time of the Stuarts. A. and C. Black, 400p. (Survey of London)

Describes the Stuart sovereigns, religion, government, trade, the plague, fire, manners and customs.

469 BESANT, W. (1904) London in the time of the Tudors. A. and C. Black, 430p. (Survey of London)

Describes the Tudor sovereigns, religion, Stow's London, government, trade and social life.

470 STEPHENSON, H. T. (1905) Shakespeare's London. Constable, 307p.

A carefully reconstructed topographical description of the City and Southwark c. AD 1600.

471 STOW, J. (1908) A survey of London, reprinted from the text of 1603, with introduction and notes by Charles Lethbridge Kingsford. Oxford: Clarendon Press, 2 vols. (Later reprints include the additional notes published as a separate pamphlet in 1927.)

This is Stow's enlarged 1603 version of his famous survey which first appeared in 1598. It contains some general description of the wall, gates, sports and customs before embarking on a topographical description of each of the wards of London. Stow describes a city still bearing its medieval imprint, much of which disappeared in the Great Fire of 1666.

472 BELL, W. G. (1920) The Great Fire of London in 1666. John Lane, The Bodley Head, 387p.

The first scholarly history of the fire based on a mass of documentary evidence. There is a description of London before the fire, the progress of the fire and an account of the rebuilding of the city.

473 CHANCELLOR, E. B. (1920) The XVIIIth century in London: an account of its social life and arts. B. T. Batsford, 271p.

Describes street topography, great houses, public buildings, churches and architectural relics as well as social life and the arts.

474 MADGE, S. J. (1921–2) Rural Middlesex under the Commonwealth. *Transactions of the London and Middlesex Archaeological Society*, New Series, Vol. 4, pp. 273–322; 403–57.

A study based on the parliamentary surveys of the confiscated royal estates. Gives detailed information on land values, place names and rural economy.

475 DAVIS, E. J. (1924) The transformation of London. In: SETON-WATSON, R. W. (ed.) Tudor studies, presented to A. F. Pollard. Longmans, pp. 247–314.

Describes how the topography of the medieval city was altered in the sixteenth century owing to a rising population and the construction of all kinds of housing and other buildings. The dissolution of the monasteries yielded many sites for building and redevelopment as well as sources of building materials.

476 BEETON, M. M. and CHANCELLOR, E. B. (1929) A tour through London about the year 1725. B. T. Batsford, 118p.

The relevant section, of Defoe's *Tour thro' the whole island of Great Britain* (1724–7) is presented along with mostly contemporary maps and drawings.

477 BRETT-JAMES, N. G. (1933) A speculative London builder of the seventeenth century, Dr Nicholas Barbon. *Transactions of the London and Middlesex Archaeological Society*, New Series, Vol. 6, pp. 110–45.

Describes the contribution of one somewhat unscrupulous builder/developer in the period of rapid urban expansion after the Great Fire of 1666.

478 BRETT-JAMES, N. G. (1935). The growth of Stuart London. London and Middlesex Archaeological Society and George Allen and Unwin, 556p.

An extensively detailed and thorough analysis of the growth of London between 1603 and 1702. Reconstructs the city c. 1600 by developing Stow's survey and then enlarges on the struggle between London's natural tendency to expand and the attempts by the Crown to prevent its growth.

The growth of the City and Westminster before and after 1660 is charted in detail, and a series of detailed street maps shows the extent of London, c. 1600, 1666 and 1708.

479 JONES, P. E. and JUDGES, A. V. (1935–6) London population in the late seventeenth century. *Economic History Review*, Vol. 6, pp. 45–63.

The authors calculate the population of London in 1695 using the returns of inhabitants compiled by parish assessors for the tax on marriages, births and deaths. This is compared with Gregory King's 1695 estimate of the population and with Rickman's retrospective calculation for 1700.

480 SPATE, O. H. K. (1936) The growth of London AD 1660–1800. In: DARBY, H. C. (ed.) An historical geography of England before AD 1800. Cambridge: University Press, pp. 529–48.

A succinct account of the transformation of London from the time of the Great Fire to the first official census. This period witnessed the expansion of the West End estates and the emergence of the poorer East End, along with the growth of trade and the incipient expansion of dock facilities. Despite improved food and water supplies, high rates of mortality only permitted a twofold population increase.

481 de BEER, E. S. (ed.) (1938) London revived, by John Evelyn. Oxford: Clarendon Press, 61p.

An edited version of John Evelyn's discourse *Londoninium Redivivium or London Restored* (1666) which describes his (unsuccessful) proposals for the rebuilding of London after the Great Fire. An introductory chapter outlines the history of the discourse and Evelyn's interest in a city of 'beauty, commodiousness and magnificance'.

482 FISHER, F. J. (1948) The development of London as a centre of conspicuous consumption in the sixteenth and seventeenth centuries. *Transactions of the Royal Historical Society*, Fourth Series, Vol. 30, pp. 37–50.

Maintains that by the early seventeenth century, the economy of London and its suburbs was adapting to a substantial seasonal immigration and permanent settlement of landowners and students who created considerable demand for goods and entertainment.

483 BELL, W. G. (1951) The great plague in London in 1665. 2nd ed. John Lane, The Bodley Head, 361p.

A history of the origins, spread and decline of the plague and its effect on the city and the countryside. The flight of the population and the removal of Parliament to Oxford is described, and a map compiled from the Bills of Mortality shows the intensity of the pestilence.

484 GEORGE, M. D. (1951) London life in the eighteenth century. 3rd ed. Kegan Paul, 452p.

A fully referenced social history of Georgian London. An overview for the period 1700–1815 is developed by selected themes on: life and death; housing and urban growth; immigrants and emigrants; trades and industries (especially watchmaking, shoemaking, silk weaving); parish supported children and apprentices; uncertainties of life and trade.

485 PHILLIPS, H. (1951) The Thames about 1750. Collins, 227p.

A detailed topographical description of the river and its adjacent buildings, wharves and streets from Woolwich to Hampton Court. Illustrated with many contemporary illustrations.

486 REDDAWAY, T. F. (1951) The rebuilding of London after the Great Fire. Edward Arnold, 333p.

A thorough and full documented account of the problems and progress in rebuilding the devastated city. The contemporary plans, rebuilding Acts and the funds raised are described along with an assessment of changes incorporated in the rebuilt city.

487 CARRIER, R. and DICK, O. L. (1957) The vanished city: a study of London. Hutchinson, 114p.

Describes and illustrates, with contemporary prints, the now largely disappeared Renaissance city of the seventeenth and eighteenth centuries. Arranged by individual buildings and more general themes, including the port of London, Whitehall, theatres, houses, squares, public buildings, and Royal Exchange, schools, hospitals, churches, the Temple, and City gates.

488 JONES, P. E. (1963) The growth of London. *Transactions of the GHA* (Guildhall Historical Association), Vol. 3, pp. 145–62.

The text of a lecture given on London's expansion beyond the walls in the sixteenth and seventeenth centuries, and the attempts to legislate against and control the growth of the city.

489 OLSEN, D. J. (1964) Town planning in London: the eighteenth and nineteenth centuries. Yale University Press, 245p.

A detailed analysis of the influence on and control of the built environment of the incipient West End of London by the great estate landlords. The Bedford Estates (Covent Garden, Bloomsbury and Figs Mead) and the Foundling Hospital Estate are treated in detail.

490 PHILLIPS, H. (1964) Mid-Georgian London. Collins, 321p.

A topographical and social survey of central and western London, c. 1750. It comprises a street-by-street description of the houses, residents and entertainments, identified for 1750. Fully illustrated with contemporary illustrations.

491 HEARSEY, J. E. N. (1965) London and the Great Fire. John Murray, 207p.

Describes the London of 1666 (generally and topographically), the origins of the fire and the devastation, and the rebuilding of the city. An account for the general reader with many contemporary quotations.

492 BEDFORD, J. (1966) London's burning. Abelard Schuman, 272p.

A detailed description of the daily progress of the Great Fire in 1666 and an account of the rebuilding of the city. Makes much use of contemporary writings and illustrations.

493 GLASS, D. V. (1966) London inhabitants within the walls 1695. London Record Society, 337p.

An Act of 1694 required an enumeration of the population of London for tax purposes and this is here published – almost 70,000 people arranged alphabetically. (A further 54,000 resided in parishes outside the walls.)

494 PRIESTLEY, H. E. (1966) London, the years of change. Muller, 243p.

This history of Tudor and Stuart London looks at the surviving medieval city, problems of population, water supply and drainage, the plague, Great Fire and the reconstruction of the city.

495 EDIE, C. A. (1967) New buildings, new taxes and old interests: an urban problem of the 1670s. *Journal of British Studies*, Vol. 6, pp. 34–63.

Outlines the conflicting claims and interests involved in attempts at regulating and taxing the rebuilding and expansion of London beyond the City boundary in the 1670s.

496 WRIGLEY, E. A. (1967) A simple model of London's importance in changing English society and economy, 1650–1750. *Past and Present*, Vol. 37, pp. 44–70. Also printed in: BAUGH, D. A. (ed.) (1975) Aristocratic government and society in eighteenth century England. New York: Franklin Watts, pp. 62–95.

London's population almost doubled between 1650 and 1750, and this paper argues that so great was the influx of people that at least one-sixth of the total adult population of England had lived in London at some stage of their lives.The demand for food and fuel stimulated new methods of production which, in turn, increased both the scale and productivity of economic activity. London thus contributed to the complex of processes that led to the industrial revolution.

497 MARSHALL, D. (1968) Dr Johnson's London. John Wiley, 293p. (New Dimensions in History: Historical Cities)

A general survey of London in the eighteenth century, largely derived from secondary sources. The main topics considered are: topography and growth; industry, commerce and banking; government and politics.

498 ROBERTSON, A. G. (1968) Tudor London. Macdonald, 126p. (Discovering London, No. 4)

Describes the main buildings of Tudor London (including outlying palaces) and the result of the dissolution of the monasteries. Background Tudor history is supplied, along with information on how to find the most important sites.

499 BRECHIN, D. (1969) Georgian London. Macdonald, 110p. (Discovering London, No. 6)

Although subtitled 'how to find the London of Queen Anne and the first two Georges in present day London', the book is mainly concerned with Georgian history, social conditions, architecture and furniture, along with accounts of places to see and museums to visit.

500 de BEER, E. S. (1969) Places of worship in London about 1738. In: HOLLAENDER, A. E. J. and KELLAWAY, W. (eds) Studies in London history. Hodder and Stoughton, pp. 391–400.

Uses W. Maitland's *The history of London (1739)* as a source for the first list of all chapels and meeting houses in addition to the Anglican churches. The distribution of these in the City, Westminster and parts of Middlesex and Surrey shows that the upper and middle classes were adequately provided for, the working class less so.

501 HILL, D. (1969) Regency London. Macdonald, 118p. (Discovering London, No. 7)

The growth of Regency London and what survives of it, set in a background of Regency history and social conditions.

502 HOLMES, M. J. R. (1969) Elizabethan London. Cassell, 123p. (Cassell's London Series)

Examines the layout and topography of Elizabethan London along with an account of commerce, home life, entertainment and crime. The position of Westminster and the court is analysed and the author shows how the City and Westminster had become geographically and socially linked by the end of the sixteenth century.

503 PEARSE, M. (1969) Stuart London. Macdonald, 128p. (Discovering London, No. 5)

A guide to the location and significance of the most important Stuart buildings in London, set in a background of seventeenth century history.

504 COWIE, L. W. (1970) Plague and fire, London 1665–6. Wayland Publishers, 128p. (Wayland Documentary History Series)

Uses contemporary writings (including Pepys and Evelyn) and illustrations to trace the spread and decline of the plague and the fire, followed by an account of the rebuilding of the city. (Mainly aimed at GCE history projects.)

505 HILL, D. (1970) A hundred years of Georgian London from the accession of George I to the heyday of the Regency. Macdonald, 126p.

Brief introductions to the Georgian world, the face of the city and social life backed up with over 100 contemporary illustrations.

506 RUDÉ, G. (1971) Hanoverian London 1714–1808. Secker and Warburg, 271p. (History of London)

A general history of eighteenth century London, arranged by topics, including: growth of the metropolis, economic life, social life, religion, government, politics, radicalism, social protest and riots.

507 POWER, M. J. (1972) East London housing in the seventeenth century. In: CLARK, P. and SLACK, P. (eds) Crisis and order in English towns, 1500–1700. Routledge and Kegan Paul, pp. 237–62.

The East End settlements of Spitalfields, Mile End, Wapping, Shadwell and Limehouse, expanded rapidly in the seventeenth century as a result of a large increase in population. Many new streets of houses were laid out and as the century progressed, the streets tended to be more regular and the houses better built.

508 ZITO, G. V. (1972) A note on the population of seventeenth century London. *Demography*, Vol. 9, pp. 511–4.

Uses a least squares analysis of population estimates for seventeenth century London to find regression equations that best satisfy contemporary estimates. It is suggested that Petty's estimate for 1686 is too high.

509 ESPENSHADE, T. J. (1973) Comment on George Zito's 'A note on the population of seventeenth century London'. *Demography*, Vol. 10, pp. 659–60 and Reply to Espenshade, by G. ZITO. *Demography*, Vol. 10, pp. 661–2.

Espenshade questions Zito's least squares analysis of seventeenth century London population estimates (*Demography*, Vol. 9, pp. 511–4) and asks if it is valid to assume a linear growth pattern for 1550–1690. Zito replies that contemporary estimates of population have to be treated with caution and Petty's estimate is highly suspect.

510 SMITH, S. R. (1973) The social and geographical origins of the London apprentices, 1630–1660. *Guildhall Miscellany*, Vol. 4, pp. 195–206.

Shows that though apprentices were drawn from all social classes and all regions of the country, they did not represent the population as a whole. Socially and geographically they were less rural than the general population and by 1660 an increasing number was being recruited from London.

511 KELSALL, A. F. (1974) The London house plan in the later seventeenth century. *Post Medieval Archaeology*, Vol. 8, pp. 80–91.

Uses known examples to establish a particular plan for small terraced houses in the period 1660–80. This was a period of experiment in domestic planning when attempts were made to adapt a traditional form to the needs of an expanding metropolis.

512 SPUFFORD, P. (1974) Population mobility in pre-industrial England: Part 2, The magnet of the metropolis. *Genealogist's Magazine*, Vol. 17, pp. 475–80.

Most migration in the sixteenth and seventeenth centuries was confined to a distance less than 20 miles. However, this paper shows that in the case of London, migrants came from all over the country and it was their presence that allowed London's population to increase by 8,000 per annum, despite a high urban death rate.

513 A supplement to the London inhabitants list of 1695. (1975–7) *Guildhall Studies in London History*, Vol. 2, pp. 77–104; 136–57.

The London Record Society's list (*see* GLASS, 1966) is incomplete because 17 parishes were omitted owing to their assessments being missing. The supplement attempts to fill the gaps from other sources.

514 de MARÉ, E. S. (1975) Wren's London. Folio Society, 128p.

An assessment of Wren's role and contribution to the rebuilding of London. His ambitious but unsuccessful new town plan is offset by his success in rebuilding St Paul's and many other City churches.

515 MINGAY, G. (1975) Georgian London. B. T. Batsford, 163p.

Commences with an examination of the expansion of London and the creation of the Georgian city, followed by a discussion of society, commerce, crime, poverty and labour. Many contemporary illustrations.

516 ELLIS, P. B. (1976) The Great Fire of London: an illustrated account. New English Library, 126p.

A day by day account of the origin, progress and demise of the fire. The suggested plot to burn the city, and the subsequent rebuilding, is discussed.

517 FISHER, F. J. (1976) London as an 'engine of economic growth'. In: CLARK, P. (ed.) The early modern town. Longmans, pp. 205–15.

Argues that London's role in the seventeenth century English economy changed from being merely influential to becoming the centre of the economy. This was due to investment growth and the fruits of the division of labour becoming increasingly channelled through London.

518 CHARTRES, J. A. (1977) The capital's provincial eyes: London's inns in the early eighteenth century. *London Journal*, Vol. 3, pp. 24–39.

Shows that inns were of great importance in the Georgian period, as not only did they represent major sites of business activity and accommodation, but they had a primary function as transport centres. Inns serving specific country areas tended to concentrate in the same part of the City, and became centres for county influences in the capital.

519 POWER, M. J. (1978) The East and West in early modern London. In: IVES, W. W. et al. (eds) Wealth and power in Tudor England. Athlone Press, pp. 167–85.

Argues that the present day contrasting social and economic environments of the East and West Ends of London can be identified during their initial expansion in the sixteenth and seventeenth centuries. In the seventeenth century the East End had poorer quality housing, more industry and a higher incidence of plague deaths than the West End.

520 SUMMERSON, J. (1978) Georgian London, 3rd ed. Barrie and Jenkins, 348p.

A detailed and fully illustrated description of the architecture and construction of churches, public buildings and great residential estates in London between 1714 and 1830. An appendix lists all the surviving Georgian buildings in the City and London generally.

521 PEARL, V. (1979) Change and stability in seventeenth century London. *London Journal*, Vol. 5, pp. 3–34.

Argues that London was not dominated by crisis, conflict and social polarization in the seventeenth century as some historians proclaim. Rich and poor lived close to each other, urban disorder was controlled, parish government was effective, while the gilds still dominated trade and industry.

522 PROCKTOR, A. and TAYLOR, R. (1979) The A to Z of Elizabethan London. London Topographical Society, 62p.

A detailed analysis of the streets, buildings and open spaces shown on three maps of London published between 1558 and 1572.

4.8 Nineteenth century London

523 SHEPHERD, T. H. (1829) London in the nineteenth century. (Reissued by Frank Graham, Newcastle-upon-Tyne, 1970, 160p.)

A nearly complete guide, with over 100 engravings, to the new buildings of early nineteenth century London, ranging in scale from Regent Street to local schools.

524 DORÉ, G. and JERROLD, B. (1872) London: a pilgrimage. Grant, 191p. (Reissued by Blom, New York in 1968, and S. R. Publishers, Wakefield)

Doré's pilgrimage is described in 21 chapters each focusing on a theme or scene, such as the docks, the river, the West End and markets. Written in a highly personal style, punctuated by Doré's famous engravings.

525 EDWARDS, P. J. (1898) History of London street improvements 1855–1897. London County Council, 312p.

A street-by-street description (with plans) of the improvements carried out by the Metropolitan Board of Works and the London County Council during the period.

526 GOMME, G. L. (1898) London in the reign of Queen Victoria (1837–1897). Blackie, 256p. (Victorian Era Series)

A wide-ranging survey of London that examines the capital in 1837 and then describes the progress during the period of trade, commerce, industries, population growth, architecture, parks, street improvements, local government and taxation.

527 BESANT, W. (1909) London in the nineteenth century. A. and C. Black, 421p. (Survey of London)

Describes the history, government, education, entertainment, societies, clubs, charitable work and open spaces of nineteenth century London with an account of improvements made to sewerage, lighting, water supply and communications.

528 REES, H. (1945) A growth map of north-east London during the railway age. *Geographical Review*, Vol. 35, pp. 458–65.

Presents a detailed growth map of the area north of the Thames and east of Ermine Street and identifies areas and periods of growth. Much of the explanation of the patterns of growth is related to the physical geography of the area.

529 BULL, G. B. G. (1956) Thomas Milne's land utilization map of the London area in 1800. *Geographical Journal*, Vol. 122, pp. 25–30.

Milne's pioneer map identified 12 categories of land use which are used as the basis for analysing land use regions in the London area in the early nineteenth century.

530 DYOS, H. J. (1957) Urban transformation: a note on the objects of street improvement in Regency and early Victorian London. *International Review of Social History*, Vol. 2, pp. 259–65.

Compares street improvements in London with those of Hausmann's Paris. London's street improvements, though initially aimed at relieving traffic congestion, were soon identified with improving public health by being routed through slum areas. The new streets of Paris, however, were created by a political decision to beautify the capital.

531 DYOS, H. J. (1961) Victorian suburb: a study of the growth of Camberwell. Leicester: University Press, 240p.

Discusses the meaning and economic and social functions of suburbs generally, followed by a detailed analysis of the processes and patterns of development in the suburbanization of Camberwell, with special reference to estate development and transport history.

532 BRIGGS, A. (1963) London: the world city. In: BRIGGS, A. Victorian cities. Odhams Press, pp. 311–60.

Traces London's rapid growth and its social polarization into East and West Ends in the second half of the nineteenth century. The personages and politics involved in creating London County Council in 1888 as an attempt to govern the growing city are outlined, along with the role of social reformers and novelists in perceiving the city.

533 PRINCE, H. C. (1964) North-west London 1814–1863 *and* North-west London 1864–1914. In: COPPOCK, J. T. and PRINCE, H. C. (eds) Greater London. Faber and Faber, pp. 80–119; 120–141.

An analysis of suburban growth in the area of the former boroughs of St Marylebone, St Pancras and Hampstead. The influence of landed estates and railways in promoting suburban development is considered, along with the relationship of the suburbs to the City.

534 DYOS, H. J. (1968) The speculative builders and developers of Victorian London. *Victorian Studies*, Vol. 11, pp. 641–90.

Shows that from the 1840s to the 1870s small firms dominated the London building market, while the great building boom of the 1870s and 1880–1 saw the emergence of larger firms who survived the succeeding recession. By the time of the next boom, very large firms dominated the market and 90 per cent of the houses were speculatively built, in anticipation of demand.

535 REEDER, D. A. (1968) A theatre of suburbs: some patterns of development in West London, 1801–1911. In: DYOS, H. J. (ed.) The study of urban history. Edward Arnold, pp. 253–71.

Discusses the process of suburban growth in West London, an area transformed in the nineteenth century by the influx of members of all social classes, largely as a result of railway transport. Paddington and Hammersmith are examined in detail, showing that suburban development was complex and variable: Paddington's reputation as a fashionable place grew, while Hammersmith, a fashionable place in eighteenth century, gradually declined.

536 NORTON, G. (1969) Victorian London. Macdonald, 127p. (Discovering London, No. 8)

An account of life in Victorian London, with sections on transport, drainage, public buildings and the growth of suburbs.

537 HOBHOUSE, H. (1971) Thomas Cubitt: master builder. Macmillan, 640p.

Definitive study of the man who gave London much of its distinctive appearance. Cubitt's plans for and building developments in Bloomsbury, Belgravia, Pimlico, Clapham and other areas are fully described.

538 MARGETSON, S. (1971) Regency London. Cassell, 155p. (Cassell's London Series)

An illustrated general history of the period 1811–20, encompassing the Regency Government, the growth of commerce, Nash's architectural and building schemes, society, entertainment, artists, writers, the ordinary people and the expanding city.

539 SHEPPARD, F. H. W. (1971) London 1808–1870: the infernal Wen. Secker and Warburg, 427p. (History of London)

A general, though detailed, history of London in a period when the capital was transformed by new transport linkages, rapid expansion and social and economic changes. The topics covered are government, finance, physical growth, transport, industry, commerce, church, school and state, public health, politics and social life.

540 de MARÉ, E. S. (1972) London 1851: the year of the Great Exhibition. Folio Society, 128p.

Uses 108 contemporary illustrations to describe the metropolitan scene in 1851. The Great Exhibition and its aftermath forms a major part of the text.

541 METCALF, P. (1972) Victorian London. Cassell, 190p. (Cassell's London Series)

Charts the progress and expansion of London from 1837 to 1901, arranged by theme-decades. There is considerable emphasis on urban form, buildings and architecture.

542 COLLINS, P. (1973). Dickens and London. In: DYOS, H. J. and WOLFF, M. (eds) The Victorian city: images and realities. Routledge and Kegan Paul, pp. 537–57.

Examines Dickens' personal image of London as portrayed in his novels.

543 de MARÉ, E. S. (1973) The London Doré Saw: a Victorian evocation. Allen Lane, 229p.

Discusses the vision and content of Doré's *London: a pilgrimage* (DORÉ, G. and JERROLD, B., 1872) and reproduces many of his illustrations. Doré's main themes are analysed and discussed in their wider setting.

544 OLSEN, D. J. (1973) House upon house: estate development in London and Sheffield. In: DYOS, H. J. and WOLFF, M. (eds) The Victorian city: images and realities. Routledge and Kegan Paul, pp. 333–57.

Compares the development of the Eton College Estate in Hampstead with the Norfolk Estate in Sheffield. The architecture and layout of both estates was uninspiring – middle class villas in Hampstead and artisan terraces in Sheffield – but both estates achieved the social and financial aims of their developers.

545 SUMMERSON, J. (1973) The London building world of the eighteen-sixties. Thames and Hudson 60p.

An illustrated summary of the work of the great building contractors during one of the busiest decades in London's building. Architects were dependent on everyday building rather than the aesthetic side.

546 SUMMERSON, J. (1973) London, the artifact. In: DYOS, H. J. and WOLFF, M. (eds) The Victorian city: images and realities. Routledge and Kegan Paul, pp. 311–22.

An illustrated analysis of the architectural developments that transformed the centre of London from 1837–1901. Unlike the Georgian period whose distinctive style dominated the city until about 1857, no single architectural type emerged in the Victorian period but en masse the varying styles contributed a distinctive Victorian ethos.

547 BINFORD, H. C. (1974) Land tenure, social structure and railway impact in North Lambeth, 1830–61. *Journal of Transport History*, Vol. 2, pp. 129–54.

Uses rate books, census returns and railway records to analyse the social impact of the construction of the Charing Cross Railway in the early 1860s. Though the railway caused the demolition of 20 per cent of the houses and the downward social drift of the area, the author argues that such processes were already in operation and the railway only accelerated them.

548 OLSEN, D. J. (1974) Victorian London: specialisation, segregation and privacy. *Victorian Studies*, Vol. 17, pp. 265–78.

Argues that the Victorians transformed London into an environment that allowed privacy for the family and separation of the working and middle classes into their own specialized neighbourhoods and suburban areas.

549 BÉDARIDA, F. (1975) Urban growth and social structure in nineteenth century Poplar. *London Journal*, Vol. 1, pp. 159–88.

Demonstrates how Poplar grew in the nineteenth century after the construction of docks and railways. Employment, however, was progressively concentrated in manufacturing industries and Poplar's mainly working class population was divided between skilled workers and casual labourers. The geographical and psychological isolation of the area, coupled with a strong socialist tradition, earned Poplar a reputation for radicalism and disorder.

550 DYOS, H. J. (1975) A castle for everyman. *London Journal*, Vol. 1, pp. 118–34.

A review article on Victorian and twentieth century suburbanization that examines a large number of diverse publications.

551 EVERSLEY, D. E. C. (1975) Searching for London's lost soul. *London Journal*, Vol. 1, pp. 103–17.

A review article on the contributions on London in DYOS, H. J. and WOLFF, M. (eds) (1973) *The Victorian city*. It is suggested that a synthesis of the findings is required.

552 WHITEHAND, J. W. R. (1975) Building activity and intensity of development at the urban fringe: the case of a London suburb in the nineteenth century. *Journal of Historical Geography*, Vol. 1, pp. 211–24.

Presents a model that explains the building process at the urban fringe, in terms of distance from the city, landholding patterns and spatial and temporal variations in plot size. The model is tested for the development of North Kensington between 1826–1869.

553 OLSEN, D. J. (1976) The growth of Victorian London. B. T. Batsford, 384p.

Starts with an analysis of Georgian London and how its ideals were rejected and destroyed in certain areas while being preserved and extended in many private estates. The arrival of new forms of transport promoted the growth of outer London for both the middle and working classes, and new and distinctive building patterns and styles emerged.

554 SUMMERSON, J. (1977) The Victorian rebuilding of the City of London. *London Journal*, Vol. 3, pp. 163–85.

Describes how in the nineteenth century the Georgian City was redeveloped as a commercial capital. The City ceased to be an important residential area, becoming instead an area dominated by daytime business.

555 ROEBUCK, J. (1979) Urban development in nineteenth century London: Lambeth, Battersea and Wandsworth, 1838–1888. Phillimore, 211p.

An analysis of the changing local governmental influences on urban growth in South London, largely derived from the minutes and reports of the local vestries. The emphasis is on the development of public utilities, but population and social changes are also outlined.

556 SERVICE, A. (1979) London 1900. Granada Publishing, 274p.

A richly illustrated review of the buildings and architectural types current in London about 1900. The buildings are arranged by function and set in their historical and social context.

4.9 General topography

557 LYSONS, D. (1792–6) The environs of London, being an historical account of the towns, villages and hamlets, within twelve miles of that capital. Cadell, 4 vols. (supplementary vol. 1811)

Vol. 1, Surrey.
Vol. 2 and 3, Middlesex.
Vol. 4, Hertfordshire, Essex and Kent.
Vol. 5, Supplement (provides additional details up to 1811).

A magnificent survey of the history of individual settlements around London. The emphasis is very much on manorial and church history, but Lysons also provides a useful description of each place in the eighteenth century, including the average number of births and deaths from the parish register and the number of houses in the parish. Descriptions of developments such as the New River, Woolwich Arsenal and the docks are also provided.

558 LYSONS, D. (1800) An historical account of those parishes in the County of Middlesex, which are not described in *The environs of London*. Cadell, 316p.

Uses the same approach as in his *The environs of London* (1792–6) for parishes lying in the south-west, west and north-west of Middlesex.

559 CUNNINGHAM, P. (1850) Handbook of London: past and present. 2nd ed. John Murray. Reissued (1978) with a new introduction by Michael Robbins. East Ardsley: E. P. Publishing, 602p.

An alphabetical guide to the streets and buildings of London drawing extensively on personal knowledge and quotations from other sources. The area covered is the City and Westminster and as far out as Hampstead, Kilburn, Hammersmith, Battersea, Dulwich, Norwood, Greenwich, Blackwall and Limehouse.

560 THORNBURY, G. W. and WALFORD, E. (1872–8) Old and New London: a narrative of its history. Cassell, Petter and Gilpin, 6 vols.

A voluminous illustrated history of London and description of its buildings, institutions and traditions, as viewed in the nineteenth century. Vols 1 and 2 (by Thornbury) are on the City, vols 3–6 (by Walford) cover Westminster and the western, northern and southern suburbs.

561 THORNE, J. (1876) Handbook to the environs of London, n.p. Reissued (1970) by Adams and Dart, Bath, 794p.

A Victorian gazetteer of the villages and places of historical interest within 20 miles of London. The rural descriptions of many places which are now urbanized are particularly evident.

562 WHEATLEY, H. B. (1891) London past and present. John Murray, 3 vols.

An alphabetical dictionary of buildings, streets and districts described in terms of 'history, associations and traditions'.

563 SEXBY, J. J. (1898) The municipal parks, gardens and open spaces of London. Elliot Stock, 646p.

Separate topographical and historical descriptions of a large number of such spaces, ranging from Hampstead Heath to Red Lion Square.

564 The Victoria history of the County of Surrey. (1902–12) Constable, 4 vols.

Vol. 1 (1902) covers the following general topics: natural history, geology, early man, Anglo-Saxon remains, Domesday Surrey.

Vol. 2 (1905) covers the following general topics: ecclesiastical history, military history, schools, industries, ecclesiastical architecture, domestic architecture, sport, forestry, topography of Farnham Hundred.

Vol. 3 (1911) includes the topographical and historical description of Kew, Kingston-upon-Thames, Malden, Peterham, (that is, Ham), Richmond.

Vol. 4 (1912) contains the topographical and historical description of the following parishes: Barnes, Battersea with Penge, Bermondsey, Camberwell, Clapham, Deptford St Paul, Lambeth, Merton, Mortlake, Newington, Putney, Rotherhithe, Streatham, Tooting, Graveney, Wandsworth, Wimbledon, Southwark, Addington, Beddington, Carshalton, Cheam, Coulsdon, Croydon, Mitcham, Morden, Sanderstead, Sutton.

565 The Victoria History of the County of Essex. (1903–78) Constable, Vols 1 and 2; Oxford University Press for Institute of Historical Research, Vols 3–7.

Vol. 1 (1903) covers the following general topics: natural history, geology, early man, ancient earthworks, Anglo-Saxon remains, Domesday Essex.

Vol. 2 (1907) covers the following general topics: religious houses, political history, maritime history, social and economic history, industries, schools, sports, forestry.

Vol. 3 (1963), Roman Essex (including gazetteer).

Vol. 5 (1966), Metropolitan Essex since 1850. Topographical and historical description of the following parishes: Chingford, Epping, Nazeing, Waltham Holy Cross, Barking, Ilford, Dagenham.

Vol. 6 (1973). Topographical and historical description of the following parishes in Greater London: East Ham, West Ham, Little Ilford, Leyton, Walthamstow, Wanstead, Woodford.

Vol. 7 (1978). Topographical and historical description of the following parishes in Greater London: Havering, Hornchurch, Romford, Cranham, North and South Ockendon, Rainham, Upminster, Great and Little Warley, Wennington.

566 The Victoria History of Hertfordshire. (1908; 1912) Constable, Vols 2 and 3.

Vol. 2 (1908) includes the topographical and historical description of Barnet and East Barnet; volume 3 (1912) includes Totteridge.

567 The Victoria History of Kent. (1908–32) Constable, 3 vols.

Provides a general history of Kent, including ecclesiastical history and religious houses, Romano-British remains, Domesday Kent, social and economic history. There are no volumes covering individual parish histories.

568 The Victoria History of London. (1909) Constable, 588p.

A general survey of the history of the County of London, covering the following aspects: Romano-British London, Anglo-Saxon remains, ecclesiastical history and religious houses (with a description of each). Only one volume has so far been published.

569 The Victoria history of the County of Middlesex. (1911–) Constable, Vols 1 and 2. Oxford University Press for the Institute of Historical Research, Vols 3– .

Vol. 1 (1969) covers the following general topics: the physique of Middlesex, archaeology, Domesday Middlesex, ecclesiastical organization, the Jews, religious houses, education.

Vol. 2 (1911) covers the following general topics: ancient earthworks, political history, social and economic history, industries, agriculture, forestry, sport. The topographical and historical description of individual parishes is commenced with: Ashford, East Bedfont and Hatton, Feltham, Hampton, Hanworth, Laleham, Littleton.

Vol. 3 (1962). Topographical and historical description of the following parishes: Shepperton, Staines, Stanwell, Sunbury, Teddington, Heston and Isleworth, Twickenham, Cowley, Cranford, West Drayton, Greenford, Hanwell, Harefield, Harlington.

Vol. 4 (1971). Topographical and historical description of the following parishes: Harmondsworth, Hayes, Northwood (including Southall), Hillingdon (including Uxbridge), Ickenham, Northolt, Perivale, Ruislip, Edgware, Harrow (including Pinner).

Vol. 5 (1976). Topographical and historical description of the following parishes: Hendon, Kingsbury, Great Stanmore, Little Stanmore, Edmonton, Enfield, Monken Hadley, South Mimms, Tottenham.

570 CHANCELLOR, E. B. (1907) The history of the squares of London: topographical and historical. Kegan Paul, Trench, Trubner, 420p.

Describes the origins, buildings and famous inhabitants of over 100 squares with brief accounts of ones that have disappeared.

571 HUTCHINGS, W. W. (1909) London town, past and present. Cassell, 2 vols.

A street-by-street survey of buildings and their literary and historical associations in the City and Westminster and all the boroughs of the former LCC area.

572 BESANT, W. (1911) London: north of the Thames. A. and C. Black, 682p. (Survey of London)

A survey of the history and topography of north London, described district-by-district.

573 RICHARDSON, A. E. and GILL, C. (c. 1911) London houses from 1660 to 1820. B. T. Batsford, 87p.

General sections describe the history, plan and decoration of town houses, followed by almost 100 photographs of houses, with descriptions, subdivided into the periods 1666–1720, 1720–1760, 1760–1820.

574 BESANT, W. (1912) London: south of the Thames. A. and C. Black, 372p. (Survey of London)

A survey of the history and topography of south London, described district-by-district.

575 ROYAL COMMISSION ON HISTORICAL MONUMENTS (ENGLAND) (1924–30) An inventory of the historical monuments in London. HMSO, 5 vols.

The five volumes cover: (1) Westminster Abbey; (2) West London; (3) Roman London; (4) City of London; (5) East London; each fully described with maps, plans and photographs (each volume is separately listed in the appropriate section of this bibliography).

576 ROYAL COMMISSION ON HISTORICAL MONUMENTS (ENGLAND) (1925) An inventory of the historical monuments of London, Vol. II West London (excluding Westminster Abbey). HMSO, 194p.

A borough-by-borough illustrated list of all the monuments and buildings (prehistoric, Roman, secular, ecclesiastical and unclassified) dating from before 1714 in the former boroughs of Battersea, Chelsea, Finsbury, Fulham, Hammersmith, Hampstead, Holborn, Islington, Kensington, Lambeth, St Marylebone, St Pancras, Stoke Newington, Wandsworth and Westminster.

577 BOYS, T. S. (1926) Original views of London as it is ... 1842, with descriptive notes and introduction by E. B. Chancellor. Architectural Press, 61p.

A collection of accurate drawings of the nineteenth century townscape.

578 CLUNN, H. P. (1927) London rebuilt 1897–1927. John Murray, 316p.

A street-by-street survey of the redevelopment of inner London, with many interesting before and after photographs.

579 ROYAL COMMISSION ON HISTORICAL MONUMENTS (ENGLAND) (1930) An inventory of the historical monuments in London, Vol. V, East London. HMSO, 149p.

An illustrated and descriptive list of all the monuments and buildings (secular and ecclesiastical) dating from before 1714 in the (former) boroughs of Bermondsey, Bethnal Green, Camberwell, Deptford, Greenwich, Hackney, Lewisham, Poplar, Shoreditch, Southwark, Stepney, Woolwich.

580 HOME, G. C. (1931) Old London Bridge. John Lane, The Bodley Head, 382p.

A detailed account, derived mainly from documentary sources of the construction, history and replacement of the stone built London Bridge 1176–1832.

581 BRIGGS, M. S. (1934) Middlesex, old and new. George Allen and Unwin, 312p.

A topographical history of Middlesex, arranged by districts.

582 BELL, W. G. et al. (1937) London wall through eighteen centuries. Council for Tower Hill Improvement, 124p.

A history of the wall from Roman construction to twentieth century uncovering, with a guide to the surviving remains.

583 BRAUN, H. (1937) Some earthworks of north-west Middlesex. *Transactions of the London and Middlesex Archaeological Society*, New Series, Vol. 7, pp. 365–92.

Describes the major earthworks of the area, including hunting parks, ancient routeways, Grim's Dyke and the supposed site of Sulloniacae at Brockley Hill.

584 ROYAL COMMISSION ON HISTORICAL MONUMENTS (ENGLAND) (1937) An inventory of historical monuments in Middlesex. HMSO, 176p.

An illustrated list of Roman, ecclesiastical, secular and unclassified monuments in Middlesex dating from before 1714, arranged by parishes.

585 KENT, W. (1947) The lost treasures of London. Phoenix House, 150p.

A building-by-building survey of the damage caused by the Second World War air raids on London.

586 SINCLAIR, R. (1950) East London. Robert Hale, 416p. (County Books Series)

Covers the old County of London, east of the City and north of the Thames, plus the urbanized area of south-west Essex. The book is the three parts: (1) the shape of east London, its people and the port; (2) the growth of settlement (from earliest times); (3) a perambulation around buildings, markets and the river.

587 COLVILLE, R. (1951) London, the northern reaches. Robert Hale, 268p. (County Books Series)

A general historical and topographical account of Hampstead, Highgate, Holloway, St Pancras, Regent's Park and Primrose Hill, St Marylebone and Willesden.

588 PEVSNER, N. (1951) The buildings of England: Middlesex. Harmondsworth: Penguin Books, 204p.

A parish-by-parish survey of churches, public buildings and houses of architectural note, with brief descriptions of each.

589 PEVSNER, N. (1952) The buildings of England: London, Vol. 2 (except the Cities of London and Westminster). Harmondsworth: Penguin Books, 496p.

Brief descriptions of the buildings of architectural importance, including street perambulations, for all the former LCC boroughs.

590 PEVSNER, N. (1953) The buildings of England: Hertfordshire. Harmondsworth: Penguin Books, 313p.

Includes descriptions of the buildings of architectural importance in the parishes of Chipping Barnet, East Barnet and New Barnet, which were transferred to Greater London in 1965.

591 REID, K. (1954) The watermills of London. *Transactions of the London and Middlesex Archaeological Society*, New Series, Vol. 11, pp. 227–36.

A brief survey of the now largely demolished stream and tide mills on the Thames and its tributaries.

592 LONDON AND MIDDLESEX ARCHAEOLOGICAL SOCIETY (1955) Middlesex Parish Churches. *Transactions of the London and Middlesex Archaeological Society*, Supplement to Vol. 18, unpaged.

A survey of the structures and furnishings of all the parish churches (new and old) in Middlesex.

593 FEIN, A. (1962) Victoria Park: its origins and history. *East London Papers*, Vol. 5, pp. 73–90.

Argues that, although the need for a public park in the overpopulated areas of the East End was agreed by Parliament, a bad choice of site and delays in constructing the access roads meant that, though well used, the park failed to achieve the ambitions of the reformers who proposed it.

594 GODFREY, W. H. (1962) A history of architecture in and around London, 2nd ed. Phoenix House, 368p.

A treatise on the history of architecture using examples from the London area, all of which are located on maps and many illustrated.

595 PRINCE, H. C. (1964) Parks and parkland. In: COPPOCK, J. T. and PRINCE, H. C. (eds) Greater London. Faber and Faber, pp. 333–57.

An historical and topographical description of Royal Parks, public parks and other open spaces. The origins and landscapes of the parks are analysed according to their location: (1) in London; (2) on the outskirts of London; (3) beyond London.

596 CHURCH, R. (1965) The Royal Parks of London. 2nd ed. HMSO, 61p.

Short histories and descriptions of London's ten parks that belong to the Crown and are open to the public.

597 PEVSNER, N. (1965) The buildings of England: Essex. Harmondsworth: Penguin Books, 482p.

Describes the principal buildings of architectural note in the county of Essex, parish by parish, including the areas transferred to Greater London in 1965.

598 WILSON, A. (1967) London's industrial heritage. Newton Abbot: David and Charles, 160p.

A beautifully illustrated collection of 58 vignettes of the surviving remains of a large variety of London industries.

599 SMITH, D. (1969) Industrial archaeology of the Lower Lea Valley. *East London Papers*, Vol. 12, pp. 83–114.

A detailed and illustrated survey of the main classes of industrial monuments in the area between the Thames and Waltham Abbey, but with most emphasis on the area south of Tottenham. The main industries covered are: shipbuilding, main drainage, public water supply, water mills and transport.

600 CLUNN, H. P. (1970) The face of London. Spring Books, 628p.

Describes in detail the streets, buildings and parks in the whole of Greater London, arranged as a number of walks and drives.

601 KENT, W. (ed.) (1970) An encyclopedia of London. 2nd ed. by Godfrey Thompson. J. M. Dent, 618p.

An historical and descriptive encyclopedia of buildings, places and institutions.

602 NEALE, K. (1970) Discovering Essex in London. Essex Countryside, 190p.

Describes the history and topography of that area of Essex which is now part of Greater London, from a general point of view as well as individually, for the metropolitan boroughs of Barking, Havering, Newham, Redbridge and Waltham Forest.

603 JACKSON, P. (1971) London Bridge. Cassell, 134p.

A pictorial history of London Bridge covering: the timber Roman and Saxon bridge, Old London Bridge 1209–1831, New London Bridge 1831–1967, and the present bridge.

604 NAIRN, I. and PEVSNER, N. (1971) The buildings of England: Surrey, 2nd ed. revised by B. Cherry. Harmondsworth: Penguin, 600p.

Describes the principal buildings of architectural note in the county of Surrey, parish by parish, including the areas transferred to Greater London in 1965.

605 SHEPHERD, C. W. (1971) A thousand years of London Bridge. John Baker, 140p.

A history of London's early bridge-point, with special emphasis on the first stone built bridge and its subsequent demolition and re-erection.

606 BETJEMAN, J. (1972) London's historic railway stations. John Murray, 126p.

A photographic and architectural survey of 16 main line termini.

607 MEE, A. (1972) London north of the Thames (except the City and Westminster) 2nd ed. by Ann Saunders. Hodder and Stoughton, 432p. (The King's England)

A descriptive topographical history of places and buildings arranged by London boroughs.

608 OLSEN, D. J. (1973) The changing image of London in *The Builder*. *Victorian Periodicals Newsletter*, No. 19, pp. 4–9.

Shows how articles in *The Builder* deprecated Georgian grey brick and Regency stucco architecture in London in favour of more colourful, sculptural and stone-built buildings. However, by the 1890s neo-classical ideas were being favoured again.

609 PEVSNER, N. (1973) The buildings of England: London. Vol. 1, The Cities of London and Westminster, 3rd ed. revised by B. Cherry. Harmondsworth: Penguin Books, 756p.

A survey of religious buildings, public buildings and street perambulations in the City, Westminster and the southern part of Holborn.

610 THURSTON, H. (1974) Royal parks for the people: London's ten. Newton Abbot: David and Charles, 163p.

Examines the evolution and distinctive character of the following parks: Hyde Park, Kensington Gardens, St James's Park, Green Park, Regent's Park and Primrose Hill, Richmond Park, Hampton Court Park and Bushey Park, and Greenwich Park. The management of Royal Parks is also considered.

611 BUSH, G. (1975) Old London. Academy Editions, 174p.

An annotated re-issue of a magnificent series of photographs of buildings in London, first published 1875–86. Many of the buildings have now disappeared as the emphasis was on the back streets rather than public buildings.

612 CRUIKSHANK, D. and WYLD, P. (1975) London: the art of Georgian building. Architectural Press, 232p.

Mainly concerned with the external appearance and building materials of terraced Georgian houses.

613 MEE, A. (1975) London: the City and Westminster. 2nd ed. by Ann Saunders. Hodder and Stoughton, 436p.

A descriptive topographical history of the principal streets, buildings and open spaces in the City and Westminster.

614 NEWMAN, J. (1976) The buildings of England: West Kent and the Weald. 2nd ed. Harmondsworth: Penguin Books, 672p.

Describes the principal buildings of architectural note in the county of Kent, parish by parish, including the areas transferred to Greater London in 1965.

615 POULSON, C. (1976) Victoria Park: a study in the history of East London. Stepney Books, 118p.

Victoria Park was established in London by the government of the 1840s in an attempt to bring fresh air and recreation to the disease ridden East End. This book traces the history and building of the park and its role in the social and political life of the East End.

616 PORT, M. H. and HALL, J. M. (1977) Two Londons: four jubilees. *Geographical Magazine*, Vol. 49, pp. 552–68.

A description of the streets and buildings of London at the time of Queen Victoria's Diamond Jubilee in 1897 (by M. H. Port) and a similar description of London at the time of Queen Elizabeth II's Silver Jubilee in 1977 (by J. M. Hall).

617 WILLIAMS, G. (1978) The Royal Parks of London. Constable, 234p.

An account of the transformation of the Royal Parks from hunting grounds and private gardens to public domains, including the historical events that have taken place in them. Covers the following: Greenwich Park, St James's Park, Green Park, Hyde Park and Kensington Gardens, Richmond Park, Bushey Park, Hampton Court Park and Regent's Park.

4.10 Local topography and history

Arranged by Greater London boroughs (with pre-1965 boroughs in brackets).

Barking (Barking, Dagenham)

618 OXLEY, J. E. (1955) Barking Vestry minutes and other parish documents. Colchester: Benham, 344p.

A social and administrative history of Barking in the eighteenth and nineteenth centuries using primary sources. Poverty and transport are major themes, and the separation of Ilford from Barking is described.

619 O'LEARY, J.G. (1964) The book of Dagenham: a history. 3rd ed. Borough of Dagenham, 156p.

A well-illustrated history of Dagenham from prehistory to the twentieth century followed by historical descriptions of manors, the river, Hainault forest, churches, schools, inns, industry and transport.

620 HOWSON, J. (1977) Barking – a major fishing port before 1860. *Port of London*, Vol. 52, pp. 127–8.

A brief account of the development of Barking from Domesday Book to the nineteenth century. Barking's role as a fishing and general river port is examined.

Barnet (Barnet, East Barnet, Finchley, Friern Barnet, Hendon)

621 BRETT-JAMES, N. G. (1931) The story of Hendon: manor and parish. Warden, 164p.

Based on local documents, this well-researched history ranges from the Romans to the early twentieth century, with considerable emphasis on manorial history.

622 HOPKINS, J. (1964) A history of Hendon. Hendon Borough Council, 112p.

A general history from the Romans to the twentieth century.

623 LAWRENCE, G. R. P. (1964) Village into borough. 2nd ed. Finchley Public Libraries Committee, 44p.

Describes the physical setting and early history of Finchley as a prologue to the transformation of the village in the second half of the nineteenth century as it became engulfed by the expansion of London.

624 LLOYD, E. (1967) Farm accounts of the manor of Hendon, 1316–1416. *Transactions of the London and Middlesex Archaeological Society*, Vol. 21, pp. 157–63.

A scholarly interpretation of the Westminster Abbey yearly account rolls for Hendon. Shows the creation of Hendon as the home farm for Westminster Abbey and its subsequent leasing to a pioneer yeoman farmer.

625 WILMOT, G. (1973) The railway in Finchley: a study in suburban development. 2nd ed. Barnet London Borough Council, 70p.

Explains how amid great organizational difficulties the Great Northern Railway built a suburban line to Finchley, encouraging rapid suburban development in Finchley and neighbouring districts.

626 GREEN, B. G. (1977) Hampstead Garden Suburb, 1907–1977. Hampstead Garden Suburb Residents' Association, 20p.

A history of the creation of Hampstead Garden Suburb and its architectural development to the sale of the estate in 1961.

Bexley (Bexley, Crayford, Erith, Sidcup)

627 HARRISS, J. (1885) The parish of Erith in ancient and modern times. Mitchell and Hughes, 88p.

A collection of vignettes including the church, Lesnes Abbey, Belvedere Heath, river Thames, and Crossness sewage works.

628 CASTELLS, F. de P. (1910) Bexley Heath and Welling. Bexley Heath: Jenkins, 164p.

Examines several aspects of the history of the area with a bias towards ecclesiastical history.

629 de BOULAY, F. R. H. (1961) Medieval Bexley. Bexley Corporation Public Libraries, 56p.

A brief but scholarly history of the village, church, and manorial organization of medieval Bexley, derived from primary sources.

Brent (Wembley, Willesden)

630 MORRIS, J. C. (1950) The Willesden Survey 1949. Corporation of Willesden, 96p.

A survey by the local authority of the following aspects of Willesden, each based on a large format map: geographical and historical background, land use, population and housing, redevelopment, industry, shopping, employment, transport, open spaces, community services.

631 ROBBINS, M. (1975) Railways and Willesden. *Transactions of the London and Middlesex Archaeological Society*, Vol. 26, pp. 309–18.

A chronology of the arrival of the different railway companies at Willesden and the creation of Willesden Junction.

Bromley (Beckenham, Bromley, Chislehurst, Orpington, Penge)

632 TRENCH, F. C. (1898) The Story of Orpington. Bromley: Bush, 87p.

A brief history of Orpington from Domesday Book to the late nineteenth century, based on documentary evidence.

633 WEBB, E. A. et al. (1899) The history of Chislehurst, its church, manors and parish. Eyre and Spottiswoode, 487p.

A history from earliest times with a detailed topographical description of the parish.

634 BORROWAN, R. (1910) Beckenham past and present. Beckenham: Thornton, 307p.

Gives a history of Beckenham from Domesday Book to 1909 with description of the church, mansion houses and landmarks.

635 HORSBRUGH, E. L. S. (1929) Bromley, Kent. Hodder and Stoughton, 494p.

A detailed history from Domesday Book to the 1920s, largely derived from primary sources. There are detailed descriptions of the manor, palace and churches, together with topographic descriptions of Bromley town and surrounding settlements.

636 WARWICK, A. R. (1972) The phoenix suburb: a south London social history. Blue Boar Press, 271p.

Examines Norwood from the sixteenth century to the present day. The account is dominated by the erection, use and destruction of the Crystal Palace, and Norwood is seen as a suburb that emerged from the ashes of the Crystal Palace fire.

Camden (Hampstead, Holborn, St Pancras)

637 PRICKETT, F. (1842) The history and antiquities of Highgate, Middlesex. The author, 174p. (Re-issued by S.R. Publishers, 1971)

A traditional history of Highgate, its buildings and important inhabitants. It paints an attractive period picture of Highgate as a select rural village.

638 MILLER, F. (1874) Saint Pancras past and present. Abel and Heywood, 355p.

A general history of St Pancras and a detailed topographical description of streets and buildings from the edge of the City to Highgate.

639 BAINES, F. E. (ed.) (1890) Records of the manor, parish and borough of Hampstead. Whittaker, 575p.

A wide-ranging history and description of Hampstead, arranged by topics, including the manor, topography, the heath, local government, services and institutions.

640 SURVEY OF LONDON, Vol. III (1912) The parish of St Giles-in-the-Fields, (Part I) Lincoln's Inn Fields, by W. E. Riley. London County Council, 135p.

An architectural and historical survey of the principal buildings adjacent to Lincoln's Inn Fields.

641 SURVEY OF LONDON, Vol. V (1914) The parish of St Giles-in-the-Fields, (Part II) by W. E. Riley. London County Council, 205p.

An architectural and historical survey of buildings in the parish, including the Bedford Square area.

642 WILLIAMS, E. (1927) Early Holborn and the legal quarter of London. Sweet and Maxwell, 2 vols.

A detailed topographical and documentary history of the buildings and landholdings in this important area that straddled the City boundary.

643 COOKE, M. E. (1932) A geographical study of a London Borough: St Pancras. University of London Press, 68p.

A physical and human geography of St Pancras, aimed at school teachers.

644 DAVIS, E. J. (1936) The university site, Bloomsbury. *London Topographical Record*, Vol. 17, pp. 1–139.

A fully researched history of part of Bloomsbury, covering its physical geography, medieval estate history, post-medieval urban development as part of the Bedford Estate, and its subsequent purchase by the University of London.

645 SURVEY OF LONDON, Vol. XVII (1936) The village of Highgate (The parish of St Pancras, Part I), by P. W. Lovell and W. McB. Marcham. London County Council, 170p.

An architectural and historical survey of buildings in part of Highgate, including Watergate House and Ken Wood.

646 SURVEY OF LONDON, Vol. XIX (1938) Old St Pancras and Kentish Town (The parish of St Pancras, Part II), by P. W. Lovell and W. McB. Marcham. London County Council, 173p.

An architectural and historical survey that includes old St Pancras church and Nash's terraces on the east side of Regent's Park.

647 SURVEY OF LONDON, Vol. XXI (1949) Tottenham Court Road and neighbourhood (The parish of St Pancras, Part III), by W. H. Godfrey and W. McB. Marcham. London County Council, 176p.

An architectural and historical survey of buildings in the area east, west and north of Tottenham Court Road, including University College and Euston Station.

648 SURVEY OF LONDON, Vol. XXIV (1957) King's Cross neighbourhood (The parish of St Pancras, Part IV), by W. H. Godfrey and W. McB. Marcham. London County Council, 181p.

An architectural and historical survey that includes Camden Town, King's Cross and St Pancras Stations, and the northern parts of the Bedford and Southampton estates.

649 LEHMANN, J. (1970) Holborn: an historical portrait of a London borough. Macmillan, 208p.

Traces the history of Holborn from medieval times to its incorporation into the London Borough of Camden in 1965. Themes include: the Inns of Court and Chancery; the development of Bloomsbury; slums and poverty; places of learning; entertainment; writers and artists.

650 IKIN, C. W. (1971) Hampstead Heath centenary, 1871–1971. Greater London Council, 24p.

A brief account of how the Hampstead Heath Act of 1871 saved the Heath for the public from the clutches of developers. Maps show the inroads made into the Heath before 1871 and additions made since then, along with an assessment of current problems in managing the Heath.

651 THOMPSON, F. M. L. (1974) Hampstead: building a borough, 1650–1964. Routledge and Kegan Paul, 459p.

A detailed study of the transformation of Hampstead from rural parish to Metropolitan Borough. The emphasis is on the contribution of landowners and builders in creating the urban environment of Hampstead.

652 Le FAYE, D. (1975) Mediaeval Camden. Camden History Society, 22p.

Described as 'a collation of information available in print' regarding Highgate, St Pancras and Hampstead during the period 1086–1485. The material is arranged by parish, with select bibliographies.

653 BORER, M. C. (1976) Hampstead and Highgate: the story of two hilltop villages. Allen, 255p.

A history of the two villages from the medieval period to the twentieth century. The transformation from purely agricultural settlements to spa villages and eventually desirable London suburbs is described from a social viewpoint.

654 THOMPSON, F. M. L. (1977) Hampstead, 1830–1914. In: SIMPSON, M. A. and LLOYD, T. H. (eds) Middle-class housing in Britain. Newton Abbot: David and Charles, pp. 86–113.

Describes the process of suburbanization in Hampstead in the nineteenth century. Improved road and rail communications and the recognition of a market for middle class houses encouraged estate owners and speculative builders to slowly transform Hampstead into one of the most desirable residential suburbs of nineteenth century London.

655 TINDALL, G. (1977) The fields beneath: the history of one London village (Kentish Town). Temple Smith, 255p.

A well researched history (with strong spatial content) of the origins and development of Kentish Town. The tale encompasses medieval manor, Tudor village, select eighteenth century suburb and the inner London district of today.

City of London

656 NORMAN, P. (1908) London City churches that escaped the Great Fire. *London Topographical Record*, Vol. 5, pp. 26–116.

Historical and architectural descriptions of the following churches: St Olave's (Hart Street); All Hallows, Barking; St Katherine Cree; St Andrew Undershaft; St Helen's (Bishopsgate); St Ethelburga (Bishopsgate); St Giles', Cripplegate; St Bartholomew the Great.

657 BESANT, W. (1910) London: City. A. and C. Black, 491p.

A detailed topographical survey of the streets of the City of London, with descriptions of the origins of the streets and the important buildings in them.

658 BELL, W. G. (1912) Fleet Street in seven centuries. Pitman, 608p.

A topographical history of the development of the western end of the City with emphasis on the transformation of the semi-open area of medieval times to the urbanized area of today.

659 HARBEN, H. A. (1918) A dictionary of London. Jenkins, 641p.

Topographical and historical notes for the streets and principal buildings of the City of London only. The earliest recorded occurrence of each name is given.

660 BELL, W. G. (1920) Surviving City houses built after the Great Fire. *Transactions of the London and Middlesex Archaeological Society*, New Series, Vol. 4, pp. 189–210.

A descriptive survey of some of the last surviving houses of the rebuilt City, after the Victorian phase of redevelopment.

661 ROYAL COMMISSION ON HISTORICAL MONUMENTS (ENGLAND) (1921) An inventory of the historical monuments in London, Vol. IV, The City. HMSO, 258p.

A list of the secular and ecclesiastical buildings (excluding Roman) in the City of London, dating from before 1714, arranged by wards.

662 SURVEY OF LONDON, Vol. IX (1924) The parish of St Helen, Bishopsgate (Part I), by M. Reddan and A. W. Clapham. London County Council, 99p.

An architectural and historical survey of St Helen's church.

663 SURVEY OF LONDON, Vol. XII (1929) The parish of All Hallows, Barking, (Part I), by L. J. Redstone. London County Council, 100p.

An architectural and historical survey of the Church of All Hallows, Barking.

664 HONEYBOURNE, M. B. (1932) The precinct of the Grey Friars. *London Topographical Record*, Vol. 14, pp. 9–51.

Reconstructs the monastic and other buildings of the Grey Friars in the area between Newgate Street and Aldersgate.

665 SURVEY OF LONDON, Vol. XV (1934) The parish of All Hallows, Barking (Part II), by members of the Surrey Committee. London County Council, 130p.

An architectural and historical survey of the buildings of the parish, including the quays, but excluding the parish church.

666 HONEYBOURNE, M. B. (1947) The Fleet and its neighbourhood in early and medieval times. *London Topographical Record*, Vol. 19, pp. 13–87.

A brief account of the topographical geography and early history of the Fleet, followed by a detailed description of the landholdings and buildings in the area between St Paul's and Temple Bar and between the Thames and Holborn.

667 HOLDEN, C. H. and HOLFORD, W. G. (1951) The City of London: a record of destruction and survival. Corporation of the City of London, 341p.

A record of the damage caused to the City during the Second World War with the full text of the consultant's report on the reconstruction of the City. The aims of the reconstruction proposals and their effect on the face and function of the City are described.

668 REDDAWAY, T. F. (1958) The London Custom House, 1666–1740. *London Topographical Record*, Vol. 21, pp. 1–25.

An historical description of the Custom House designed by Wren to replace that destroyed in the 1666 fire. Accidental damage and possible poor quality building and foundations led to its replacement in 1740 to a design by Ripley.

669 COBB, G. (1971) London City churches. Corporation of London, 52p.

Gives a brief history and description of all the surviving churches in the City of London, along with lists of those that have disappeared.

670 HONEYBOURNE, M. B. (1971) The City of London's historic streets. *Transactions of the Ancient Monuments Society*, Vol. 18, pp. 77–94.

A descriptive history of the alignments of the streets, gates and bridges in London from Roman London to the present day. The survival of Roman streets, the deviations made by the Saxons, and the names and functions of medieval streets are discussed.

671 MEGARRY, R. (1972) Inns ancient and modern: a topographical and historical introduction to the Inns of Court, Inns of Chancery and Sergeants' Inns. Seldon Society, 57p.

An account of the creation of the Inns as London's early legal quarter and the later demise of many of the Inns and survival of a few.

672 COWIE, L. W. (1974) Blackfriars in London. *History Today*, Vol. 24, pp. 846–53.

A history of the medieval priory and its post-dissolution uses in the sixteenth and seventeenth centuries.

673 COWIE, L. W. (1975) The Steelyard of London, from the twelfth century. *History Today*, Vol. 25, pp. 776–81.

A history of the market hall and community maintained by Hansard merchants in London.

674 COWIE, L. W. (1975) Whitefriars in London. *History Today*, Vol. 25, pp. 436–41.

A history of the medieval priory of the Whitefriars and the post-dissolution use of the area as a sanctuary for criminals.

675 CRAWFORD, D. (1975) Modern building in the City. *Building*, No. 229, pp. 76–85.

A photo essay on the changing face of the City of London since the war with comment on some of the major developments still to come.

676 COWIE, L. W. (1976) The London Greyfriars from 1224. *History Today*, Vol. 26, pp. 462–7.

A history of the medieval priory and of the post-dissolution development of the area as Christ's Hospital.

677 CRAWFORD, D. (1976) The City of London its architectural heritage: the book of the City of London's heritage walks. Cambridge: Woodhead Faulkner, 143p.

A guide to the history and architecture of the buildings located on the two heritage walks laid out in the City of London.

678 FISHER, J. K. (1976) City of London past and present: a pictorial record of the City of London. Oxford: Oxford Illustrated Press, 64p.

A selection of photographs (from the Guildhall collection) of streets and buildings in the City of London dating from the 1860s to the 1950s compared with the same scenes photographed in 1975.

Croydon (Coulsdon and Purley, Croydon)

679 ANDERSON, I. C. (1889) Plan and award of the commissioners appointed to enclose the commons of Croydon ... (no publisher named), 240p.

A published enclosure award that includes a map of the new enclosures (the enclosure was 1793–1803) and a list of occupiers of land, and provides a perspective on the former rural parish before urbanization.

680 RESKER, R. R. (1916) The history and development of Purley. Cassell, 99p.

Describes the geology, river Bourne, transport, mansion houses, religion, education and local government of Purley.

681 CROYDON NATURAL HISTORY AND SCIENTIFIC SOCIETY (1936–) Regional Survey Atlas of Croydon and District. Croydon Natural History and Scientific Society, (irregular).

An authoritative and well produced collection of maps with interpretative notes.

682 CROYDON NATURAL HISTORY AND SCIENTIFIC SOCIETY (1970) Croydon: the story of a hundred years. Croydon Natural History and Scientific Society, 60p.

An illustrated review of Croydon's urban, commercial, industrial and municipal progress after 1870.

683 COX, R. C. W. (1973) The old centre of Croydon: Victorian decay and redevelopment. In: EVERITT, A. (ed.) Perspectives in English urban history. Macmillan, pp. 184–212.

Describes how the old centre of Croydon had become physically and morally run down by the mid-nineteenth century, despite a rapidly growing population. Attempts to redevelop the centre were frustrated for half a century by municipal indifference and private greed until the Croydon Improvement Act of 1890 permitted the council to replan and rebuild the central area.

Ealing (Acton, Ealing, Southall)

684 JACKSON, E. (1898) Annals of Ealing, from the twelfth century to the present time. Phillimore, 348p.

A documentary history of the church, the manor and important houses. Brentford and Gunnersbury are also included.

685 NEAVES, C. M. (1931) A history of Greater Ealing. 2nd ed. Brentford Printing Co., 213p. (Reissued by S. R. Publishers, 1971)

Provides a general history of the area, followed by individual histories of

Ealing, Hanwell, Greenford, Perivale, West Twyford, Northolt and Brentford.

686 ALLISON, K. J. (1962) An Elizabethan 'Census' of Ealing. Ealing Local History Society, 20p. (ELHS Members' Papers, No. 2). Also in *Bulletin of the Institute of Historical Research*, (1963), Vol. 36, pp. 91–103.

Describes and transcribes a census of all inhabitants in Ealing in 1599, along with their ages, relationships and occupations.

687 KIRWAN, P. (1965) Southall: a brief history. Southall Public Libraries, 54p.

Covers the geology, physical geography and history of Southall from prehistory to the twentieth century.

688 KEENE, C. H. (1975) Field monuments in the London Borough of Ealing. London Borough of Ealing Technical Services Group 28p.

Lists, maps and discusses the different kinds of field monuments and find sites in Ealing.

Enfield (Edmonton, Enfield, Southgate)

689 ROBINSON, W. (1823) The history and antiquities of Enfield. J. Nichols, 2 vols.

A comprehensive history of the parish, church, manor, buildings and Enfield Chase.

690 GILLAM, G. R. (1953) A Romano-British site at Edmonton, Middlesex. Edmonton Hundred Historical Society, 23p.

An account of excavations carried out at Churchfield, Edmonton, in 1951.

691 EDMONTON HUNDRED HISTORICAL SOCIETY (1962–) Occasional Papers, New Series:

2. CRAKE, K. M. (1962) The 1777 division of Enfield Chase. 9p.
5. PAM, D. O. (1963) The Stamford Hill, Green Lanes' Turnpike Trust. Part 1. 23p.
6. AVERY, D. (1963) Manorial systems in the Edmonton Hundred in the late medieval and Tudor periods. 19p.
7. PAM, D. O. (1965) The Stamford Hill, Green Lane Turnpike Trust. Part 2. 41p.
9. AVERY, D. (1965) The irregular common fields of Edmonton. 54p.
13. AVERY, D. (n.d.) The Tudor Hundred of Edmonton. 15p.
18. PAM, D. O. (n.d.) Tudor Enfield; the maltmen and the Lea navigation.
22. LEWIS, T. and PAM, D. (n.d.) William and Robert Cecil as landowners in Edmonton and Southgate 1561–1600; two essays. 25p.

27. PAM, D. O. (1974) The fight for common rights in Enfield and Edmonton, 1400–1600. 15p.
30. PAM, D. O. (1975) Elizabethan Enfield 1572. 17p.
35. BOWBELSKI, M. (1977) The Royal Small Arms factory. 27p.

692 ENFIELD ARCHAEOLOGICAL SOCIETY (1971) Industrial Archaeology in Enfield. Enfield Archaeological Society, 46p. (EAS Research report, No. 2)

A survey of industrial monuments in the London Borough of Enfield, arranged by topic.

693 GILLAM, G. R. (1973) Prehistoric and Roman Enfield. Enfield Archaeological Society, 32p. (EAS Research report, No. 3)

A summary of the archaeology of the London Borough of Enfield, arranged by periods, and with a gazetteer of Romano-British finds and sites.

694 IVENS, J. and DEAL, G. (1977) Finds and excavations in Roman Enfield. *London Archaeologist*, Vol. 3, pp. 59–65.

Provides a gazetteer of Roman find sites in Enfield, with a description of the excavation of an important occupation site in Lincoln Road. It is suggested that Enfield may have been a posting station on Ermine Street.

Greenwich (Greenwich, Woolwich)

695 VINCENT, W. T. (1888–90) The records of the Woolwich district. Woolwich: Jackson, 2 vols.

Volume 1 presents a large number of short articles on almost every aspect of the history of Woolwich and district, including the dockyard and the arsenal. Volume 2 contains similar vignettes for Plumstead, Eltham, East Wickham, Welling, Charlton, Lesnes, Erith, Bexley and Bexley Heath.

696 GRINLING, C. H. et al. (eds) (1909) A survey and record of Woolwich and West Kent. Woolwich: Labour Representation Printing Co., 526p.

Contains accounts of the geology, botany, zoology, archaeology and industries of the area.

697 MILLWARD, R. and ROBINSON, A. (1971) The landscape of Greenwich. In: MILLWARD, R. and ROBINSON, A. South East England: Thames-side and the Weald. Macmillan, pp. 65–76.

The physical geography of Greenwich and the evolution of the present cultural landscape (for secondary schools).

698 LIPMAN, V. D. (1973) Greenwich: palace, park and town. *Transactions of the Ancient Monuments Society*, Vol. 20, pp. 25–48.

Given originally as an address to the Ancient Monuments Society, this article presents short histories (with long footnotes) of the palace, Queen's House, Royal Hospital, Ranger's House, Royal Park, Greenwich Castle, Observatory, the old town, Croom Hill, Maze Hill and the riverside.

699 PLATTS, B. (1973) A history of Greenwich. Newton Abbot: David and Charles, 231p.

A history from the Roman period to the early nineteenth century but with special emphasis on the early periods and Greenwich palace. The first chapter argues that Greenwich originated as the pre-Roman settlement of Trinovantum and that its importance as a Roman town has been confused by being identified with London.

Hackney (Hackney, Shoreditch, Stoke Newington)

700 ROBINSON, W. (1820) The history and antiquities of the parish of Stoke Newington. Nichols, 296p.

A traditional history emphasizing church and manor. There are useful maps of the parish and the manorial estate.

701 ROBINSON, W. (1842–3) The history and antiquities of the parish of Hackney. Nichols, 2 vols.

A comprehensive work, based on original sources, describing the topography, buildings, manors and churches. Maps of the fields before and after enclosure are reproduced.

702 SURVEY OF LONDON, Vol. VIII (1922) The Parish of St Leonard, Shoreditch, by G. T. Forrest. London County Council, 211p.

An architectural and historical survey of the principal buildings of the parish.

703 SURVEY OF LONDON, Vol. XXVIII (1960) Parish of Hackney, (Part 1). Brooke House: a monograph. London County Council, 90p.

Brooke House had been surveyed earlier by the Survey of London, but as a result of bomb damage and its demolition in 1955, it was possible to undertake a more detailed architectural and archaeological survey.

704 BLACK, G. (1978) The archaeology of Hackney. Inner London Archaeological Unit, 16p.

Briefly describes the main archaeological sites of Hackney from prehistoric to the post-medieval period.

Hammersmith and Fulham (Fulham, Hammersmith)

705 FERET, C. J. (1900) Fulham, old and new. Leadenhall Press, 3 vols.

A comprehensive and detailed history of Fulham and its streets, buildings and open spaces.

706 SURVEY OF LONDON, Vol. VI (1915) The parish of Hammersmith London County Council, 143p.

An architectural and historical survey of buildings in the parish.

707 WHITTING, D. (ed.) (1965) A history of Hammersmith based upon that of Thomas Faulkner in 1839. Hammersmith Local History Group, 273p.

A collection of 19 essays by local historians on aspects of the history of the former Metropolitan Borough of Hammersmith. Topics include: backcloth to Hammersmith, the manor of Fulham, the parish church, religion, farming, parishes, public houses, transport, industry, entertainment, hospitals.

708 WHITTING, P. D. (ed.) (1970) A history of Fulham to 1965. Fulham History Society, 330p.

A collection of 14 essays by local historians covering the following aspects of Fulham: early Fulham, the older settlements, local administration and government, poor law and hospitals, growth and development of Fulham 1851–1900, churches, charities, education, farms, gardens, industry, Fulham in the twentieth century, bibliography.

709 WHITEHOUSE, K. (1972) Early Fulham. *London Archaeologist*, Vol. 1, pp. 343–7.

Outlines the origins of Saxon Fulham as a manor belonging to the Bishops of London who built Fulham Palace. The author suggests that the moat round the palace may have originated as a Viking defensive construction or as a Roman fort.

710 HASELGROVE, D. (1972) Early Fulham – a rejoinder. *London Archaeologist*, Vol. 2, pp. 18–21.

Presents new information and re-interprets some of the suggestions put forward by K. Whitehouse's article 'Early Fulham' in *London Archaeologist*, Vol. 1, pp. 343–7.

711 GREEVES, T. A. (1975) Bedford park: the first garden suburb – a pictorial survey. Anne Bingley, 62p.

Bedford park was built between 1875 and 1886 on the garden suburb principle and with its architecture in keeping with the aesthetic movement of the 1870s. This book provides a brief outline of the architectural history of the estate, with over 100 illustrations of houses and house plans.

712 BOLSTERLI, M. J. (1977) The early community at Bedford Park. Routledge and Kegan Paul, 136p.

Describes the garden suburb movement generally and the creation and architecture of Bedford Park. The idea of the estate as a middle class utopia and the evolution and practice of community life is considered.

Haringey (Hornsey, Tottenham, Wood Green)

713 ROBINSON, W. (1840) The history and antiquities of the parish of Tottenham. Nichols, 2 vols.

A comprehensive history of Tottenham, including the church, manors, water supply, river Lea, manufactories, and principal buildings.

714 MARCHAM, W. McB. (1937) The village of Crouch End, Hornsey. *Transactions of the London and Middlesex Archaeological Society*, New Series, Vol. 7, pp. 393–435.

A largely documentary history of the manor and other houses and landholdings surrounding the Broadway.

715 MADGE, S. J. (1938) The early records of Harringay alias Hornsey, from prehistoric times to AD 1216. Hornsey Public Libraries Committee, 99p.

A brief description of archaeological remains and a discussion of the documentary sources for the early medieval period.

716 MADGE, S. J. (1939) The medieval records of Harringay alias Hornsey, from 1216 to 1307. Hornsey Public Libraries Committee, 138p.

Presents excerpts from the wealth of medieval documents on Hornsey resulting from its ecclesiastical domination by the Bishops of London.

717 BROWN, A. E. and SHELDON, H. L. (1969–71 and 1974) Highgate Wood. *London Archaeologist*, Vol. 1, pp. 39–44, 150–4 and 300–3, and Vol. 2, pp. 222–31.

Describes the excavation of ten Roman pottery kilns in Highgate Wood and illustrates their products.

718 ENFIELD HUNDRED HISTORICAL SOCIETY (1971–) Occasional Papers, New Series.

No. 11. BOLITHO, J. R. (1974) Tudor Tottenham. 10p.
No. 21. HOARE, E. (1971) Eighteenth-century Tottenham. 23p.
No. 29. POLLOCK, J. (1974) Tottenham 1800–1850. 22p.

719 MOSS, D. and MURRAY, I. (1973) Land and labour in fourteenth century Tottenham. *Transactions of the London and Middlesex Archaeological Society*. Vol. 23, pp. 199–220.

Uses a variety of documents such as court rolls, extents, accounts and rentals, to show the manorial structure, landholdings, finances and agricultural methods operating on the four manors that were created from the Domesday Manor of Tottenham.

720 MOSS, D. and MURRAY, I. (1974) A fifteenth century Middlesex terrier. *Transactions of the London and Middlesex Archaeological Society*, Vol. 25, pp. 285–94.

Uses a terrier of 1459 to reconstruct the pattern of land ownership in fifteenth century Tottenham. From the 24 divisions of the terrier, it is suggested that much of the land was already consolidated and enclosed.

721 CARRINGTON, R. (1975) Alexandra Park and Palace: a history. Greater London Council, 215p.

Details the building of Alexandra Palace and its history as a place of entertainment, and outlines a continuing saga of financial difficulty.

722 MOSS, D. and MURRAY, I. (1976) Signs of change in a medieval village community. *Transactions of the London and Middlesex Archaeological Society*, Vol. 27, pp. 280–7.

A documentary analysis of the pattern of land transfers in Tottenham in the late fourteenth and early fifteenth centuries. Agricultural land was accumulating in the hands of the more substantial tenants by the early fifteenth century, many of them London citizens who were investing in farming and whose commercial background was undermining the traditional village community.

Harrow

723 DRUETT, W. W. (1937) Pinner through the ages. Uxbridge: King and Hutchings, 226p.

A general history of Pinner from prehistory to the twentieth century.

724 DRUETT, W. W. (1938) The Stanmores and Harrow Weald through the ages. Uxbridge: King and Hutchings, 263p.

A general history of the area from prehistory to the twentieth century.

725 DRUETT, W. W. (1956) Harrow through the ages. 3rd ed. Uxbridge: King and Hutchings, 201p. (Reprinted with new foreword and appendix by S. R. Publishers, 1971)

A social, political and ecclesiastical history of Harrow from the Romans to 1955.

726 BALL, A. W. (1978) Paintings, prints and drawings of Harrow-on-the-Hill, 1562–1899. London Borough of Harrow, 160p.

Covers distant views of Harrow, Harrow village and the School.

Havering (Hornchurch, Romford)

727 WESTLAKE, H. F. (1923) Hornchurch Priory: a kalendar of documents in the possession of the warden and fellows of New College, Oxford. Philip Allan, 152p.

Reproduces excerpts from twelfth and thirteenth century documents that illustrate the life and economy of the priory and its surrounding area.

728 ROBERTS, S. (1969) Romford in the nineteenth century. Local History Reprints, 72p.

Describes the nineteenth century development of Romford before and after the railways with brief histories of the market, industries and local government.

Hillingdon (Hayes and Harlington, Ruislip and Northwood, Uxbridge, Yiewsley and West Drayton)

729 de SALIS, R. (c. 1926) Hillingdon through eleven centuries. Uxbridge: Lucy and Birch, 103p.

A history of Hillingdon and Uxbridge from Roman times to the nineteenth century with an emphasis on manorial and church history.

730 MORRIS, L. E. (1956) A history of Ruislip. Ruislip Residents' Association, 53p.

A short but thorough history of the manor, land ownership, peasantry, agriculture, enclosure and woods of Ruislip, before suburbanization.

731 BOWEN, M. (1977) The archaeology of the Colne Valley Park. Greater London Council, 48p. (GLC Research Memorandum, No. 516)

Reports on an archaeological survey of the Colne Valley Park undertaken as part of the consideration of alternative routes for the North Orbital Motorway. Each period from palaeolithic to medieval is discussed along with lists and maps of finds, sites and threatened area.

Hounslow (Brentford and Chiswick, Feltham, Heston and Isleworth)

732 FAULKNER, T. (1845) The history and antiquities of Brentford, Ealing and Chiswick. Simpkin and Marshall, 504p.

Provides traditional histories and contemporary descriptions for each of the three parishes.

733 BATE, G. E. (1948) And so make a city here: the story of a lost heathland. Hounslow: Thomasons, 463p.

A collection of topographical and historical essays on areas now in the London Borough of Hounslow.

734 DRAPER, W. (1973) Chiswick. 2nd ed. Anne Bingley, 236p.

A detailed history of Chiswick from prehistory to the twentieth century, with special emphasis on estates, people and buildings. Originally published in 1923, the new edition has additional illustrations and a new preface that summarizes Chiswick's development, 1923–73.

735 CANHAM, R. (1978) 2000 years of Brentford. HMSO, 158p.

Describes the outcome of several excavations in Brentford which have provided evidence of habitation from neolithic to medieval. There is a considerable emphasis on typologies of pottery, metal objects and other small finds.

Islington (Finsbury, Islington)

736 TOMLINS, T. E. (1858) A perambulation of Islington. J. S. Hodson, 214p.

A documentary history of the main estates and buildings in the parish.

737 PINKS, W. J. (1881) The history of Clerkenwell. Herbert, 800p.

A history of the parish of Clerkenwell and a detailed street-by-street survey of the history of individual buildings and sites, including the New River Head and the original Clerk's well. Useful maps.

738 RYAN, E. K. W. (1917) A short history of Cripplegate, Finsbury and Moorfields. Adams, 31p.

Uses a number of contemporary illustrations to describe these three areas which lay outside the City and served as open space until urbanized in the eighteenth century.

739 ZWART, P. (1973) Islington: a history and guide. Sidgwick and Jackson, 183p.

A guide to the history, buildings, character, people and happenings of the London Borough of Islington, arranged by sub-districts.

740 ROBERTS, S. (1975) The story of Islington. Hale, 237p.

A general history of Islington arranged by topics, including religion, entertainment, militarism, water works, industry, commerce, communications and Islington today.

741 SCHWAB, I. (c. 1979) The archaeology of Islington. Inner London Archaeological Unit, 16p.

Briefly describes the main archaeological sites in Islington from prehistoric to post-medieval.

Kensington and Chelsea (Chelsea, Kensington)

742 FAULKNER, T. (1820) History and antiquities of Kensington. Egerton, 624p.

A traditional history emphasizing manorial, aristocratic, and ecclesiastical history. A useful introductory chapter describes highways, sewers, nurseries and water supply in 1820.

743 FAULKNER, T. (1829) An historical and topographical description of Chelsea and its environs. Privately published, 2 vols.

Emphasizes manorial and ecclesiastical history along with a description of the chief buildings in the parish.

744 BEAVER, A. (1892) Memorials of Old Chelsea. Elliott Stock, 428p.

A detailed survey of the history of Chelsea, together with topographical descriptions of the principal historic buildings and streets.

745 PRIDEAUX, W. F. (1906 and 1908) Kensington Turnpike Trust plans, 1811. *London Topographical Record*, Vol. 3, pp. 21–63 and Vol. 5, pp. 138–44.

Describes the plans made by Salway for the construction of the road from Hyde Park Corner to Counter's Bridge. (The plans were published in 30 sheets by London Topographical Society, 1899–1903.)

746 SURVEY OF LONDON, Vol. II (1909) The parish of Chelsea, (Part I), by W. H. Godfrey. London Survey Committee, 103p.

An architectural and historical survey of the buildings in Paradise Row and Cheyne Walk.

747 SURVEY OF LONDON, Vol. IV (1913) The parish of Chelsea, (Part II), by W. H. Godfrey. London County Council, 101p.

An architectural and historical survey of Chelsea (except Paradise Row, Cheyne Walk, the Old Church and the Royal Hospital).

748 SURVEY OF LONDON, Vol. VII (1921) The parish of Chelsea, (Part III), by W. H. Godfrey. London County Council, 91p.

An architectural and historical survey of Chelsea Old Church.

749 SURVEY OF LONDON, Vol. XI (1927) The parish of Chelsea, (Part IV), by W. H. Godfrey. London County Council, 133p.

An architectural and historical survey of the Royal Hospital.

750 CURLE, B. R. and MEARA, P. (1968) Kensington and Chelsea street names: a progress report. Kensington and Chelsea Public Libraries, 28p.

An alphabetical list explaining the development of the street names followed by a short history of the principal estates in the area.

751 GLADSTONE, E. F. M. (1969) Notting Hill in bygone days. 2nd ed. with reassessment by A. Barker. Anne Bingley, 280p.

Provides a history of the North Kensington estate from the fifteenth century to suburbanization between 1825 and 1870, with descriptive histories of the component hamlets.

752 CURLE, B. R. and MEARA, P. (1971) An historical atlas of Kensington and Chelsea. Kensington and Chelsea Public Libraries, 30p.

Reproduces ten historical maps of Kensington (1734–1896) and ten historical maps of Chelsea (1717–1901) with short descriptions of each.

753 SURVEY OF LONDON, Vol. XXXVII (1973) Northern Kensington. Greater London Council, 415p.

An architectural and historical survey of the middle class estates that were rapidly built up between 1820 and 1880.

754 GAUNT, W. (1975) Kensington and Chelsea. 2nd ed. B. T. Batsford, 263p.

Although the two places are treated separately, a contrast is obvious between Chelsea, the village retreat of artists and writers, and Kensington the Royal suburb. Much of the description concerns buildings and famous former inhabitants.

755 STROUD, D. (1975) The South Kensington estate of Henry Smith's Charity, its history and development. Trustees of Henry Smith's Charity, 64p.

Describes the purchase of the estate and its subsequent development between 1830 and 1890 as an exclusive residential area between Fulham Road and Brompton Road.

756 SURVEY OF LONDON, Vol. XXXVIII (1975) The museums area of South Kensington and Westminster. Greater London Council, 465p.

An architectural and historical survey of the cultural area created with the surplus funds from the 1851 Great Exhibition. The museums and the Albert Hall are important Victorian monuments, and the nearby domestic buildings are prime examples of Italianate stuccoed terraces.

757 WHIPP, D. (c. 1979) The archaeology of Kensington and Chelsea. Inner London Archaeological Unit, 16p.

Briefly describes the main archaeological sites in Kensington and Chelsea from prehistoric to medieval.

Kingston (Kingston, Malden and Coombe, Surbiton)

758 BIDEN, W. D. (1852) The history and antiquities of Kingston-upon-Thames. Lindsey, 128p.

A traditional history derived from parish documents, with emphasis on manors and church.

759 LAMBERT, H. (1933) Some account of the Surrey manors held by Merton College and Corpus Christi College, Oxford, in the seventeenth century. *Surrey Archaeological Collections*, Vol. 41, pp. 34–49.

Merton College's Surrey endowments included the manor of Malden (and part of Chessington) and excellent documents and estate maps survive. This article briefly reconstructs the extent of the manor and its demesnes and shows that most was enclosed by the mid-seventeenth century.

760 HARPER, F. M. H. (ed.) (1974) Chessington's eighteenth and nineteenth century papers and documents. New Malden: The author, 26p.

Lists documents relating to, inter alia, highways, buildings, constables, burials, voting and poor law administration.

Lambeth (Lambeth, part of Wandsworth)

761 BURGESS, J. H. M. (1929) The chronicles of Clapham (Clapham Common). Privately published, 121p.

A description of the houses and inhabitants adjacent to Clapham Common, with a history and natural history of the common.

762 SURVEY OF LONDON, Vol. XXIII (1951) South Bank and Vauxhall, The parish of St Mary, Lambeth, (Part I). London County Council, 170p.

An architectural and historical survey that includes County Hall and Lambeth Palace.

763 SURVEY OF LONDON, Vol. XXVI (1956) The Parish of St Mary, Lambeth, Part II: southern area. London County Council, 226p.

An architectural and historical survey of buildings in Kennington, Vauxhall, South Lambeth, Stockwell, Brixton, Denmark Hill, Herne Hill, Tulse Hill, Brockwell Park and Norwood.

764 SHORT, M. (1971) Windmills in Lambeth: an historical survey. London Borough of Lambeth, 96p.

Reviews the documentary and pictorial evidence for the existence of 12 windmills in Lambeth. Historical accounts are provided for each mill (along with 70 illustrations) and much confusion concerning names and locations of the mills is clarified.

765 DENSEM, R. and DOIDGE, A. (1979) The topography of North Lambeth. *London Archaeologist*, Vol. 3, pp. 265–9.

Describes and maps the result of borehole surveys in North Lambeth and part of Southwark. The significance of the early topography in the settlement of Lambeth is assessed.

Lewisham (Deptford, Lewisham)

766 DEWS, N. (1884) The history of Deptford. 2nd ed. Simpkin and Marshall, 328p.

A wide-ranging history that includes manorial, commercial and industrial history. A map of Deptford made in 1623 is reproduced.

767 DUNCAN, L. (1908) History of the Borough of Lewisham. North, 173p. (Reprinted by London Borough of Lewisham, 1973)

Describes the solid and superficial geology of Lewisham and outlines the history of the borough up to the present day. Topographical description is arranged as an itinerary.

768 RHIND, N. (1971) Blackheath centenary, 1871–1971. Greater London Council, 32p.

A brief illustrated history of Blackheath Heath, which was preserved as a public open space by an Act of 1871. The use of the heath as a place for sport, recreation and defence is described, along with an account of important houses adjoining the Heath.

769 RHIND, N. (1976) Blackheath village and its environs: Vol. 1, The village and Blackheath Vale. Bookshop Blackheath, 251p.

Presents descriptive histories of almost every developed site in Blackheath since the subdivision of the original estates in the eighteenth century.

Merton (Merton and Morden, Mitcham, Wimbledon)

770 BARTLETT, W. A. (1865) The history and antiquities of the parish of Wimbledon, Surrey. Simpkin and Marshall, 222p. (Re-issued by S. R. Publishers, 1971)

A traditional history concentrating on manors, church and inhabitants, with short accounts of geology, physical geography and botany. The re-issue briefly mentions the growth of Wimbledon after the arrival of the railway.

771 JOHNSON, W. (1912) Wimbledon Common: its geology, antiquities and natural history. Fisher and Unwin, 304p.

Describes the physical geography, geology, prehistory, history and biology of both Wimbledon Common and Putney Heath.

772 JOWETT, E. M. et al. (1951) An illustrated history of Merton and Morden. Merton and Morden Festival of Britain Local Committee, 150p.

A general account from prehistory to the twentieth century.

773 DENBIGH, K. (1975) History and heroes of Old Merton. Charles Skilton, 171p.

Covers the history of Merton from the Romans to the twentieth century, along with a brief account of Merton Park conservation area.

Newham (East Ham, West Ham)

774 OUTER LONDON ENQUIRY COMMITTEE (1907) West Ham: a study in social and industrial problems. J. M. Dent, 423p.

A detailed survey of housing, employment and local government in West Ham at the beginning of the twentieth century. The employment of casual and irregular labour, especially in the docks, is held responsible for most of the social and economic problems of West Ham.

775 PAGENSTECHER, G. (1909) History of East and West Ham. Wilson and Whitworth, 231p.

A general history that presents brief accounts of most aspects of East and West Ham.

776 STOKES, A. (1933) East Ham: from a village to County Borough. 3rd ed. Wilson and Whitworth, 312p.

Uses local sources to reconstruct the municipal development of East Ham, with a section on the docks.

777 LONDON BOROUGH OF NEWHAM, Department of Planning and Architecture (1974) Buildings in Newham: a survey of buildings of architectural, historic and local interest recorded in the Borough in 1973. London Borough of Newham, 98p.

Provides photographs and descriptions of historic buildings in Newham, with an emphasis on the nineteenth century.

Redbridge (Ilford, Wanstead and Woodford)

778 TASKER, G. E. (1901) Ilford past and present. Ilford: S. W. Hayden, 160p.

Examines the growth and development of Ilford from prehistory to nineteenth century suburbanization and industrialization. Nearby places, including Barkingside and Hainault Forest, are also described.

779 TUFFS, J. E. (1962) The story of Wanstead and Woodford from Roman times to the present day. The author, 151p.

A collection of factual statements that illustrate the history of Wanstead and Woodford (separately) period-by-period from the Romans to 1961.

Richmond (Barnes, Richmond and Twickenham)

780 GARSIDE, B. (1951) The manor, lordship and great parks of Hampton Court during the sixteenth and seventeenth centuries. Privately published, 72p.

Derived from original sources and includes a description of Hampton Wick.

781 GARSIDE, B. (1953) The lanes and fields of Hampton Town during the seventeenth century. Privately published, 71p.

Uses documentary sources to show the layout and organization of the fields.

782 ROSE, C. M. (1961) Nineteenth century Mortlake and East Sheen. Privately published, 157p.

Uses vestry minutes as a basis for the examination of the development of public services, sewers, highways, education and transport.

783 URWIN, A. C. B. (1965) Twicknam Parke. Privately published, 130p.

A history of Twickenham Park and its residents and owners, from its creation in 1227 to the sale of the park in 1805 and its subsequent suburban development as the St Margaret's Estate after 1854.

784 JONES, P. M. (1972) Richmond Park: portrait of a royal playground. Phillimore, 75p.

A history of Richmond Park from its creation by Charles I as a hunting park to the present day. The development of the internal landscape of the park is analysed, with an account of its unique flora and fauna.

785 DUNBAR, J. (1973) A prospect of Richmond. 2nd ed. White Lion Publishers, 223p.

A general history of Richmond, the palace, park, town and river, plus a guide to buildings and streets.

Southwark (Bermondsey, Camberwell, Southwark)

786 BLANCH, W. H. (1877) Ye parish of Camberwell. Allen, 486p. (Reprinted by Stephen Marks for the Camberwell Society, 1976.)

A comprehensive history covering such topics as geology, population, parochial history and administration, churches, manorial history and Dulwich College.

787 PENDLE, W. (1878) Old Southwark and its people. Drewett, 333p.

A detailed and documented history of the principal buildings and streets in Southwark. Reproduces a manuscript plan of Southwark, c. 1542.

788 CLARKE, E. T. (1902) Bermondsey: its historic memories and associations. Elliott Stock, 270p.

Covers monastic Bermondsey, Bermondsey House and the growth of Bermondsey and its industries.

789 BECK, E. J. (1907) A history of the parish of St Mary, Rotherhithe. Cambridge: University Press, 270p.

A wide ranging history covering geology, Roman, Saxon, monastic and ecclesiastical history, docks and the Thames Tunnel.

790 JOHNSON, P. M. (1917, 1918) Old Camberwell. *Transactions of the London and Middlesex Archaeological Society*, New Series, Vol. 3, pp. 124–84; 331–50, *and* New Series, Vol. 4, pp. 1–16.

A series of three articles covering the church, history and topography of Camberwell, from Roman times to the early nineteenth century.

791 SURVEY OF LONDON, Vol. XXII (1950) Bankside (the parishes of St Saviour and Christchurch, Southwark). London County Council, 152p.

An architectural and historical survey of the principal historic buildings in the northern part of Southwark, including Southwark Cathedral.

792 SURVEY OF LONDON, Vol. XXV (1955) St George's Fields; the parishes of St George the Martyr, Southwark and St Mary, Newington, by Ida Darlington. London County Council, 150p.

An architectural and historical survey of the principal historic buildings of the two parishes.

793 KENYON, K. M. (1959) Excavations in Southwark, 1945–1947. Surrey Archaeological Society, 112p. (SAS Research Paper, No. 5)

A report of five excavations on or near the line of Roman Stane Street, providing evidence of Roman and medieval settlement.

794 JOHNSON, D. J. (1969) Southwark and the City. Oxford: University Press for the Corporation of London, 441p.

A detailed history of the origins and growth of Southwark and how the City obtained royal charters from AD 1327 to exert control over the borough, until interest began to wane in the nineteenth century.

795 PLOUVIEZ, J. (1973) Roman Southwark. *London Archaeologist*, Vol. 2, pp. 106–13.

Reviews the geographical and archaeological evidence for Roman settlement in Southwark. Roman Southwark is seen as an urban settlement sharing in the wealth of the City, with its nucleus around the bridgehead and approach roads. A detailed map shows the location of Roman finds.

796 SHELDON, H. (1974) Excavations at Topping's and Sun Wharves, Southwark, 1970–1972. *Transactions of the London and Middlesex Archaeological Society*, Vol. 25, pp. 1–116.

Excavations showed that there were a number of early Roman buildings on the site, possibly associated with the early Roman bridgehead. Flooding and erosion appears to have precluded the area's use until the twelfth and thirteenth centuries when a substantial building was erected, followed in the fourteenth century by the reclamation of the river bank and the construction of a dock and jetty.

797 SOUTHWARK AND LAMBETH ARCHAEOLOGICAL EXCAVATION COMMITTEE (1978) Southwark excavations 1972–1974. London and Middlesex Archaeological Society and Surrey Archaeological Society, 2 vols.

A full excavation report, arranged in three parts: (1) the significance of the excavations and their contribution to the history of Southwark; (2) detailed descriptions of the sites; (3) descriptions of the finds of pottery, coins, organic remains, etc. Most of the evidence refers to Roman Southwark.

Sutton (Beddington and Wallington, Carshalton, Sutton and Cheam)

798 JONES, A. E. (1970) From medieval manor to London suburb: an obituary of Carshalton. The author, 126p.

Based on local documents, but concentrates on the earlier history rather than on suburban growth.

799 BRIGHTLING, G. B. (1978) History of Carshalton. Sutton Libraries and Arts Services, 128p. (Reprint of 1882 edition.)

This traditional history with its emphasis on the church and tombs provides an interesting commentary on mid-Victorian small town attitudes to local history.

Tower Hamlets (Bethnal Green, Poplar, Stepney)

800 SURVEY OF LONDON, Vol. I (1900) The Parish of Bromley-by-Bow, edited by C. R. Ashbee. King, 53p.

An architectural survey of the principal buildings in the parish.

801 SMITH, H. L. (1939) The history of East London from the earliest times to the end of the eighteenth century. Macmillan, 308p.

A history of the area now equated with Tower Hamlets, from the Romans onwards. A chronological introductory section is followed by a discussion of such topics as the medieval manor of Stepney, parish creation, religion, early suburbanization, economic and social history.

802 WYLD, P. (1952) Stepney story. St Catherine's Press, 63p.

A brief history of St Dunstan's parish from prehistory to the present day.

803 SURVEY OF LONDON, Vol. XXVII (1957) Spitalfields and Mile End New Town. London County Council, 348p.

An architectural and historical survey of a wide variety of buildings including Spitalfields markets, silk weavers' houses and working class housing.

804 GIBSON, A. V. B. (1958) Huguenot weavers' houses in Spitalfields. *East London Papers*, Vol. 1, pp. 3–14.

Outlines the history of silk weaving in Spitalfields from the immigration of Huguenot weavers to the final demise of the industry on the eve of the Second World War. The architecture of the earliest wooden houses (none surviving) and the later brick built houses is described.

805 IMRAY, J. M. (1966) The Mercers' Company and east London: an exercise in urban development. *East London Papers*, Vol. 9, pp. 3–25.

An account of the transformation of 90 rural acres in Stepney into a building estate in the period 1817–50.

806 EAST LONDON HISTORY GROUP (1968) The population of Stepney in the early seventeenth century. *East London Papers*, Vol. 11, pp. 75–84.

A detailed analysis of the parish registers of Stepney, 1606–10, looking at total population fluctuations in births, deaths and marriages. The home areas of deceased immigrants, and occupations in the eight hamlets of Stepney are also analysed.

807 BOLTON, J. L. and HALL, J. M. (1973) The changing face of the East End: a photographic survey. *East London Papers*, Vol. 15, pp. 43–68.

Illustrates old and new buildings in Tower Hamlets with maps showing their location.

808 RAVENHILL, W. (1976) Joel Gascoyne's Stepney: his last years in pastures old yet new. *Guildhall Studies in London History*, Vol. 2, pp. 200–12.

Provides an illustrated list of the printed and manuscript maps made of Stepney and some of its hamlets, by Gascoyne in 1703.

809 BLACK, G. (c. 1978) The archaeology of Tower Hamlets. Inner London Archaeological Unit, 16p.

Briefly describes the main archaeological sites in Tower Hamlets from prehistoric to post-medieval times.

810 POWER, M. (1978) Shadwell: the development of a London suburban community in the seventeenth century. *London Journal*, Vol. 4, pp. 29–48.

Describes how the speculator, Thomas Neale, created Shadwell as a planned growth and unified parish with its own church, market and water-works.

Waltham Forest (Chingford, Leyton, Walthamstow)

811 SMITH, R. S. (1938) Walthamstow in the early nineteenth century. Walthamstow Antiquarian Society, 36p. (WAS Occasional Publications, No. 2)

Describes industry, population and emergent local government.

812 NEALE, K. (1968) Chingford enumerated: the village community at the Census of 1851. Chingford Historical Society, 12p. (CHS Occasional Publication, No. 2)

A brief descriptive essay on the categories of persons enumerated in the 1851 Census, showing Chingford's rural employment structure.

813 NEALE, K. (1974) Chingford in history: the story of a forest village. 2nd ed. Chingford Historical Society, 22p.

A brief account from prehistory to the present day of the origin of Chingford, the manor, church and Epping Forest.

Wandsworth (Battersea, Wandsworth)

814 ARNOLD, F. (1886) The history of Streatham. Elliot Stock, 222p.

A traditional approach emphasizing manorial history (Streatham, Tooting Bec, Leigham Court and Balham manors) and ecclesiastical history. Mineral wells and commons are also described.

815 MORDEN, W. E. (1897) The history of Tooting-Graveney. Searle, 378p.

Presents many extracts from local records to illustrate the history of manor, church, vestry administration and education.

816 TAYLOR, J. G. (1925) Our Lady of Batersey: the story of Battersea church and parish. White, 442p.

Although largely dealing with the history of the church, the book provides information on the whole parish.

817 FARRANT, N. (1972) The Romano-British settlement at Putney. *London Archaeologist*, Vol. 1, pp. 368–71.

Analyses the results of several excavations which together prove that a late Romano-British settlement existed at Putney from the first century AD to at least the end of the fourth century.

818 METCALF, P. (1978) The Park Town estate and the Battersea tangle. London Topographical Society, 61p.

A descriptive account of the creation of the Park Town estate from 1863, in an area of South London hemmed in with roads and railways.

Westminster (Paddington, St Marylebone, Westminster)

819 ROBINS, W. (1853) Paddington: past and present. Privately published, 200p.

Traces the evolution of landholdings in Paddington and the progress of urbanization up to 1850. Churches, schools and social conditions are also described.

820 CLINCH, G. (1892) Mayfair and Belgravia: being an historical account of the parish of St George, Hanover Square. Truslove and Shirley, 183p.

A history of the Manor of Ebury is followed by topographical histories of parts of Mayfair and Belgravia, including Hyde Park, St George's Church and Buckingham Palace.

821 BESANT, W. (1897) Westminster. 2nd ed. Chatto, 312p.

An illustrated history concentrating on the Abbey and Palace of Westminster.

822 SPIERS, W. L. (1912) Explanation of the plan of Whitehall. *London Topographical Record*, Vol. 7, pp. 56–66.

Describes the compilation of a plan of Whitehall showing the position of the principal buildings of the old palace. London Topographical Society also published (1900) *Comparative plan of Whitehall 1680/1896* as a separate item. The whole of volume 7 of the *London Topographical Record* is devoted to Whitehall and Westminster.

823 GODFREY, W. H. (1920) The Strand in the seventeenth century: its river front. *Transactions of the London and Middlesex Archaeological Society*, New Series, Vol. 4, pp. 211–27.

A description of the buildings between the Strand and the river, including The Savoy Palace and Somerset House.

824 ROYAL COMMISSION ON THE HISTORICAL MONUMENTS (ENGLAND) (1924) An inventory of the historical monuments in London, Vol. 1, Westminster Abbey. HMSO, 142p.

Describes the architecture, monuments, heraldry and stained glass of the Abbey dating from before 1714.

825 KINGSFORD, C. L. (1925) The early history of Piccadilly, Leicester Square, Soho and their neighbourhood. Cambridge: University Press, 178p.

A detailed description of the area before London's expansion, based on a plan of 1585. The subsequent urbanization is described street-by-street. *A London plan of 1585* was published by the London Topographical Society in 1925. *Drawings of buildings in the area described in the early history of Piccadilly* was published by the London Topographical Society in 1926.

826 SURVEY OF LONDON, Vol. X (1926) The parish of St Margaret, Westminster (Part 1), by G. T. Forrest. London County Council, 159p.

An architectural and historical survey of the principal buildings of Great George Street, Queen Anne's Gate and Old Queen Street.

827 SURVEY OF LONDON, Vol. XIII (1930) The parish of St Margaret, Westminster (Part II); Neighbourhood of Whitehall, Vol. 1, by M. H. Cox and G. T. Forrest. London County Council, 279p.

An architectural and historical survey of important buildings including Whitehall Gardens and the Banqueting House.

828 SURVEY OF LONDON, Vol. XIV (1931) The parish of St Margaret, Westminster (Part III); Neighbourhood of Whitehall, Vol. II, by M. H. Cox and G. T. Forrest. London County Council, 183p.

An architectural and historical survey of important buildings, including Downing Street and the Treasury.

829 SURVEY OF LONDON, Vol. XVI (1935) Charing Cross (The parish of St Martin in-the-Fields, Part 1), by G. H. Gater and E. P. Wheeler. London County Council, 296p.

An architectural and historical survey of the principal buildings in Cockspur Street, Charing Cross and the northern part of Whitehall, including The Admiralty and Horseguards.

830 SURVEY OF LONDON, Vol. XVIII (1937) The Strand (The parish of St Martin in-the-Fields, Part II), by Sir G. Gater and E. P. Wheeler. London County Council, 163p.

An architectural and historical survey of buildings in Adam Street, John Street, York Buildings and Buckingham Street, including Northumberland House.

831 SURVEY OF LONDON, Vol. XX (1940) Trafalgar Square and neighbourhood, (The parish of St Martin in-the-Fields, Part III), by Sir G. Gater and F. R. Hiorns. London County Council, 147p.

An architectural and historical survey of buildings around the square, including St Martin in-the-Fields Church, Carlton House and Haymarket.

832 DUGDALE, G. S. (1950) Whitehall through the centuries. Phoenix House, 192p.

A topographical history of the Palace of Whitehall to its destruction by fire in 1698 and the buildings that followed.

833 JOHNSON, B. H. (1952) Berkeley Square to Bond Street: the early history of the neighbourhood. John Murray, in association with the London Topographical Society, 240p.

A fully researched history of the pre-urban estates in the area and their subsequent development as building estates as part of London's fashionable West End.

834 HONEYBOURNE, M. B. (1958) Charing Cross riverside. *London Topographical Record*, Vol. 21, pp. 44–78.

A detailed historical description of the landholdings and buildings in the area between the river, the Strand and Whitehall. Reconstructed maps show the topography c. 1500 and c. 1690.

835 SHEPPARD, F. H. W. (1958) Local government in St Marylebone, 1688–1835; a study of the Vestry and the Turnpike Trust. Athlone Press, 326p.

Uses mainly local records to show the operation of primitive local government. The main emphasis is on poor relief and road maintenance and construction.

836 ASHFORD, E. B. (1960) Lisson Green: a Domesday village in St Marylebone. St Marylebone Society, 28p. (St Marylebone Society Publication, No. 3)

A brief description of the principal buildings and urban development of the manor of Lilestone, or Lisson.

837 SOMERVILLE, R. (1960) The Savoy; manor, hospital, chapel. Duchy of Lancaster, 278p.

A topographical and historical account of the area known as the 'Liberty of the Strand' and the creation and demise of the hospital there.

838 SURVEY OF LONDON, Vols. XXIX, XXX (1960) The parish of St James Westminster (Part 1), south of Piccadilly. London County Council, 646p.

An architectural and historical survey of buildings, including St James's Church, St James's Square, Pall Mall, St James's Street, etc., covering famous houses and clubs.

839 SURVEY OF LONDON, Vols. XXXI, XXXII (1963) The parish of St James, Westminster (Part II), north of Piccadilly. London County Council, 648p.

An architectural and historical survey that includes Burlington House, Piccadilly Circus, Regent Street quadrant and the area north of Oxford Street.

840 SURVEY OF LONDON, Vols. XXXIII, XXXIV (1966) The parish of St Anne, Soho. Greater London Council, 589p.

An architectural and historical survey that includes St Anne's Church, Soho Square and Leicester Square.

841 SAUNDERS, A. (1969) Regent's Park: a study of the development of the area from 1086 to the present day. Newton Abbot: David and Charles, 244p.

A well-researched history of Regent's Park from its creation as a royal hunting park by Henry VIII to the emergence of the public park of today. The attempts to develop it as building land and the execution of Nash's modified scheme for urban residences in a rural setting are treated in detail.

842 SURVEY OF LONDON, Vol. XXXV (1970) The Theatre Royal, Drury Lane, and The Royal Opera House, Covent Garden. Greater London Council, 132p.

Describes the earlier theatre buildings as well as the present ones.

843 SURVEY OF LONDON, Vol. XXXVI (1970) The parish of St Paul, Covent Garden. Greater London Council, 388p.

Surveys St Paul's Church and the Covent Garden Market area, along with an historical account of the early Bedford Estate.

844 MACKENZIE, G. (1972) Marylebone: great city north of Oxford Street. Macmillan, 320p.

Charts the transformation of Marylebone from manors and village to aristocratic housing estate. Sub-areas, including Oxford Street, Harley Street, Cavendish and other squares, Regent's Park and St John's Wood are described, with much emphasis on artists, writers and other notables who have lived in Marylebone.

845 HOBHOUSE, H. (1975) A history of Regent Street. Macdonald and Jane's Press, 166p.

Examines Nash's design and its execution, and the rebuilding of the Street 1904–28. The present day pattern is examined, including Piccadilly Circus and the traffic problem.

846 MACE, R. (1976) Trafalgar Square, emblem of empire. Lawrence and Wishart, 338p.

The first part of this book describes the original topography of the area and the events leading to the construction of the first open public square in London. The second part deals with Trafalgar Square as a place for assemblies and demonstrations.

847 PORT, M. H. (1976) Pride and parsimony: influences affecting the development of the Whitehall quarter in the 1850s. *London Journal*, Vol. 2, pp. 171–99.

Outlines the plans and problems involved in the development of Whitehall as a quarter for governmental administration.

848 SURVEY OF LONDON, Vol. XXXIX (1977) The Grosvenor Estate in Mayfair (Part 1), general history. Greater London Council, 236p.

A general account of the history and architecture of one of the most valuable estates in the world. The estate was mainly built in the period 1720–80, but there have been many important developments since then.

849 RICHARDSON, J. (1979) Covent Garden. Historical Publications, 112p.

Presents a history of Covent Garden and the former market along with a street-by-street survey of buildings and associations for the area bounded by Kingsway, St Martin's Lane, Holborn and The Strand.

850 WALKER, R. J. B. (1979) Old Westminster Bridge. Newton Abbot: David and Charles, 319p.

Describes the need for a bridge at Westminster and the early attempts at providing one. The bridge was much delayed by financial and other problems before its construction during 1735–50. A final section examines drawings and paintings of the bridge.

Section 5: Economic structure and patterns

The structure and performance of the economy of the London region is of central importance to the economic prosperity of the south-east region and of major concern to the United Kingdom economy. The London region remains the single most important centre of economic activity within the United Kingdom, dominating the rest of the country in terms of the size and significance of its manufacturing, office and retail sectors. It is the major centre in the country for overseas tourists. The aim of this section is to present some of the extensive range of literature published on the economy of the London region, and consequently, the bulk of the entries are concerned with manufacturing and office activities; separate subsections are included on retailing and tourism.

The economic literature on the London region is extensive in scope and varied in content. Most of the entries included in this section were published during the 1960s and 1970s, reflecting the increased post-war interest in regional economic analysis and concern with economic issues and problems. The late 1940s and early 1950s were involved with reconstruction following the damage of the Second World War, but in the 1960s, London's booming economy gave rise to the problems of congestion and expansion, with planning policies geared to economic constraint in London and to decentralization at the intra-regional and the national scales. The 1970s saw increased evidence of the decline in Greater London's industrial base and led, at the end of the decade, to a change in the policy of encouraging industrial decentralization from Greater London.

The coverage of these issues in the literature has been patchy, spread throughout a great variety of periodicals and books and it is rather surprising that there is no single text dealing in a comprehensive way with London's post-war economic development. A number of general reviews (subsection 5.2) have been published as part of larger works; for example two chapters on the south-east region (KEEBLE, D. E., 1972) or the chapter on jobs in *London 2000* (HALL, P. G., 1963). A deeper specialized analysis of London's economy is still awaited.

In the manufacturing sector (subsections 5.5, 5.6 and 5.7) there are more substantive general works. HALL, P. G. (1962) and MARTIN, J. E. (1966) have written important basic studies of Greater London's manufacturing industries and areas, including a great deal of material on the historical evolution of manufacturing in London. These books do not cover the decline of manufacturing industry in Greater London which has occurred during the 1970s, but these recent trends are examined in a variety of articles including STONE, P. A. (1977) and DENNIS, R. (1978 and 1979).

Office activities in the United Kingdom are heavily concentrated in the London region, so much of the general literature on this topic is relevant to this bibliography. The major references include GODDARD, J. (1967 onwards) and

DANIEL, P. (1969 onwards), both of whom have written widely on office location. DUNNING, J. H. and MORGAN, E. V. (1971) have written an important economic study of the City of London, while the planning agencies like the Location of Offices Bureau and the Department of the Environment have produced valuable material on London's office activities. The recent GLC research memorandum (WEATHERITT, L. and JOHN, O. N., 1979) *Office development and employment in Greater London, 1967–1976*, is a useful introduction to the subject.

Retailing and market patterns (subsection 5.9) have been analysed by academics and developed by planners. The basic geographical distribution of service centres was examined in the 1960s by SMAILES, A. E. and HARTLEY, G. (1961) and CARRUTHERS, W. I. (1962) and subsequent developments in the pattern of planned centres such as at Brent Cross have been analysed and monitored by a number of commentators. The move of Covent Garden market to Nine Elms in 1974 stimulated a good deal of published material and the reader is advised to consult the planning section (section 8) and the index.

The growth of tourism (subsection 5.10) and its impact on London's economy and environment has been considered in a number of publications produced since 1971. The planning authorities have produced studies, discussion papers and plans and the Tourist Boards have also published in this field. The GLC (1971 onwards) and certain London boroughs such as Kensington and Chelsea and the City of Westminster have been concerned with the impact of tourism upon their areas.

The literature on London's economy has been produced by a variety of individuals and organizations. Economists, geographers and planners provide the bulk of the individual contributions. Private developers interested in the markets for factories, offices and shops have published data and analyses. Planning agencies and authorities have produced a large volume of literature on economic matters – especially the Greater London Council. As the strategic planning authority the GLC, since its inception in 1965, has undertaken research and contributed to numerous studies of economic activities in London. For material on employment published by individual London boroughs, reference should be made to COCKETT, I. (1976). The Standing Conference on London and South East Regional Planning and the South East Economic Planning Council have also been active in this field during the 1970s. Central government departments such as the Department of the Environment, special agencies like the Location of Offices Bureau and the London Tourist Board have contributed to the literature. Finally, action groups, such as the Islington Economy Group, have written on the economic problems of various parts of London. As economic considerations affect most aspects of the London region, relevant material occurs in other sections of this bibliography, especially in the transport, social and planning sections. Most of the official plans for the region contain sections on economic structure and problems, and the reader interested in economic issues is advised to consult the planning section (section 8) and the index. References to transport (section 6) are only included in this section where they relate to the regional economic impact of transport developments.

For further references the GLC Research Library bibliographies and *London topics* are useful bibliographic sources. On offices see LOCATION OF OFFICES

BUREAU (1972) *Offices: a bibliography*, and for retailing THORPE, D. and KIVELL, P. T. (1974) *Atlas of Greater London shopping centres* is a valuable source of material. For statistical sources COCKETT, I. (1977) provides statistics concerned with employment in London. The census volumes and the *Annual Abstract of Greater London Statistics* (1966 onwards) are also important statistical sources. Recent references can be obtained by consulting *Urban Abstracts* and up-to-date surveys on office, factory, warehouse and shop markets can be obtained from the *Estates Gazette* and from regular surveys undertaken by bodies such as the London Chamber of Commerce and Industry.

5.1 Bibliographies, statistics and atlases

851 MINISTRY OF HOUSING AND LOCAL GOVERNMENT (1954) Industry in Greater London. Department of the Environment Library, 4p. (DOE Library Bibliography, No. 114)

A basic bibliography on London industry.

852 LOCATION OF OFFICES BUREAU (1972) Offices: a bibliography. Location of Offices Bureau, 22p.

A list (unannotated) of publications, unpublished theses and documents on office development and location in London and elsewhere.

853 REDPATH, R. U. et al. (1972) Surveys of personal income in London. Greater London Council, 27p. (GLC Research Memorandum, No. 350)

The advantages and limitations of the various government sources of regular income statistics for London are discussed. The four main sources are the Department of Employment's *Family expenditure survey* and its *New earnings survey*, the Department of Health and Social Security's *Regional statistics of earnings* and the Inland Revenue's *Survey of personal income*.

854 GREATER LONDON COUNCIL, Department of Planning and Transportation, (1973) Offices in London. Greater London Council, 16p. (GLC Research Bibliography, No. 48)

An annotated bibliography of books, pamphlets and articles on offices in London, published since 1960. Contains 109 entries.

855 THORPE, D. and KIVELL, P. T. (1974) Atlas of Greater London shopping centres. Manchester Business School, 38p. (MBS Research Report, No. 7)

Data are included on shopping floorspace, the distribution of centres by borough, distribution of supermarkets and shop rents.

856 GOMERSALL, A. (1975) Industrial relocation. Greater London Council, 21p. (GLC Research Bibliography, No. 61)

A selective annotated list of material on industrial relocation (not office location) with a separate section on Greater London and south-east England.

857 LOCATION OF OFFICES BUREAU (1975) Office relocation facts and figures: statistical handbook. Location of Offices Bureau, 60p.

A 12 year series of statistics from the Location of Offices Bureau, with cross tabulations of the most significant characteristics of moves which have taken place. Reasons for relocation or decisions against relocation are given.

858 MURRAY, N. (1975) Shopping in urban areas. Greater London Council, 29p. (GLC Research Bibliography, No. 71)

A list of general references on shopping in urban areas, with a specific section on shopping provision in London and the South East.

859 *Trend survey of manufacturing industry in the South East.* (1975–) London Chamber of Commerce and Industry, thrice yearly.

Statistics prepared regularly by the Economic Committee of the London Chamber of Commerce and Industry are presented. Data are presented on the business climate, restrictions on output, employment, inflation, the factors inflating costs, the prospects for the next 12 months, the employment situation and the initiative taken to meet increased export demand.

860 COCKETT, I. (1976) Research projects on employment within the London boroughs, 1966–1976. Greater London Council, 10p. (GLC Research Bibliography, No. 76)

A list of specific employment studies conducted by individual London Boroughs.

861 ECONOMIST INTELLIGENCE UNIT (1976) An analysis of commercial property values 1962–1975. Economist Intelligence Unit, 14p.

A survey covering offices, shops and industrial warehouses in the Greater London Council area, the South East and East Anglia.

862 LOCATION OF OFFICES BUREAU (1976) Offices: a bibliography. Location of Offices Bureau, 23p.

References, without abstracts, covering employment in offices, layout and organization of office buildings, communications and technology, office development, office location, decentralization, office location in other countries, transport and the journey to work and office planning policies. Special emphasis upon London and the South East.

863 SCOTT, G. (1976) Industry and employment in London. Greater London Council, 15p. (GLC London Topics, No. 14)

A selective guide to some of the material on industry and employment which may be useful to someone looking at the subject for the first time. Focuses upon problems of employment in the London area, with a summary of central government and local government policies on industry in London. A chronology of strategic planning in Greater London, 1965–76 is included, along with an annotated list of sources of information on employment, including Council reports and publications and other sources of data.

864 SMITH, G. M. (1976) The financial activities of the City of London: a select bibliography. Business School Press, 27p.

A list of publications describing and analysing the functions and effectiveness of the various organizations collectively known as 'The City'. Intended as a guide to the relevant literature for academics, students, politicians, practitioners, laymen and librarians.

865 The office market in 1976. (1977) *Estates Gazette*, Vol. 241, pp. 16–22.

Presents the results of an overall survey of the office property market in 1976 covering London and the South East and the major areas of the United Kingdom.

866 COCKETT, I. (1977) Statistics concerning employment in London. Greater London Council, 30p. (GLC Research Bibliography, No. 85)

A list, with abstracts, of published and unpublished reports, abstracted data, conference papers, literature reviews, research registers and committee papers concerned with statistical data on the employment situation in London. Documents are arranged under headings, including incomes and earnings, migration and land use.

867 Factories and warehouses in 1976. (1977) *Estates Gazette*, Vol. 241, pp. 145–51.

An overall survey of the factory and warehouse market in 1976 covering London and south-east England and the major areas of the United Kingdom.

868 KRISHNA-MURTY, R. (1977) Earnings and incomes data for Greater London. Greater London Council, 161p. (GLC Research Memorandum, No. 504)

A number of different sources are used to compile these data on individual earnings and household incomes in London up to the end of 1976. Some comparisons are made with the rest of the south-east region and with the United Kingdom as a whole.

869 LONDON TOURIST BOARD (1977) London's tourist statistics, 1976. London Tourist Board, 22p.

Statistics are given for volume of tourism, value of tourism, seasonal distribution of visitor arrivals, purpose of visit, type of accommodation used, hotel occupancy and the growth in the number of visitors to selected tourist attractions.

870 BERNARD THORPE AND PARTNERS (1977) Industrial property markets in Greater London and the South East: a study. Bernard Thorpe and Partners, 24p.

This study surveys activity in the factory and warehouse markets in the South East, and then makes a more detailed survey of five places in the region: Greater London, south of London (Crawley/Gatwick and South Hampshire), west of London (Reading, Wokingham and Basingstoke), east of London (the Chelmsford area and the Maidstone, Tonbridge and Malling area) and north of London (Milton Keynes).

871 ENGLISH TOURIST BOARD (1977) English heritage monitor. English Tourist Board, 49p.

The second annual compilation and analysis of data on the conservation, preservation and public use of England's architectural heritage. London is well represented.

872 FROST, D. (1978) Office property in the South East. *Commerce International*, March, pp. 15–9.

Despite extensive developments in the suburbs and in towns like Reading, the central London market remains buoyant. This article surveys the office property scene in London and the South East.

873 GREATER LONDON COUNCIL (1978) Some data on London's unemployment. *Greater London Intelligence Journal*, Vol. 41, p. 27.

Unemployment rates and numbers unemployed are given for London. Figures are for April 1978, with past data for comparison.

874 KENNINGTON, D. and MACLEOD, K. (1978) Tourism in London. Greater London Council, 24p. (GLC Research Bibliography, No. 96)

A selective list of the literature on tourism of direct relevance to London's particular problem.

875 SINCLAIR GOLDSMITH (1979) City of London space analysis: a survey of present and future supply of substantial office buildings up to 1982. Sinclair Goldsmith, 8p.

This survey shows office units over 15,000 square feet which will be available for occupation by 1982. The breakdown of supply over the next three years in each of the City postal districts is given.

5.2 General

876 FOGARTY, M. P. (1945) London and the south eastern counties. In: FOGARTY, M. P. Prospects of the industrial areas of Great Britain. Methuen, pp. 389–451.

A study of the post-war economic prospects of the main industrial regions of Great Britain. It is concerned with patterns of community living and social service as well as with questions of industrial and commercial development.

877 CHISHOLM, M. (1964) Must we all live in the South East? The location of new employment. *Geography*, Vol. 49, pp. 1–14.

Seeks to explore some of the reasoning used to explain the progressive concentration of industry and population in Great Britain. Suggests that the arguments advanced and questioned do not necessarily account for an inexorable tendency for industry and population to be concentrated in south-east England.

878 HOLMANS, A. E. (1964) Industrial development certificates and control of the growth of employment in south-east England. *Urban Studies*, Vol. 1, pp. 138–52.

Examines data on the growth of employment in manufacturing industry and the amount of industrial building for which industrial development certificates were issued during the period 1950–62. Contends that without new control methods, a substantial increase in employment even in manufacturing industry must be expected to take place in south-east England in the foreseeable future.

879 HOLMANS, A. E. (1964) Restriction of industrial expansion in south east England: a reappraisal. *Oxford Economic Papers*, Vol. 16, pp. 235–61.

Argues that the disadvantages from the standpoint of the national economic interest of further growth of industrial capacity and employment in south-east England are not so evident as maintained. The case for more drastic reductions on growth in the South East as part of a national policy for faster growth, has not been made out.

880 HOLMANS, A. E. (1965) Restrictions on industrial expansion in south east England: a rejoinder. *Oxford Economic Papers*, Vol. 17, pp. 343–5.

Replies that Thirlwall (1965) has not demonstrated any inconsistency between a rapid growth of population and employment in the South East and lower unemployment rates in the north.

881 THIRLWALL, A. P. (1965) A reply to Mr Holmans on restriction of expansion in south east England. *Oxford Economic Papers*, Vol. 17, pp. 337–41.

Disagrees with Holmans' (1964) analysis that still more stringent restrictions on industrial expansion in south-east England might be inimical to faster economic growth in the nation as a whole. Maintains that if south-east England is allowed to expand at a rapid rate, there is little prospect of success in reducing inter-regional differences in unemployment rates.

882 STANDING CONFERENCE ON LONDON AND SOUTH EAST REGIONAL PLANNING (1966) Population and employment in the conference area: report by the Technical Panel on recently available data. Standing Conference on London and South East Regional Planning, 14p. (LRP, 721)

A first analysis based upon the 1961 Census of numbers of people in employment, where they worked and where they lived, with data about changes since 1951 and migrational movements in 1960–1.

883 HALL, P. G. (1969) Jobs 2000. In: HALL, P. G. London 2000. 2nd ed. Faber and Faber, pp. 45–84.

First produced in 1963, the second edition is unaltered except that postscripts have been added to each chapter dealing with changes in the problems, plans and ideas over six years. The earlier chapter on jobs 2000 contains sections on the interim years; Barlow policy in action; the geography of post-war growth; the development of diversion policies and a final section on planning London's jobs. The postscript focuses on the trend towards a new polycentric region in which London's former absolute dominance is becoming steadily eroded.

884 LUTTRELL, W. F. (1970) Employment in Greater London. Town and Country Planning Association, 12p.

A discussion of employment policy in Greater London, based upon three divisions – the central area, the rest of the inner boroughs and the outer boroughs.

885 CRAWFORD, D. (1971) What about the workers? London's employment problems. *Commerce International*, Vol. 102, pp. 17–9.

Describes the decline in London's population and employment, attributing a major part of it to development control and decentralization policies. An examination of the Greater London Council's plans to create more jobs and maintain employment balance in the three concentric rings is followed by a discussion of schemes to promote industrial growth in the boroughs.

886 EVERSLEY, D. E. C. (1972) Rising costs and static incomes: some economic consequences of regional planning in London. *Urban Studies*, Vol. 3, pp. 347–68.

Examines the reasons for the rapid decline of population and employment in Greater London and the contrast between rising urban costs and falling urban incomes. The paper argues that migration from London should be slowed down and London should receive greater financial support from central government.

887 KEEBLE, D. E. (1972) The South East and East Anglia. I The Metropolitan Region. II The Zonal Structure. In: MANNERS, G. et al., Regional development in Britain. John Wiley, pp. 70–125.

Describes and interprets the trends, problems and uncertainties associated with the evolving patterns of regional economic change. It evaluates the political response to them and affords a bench mark for contemplating something of the prospective geography of this region in the 1970s. The first chapter examines the importance of south-east England to the British economy and the planning problems of growth; it discusses the traditional decentralizing policy and considers the links binding together the London city region. The second chapter discusses the human geography, economic character and planning problems of different parts of the region.

888 MORTLOCK, D. (1972) Employment changes in Greater London. *Quarterly Bulletin of the Intelligence Unit of the Greater London Council*, 20, pp. 16–26.

Analyses the movement levels in Greater London for manufacturing and service sectors. Local trends, national trends and national and local cyclical factors are all considered. It concludes that cyclical factors have played little part in determining London's employment level, while local and national trends have been dominant.

889 FOSTER, C. D. and RICHARDSON, R. (1973) Employment trends in London in the 1960s and their relevance for the future. In: DONNISON, D. and EVERSLEY, D. (eds) London: urban patterns, problems and policies. Heinemann, pp. 86–118.

This chapter considers the problems raised for Londoners by the decline in London's resident and working populations. It concludes that there is no clear evidence that London's population has suffered notable hardship as a result of the decline in the size of the city.

890 GREATER LONDON COUNCIL (1973) London the future and you: population and employment. Greater London Council, 20p.

Contains brief sections on population, employment, local employment problems, prices and incomes, followed by a discussion of population and employment trends. A final section lists the main points of a possible economic policy and measures for meeting the situation described.

891 GUEST, D. (1974) Turn again Whittington. *New Society*, Vol. 28, pp. 374–6.

The results of a survey into attitudes towards the London weighting allowance paid to workers in London. It is concluded that the higher costs of housing and travel in London are the main reasons behind the demand for a higher London allowance.

892 KIRWAN, R. (1974) The contribution of public expenditure and finance to the problems of inner London. In: DONNISON, D. and EVERSLEY, D. (eds) London: urban patterns, problems and policies. Heinemann, pp. 119–55.

Looks at the distribution of public expenditure and the way it is financed in Britain, with special reference to the problems of inner London. It concludes that in the medium term, if not the short term, the prospects for inner London (a very prosperous area in national terms) are somewhat bleak.

893 LOMAS, G. M. (1974) Labour and life in London. In: DONNISON, D. and EVERSLEY, D. (eds) London: urban patterns, problems and policies. Heinemann, pp. 51–85.

Examines aspects of the inter-relationship between job opportunities, the housing situation and personal incomes in London, set against the debate over the benefits and disbenefits of the decentralization of people and jobs from London, particularly for inner London.

894 O'CLEIREACAIN, C. C. (1974) Labour market trends in London and the rest of the South East. *Urban Studies*, Vol. 11, pp. 329–39.

Male employment in the GLC area fell quite substantially in the 1960s and GLC policy shifted to discourage the continuance of the trend. Evidence is presented about the level of earnings of Londoners compared with the rest of the south-east region. Problems of using highly aggregated data are outlined and conclusions about the population which may be relatively disadvantaged by remaining in London are tentative.

895 BALINT, M. (1975) Greater London's economically active population. Greater London Council, 89p. (GLC Research Memorandum, No. 441)

Data from the 1961 and 1971 Censuses and the 1966 Sample Census are used in this analysis of the size of the economically active population and economic activity rates in Greater London and the London boroughs.

896 GREATER LONDON COUNCIL (1975) Industrial policy and employment in London: submission by the GLC to the Secretary of State for Industry. Greater London Council, 16p.

Looks at the decline in both manufacturing and service industry and employment since 1961 and identifies those areas and groups of workers suffering the heaviest rates of unemployment. It recommends a range of policies to deal with these problems.

897 JAROSZEK, J. (1975) Earnings in relation to employment changes. Greater London Council, 42p. (GLC Research Memorandum, No. 500)

An examination of differences between earnings in Greater London and the rest of the south-east region. Disaggregates differences in average earnings into structural and earnings components and earnings on a residence basis together with the cost of living.

898 LATTA, E. (1975) London to-day – the South-East tomorrow? *Commerce International*, November, pp. 9–14.

Attributes London's industrial plight largely to official restrictions preventing companies from expanding or relocating in the area. It argues that similar problems could affect the whole region if policies are not relaxed.

899 MINNS, R. (1975) An alternative employment policy for the Greater London Council. *Town and Country Planning*, Vol. 43, pp. 483–5.

Looks at two employment problems – how to overcome falling manufacturing employment and how to find more money for local services. It concludes that GLC and other regional authorities should be given powers akin to the proposed Scottish and Welsh Development Agencies.

900 CORPORATION OF LONDON Department of Architecture and Planning (1976) City of London development plan background paper: economic activity. Corporation of London, 2 vols.

Examines activity patterns; current policies on floor space and employment and their background; employment; office floor space; manufacturing industry and warehousing; and similar activities.

901 GREATER LONDON COUNCIL (1976) Development of the Strategic Plan for the South East: employment in London and the region. Greater London Council, 10p.

A discussion of industrial decline in London, looking at changes in employment, manufacturing employment, the range of employment in occupation groups, unemployment, earnings and incomes.

902 GRIPAIOS, P. (1976) The end of decentralization policy in London: some comments. *Town and Country Planning*, Vol. 44, pp. 426–8.

Notes the Greater London Council announcement of its intention to abandon its decentralization policy which has been operational for 25 years. Examines the effects of dispersal policy on employment and questions whether the changed policy is likely to have much effect.

903 GRIPAIOS, P. (1976) A new employment policy for London? *National Westminster Bank Review*, Vol. 8, pp. 37–45.

Examines changes in employment in London and the role of policy in causing these changes, considering whether continuation of dispersal policy is justified; special use is made of data from Greenwich, Lewisham and Southwark. Finally, the need for a positive policy to stem the loss of jobs and encourage suitable employment in London is suggested.

904 HALL, P. G. (1976) The South East: Britain's tarnished golden corner. *New Society*, Vol. 37, pp. 228–31.

A survey of the south-east of England shows that although it is still the most economically successful region, it is stagnating and comes only 17th in a table of incomes among 49 EEC regions. Population is falling, particularly in Greater London where there is a spectacular drop in manufacturing employment; the reasons for this decline are examined. The article looks at ROSE (Rest of the South East), examines growth patterns and points out the government's dilemma in allocating funds.

905 OAKESHOTT, J. J. (1976) Unemployment in London. Greater London Council, 48p. (GLC Research Memorandum, No. 499)

Examines unemployment in relation to vacancy rates, occupation, age and duration and refers to similar data for the whole region. The paper describes the strategic planning context and discusses the usage of unemployment as an economic indicator.

906 DOBSON, N. (1977) Employment and industry in Greater London: a background document. London Council of Social Service, 50p.

Looks at employment and industry in Greater London in a planning context.

907 KNIGHT, D. R. W. et al. (1977) The structure of employment in Greater London, 1961–81. Greater London Council, 53p. (GLC Research Memorandum, No. 501)

Analyses the changing structure of employment by broad industry and occupation groups in Greater London and nine sectors of London from 1961–74, with a consideration of some of the issues arising from the formulation of strategic employment policy in the medium term.

908 LAPPING, A. (1977) London's burning! London's burning! a survey. *Economist*, Vol. 262, pp. 17–35.

A wide-ranging survey of the reasons for London's decline, covering the loss of manufacturing and office jobs, the decrease in population, decaying housing, decreasing shopping facilities, the polarization of social classes, the role of the Greater London Council and the conflicting remedies that have been put forward.

909 MANNERS, G. (1977) The 1976 Review of the Strategic Plan for the South East: some outstanding economic issues. *Planning Outlook*, Vol. 20, pp. 2–8.

Looks at the place of the South East in the UK economy, economic growth and spatial patterns of development and the style of regional planning in Britain, in an attempt to assess critically the *Review of the Strategic Plan.*

910 SIMON, N. (1977) The relative level and changes in earnings in London and Great Britain. *Regional Studies*, Vol. 11, pp. 87–98.

Examines the question of earnings within London, their present level and relative position over time, compared with the rest of the country.

911 CARMICHAEL, M. (1978) Incomes in London. Greater London Council, 12p. (GLC Research Memorandum, No. 561)

Describes and comments on data from the *Family expenditure survey*, the *New earnings survey*, the *Inland revenue survey of personal incomes* and the *General household survey*, comparing London with the rest of the south-east region, and with the country as a whole. The main conclusion is that, in broad financial terms, London's households are no worse off (or better off) than households in the country at large.

912 LEAN, B. (1978) Employment policies for inner London. *Greater London Intelligence Journal*, Vol. 41, pp. 19–23.

Briefly discusses certain features of employment in inner London, in particular the relationship between work place and residence, and draws conclusions about the policies needed to bring employment to the inner areas.

913 SALT, J. (1978) Population and employment in London. In: CLOUT, H. (ed.) Changing London. University Tutorial Press, pp. 15–26.

Reviews the main changes in population and employment in London giving cause for concern. There are sections on problems in the inner city, social composition, overseas immigrants, with a summary conclusion.

914 BLACKBURN, J. (1979) Employment strategies: a local perspective. *Greater London Intelligence Journal*, Vol. 42, pp. 9–12.

Discusses the problems facing London's inner areas, especially Newham, in their efforts to increase employment and restore vitality.

5.3 Patterns of economic development

915 MIDDLETON, J. (1807) View of the agriculture of Middlesex. 2nd ed. Nicol and Board of Agriculture, 597p.

A survey of agricultural practice and the progress of improvements in Middlesex, including a map of land use. Climate, soils, farm buildings and the landholding system are also described.

916 SHAW, C. W. (1879) The London market gardens. The Garden, 222p.

Describes the market gardeners of the latter part of the nineteenth century in London, with an emphasis upon their practices in the culture of fruit, flowers and vegetables.

917 HAZLITT, W. C. (1892) The livery companies of the City of London: their origin, character, development and social and political importance. Sonnerschein, 692p.

A comprehensive and detailed account of the livery companies of the City of London.

918 SMITH, H. L. (1931–4) The new survey of London life and labour. Vols 2, 5, 8. London industries. P. S. King.

These three volumes form part of a wider survey of life and labour in 1928. They depict conditions and tendencies in each group of London industries, covering in total approximately four fifths of the occupied population of Greater London. Each volume has a general introduction followed by chapters on individual industries.

919 SMITH, D. H. (1933) The industries of Greater London. P. S. King, 188p.

An analysis of industrial development with special reference to the northern and western sectors of Greater London.

920 THRUPP, S. L. (1933) The grocers of London: a study in distributive trades. In: POWER, E. and POSTAN, M. M. (eds) Studies in English trade in the fifteenth century. Routledge, pp. 247–92.

Uses the records of the Grocers' Company to show that the livery companies were not necessarily limited to their appointed trade or craft. The grocers were engaged in the wool and cloth trade and general victualling as well as grocering.

921 FISHER, F. J. (1935) The development of the London food market, 1540–1640. *Economic History Review*, Vol. 5, No. 2, pp. 46–64; also in CARUS-WILSON, E. M. (ed.) (1954) Essays in economic history, Edward Arnold.

Shows that as the population grew, so did the demand for food, outstripping supply and causing a food crisis. However, the city retailers gained more and more control of the supply, revolutionizing suburban farming and bringing prosperity to the Home Counties and South Midlands.

922 SPATE, O. H. K. (1938) Geographical aspects of the industrial evolution of London till 1850. *Geographical Journal*, Vol. 92, pp. 422–32.

Discusses the distribution and locational factors of the main industries of London between the late eighteenth century and the 1851 Census.

923 THRUPP, S. L. (1948) The merchant class of medieval London, 1300–1500. Ann Arbor: University of Michigan Press, 401p.

An in-depth study of the social and economic context of medieval merchant life in London. The author shows that the merchant class was a stable unit through the period and was able to exert control over the government of the City.

924 EVERARD, S. (1949) The history of the gas light and coke company, 1812–1949. Benn, 428p.

A detailed history of the various companies and their competitive battles to supply gas to different parts of London. A useful series of maps shows the areas served and the gradual domination of central London by the gas light and coke company.

925 ESTALL, R. C. (1958) The London coal trade. *Geography*, Vol. 43, pp. 75–85.

An analysis of the areas of supply, types of coal and transport costs of coal brought to London in the early 1950s. It is shown that the proportion of London's coal coming by rail from the East Midlands was rising steadily at this time.

926 ROBERTSON, A. B. (1959) The suburban food markets of eighteenth century London. *East London Papers*, Vol. 2, pp. 21–6.

Reviews the rapid growth of population in the eighteenth century and how the resulting demand for foodstuffs put pressure on the City and its markets, which had a monopoly over a seven mile radius. However, private applications for markets were sometimes successful, especially at the periphery of the area, and mainly for retail rather than wholesale goods.

927 SMITH, R. (1961) Sea coal for London: history of the coal factors in the London market. Longman, 388p.

Based on the records of the Society of Coal Factors who dominated the distribution of coal arriving by sea from Newcastle in the eighteenth and nineteenth centuries. The transport of coal to London and the market for coal are also examined.

928 UNWIN, G. (1963) The gilds and companies of London. 4th ed. Frank Cass, 401p.

First published in 1908, this book presents a history of the continuous development of the gilds and companies from the Saxon period to the nineteenth century. Largely based on the documentary sources of the gilds, it stresses the significance the gilds and companies have had for the constitutional history of the City and for the social and economic development of the nation.

929 WHETHAM, E. H. (1964) The London milk trade, 1860–1900. *Economic History Review*, Vol. 17, pp. 369–380.

Examines the changes in the London milk trade in the 40 years before 1900. Topics covered include: the growth in demand for liquid milk from urban consumers; changes in the transport and presentation of milk; increasing public concern over milk quality.

930 VEALE, E. M. (1966) The English fur trade in the later middle ages. Oxford: Clarendon Press, 251p.

Largely based on the records of the Worshipful Company of Skinners, this book examines the medieval and Tudor trade in skins, and the structure and organization of the skinners in London.

931 WEBBER, R. (1968) The early horticulturists. Newton Abbot: David and Charles, 224p.

This study of early horticulturists includes gardeners, most of whom lived in or near London. Each represents a period or a particular branch of horticulture.

932 WHETHAM, E. H. (1970) The London milk trade 1900–1930. Reading: University of Reading, Institute of Agricultural History. 16p. (IAH Research paper, No. 3)

Summarizes the main features of the London milk market as it existed in the railway age and describes the changes which took place in the London milk trade between 1900 and 1930, up to the creation of the Milk Marketing Board.

933 SCHWARZ, L. D. (1972) Occupations and incomes in late eighteenth century east London. *East London Papers*, Vol. 14, pp. 87–100.

Uses land tax records and parish registers to establish occupations and incomes in east London. These show that the area was a poor and overwhelmingly working class area, marginal to London society and mainly providing goods and services.

934 HARVEY, J. H. (1973) The nurseries on Milne's land use map. *Transactions of the London and Middlesex Archaeological Society*, Vol. 24, pp. 177–98.

Thomas Milne's map of the London area, published in 1800, was the first to show detailed land use. This paper uses the mapped information to list and identify all the nurseries and reconstruct the nursery trade of London, 1795–1800.

935 WEBBER, R. (1973) London's market gardens. *History Today*, Vol. 23, pp. 871–8.

Outlines the history of market gardening around London from the sixteenth to the nineteenth centuries, when advancing urbanization and improved rail transport pushed the market gardening belt further afield.

936 MASTERS, B. R. (1974) The public markets of the City of London, surveyed by William Leyburn in 1677. London Topographical Society, 44p.

Reproduces Leyburn's detailed plans of the layout of Leadenhall, Honey-Lane, Stocks and Newgate Markets, along with a description of the operation and history of the markets.

937 HARVEY, J. H. (1975) Mid-Georgian nurseries of the London region. *Transactions of the London and Middlesex Archaeological Society*, Vol. 26, pp. 293–308.

This article corrects and supplements the list of nurseries given by Harvey (1973) and extends the period to 1750–1800, and widens the area under consideration.

938 SCHMIECHAN, J. A. (1975) State reform and the local economy: an aspect of industrialization in late Victorian and Edwardian London. *Economic History Review*, 2nd series, Vol. 28, pp. 413–28.

Considers the impact of the Factory and Workshop Acts on the structure and location of industry in London. Though industry was displaying decentralization tendencies due to economic change, the role of factory inspectors in closing insanitary or dangerous workplaces in the central areas promoted the outwards movement.

939 JARVIS, R. C. (1976) The early customs and custom houses in the port of London. *Transactions of the London and Middlesex Archaeological Society*, Vol. 27, pp. 271–9.

Reviews the history of the administration of port customs in London from the Roman period to the nineteenth century.

940 JONES, P. E. (1976) The butchers of London. Secker and Warburg, 246p.

Examines the history of meat markets, slaughterhouses, prices and weights, as well as the organization of the Company of Butchers.

941 ATKINS, P. J. (1977) London's intra-urban milk supply circa 1790–1914. *Transactions of the Institute of British Geographers*, New Series, Vol. 2, pp. 383–99.

Argues that before the 1860s, must of London's milk was produced within the capital in urban cowsheds. Encroaching urbanization often meant the destruction of cow pastures and the loss of leases on cowsheds. Country milk became increasingly the mainstay of the trade, though the quality of the milk was sometimes poorer than before.

942 ATKINS, P. J. (1978) The growth of London's railway milk trade c. 1845–1914. *Journal of Transport History*, New Series, Vol. 4, pp. 208–26.

Shows that the growth of London's rail transported country milk supply in the nineteenth century was a complex process. Between 1875 and 1885 a number of factors operated that spelt the demise of intra-urban supply and the dominance of railway milk.

5.4 Primary activities

943 BENNETT, L. G. (1952) The horticultural industry of Middlesex. Reading: University of Reading, 70p; (Department of Agricultural Economics, Miscellaneous Studies No. 7)

Considers the horticultural industry of rural Middlesex, how the industry was established and the forces which moulded its development and importance in the early post-war period.

944 SOUTH EAST JOINT PLANNING TEAM (1971) Minerals. In: SOUTH EAST JOINT PLANNING TEAM, Strategic plan for the South East: social and environmental aspects, Studies Volume, No. 2. HMSO, pp. 90–100.

Looks at the supply and demand position in the wide variety of mineral resources found in the south-east region – especially sand and gravel and cement associated with the construction industry.

945 STANDING CONFERENCE ON LONDON AND SOUTH EAST REGIONAL PLANNING (1974) Sand and gravel extraction. Standing Conference on London and South East Regional Planning, 73p.

Includes reports from the various gravel regions – Wessex; middle Anglia; middle and upper Thames; northern London and the Western and Maidenhead service areas.

946 BOSTON, A. (1978) City farms. *Vole*, No. 6, pp. 32–7.

Describes farms within the city at Kentish town, in the Surrey Docks at Rotherhithe and the Newham city farm.

947 UNIVERSITY COLLEGE (London) et al. (1979) Land use conflicts in the urban fringe: a case study of aggregate extraction in the London Borough of Havering. Countryside Commission, 40p. (CC Working Paper, No. 11)

Looks at land use changes in Havering, the interests concerned with land use changes and causes of the inefficient use of land in the area.

5.5 Manufacturing industries: general

948 WISE, M. J. (1956) The role of London in the industrial geography of Great Britain. *Geography*, Vol. 41, pp. 219–32.

Examines the background to London's industrial growth in the context of the role of the metropolis in the industrial geography of Great Britain. A final section looks at problems of the decentralization of industry and population and the stabilization or reduction of the size of Greater London.

949 ESTALL, R. C. and MARTIN, J. E. (1958) Industry in Greater London: a survey of trends in the new factory building and industrial employment in the London area. *Town Planning Review*, Vol. 28, pp. 261–77.

Examines trends in manufacturing activity in the London area in the period 1950–6 with special reference to new factory floor space and industrial employment changes. Regional aspects of change are considered – the growth of employment in central London and the outer ring areas with a decline in industrial employment in the inner ring districts; the engineering and vehicle industries are shown as major industries. Argues that London must retain a sound industrial structure, and industrial location controls must be selective, and modernization, and even new growth must be allowed if London itself is not to become a stagnant pool of outmoded industries.

950 HALL, P. G. (1962) The industries of London since 1861. Hutchinson, 192p.

Studies the industrial geography of Greater London in a historical way concentrating upon manufacturing industry. The core chapters cover the older industries – clothing, furniture and printing as well as the newer industries – general engineering, electrical engineering and vehicles. A final chapter assesses the effect of planning policy on London industry since 1940.

951 HALL, P. G. (1964) Industrial London: a general view. In: COPPOCK, J. T. and PRINCE, H. C. (eds) Greater London. Faber and Faber, pp. 225–45.

A general view of the evolution of manufacturing industry in London. The industries in the main manufacturing areas are examined as follows: the Victorian belt, the Riverside industrial belt, the west London industrial belt, the Lea valley belt. A final section looks at the growth since 1948 in London's outer ring.

952 MARTIN, J. E. (1964) Three elements in the industrial geography of Greater London. In: COPPOCK, J. T. and PRINCE, H. C. (eds) Greater London. Faber and Faber, pp. 246–64.

A study of three groups of industries clothing, precision instruments and electrical engineering, and waterside industries. Each provides a distinct element in the geography of London manufacturing and each has a component in inner north-east London – the detailed focus of this chapter.

953 MARTIN, J. E. (1964) The industrial geography of Greater London. In: CLAYTON, R. (ed.) The geography of Greater London: a source book for teacher and student. George Philip, pp. 111–42.

Examines the overall size, structure, evolution and geographical distribution of manufacturing industries in Greater London. Presents detailed studies of two industrial areas: inner north-east London and outer north-west London.

954 BROWN, C. M. (1966) The structure of manufacturing industry in London's new towns. *Tijdschrift voor Economische en Sociale Geografie*, Vol. 57, pp. 121–4.

Examines the industrial structure of the eight new towns around London. The structure is dominated by manufacturing firms located in one or two industrial estates. It is generally clean and light, with general engineering, electrical goods and vehicles prominent among the new firms; there are many branch factories.

955 MARTIN, J. E. (1966) Greater London: an industrial geography. G. Bell, 292p.

Traces and analyses the location of manufacturing industry in Greater London. A survey of nineteenth century and twentieth century developments is followed by detailed studies of the engineering, chemicals, food, clothing and other industries. Recent trends in the London region, especially the new towns, are examined with a summary chapter on the metropolitan industrial city.

956 MANNERS, G. (1970) Greater London development plan: location policy for manufacturing industry. *Area*, Vol. 3, pp. 54–6.

Examines the forecast of the future size and structure of manufacturing employment contained in the *Greater London Development Plan* (1969).

957 SOUTH EAST JOINT PLANNING TEAM (1971) Strategic plan for the South East, Studies Volume No. 5; report of Economic Consultants Ltd. HMSO, 102p.

A detailed examination of manufacturing industry in the South East, with particular reference to locational factors affecting efficiency and profitability and special attention to industrial movement. Small firms (less than 100 employees) were excluded and emphasis placed upon firms in the mechanical and electrical engineering industries. Concludes that the requirements of manufacturing industry in general do not dictate or indeed act as a major constraint on the future strategy of development in the South East.

958 WOOD, P. A. (1974) Urban manufacturing: a view from the fringe. In: JOHNSON, J. H. (ed.) Suburban growth; geographic processes at the edge of the western city. John Wiley, pp. 129–54.

A critical view of much of the systematic literature concerned with urban manufacturing location. Suggests a focus on manufacturing growth processes and illustrates this with empirical evidence from studies of south-east England.

959 WEATHERITT, L. and LOVETT, A. F. (1975) Manufacturing industry in Greater London. Greater London Council, 26p. (GLC Research Memorandum, No. 498)

Analyses some of the reasons for the decline in London's manufacturing industry. It suggests certain factors that increase costs are responsible and describes a planned research programme which may contribute positively to the formulation of policies for manufacturing industry.

960 SIEVE, Y. (1976) The GLC attitude to manufacturing industry. Regional Studies Association, 10p.

A paper on the economic crisis and local manufacturing employment. Discusses the scale, effect and causes of industrial decline in London and the policies being implemented by the GLC to counter it.

961 GREATER LONDON COUNCIL, Freight Unit (1977) Factors affecting the location of industry: report of survey of 88 companies. Greater London Council, 68p.

Describes the first phase of the unit's work in establishing how transport, and in particular freight policy, can best assist industry and attract industry to London. It is based upon a series of case studies of manufacturing companies from which relocation factors have been identified and their relative importance assessed.

962 STONE, P. A. (1977) Policies for manufacturing industry in London. *Greater London Intelligence Journal*, Vol. 37, pp. 9–21.

Discusses London's economy and, in particular, the decline of manufacturing industry and the consequent reduction in job opportunities. It argues that London is in danger of suffering a self-perpetuating decline in its economy. The article considers the nature of conditions leading to industrial decline, the policies to be pursued to make London attractive to industrial firms and the action needed from each level of authority.

963 DENNIS, R. (1978) The decline of manufacturing industry in Greater London, 1966–74. *Urban Studies*, Vol. 15, pp. 63–75.

Uses data on closures and transfers to resolve the decline of manufacturing employment in Greater London into its components. The largest component is the net decline due to the difference between closures and openings (44 per cent total job loss), movement (27 per cent) and in situ shrinkage (22 per cent); inner London suffers more from closures, outer London from movement. The major cause of movement from London was restrictions upon expansion caused by site congestion, obsolescent premises and labour shortages.

964 HAYDEN, F. W. (1978) Factors influencing the location of manufacturing. Greater London Council, 75p. (GLC Research Memorandum No. 528)

Three surveys undertaken with the co-operation of the London Chamber of Commerce in 1976–7, serve as pilot studies in identifying factors which influence the location of industry in London.

965 DENNIS, R. (1979) London's industrial decline: causes and prospects. Regional Studies Association, 10p.

Reports the findings of the Department of Industry research projects into the causes of London's decline in manufacturing industry and tentatively considers future prospects. Concludes that the scale of the decline is so great that the only realistic assumption must be that industrial decline is likely to continue.

966 LONDON INDUSTRY AND EMPLOYMENT RESEARCH GROUP (1979) Economic policies and powers in London Boroughs. Middlesex Polytechnic, 60p. (MP Occasional Paper No. 1)

Based upon interviews with council officers, the paper describes the policies pursued by 11 London boroughs and the powers available for the pursuit of their economic aims. There are sections on the local economic context and planning framework, the various approaches to state intervention and the policy approaches evident in London under the powers held by the boroughs.

5.6 Manufacturing industries: specific

967 POLLARD, S. (1950) The decline of shipbuilding on the Thames. *Economic History Review*, 2nd Series, Vol. 3, pp. 72–89.

An analysis of the decline of shipbuilding on the river Thames from the 1840s to the launching of the last large ship in 1912.

968 CRACKNELL, B. E. (1952) The petroleum industry of the lower Thames at Medway. *Geography*, Vol. 37, pp. 79–88.

Traces the growth of the petroleum industry in the Thames-Medway area from its early beginning in 1875 to the rapid growth of oil refineries in the early 1950s.

969 HALL, P. G. (1960) The location of the clothing trades in London. *Transactions of the Institute of British Geographers*, Vol. 28, pp. 155–78.

Describes and analyses the geographical distribution of the clothing trades in London. It is argued that similar principles govern the location of other old established London trades.

970 OLIVER, J. L. (1961) The east London furniture industry. *East London Papers*, Vol. 64, pp. 88–101.

A detailed analysis of the spread and distribution of furniture makers in Bethnal Green, Shoreditch and Hackney in the nineteenth century. The industry was characterized by a large number of small, specialized, though interdependent, manufacturers usually living and working in the same premises.

971 HALL, P. G. (1962) The east London footwear industry: an industrial quarter in decline. *East London Papers*, Vol. 5, pp. 3–21.

Analyses the distribution of footwear makers in 1800, 1901 and 1951, chiefly in the areas of Hackney, Stepney and Bethnal Green. The emphasis was on poorer quality footwear, often made in sweatshops, but competition from provincial factories contributed to a progressive decline after the industry reached its peak in the last quarter of the nineteenth century.

972 OLIVER, J. L. (1964) The location of furniture manufacture in England and elsewhere. *Tijdschrift voor Economische en Sociale Geografie*, Vol. 55, pp. 49–53.

Examines the geographical distribution of furniture manufacture in Great Britain. In 1958, over 61 per cent of manufacture took place in two regions – London and South East and Southern; Greater London was the leading area of production (mainly High Wycombe). Special attention is given to the evolution and recent changes in the locational patterns within these regions.

973 MARTIN, J. E. (1969) Size of plant and location of industry in Greater London. *Tijdschrift voor Economische en Sociale Geografie*, Vol. 60, pp. 369–74.

Examines the location of different sizes of plant within Greater London, using detailed 1954 information.

974 GREATER LONDON COUNCIL (1974) Power generation in Greater London: strategic issues. Greater London Council, 17p.

Reviews the background to GLC discussions with the Central Electricity Generating Board, on London's future electricity needs and gives estimates of needs and possible supply sources. The environmental effects of power station development are considered.

975 HOARE, A. G. (1974) International airports as growth poles: a case study of Heathrow Airport. *Transactions of the Institute of British Geographers*, Vol. 63, pp. 75–96.

Investigates certain aspects of the possible growth-pole effects of Heathrow Airport on surrounding opportunities for employment. Field surveys of manufacturing and office firms attempt primarily to determine the nature and strength of indirect linkages associated with the airport. An attempt is made to determine the role of the airport in the regional and sub-regional geography of manufacturing growth.

976 SHAH, S. (1975) Immigrants and employment in the clothing industry: the rag trade in London's East End. Runnymede Trust, 42p.

Describes and explains the factors affecting homeworking in the clothing industry. Homeworking is examined in detail, especially wages, hours and working conditions.

977 WHITE, D. (1976) Underneath the arches. *New Society*, Vol. 36, pp. 221–2.

Looks at the small businesses and workshops attracted to the 6,500 railway arches in London.

978 McDERMOTT, P. J. (1978) Changing manufacturing enterprise in the metropolitan environment: the case of electronics firms in London. *Regional Studies*, Vol. 12, pp. 541–50.

Examines variations in the recent growth records of electronics manufacturing firms in inner London, outer London and the outer metropolitan area. The very poor performance of inner London firms may be attributable in part to the spatial sorting of enterprise, whereby the movement of growing organization from the city centre has reduced the growth prospects of remaining establishments. Argues that differences in the process of firm creation, growth and decline may account for some of the contrasts in manufacturing activity within the metropolitan region.

979 STANDING CONFERENCE ON LONDON AND SOUTH EAST REGIONAL PLANNING (1979) Electricity supply in south east England: a review. Standing Conference on London and South East Regional Planning, 30p.

A report from a joint working party (SCLSERP and CEGB) presenting the prospect over the next ten years of demand for electricity in south-east England and of the power stations that may be needed.

5.7 Industrial areas

980 ALLEN, G. R. (1951) The growth of industry on trading estates, 1920–39 with special reference to Slough Trading Estate. *Oxford Economic Papers*, Vol. 3, pp. 272–300.

Examines the economic aspects of trading estates. This paper deals with the growth and location of new manufacturing enterprise on trading estates and the rates of failure of firms in the process of this development with a special detailed study of the growth of industry at Slough Trading Estate.

981 MUNBY, D. L. (1951) Industry and planning in Stepney: a report presented to the Stepney Reconstruction Group. Oxford University Press, 466p.

A detailed account of Stepney's industries and the implications of post-war reconstruction in the light of the *County of London plan (1943).* Many industrial units were considered suitable for decentralization, except the clothing trades (dependent on outworkers); housing densities needed to be reduced and transport facilities improved.

982 BIRD, J. (1952) The industrial development of lower Thameside. *Geography*, Vol. 37, pp. 89–97.

Describes and analyses the twentieth century development of industries along the river Thames from Woolwich to Shellhaven.

983 GRIFFITH, E. J. L. (1955) Moving industry from London. *Town Planning Review*, Vol. 26, pp. 51–63.

A study of the progress in the decentralization of industry from London to out-county estates, new towns and expanded towns. Analyses the character of 145 firms moving up to the end of 1953, the factors persuading a firm to move and the problems for the planning authority of identifying mobile firms and controlling vacated premises.

984 MARTIN, J. E. (1957) Industry in inner London. *Town and Country Planning*, Vol. 25, pp. 125–8.

A report of a study of the localization of particular types of manufacturing industry in certain quarters of the north-eastern parts of inner London (clothing, printing, precision engineering and furniture).

985 HOOSON, D. M. J. (1958) The recent growth of population and industry in Hertfordshire. *Transactions of the Institute of British Geographers*, Vol. 25, pp. 197–208.

A case study of geographical changes in the population of Hertfordshire during the twentieth century. It includes sections on the distribution of employment in 1951 and the development of garden cities, dormitories and new towns.

986 LONSDALE, G. (1962) The changing character of east London industry. *East London Papers*, Vol. 5, pp. 91–102.

Analyses, with maps, the distribution of the clothing and wood and timber industries in Stepney, Bethnal Green and Poplar in 1938 and 1958. The area as a whole experienced great social and economic changes during the period, and the workshop industries are shown to have contracted in number and become more concentrated in space. It is suggested that the industry will eventually become concentrated into factories as the functional re-zoning of the area progresses.

987 DUNNING, J. H. (1963) Economic planning and town expansion. Basingstoke: Workers' Educational Association, 168p.

A case study in the economic planning of Basingstoke – an expanded town. The book attempts to apply economic principles and statistical techniques to problems associated with urban growth. The core chapters are concerned with population, industry and employment and the service trades.

988 KEEBLE, D. E. (1965) Industrial migration from north west London, 1940–64. *Urban Studies*, Vol. 2, pp. 15–32.

Describes the movement of industry out of north-west London, 1940–64, creating some 33,000 jobs in the provincial zone (outside the London metropolitan area). More firms have, however, moved from north-west London to new locations within the metropolitan zone, chiefly in the north-west sector, encouraged by radial transport routes.

989 BROWN, C. M. (1966) The industry of new towns of the London region. In: MARTIN, J. E. Greater London: an industrial geography. G. Bell, pp. 238–52.

Analyses the growth of manufacturing industry in London's new towns, with special emphasis upon the industrial structure and the factors affecting firms locating in the eight towns.

990 BROWN, C. M. (1966) Successful features in the planning of new town industrial estates. *Journal of the Town Planning Institute*, Vol. 52, pp. 15–8.

Describes the planning of new town industrial estates. Four criteria for the further successful development of industrial estates in new towns are suggested as follows: more than one estate; easy access to town centre and neighbourhoods; one estate at a distance to accommodate unpleasant and heavy industries; provision for small estates of light industry within the neighbourhoods.

991 EVANS, A. W. (1967) Myths about employment in central London. *Journal of Transport Economics and Policy*, Vol. 1, pp. 214–25.

The *Administrative County of London development plan: first review* (LONDON COUNTY COUNCIL, 1960) and the *London traffic survey* (FREEMAN FOX AND PARTNERS, 1964, 1966) have forecast a large and continuing increase in employment in central London. This article marshals evidence to show that employment in central London has passed its peak and is likely to decline in future.

992 KEEBLE, D. E. (1968) Industrial decentralisation and the metropolis: the north west London case. *Transactions of the Institute of British Geographers*, Vol. 44, pp. 1–54.

Post-war movement of industry from north-west London is evaluated by means of government statistics and questionnaire surveys. It is demonstrated that the majority of migrant firms are those which are expanding and are forced to move as a result of various restrictions on growth in London.

993 KEEBLE, D. E. (1968) Airport location, exporting and industrial growth. *Town and Country Planning*, Vol. 36, pp. 209–14.

The author's work on industrial growth in north-west London suggests that the siting of a major international airport could generate pressure by many manufacturing concerns for locations in a 20 to 30 mile belt around the airport. In the case of Stansted, little attention has been given to the implications of this kind of pressure for regional planning in the South East.

994 DUNNING, J. H. (1969) The City of London: a case study in urban economics. *Town Planning Review*, Vol. 40, pp. 207–32.

Presents some of the findings of an economic study of the City of London commissioned by the Corporation of London. Focuses on the spatial structure of activities in the Central Business District (CBD), the nature and extent of interdependence between these activities and methods of forecasting economic growth in the CBD.

995 KEEBLE, D. E. (1969) Local industrial linkage and manufacturing growth in outer London. *Town Planning Review*, Vol. 40, pp. 63–88.

A detailed survey of north-west London manufacturing firms in 1963 showed that local industrial linkage had not been dominant in the area's industrial organization and growth – unlike the inner London position. Regional and national scale linkage was of great importance. Larger firms, important in migration, showed a tendency to a radial, sectoral movement pattern to the north-west of London.

996 TULPULE, A. H. (1969) Dispersion of industrial employment in the Greater London area. *Regional Studies*, Vol. 3, pp. 25–40.

Census data are used in an examination of the differing growth rates of industries in different zones of London.

997 HOARE, A. G. (1971) Heathrow airport: a spatial study of its economic impact. British Airports Authority, 14p.

Reports a study into the impact of a major airport upon its surrounding economy, specifically upon businesses concerned with supplying services for airport users, and firms that make use of the airport as part of their normal commercial activities. The paper also discusses the effect on the economy of aircraft noise and of competition for labour.

998 KEEBLE, D. E. and HAUSER, D. P. (1971) Spatial analysis of manufacturing growth in outer south east England, 1960–1967, I, Hypotheses and variables. *Regional Studies*, Vol. 5, pp. 229–62.

Discusses the hypotheses and variables used in multiple regression analyses of the inter-urban spatial pattern of manufacturing change in south-east England outside Greater London – the most rapidly growing major industrial region of Britain.

999 KEEBLE, D. E. and HAUSER, D. P. (1972) Spatial analysis of manufacturing growth in outer south east England 1960–1967, II, Method and results. *Regional Studies*, Vol. 6, pp. 11–36.

Reports the main findings of multiple regression analyses of the inter-urban spatial pattern of recent manufacturing growth in south-east England outside Greater London. Technical problems are discussed and special analyses presented for the new and expanded towns.

1000 KEEBLE, D. E. (1972) Modelling industrial movement: the south east England case. In: ADAMS, W. P. and HELLEINER, F. M. (eds) International Geography, Vol. 1. Toronto: University of Toronto Press, pp. 609–11.

Suggests the value of a modelling approach to the study of the movement of manufacturing firms at the intra-regional scale. Reports on tests of four hypotheses related to movement and concludes that gravity models and associated multiple regression procedures provide a useful hypotheses testing framework. In the south-east England case, labour availability and distance variables are of key importance in influencing the geography of recent manufacturing movement.

1001 HOARE, A. G. (1973) The detrimental economic effects of a major airport:

the consequences of Heathrow upon factory and office firms. *Tijdschrift voor Economische en Sociale Geografie*, Vol. 64, pp. 339–51.

A case study of the negative effects of Heathrow Airport upon local factories and offices. The incidence of problems (especially labour competition and noise interference) emanating from the airport are examined in detail and attempts made to assess their importance for local economic growth and through feedback on the airport itself. Evidence suggests that the initial negative effects may be appreciable locally, but their feedback consequences for the airport are insignificant.

1002 CENTRAL LONDON PLANNING CONFERENCE (1974) Economic activity. Central London Planning Conference, 65p. (Advisory Plan for central London Topic Paper 1)

An examination of the existing situation and main trends in four sectors: offices, industry, tourism and shopping. The goals and objectives for the Central Area economy are considered and the main issues and possible action choices are discussed.

1003 WRAY, M. et al. (1974) Location of industry in Hertfordshire: planning and industry in the post-war period. Hatfield Polytechnic, 312p.

Based upon a survey of industrial firms in the early 1970s, this report examines the impact of local authority planning on industrial development in Hertfordshire in the post-war period.

1004 CANNING TOWN COMMUNITY DEVELOPMENT PROJECT (1975) Canning Town to North Woolwich: the aims of industry? Canning Town Community Development Project, 75p.

Examines the decline of industry in Canning Town with proposals for policy and local change.

1005 DOCKLANDS JOINT COMMITTEE (1975) Work and industry in East London. Docklands Joint Committee, 44p. (DJC Working Paper for Consultation, No. 2)

This paper produced by the Docklands Development Team, and approved by the committee, examines the changes in the types of goods and services being produced in east London, the changes in the resident labour force, the local, national and regional policies influencing economic conditions in east London and possible future trends. The committee's proposals for improving the economic situation in the docklands includes the establishment of three or four large industrial areas as centres of employment serving the whole of docklands, and positive action to attract firms to the area.

1006 JOINT DOCKLANDS ACTION GROUP (1975) Tower Hamlets and the fight for jobs. Joint Docklands Action Group, 36p.

Bethnal Green and Stepney Trades Council and others, look at the basic problem underlying the fundamental reasons for the industrial decline in Tower Hamlets, taking some case studies, such as brewing, the docks and the clothing industry to illustrate the theme. A final section stresses the need for action to halt the process of decline.

1007 HOARE, A. G. (1975) The sphere of influence of industrial location factors: a case study of the use of Heathrow Airport. *Geoforum*, Vol. 6, pp. 219–30.

Examines the concept of sphere of influence with a case study of Heathrow based upon the use made of the airport for commercial purposes. In practice a number of different use types can be distinguished for which the related spheres of influence also vary spatially and in intensity of industrial involvement.

1008 HOARE, A. G. (1975) Foreign firms and air transport: the geographical effect of Heathrow Airport. *Regional Studies*, Vol. 9, pp. 349–68.

Tests the idea that the geography of foreign investment in the UK is partly related to the desire of overseas companies to be accessible to international airports. General support is found for the idea, at the scale of the UK as a whole and of south-east England. The need to be within a certain time distance of Heathrow created a noticeable clustering of foreign firms within the South East.

1009 HOARE, A. G. (1975) Linkage flows, locational evaluation and industrial geography: a case study of Greater London. *Environment and Planning*, Vol. 7, pp. 41–58.

Examines a series of hypotheses set up and tested for Greater London concerning the interrelationships between linkage flows, spatial patterns and locational evaluation.

1010 MARTIN, J. E. and SEAMAN, J. M. (1975) The fate of the London factory: 20 years of change. *Town and Country Planning*, Vol. 43, pp. 492–5.

Examines the contemporary use of redundant factories based on a 1974 field study of factories, with 200 plus workers in 1954 in Ealing and Southwark, contrasting London boroughs. Concludes that the factories of inner London are ill adapted for modern manufacturing without conversion. The provision of modern, more functionally efficient space, at economic rents might attract new industry.

1011 SOCIAL AUDIT (1975) Newham – prosperity or decline. Social Audit, 18p.

The causes and effects of industrial decline in Newham are outlined. New jobs are necessary to revitalize the borough.

1012 WHITE, D. (1975) Newham – an example of urban decline. *New Society*, Vol. 34, pp. 201–4.

Illustrates the decline of this dockland London borough in terms of the fall in population, demise of traditional industry, emigration of the middle class and their replacement by poorer groups (notably immigrants), and poor educational standards. Community spirit is broken up although there is some evidence of grass roots militancy.

1013 CANNING TOWN COMMUNITY DEVELOPMENT PROJECT (1976) Canning Town's declining community income: case study, Tate and Lyle. Canning Town Community Development Project, 33p.

Calculates the potential effect of a single major industrial closure on a low income community. Tate and Lyle employs 3,000 and controls 10 per cent of local industrial jobs in an area where the community's collective income has declined by 10 per cent in the last ten years. Cutting jobs to improve the company's profitability would, therefore, involve a huge social cost and increased public spending to alleviate the problems.

1014 ROBERTS, J. C. (1976) Employment in Southwark: a strategy for the future. Southwark Trades Council, 80p.

Southwark is an area of economic decline which is suffering from structural unemployment as manufacturing industry contracts and office-based employment increases. This report recommends revisions in the Industrial Development Certificate system, increased intervention by local authorities to arrest the decline of manufacturing industry, the meeting of the full social costs of relocation by the firms which choose to move, and an increased use of water transport services to increase jobs and minimize the anti-social aspects of road traffic.

1015 GRIPAIOS, P. (1977) Industrial decline in London: an examination of its causes. *Urban Studies*, Vol. 14, pp. 181–9.

Examines the reasons for industrial decline in London by the analysis of data relating to south-east London (a quadrant of inner London). Concludes that locational influences, the decline of the docks, merger and rationalization and urban and regional policy, have probably all contributed significantly to industrial decline in inner London. The accelerated rate of decline of the last decade has probably reinforced the locational disadvantages of the area and increased the long term impact of past decentralization on firms that remain.

1016 GRIPAIOS, P. (1977) The closure of firms in the inner city: the south east London case 1970–75. *Regional Studies*, Vol. 11, pp. 1–7.

Within the context of a significant fall in employment in inner London and a reduction in the stock of industry, this paper examines 359 establishment closures in south-east parts of inner London since 1970.

1017 JOINT DOCKLANDS ACTION GROUP (1977) Industrial estates in docklands: what jobs for the future? Joint Docklands Action Group, 20p.

Presents findings of a survey of seven industrial estates in the dockland boroughs and questions whether a strategy based only on the promotion of industrial estates is sufficient for the economic revival of the area.

1018 HOWICK, C. and KEY, T. (1978) The local economy of Tower Hamlets: an inner city profile. Centre for Environmental Studies, 88p.

A preliminary examination of the economic problems facing Tower Hamlets. It assembles specific information on the borough's economy placing it in the context of general theories about the economic problems of inner cities.

1019 WOOD, P. (1978) Industrial changes in inner London. In: CLOUT, H. (ed.) Changing London. University Tutorial Press, pp. 38–48.

Examines the character of inner London industry and the reasons for its decline, with an analysis of some of the consequences of manufacturing decline in inner London.

1020 ISLINGTON ECONOMY GROUP (1979) Islington's multinationals and small firms: magic or myth? Islington Economy Group, 30p.

A study of 50 industrial firms closed between 1971 and 1976. Focuses upon the activities of the multinational companies in closing down subsidiaries in the London Borough of Islington.

5.8 Office activities

1021 JOHN, A. H. (1953) Insurance investment and the London money market of the eighteenth century. *Economica*, Vol. 20, pp. 137–58.

Examines the growth of London as an international monetary centre in the eighteenth century after the decade 1720–30, with special reference to insurance offices and friendly societies.

1022 AUCOTT, J. V. (1960) Dispersal of offices from London. *Town Planning Review*, Vol. 31, pp. 37–52.

A general review of the problem of office dispersal from London.

1023 MORGAN, W. T. W. (1961) The two office districts of central London. *Journal of the Town Planning Institute*, Vol. 47, pp. 161–6.

Describes the financial district in the City and the general office district of the West End and distinguishes between their functions.

1024 MORGAN, W. T. W. (1961) A functional approach to the study of office distributions: internal structures in London's central business district. *Tijdschrift voor Economische en Sociale Geografie*, Vol. 52, pp. 207–10.

The degree of concentration of different categories of offices in London's central business district is discussed and mapped. Many offices are shown to be highly specific in their location, according to the nature of their business.

1025 MORGAN, W. T. W. (1961) Office regions in the West End of London. *Town and Country Planning*, Vol. 29, pp. 257–9.

Based upon a door-to-door survey of offices in the West End of London, undertaken in 1954–5. The author divides the office concentrations into 11 office regions and describes the characteristics of each region.

1026 BORER, M. C. (1962) The City of London: its history, institutions and commercial activities. London Museum Press, 125p.

Describes the port, food markets, the commodity markets, the Baltic Exchange, Lloyds, the Stock Exchange, the Banks, the Foreign Exchange market, insurance and hire purchase houses.

1027 MORGAN, W. T. W. (1962) The geographical concentration of big business in Great Britain. *Town and Country Planning*, Vol. 30, pp. 122–4.

A study of the location of offices with specific references to concentrations in London.

1028 TOWN AND COUNTRY PLANNING ASSOCIATION (1962) The paper metropolis: a study of London's office growth. Town and Country Planning Association, 87p.

Provides data about the concentration of offices in the City, examines the social and economic consequences of the two-way movement of London's working population and makes recommendations for solving the problems identified.

1029 ECONOMIST INTELLIGENCE UNIT (1964) A survey of factors governing the location of offices in the London area. Economist Intelligence Unit, 226p.

A study commissioned by the Location of Offices Bureau to research into factors governing the location of offices in the Greater London area (up to 30 miles from Charing Cross). Excludes central government agencies.

1030 STANDING CONFERENCE ON LONDON AND SOUTH EAST REGIONAL PLANNING (1964) Office employment in the conference area: report by technical panel. Standing Conference on London and South East Regional Planning, 8p. (LRP, No. 279)

Looks generally at the decentralization of office employment from London and the establishment of new office centres outside. The significance of estimates of employment growth in the conference area are examined and a final section suggests that new measures to restrict office developments in the Greater London conurbation are required.

1031 DEPARTMENT OF ECONOMIC AFFAIRS (1965) Office decentralisation: an empirical study. Department of Economic Affairs, 20p.

Report of a survey of Location of Offices Bureau records to investigate factors affecting the location of offices and their decentralization from London.

1032 WILLIAMS, K. B. (1965) The factual basis of office policy in south east England. Department of Economic Affairs, 71p.

Reviews office location policy since 1948 and describes the office problem as a cause of congestion in London and imbalance in the regions. A policy for office growth is outlined.

1033 LOCATION OF OFFICES BUREAU (1966) Commuters to the London office: a survey. Location of Offices Bureau, 16p.

A sample of commuters have been questioned regarding their method of travel, reasons for working in central London, and whether or not they would be prepared to leave the south-east if suitable jobs were available.

1034 WABE, J. S. (1966) Office decentralisation: an empirical study. *Urban Studies*, Vol. 3, pp. 35–55.

Using some of the records of the Location of Offices Bureau, this article examines the behaviour of 207 different sized firms, exploring the possibility of decentralizing their offices from London.

1035 SALISBURY, T. P. (1962) A head office moves to Crawley. *Town and Country Planning*, Vol. 30, pp. 48–50.

The director of a multiple leather goods retail firm unable to find satisfactory accommodation for expansion in north London describes the move of his head office to Crawley new town, and the advantages and disadvantages of the move.

1036 GODDARD, J. B. (1967) Changing office location patterns within central London. *Urban Studies*, Vol. 4, pp. 276–85.

Using Kelly's Directories, the author has examined the location of various office activities in London during the twentieth century. He concludes that these firms tend to be less tied to particular parts of the centre than formerly, and prefer locations within 11 square miles of the central area, with a good working environment and a prestige address.

1037 HAMMOND, E. (1967) Dispersal of government offices: a survey. *Urban Studies*, Vol. 4, pp. 258–75.

Provides details of the various offices involved in the programme of dispersal of government jobs from central London and assesses the value of the programme in relation to the problems of London and regional policies.

1038 HARTLEY, R. (1967) No mean city: a guide to the economic City of London. Queen Anne Press, 110p.

An attempt to show how the City works – from the Port of London to the Commodity Markets.

1039 INTERSCAN LTD. (1967) The non-movers. Interscan, 58p.

A report for the Location of Offices Bureau on those companies who consulted LOB, but had not subsequently moved out of London. Attempts to assess the factors influencing the decision.

1040 LIND, H. G. (1967) Location by guesswork. *Journal of Transport Economics and Policy*, Vol. 1, pp. 154–63.

The government policy of encouraging the movement of offices from London and the south-east to the north and west is criticized on the grounds that insufficient is known about the economic advantages or disadvantages of such movement.

1041 LOCATION OF OFFICES BUREAU (1967) White collar commuters: a second survey. Location of Offices Bureau, 79p. (LOB Research paper, No. 1)

Report of a survey of office workers living outside but working in central London, including information on mode of travel to work, attitudes to work and attitudes to the possibility of moving out of the area.

1042 WABE, J. S. (1967) Dispersal of employment and the journey to work. *Journal of Transport Economics and Policy*, Vol. 1, pp. 345–61.

A study of one firm, with 600 employees, moving from Victoria to Epsom.

1043 WRIGHT, M. (1967) Provincial office development. *Urban Studies*, Vol. 4, pp. 218–57.

A comparison between the provinces and London, of location factors, with costings and policy recommendations.

1044 BRITISH MARKET RESEARCH BUREAU LTD. (1968) Communications and the relocation of offices: report on a depth investigation among manufacturing companies prepared for the Location of Offices Bureau. British Market Research Bureau, 46p.

A report on communication problems created by decentralization.

1045 GODDARD, J. B. (1968) Multivariate analysis of office location patterns in the city centre. *Regional Studies*, Vol. 2, pp. 69–85.

Considers the problem of defining the linkages binding firms together in the City of London and suggests that the linkages contribute a definable activity system.

1046 HAMMOND, E. (1968) London to Durham: a study of the transfer of the Post Office Savings Certificate Division. University of Durham, Rowntree Research Unit, 161p.

Discusses a transfer carried out between 1963 and 1969 with chapters on the administrative aspects of the move, the attitudes of the workers to the move and, a broad appraisal of the move.

1047 CAREY, S. J. (1969) Relocation of office staff: a study of the reactions of office staff decentralised to Ashford. Location of Offices Bureau, 64p. (LOB Research Paper, No. 4)

Summarizes results of a survey of a firm decentralizing in 1967 and covers social breakdown, housing changes, cost of living, journeys to work, social amenities and job attitudes.

1048 COWAN, P. et al. (1969) The office: a facet of urban growth. Heinemann, 280p.

Looks at many aspects of the growth and evolution of office accommodation and activities in post-war London – the occupiers, the provision of office space, the legislative background to controls and the measure of growth. Finally, an attempt is made to use mathematical models to describe the patterns of growth and change.

1049 DANIELS, P. W. (1969) Office decentralization from London: policy and practice. *Regional Studies*, Vol. 3, pp. 171–8.

Looks at office decentralization from central London, with particular reference to the national distribution of decentralized offices. A critical analysis of policy and control is related to south-east England.

1050 MARRIOTT, O. (1969) The property boom. Pan Books, 325p.

Considers the evolution of the new great estates and the activities of the property developers in London's post-war property market.

1051 REES, G. and WISEMAN, J. (1969) London's commodity markets. *Lloyds Bank Review*, Vol. 91, pp. 22–46.

A brief description of the physical evolution and location of the market in specific types of products is given, followed by a discussion of the volume of business, public and private sales and problems of the enterprises involved.

1052 DENMAN, D. R. (1970) Frustrated development. *National Westminster Bank Quarterly Review*, November, pp. 45–54.

Attacks the Control of Office and Industrial Development Act, as failing to shift demand for offices from London and the South East to other parts of Britain. Favours a free market rather than restrictive controls.

1053 GODDARD, J. B. (1970) Functional regions within the city centre: a study by factor analysis of taxi flows in central London. *Transactions of the Institute of British Geographers*, No. 49, pp. 161–82.

A study of the problem of measuring the relationship between movement patterns and the location of activities in central London. This paper analyses the complex linkage systems in the city centre, using data on taxi flows as an indicator and suggests that the central area of London contains a number of remarkably self-contained functional regions with strong internal bonds.

1054 HALL, R. (1970) The vacated offices controversy. *Journal of the Town Planning Institute*, Vol. 58, pp. 298–300.

A report of the results of a survey undertaken to discover what happened to premises vacated by firms moving out of London. A number of recommendations are made.

1055 BATEMAN, M. et al. (1971) Office staff on the move. Location of Offices Bureau, 102p. (LOB Research Paper No. 6)

Examines questions concerned with moving existing office staff to new locations with substantial case studies of company moves to Ipswich and Portsmouth.

1056 BATEMAN, M. and BURTENSHAW, D. (1971) Sponsored white collar migration. *Town and Country Planning*, Vol. 39, pp. 554–8.

A study of an international company which moved its offices from London to Portsmouth.

1057 CHILD, P. (1971) Office development in Croydon: a description and statistical analysis. Location of Offices Bureau, 118p. (LOB Research Paper, No. 5)

Studies the growth of Croydon as a suburban office centre, investigates features of the office market in Croydon and attempts to establish the terms on which Croydon competes with other locations in the London labour market.

1058 COWAN, P. (1971) Employment and offices. In: HILLMAN, J. (ed.) Planning for London. Harmondsworth: Penguin Books, pp. 64–76.

This book is concerned with aspects of life and planning in London. The chapter concentrates upon the growth and location of office employment in the post-war years.

1059 DIAMOND, D. (1971) The location of offices. *Town and Country Planning*, Vol. 39, pp. 106–9.

Examines the direct effect of raising the exemption limit for office development permits in Greater London, and the difficulties in deciding how much office space is appropriate in any given location at any given time, and the lack of information about the office activity system.

1060 DUNNING, J. H. and MORGAN, E. V. (eds) (1971) An economic study of the City of London. George Allen and Unwin, 460p.

A study commissioned by the Common Council of the City of London to serve as a basis for a development plan. Describes and analyses the structure and functioning of the City's economic activity, examines the main operating factors and looks in detail at the prospective development in individual fields (finance, insurance, trade, manufacturing and so on). A final chapter summarizes the findings of the study.

1061 GODDARD, J. B. (1971) Office linkages in central London. South East Economic Planning Council, 77p.

A report commissioned by the Department of Economic Affairs on behalf of the South East Economic Planning Council and covering office location and employment.

1062 GODDARD, J. B. (1971) Office communications and office location: a review of current research. *Regional Studies*, Vol. 5, pp. 263–80.

Looks at the growth of office type activities, the role of communication in office location and the problem of assessing the likely impact of future telecommunication systems on location. Examines research into telecommunication systems with specific reference to person-to-person contact, communication patterns within and between existing organizations and decisions on office location. The bulk of the examples are taken from the London area.

1063 RHODES, J. R. and KAN, A. (1971) Office dispersal and regional policy. Cambridge: University Press, 132p. (University of Cambridge, Department of Applied Economics, Occasional Paper No. 30)

Examines the regional distribution of non-manufacturing activities in the economy, and government policies designed to influence that distribution. There are three parts: (1) the historical background to office development and movement (especially London and South East Region); (2) the results of an empirical survey to explain the patterns of office movement; (3) the effects of government policies in recent years to influence the location of office and other non-manufacturing activities.

1064 STANDING CONFERENCE ON LONDON AND SOUTH EAST REGIONAL PLANNING (1971) Office development in Greater London and the South East: technical panel report. Standing Conference on London and South East Regional Planning, 10p. (LRP No. 852)

Deals with office employment and office accommodation in the region, together with the evolution of objectives, policies and proposals relating to office development. Suggests setting down of objectives, context and criteria for office development and the improvement of information on office activities.

1065 BAKER, L. L. H. and GODDARD, J. B. (1972) Inter-sectoral contact flows and office location in central London. In: WILSON, A. G. (ed.) Patterns and processes in urban and regional systems. Pion, pp. 243–80.

The concept of linkage is a possible criterion for assessing which types of office activity most need to remain in central locations within cities. Using details of the telephone and meeting contacts of a sample of business executives from a selection of commercial offices in central London, groups of strongly connected office types are identified using factor analysis.

1066 ECONOMISTS ADVISORY GROUP (1972) Office rents in the City of London and their effect on invisible earnings: a study carried out for the committee on invisible exports. Committee on Invisible Exports, 46p.

Recommends the removal of office development permits and a rise in GLC office floor space targets for central London to counteract the outflow of business from London.

1067 EVERSLEY, D. E. C. (1972) Is the property boom in central London over? *Property and Investment Review*, December, pp. 18–21.

Considers the state and prospects of the London office market and suggests that 10 million square feet of vacant office space is deliberately kept empty by property developers.

1068 HALL, R. (1972) The movement of offices from central London. *Regional Studies*, Vol. 6, pp. 385–92.

Reports the results of a survey of office movement from central London between 1963 and 1969, showing that about 24,000 jobs per year were decentralized. The main cause of the movement lies in the spatial structure of office costs.

1069 KRALL, H. (1972) Offices: the issues restated. *Built Environment*, Vol. 1, pp. 468–9.

Argues that office development in London is necessary because of the activities in the city and criticizes rigid controls as hampering natural growth. Suggests methods of solving the traffic congestion problem.

1070 LOCATION OF OFFICES BUREAU (1972) Moving your office. Location of Offices Bureau, various paging.

Discusses the pattern of decentralization and its advantages and disadvantages.

1071 LOCATION OF OFFICES BUREAU (1972) Case studies of decentralized firms: a summary. Location of Offices Bureau, 17p.

Looks at 20 firms from LOB's record of firms moved from London, to provide evidence of the way firms approach the problem of office movement, and the problems and benefits the move produced.

1072 MANNERS, G. (1972) On the mezzanine floor: some reflections on contemporary office location policy. *Town and Country Planning*, Vol. 40, pp. 210–3.

Argues that the policy for moving offices from central London is ineffective and out-of-date.

1073 MOOR, N. (1972) A policy conflict. *Built Environment*, Vol. 1, pp. 455–7.

The conflict between the Location of Offices Bureau, Department of the Environment, office developers and the Greater London Council, over office location policy in London, produces problems which are discussed in terms of *Greater London development plan* (1969) policy.

1074 YANNOPOLOUS, G. (1972) Reasons for London's dominance. *Built Environment*, Vol. 1, pp. 445–7.

As the number of office workers increases, efforts to redirect them away from London have largely failed. The City's importance as a national and international centre has grown.

1075 PRENDERGAST, C. A. (1972) Decentralisation: the pattern changes. *Investors Chronicle*, pp. 49–53.

Suggests that the task of the Location of Offices Bureau is changing as the size and type of company alters – with a wider variety of company considering decentralization possibilities and with more interest in areas beyond the suburbs.

1076 EVANS, A. W. (1973) The location of the headquarters of industrial companies. *Urban Studies*, Vol. 10, pp. 387–95.

In an investigation of the locations of head offices of the 100 largest firms in the UK, over one third were found to be in central London. As the size of the firms declined, so also did the proportion with their head offices in London. The main reason for this is said to be that the head office is located to minimize communications for several locations.

1077 GODDARD, J. B. (1973) Office linkages and location: a study of communications and spatial patterns in central London. *Progress in Planning*, Vol. 1, pp. 111–232.

Outlines the structure, localization and spatial linkages of office employment in central London and investigates the nature of communication ties that appear to confine many office jobs to central London.

1078 INTER-BANK RESEARCH ORGANISATION (1973) The future of London as an international finance centre: report by the Inter-Bank Research Organisation. HMSO, 218p.

Looks into questions concerning London's future as a world financial centre, 10–15 years ahead. It examines features that must be possessed by an international financial centre. Changing functions and government policies are discussed and the needs of a financial centre such as London – office accommodation, labour and telecommunications are estimated.

1079 The dispersal of government work from London. (1973) HMSO, 228p. (Cmnd. 5322) Chairman: Sir Henry Hardman.

A review of the possibility of dispersing more government work from London. Recommends dispersal from London of some 31,000 jobs.

1080 SLADEN, C. (1973) Banks move head offices out of London. *Administrative Management*, Vol. 27, pp. 23–5.

Discusses the trend towards decentralization among banks and finance companies for reasons including increasing rents, high rates and salaries, rapid staff turnover, absenteeism and the need for expansion. Argues that movements in road, rail and electronic communications, have encouraged the mobility of executive personnel already inherent in the banks' staff structure.

1081 WAREHAM, C. R. (1973) The pattern of office development. *Chartered Surveyor*, Vol. 105, pp. 350–6.

Discusses the pattern of office development in the nation and London, government measures and their effect and future developments.

1082 BURTENSHAW, D. et al. (1974) Office decentralisation – ten years experience. *Surveyor*, Vol. 143 (4263) pp. 22–5.

Summarizes some of the post-war research on the effects of office location. Much of it supports the Location of Offices Bureau's views on the high costs of offices in London and the economic and social advantages of relocation but some see the decentralization policy as harmful to London.

1083 BURTENSHAW, D. et al. (1974) Office decentralization – ten years experience. *Surveyor*, Vol. 143 (4264) pp. 21–23.

The second part of an article on decentralization giving examples of moves out of London and the likely effects on office employees.

1084 SIDWELL, E. (1974) The attitudes of firms and local authorities to office decentralization and office development. London School of Economics, 18p. (London School of Economics, Graduate Geography Department Discussion Paper, No. 53)

A pilot study analysing the overall pattern of office dispersal. Examines the attitudes of 28 firms decentralizing from London since July 1972, and the attitudes and policies of 39 local planning departments to office development. Suggests a policy focus on flexible firms with flexible location requirements, a sustained advertising campaign on areas north of London and the dissemination of planning authority attitudes to office development.

1085 DANIELS, P. W. (1975) Employment planning 3: strategic offices in London. *Town and Country Planning*, Vol. 43, pp. 209–14.

Suggests the need for a well defined strategy for suburban office development. Argues that more positive planning of office concentrations could help reduce the volume of long, uncomfortable and arduous journeys to work.

1086 DANIELS, P. W. (1975) Office location study. G. Bell, 240p.

Concerned with factors which have influenced the growth and location of office activities and their consequences for contemporary and future urban development. Many examples from Greater London.

1087 GODDARD, J. B. (1975) Office location in urban and regional development. Oxford: University Press, 60p.

Examines the geography of office activities, outlines recent trends in the location of office employment and describes research into the factors behind these trends. Finally, it sets this research in the context of office location policy. Uses London as a major example throughout the book.

1088 STEPHEN, F. H. (1975) The Hardman report: a critique. *Regional Studies*, Vol. 9, pp. 111–6.

Criticizes the Hardman Report (1973) recommending the dispersal of Civil Service jobs from London. Argues that there were flaws in the methodology – the report falls down because of its reluctance to use monetary values overtly; the treatment of the social desirability of dispersing jobs is shown to create problems.

1089 BAILEY, A. (1976) Property people: C. A. Prendergast. *Estates Gazette*, Vol. 239, pp. 184–7.

An interview with the Chairman of the Location of Offices Bureau who gives his views on London and its future, its key activities, commuting, office location and the future for the City of Westminster.

1090 DEPARTMENT OF THE ENVIRONMENT Urban Affairs and Commercial Property Directorate. (1976) The office location review. HMSO, 100p.

Review of national policy objectives concerned with the location of new commercial development related to offices. There is a brief historical review of government measures, an analysis of the present position and forecasts and discussion of objectives generally towards assisted areas. A final section reviews existing and possible alternative policy instruments.

1091 BLAKE, J. (1977) 'Backhand LOB'. *Town and Country Planning*, Vol. 7, pp. 344–7.

Reviews the new role announced for the Location of Offices Bureau – including the promotion of office employment in inner urban areas (including London) following 12 years busily persuading firms to move their premises out of London. Suggests that the shot by Mr Shore is likely to fall wide of the mark.

1092 DANIELS, P. W. (1977) Office location in the British conurbations; trends and strategies. *Urban Studies*, Vol. 14, pp. 261–74.

Examines trends and strategies for the location of the office industry in the British conurbations since 1951. Suburbanization of office employment has been a feature of changing patterns, and, intra-conurbation movements, it is argued, will continue in London itself. Past efforts to shift office employment from London to other regions have not been successful.

1093 DANIELS, P. W. (1977) Office policy problems in Greater London. *Planner*, Vol. 63, pp. 102–5.

Discusses Greater London's experience in devising an effective office location policy and its wider relevance to the proportion of land use policies in the more recently designated metropolitan counties, and to activities such as transportation planning.

1094 KERR, D. (1977) An assessment of development potential in London's commercial property sector. Greater London Council, 44p. (GLC Research Memorandum, No. 505)

Investigates the factors affecting the funding of commercial development and the acceptability of commercial property of various kinds of investors. Discusses, the effects of the Community Land Act and Development Land Tax on this market, detailing them in appendices. Uses data sources from business as well as local and central government.

1095 MANNERS, G. (1977) New tactics for LOB. *Town and Country Planning*, Vol. 9, pp. 444–6.

Discusses the new role for the Location of Offices Bureau – 'to promote the better distribution of office employment ... and to take such steps as may be necessary for this purpose, including ... the provision of information and publicity and the promotion of research'. Discusses the international, inter-regional and intra-regional tactics involved with this new role.

1096 PYE, R. (1977) Office location and the cost of maintaining contact. *Environment and Planning*, A, Vol. 9, pp. 149–68.

Examines the costs and benefits of moving an office out of London; the greatest net saving is usually obtained by a move to a town in the outer South East. Few offices would obtain greater savings by moving to the assisted areas, despite government incentives.

1097 GODDARD, J. B. and SMITH, I. J. (1978) Changes in corporate control in the British urban system 1972–1977. *Environment and Planning*, A, Vol. 10, pp. 1073–84.

Analysis of changes in the locations of the headquarters of the leading 1,000 companies in the UK, both at the regional and urban levels, reveals evidence of increasing centralization of control in the South East. The main elements of change are analysed to assess their relative significance in the process.

1098 MORRIS, D. (1978) Trends in office employment: are planners initiating the right strategies? *Planner*, Vol. 64, pp. 106–7.

An analysis concentrating on employment in the private office sector.

1099 WEHRMANN, G. (1978) A policy in search of an objective. *Public Administration*, Vol. 56, pp. 425–37.

Traces the introduction of Office Development Permits (ODPs) in London and the effect of the policy on the capital. Concludes that 'few tears would be shed' if a Conservative government carries out its promise to abolish them. Chronicles the changes in ODP control from 1964–77.

1100 DAMESICK, P. (1979) Putting LOB in perspective. *Town and Country Planning*, Vol. 47, pp. 16–9.

The author discusses Location of Offices Bureau's past activities and future tasks. He finds no support for the view that LOB's new objectives are mutually contradictory.

1101 DANIELS, P. W. (1979) Confusing LOB? *Town and Country Planning*, Vol. 46, pp. 414–8.

Examines the recent changes in the work of the Location of Offices Bureau and suggests that confusion and internal contradication exist concerning its new role and objectives.

1102 GIBSON-JARVIE, R. R. (1979) The City of London: a financial and commercial history. Cambridge: Woodhead-Faulkner, 128p.

Examines the historical evolution of the various financial and commercial activities of the City of London.

1103 WEATHERITT, L. and JOHN, O. N. (1979) Office development and employment in Greater London, 1967–1976. Greater London Council, 88p. (GLC Research Memorandum, No. 556)

Examines the nature and significance of office activity in London and its place in the economic development of the city.

5.9 Retailing and markets

1104 BIRD, J. (1958) Billingsgate: a central metropolitan market. *Geographical Journal*, Vol. 124, pp. 64–75.

A study of London's wholesale fish market at Billingsgate; its historical development, its market area and its locational problems and advantages are analysed.

1105 SMAILES, A. E. and HARTLEY, G. (1961) Shopping centres in the Greater London area. *Transactions of the Institute of British Geographers*, Vol. 29, pp. 201–13.

A study of the geographical distribution of 269 shopping centres within Greater London (20 mile radius of Charing Cross) classifying the centres into three main levels in the urban hierarchy – regional centres, suburban centres and minor suburban centres.

1106 CARRUTHERS, W. I. (1962) Service centres in Greater London. *Town Planning Review*, Vol. 33, pp. 5–31.

A study of service centres ('clusters of shops, along with banks, cinemas') and service areas in Greater London. The provision and use of central area facilities is described on maps using various indicators and service centres and service areas are classified and mapped into a hierarchy of orders. Concludes that the intensity of use of existing services is a more telling and sensitive index of the importance of service centres than is the provision and availability of services.

1107 WEBBER, R. (1969) Covent Garden mud salad market, J. M. Dent, 178p.

Tells the story of the development of Covent Garden from the Convent garden to the late 1960s decision to move the market to Nine Elms.

1108 BLAKE, J. (1971) Future pattern of London's shopping (1). *Surveyor*, Vol. 138, pp. 16–9.

A summarized account of the enquiry into the *Greater London development plan's* policy for shopping and town centres, including the proposed shopping 'hierarchy' in London.

1109 BLAKE, J. (1971) Future pattern of London's shopping (2). *Surveyor*, Vol. 138, pp. 16–7.

Examines in a critical way the proposed creation by the Greater London Council of 'strategic' and 'major strategic' shopping centres.

1110 BLAKE, J. (1971) Future pattern of London's shopping: an elaborate plan that almost isn't (3). *Surveyor*, Vol. 138, pp. 32–4.

Looks critically at the floor space allocation to strategic centres proposed for London by the Greater London Council and argues that 'the GLC have come as near as they decently could to producing a non plan for shopping'.

1111 BLAKE, J. (1971) Shopping and suburban development. In: HILLMAN, J. (ed.) Planning for London. Penguin Books, pp. 77–85.

Part of a book concerned with aspects of life and planning in London. Examines the distribution and decentralization of shopping centres in London since 1945 and links with transportation plans.

1112 SCHILLER, R. K. (1971) Location trends of specialist services. *Regional Studies*, Vol. 5, pp. 1–10.

Evidence is produced to show that specialist services in the outer metropolitan area around London are less centrally located than would be expected by current theory. High income, car ownership and population dispersal are suggested as causes. It is argued that specialist services will tend to polarize in future between non-central locations serving car-based local consumers and metropolitan central business districts serving public transport-based commuters, tourists and distant visitors.

1113 Shopping in London (1971) *Shop Property*, supplement, June, 24p.

A series of articles on the existing provision of shopping in London, shop property values and likely future developments in various parts of London.

1114 CROFTS, C. (1972) Warehousing and distribution: cash and carry wholesalers. Greater London Council, 8p. (GLC Research Memorandum No. 376)

Describes the development and growth of cash and carry wholesalers in London.

1115 BURGESS, J. (1974) Street markets – an extendible resource, *Built Environment*, Vol. 3, pp. 403–5.

The fascination of markets for tourists is briefly discussed with particular reference to Portobello Road and Petticoat Lane markets.

1116 The new Covent Garden (1974) *Fruit Trades Journal*, supplement, 152p.

A supplement to mark the opening of the new Covent Garden horticultural market at Nine Elms on the South Bank of the Thames west of Vauxhall Bridge.

1117 ROCK, D. (1974) Long live Covent Garden. *Building*, Vol. 227, pp. 112–19.

A nostalgic look at the old Covent Garden to mark the departure of the market to Nine Elms on 11 November, 1974.

1118 BRITISH TOURIST AUTHORITY (1975) Shopping in London. British Tourist Authority, 56p.

Guide to the main shopping centres of the West End, with special sections on shopping for the disabled, street markets and speciality shops.

1119 CITY OF WESTMINSTER (1975) Shopping in Westminster. City of Westminster, 104p. (City of Westminster, Development Plan Topic Paper No. 5)

Outlines national and regional policies and trends related to shopping as a background to Westminster's planning powers and policies in the field. Describes shopping provision in the West End, secondary and intermediate centres, local centres and street markets and includes a chapter on pedestrianization and servicing.

1120 DOCKLANDS JOINT COMMITTEE (1975) Shopping. Docklands Joint Committee, 24p. (DJC Working Papers for Consultation, No. 5)

Recommends that no new major shopping centre be developed in docklands, but small centres for food shopping using pedestrian precincts were needed.

1121 THORNE, V. (1975) Less of a warehouse, more of a transit camp. *Property and Investment Review*, January, pp. 39–41.

A study of the Nine Elms site of the new Covent Garden, examining the prospect from its birth in 1961 to the establishment of the new fruit, vegetable and flower market. Documents the improvements and benefits in the speed of throughput of produce.

1122 WILLIAMS, A. (1975) New Covent Garden Nine Elms. *Building*, Vol. 228, pp. 71–86.

Building dossier on the new Covent Garden Market with a brief history of the market costs of the new buildings, maps showing their exact location, photographs of the accommodation, construction details and facilities and plans of the trading units in both the fruit and vegetable market and the flower market.

1123 BERMAN, J. (1976) Sunday markets. *Town and Country Planning*, Vol. 44, pp. 225–30.

Discusses the growth, legal status, planning problems and the costs and benefits to the community of Sunday markets in London.

1124 BLAKE, J. (1976) Brent Cross shopping centre. *Town and Country Planning*, Vol. 44, pp. 231–6.

Describes in detail the large new shopping centre opened in March 1976 in north-west London at the junction of the M1 and the North Circular Road and assesses its implications for future centres.

1125 CORPORATION OF LONDON Department of Architecture and Planning (1976) City of London development plan background study: shopping. Corporation of London, 2 vols.

Examines the historical development and composition of trade in the city: trends in consumer demand, the supply of retail units, street markets and external centres, existing policies, individual centre studies, and the issues involved.

1126 DOWNEY, P. (1976) Brent Cross shopping centre impact study: results of the first diary study of household shopping trips, preliminary analysis of shop vacancies and changes in NW London, 1971–76. Greater London Council, 13p. (GLC Research Memorandum, No. 525)

Presents an analysis of shop vacancies and changes in north-west London between 1971–6 before the Brent Cross centre opened. It covers four central shopping centres outside north-west London, so that shopping centres likely to be affected by Brent Cross can be compared with those not likely to be affected, but which are similarly under the influence of London-wide factors.

1127 O'TOOLE, F. (1976) The prosperous Home Counties. *Shop Property*, June, pp. 14–23.

One-third of the country's population live and work in the South East and this proportion has an above average level of prosperity. Retail companies have their headquarters in the area and thus there is current pressure for development. The large suburban centres of London and nearby towns appear to be the most prosperous centres.

1128 WILLSON, J. M. (1976) Note on the 1971 Census of distribution for London and the South East. Greater London Council, 20p. (GLC Research Memorandum No. 486)

By using the 1971 Census of Distribution and comparing it with the 1961 figures, presents some pointers to changes in the pattern of retail trade in London and the south-east region.

1129 BRUCE, A. J. and MANN, H. R. (1977) The Brent Cross shopping centre impact study: results of the first diary study of household shopping trips. Greater London Council, 59p. (GLC Research Memorandum, No. 522)

In this household survey carried out before the centre opened, shoppers filled in diaries to record the trips that they made to all shopping centres in the area. Data about these trips are presented and analysed according to various characteristics of shoppers and householders.

1130 LEE, M. and KENT, E. (1977) Brent Cross study. Donaldsons, 73p.

Describes the operations and functions of Brent Cross Centre against the background of planning and shopping in north-west London. The work is based mainly on a survey of shoppers at Brent Cross undertaken in October, 1976, some six months after the Centre's opening.

1131 MANN, H. R. (1977) The Brent Cross shopping centre impact study: the first home interview survey, Greater London Council, 85p. (GLC Research Memorandum, No. 510)

Describes the planning, execution and limitations of the first interview survey undertaken by the GLC and the Department of the Environment. A summary of the survey's findings is given in two appendices, one of which is concerned with shoppers' attitudes to those shopping centres likely to face competition from Brent Cross.

1132 WILLSON, J. M. (1977) Brent Cross Shopping Centre impact study: household income and expenditure in 1976. Greater London Council, 78p. (GLC Research Memorandum, No. 526)

Examines the relationship between household type and size with household income and expenditure as revealed by home interviews and diaries.

1133 BLAKE, J. (1978) Brent Cross: a regional shopping centre. *Planner*, Vol. 64, pp. 115–7.

Describes the development and performance of the new shopping centre at Brent Cross – unique as an in-situ regional centre located in an already built up area.

1134 SHEPHERD, I. D. H. and NEWBY, P. T. (1978) The Brent Cross regional shopping centre – characteristics and early effects. Retailing and Planning Associates, 88p.

Traces the development of the centre and describes its trading characteristics; the spatial impact on the shopper and other shopping centres is examined. Patterns of access to Brent Cross are looked at with a final chapter on the wider implications of the Brent Cross experience.

1135 NEWBY, P. T. and SHEPHERD, I. D. H. (1979) Brent Cross: a milestone in retail development. *Geography*, Vol. 64, pp. 133–7.

Describes the Brent Cross regional shopping centre in north-west London, opened in 1976 and the subsequent development of the site. The impact of the centre upon north-west London and the planning implications of such centres are examined.

1136 OMAND, E. (1979) Retail investment: a study based on interviews with some major retail and property companies. Greater London Council, 18p. (GLC Research Memorandum, No. 546)

Reports on meetings held between GLC officers and representatives of some major retail and property companies in 1976–7 with emphasis on town centre redevelopments in London and future policies.

5.10 Tourism

1137 GREATER LONDON COUNCIL (1971) Tourism and hotels in London: a paper for discussion. Greater London Council, 40p.

A discussion paper highlighting some of the opportunities and dangers inherent in the recent growth of the London tourist trade and the implications for planning policies of the acceptance of further tourist growth.

1138 MULLIN, S. (1971) Change, conservation and the tourist trade. In: HILLMAN, J. (ed.) Planning for London. Harmondsworth: Penguin Books, pp. 112–23.

Examines recent changes in the conservation of London's built environment in the light of the spectacular growth of London's tourist trade. Suggests that the GLC should take a more forceful role 'to promote innovation and change both for the voiceless and underprivileged and for the prosperous and sophisticated transient ...'.

1139 SOUTH EAST JOINT PLANNING TEAM (1971) Tourism. In: SOUTH EAST JOINT PLANNING TEAM Strategic Plan for the South East, Studies Volume No. 2: social and environmental aspects. HMSO, pp. 147–53.

Reviews the scarce data on the tourist industry in the south-east region, against the better quantified national background. Examines the changing pattern of demand for holidays at home and overseas, the place of overseas visitors to Britain and the significance of London in their itineraries. Emphasizes the importance of high standards of accommodation and tourist recreational and service facilities, especially in and near London and the value of the earnings of foreign currency from visitors.

1140 CITY OF WESTMINSTER (1972) Tourism and hotel development. City of Westminster, 89p. (City of Westminster Tourist Development Plan, Topic Paper No. 1)

The results of a study aimed at identifying the problems and potentialities for tourism as part of the production of the City of Westminster development plan.

1141 GREATER LONDON COUNCIL (1973) Tourism in London: towards a short term plan. Greater London Council, 55p.

A consultation document on tourism in London, composed by a steering group appointed by the GLC, English and London Tourist Boards and the London Boroughs Association.

1142 LONDON CONVENTION BUREAU (1973) Convention London: handbook of the London Convention Bureau. London Convention Bureau, 124p.

Guide to London's conference and meeting facilities with short sections on post-conference tours, London's night life and cultural activities.

1143 BROWN, A. (1974) After eight – West End choc-a-bloc? *Built Environment*, Vol. 3, pp. 400–2.

Discusses the comparatively poor provision of entertainment for the night-time visitor to London and calls for a more flexible attitude to such things as licensing laws and greater publicity for established and 'fringe' night-time activities.

1144 GREATER LONDON COUNCIL (1974) Tourism in London: plan for management. Greater London Council, 44p.

An advisory plan suggesting guidelines for the provision, maintenance and improvement of facilities for tourists and sightseers.

1145 HALL, J. M. (1974) The capacity to absorb tourists. *Built Environment*, Vol. 3, pp. 392–7.

Examines the activities of both tourists and residents in various areas with particular mention of the GLC's role in planning for increased tourism in London.

1146 BRITISH TOURIST AUTHORITY (1975) Survey among visitors to London – Summer, 1974. British Tourist Authority, 28p.

Covered personal particulars, purpose of visit, length of stay, type of accommodation, and transport used whilst in the city. Additional topics were attitudes towards London, their expenditure in it and their use of various sources of information on it.

1147 A capital solution? (1975) *British Travel News*, Vol. 51, pp. 10–5.

An interview with the chairman of the Greater London Council Policy and Resources Committee on the role tourism could play in helping to restore London's economy.

1148 LAVERY, P. (1975) Is the supply of accommodation inhibiting the growth of tourism in Britain? *Area*, Vol. 7, pp. 289–96.

A study of the 1961 and 1971 Census returns for south-east England shows a disturbing decline in the proportion of small hotel accommodation that is available. Some possible attributing factors are examined and the implications of this need are considered. Recent developments affecting the hotel industry are also examined.

1149 ROYAL BOROUGH OF KENSINGTON AND CHELSEA Development Plan Group (1975) Hotels and tourism. Royal Borough of Kensington and Chelsea, 45p. (RBKC Context Paper No. 4)

A consultation paper looking at current trends in hotels and tourism in London and the borough in particular, with their secondary effects on transport, employment, housing, social life, environment and finance.

1150 CORPORATION OF LONDON Department of Architecture and Planning (1976) City of London development plan background study: tourism. Corporation of London, unpaged.

Examines existing policies, overseas visitors to Britain, tourist attractions in the City, employment in the tourist industry, problems and development opportunities.

1151 LONDON TOURIST BOARD (1963–) Annual report. London Tourist Board.

Reviews the role of the LTB and the London Convention Centre, ground services, publications enquiries, guides, facts about tourism in London, income and expenditure and LTB objectives.

1152 EVERSLEY, D. E. C. (1977) The ganglion of tourism: an unresolvable problem for London? *London Journal*, Vol. 3, pp. 186–211.

Presents the case for more powers for the GLC in controlling the effects of a rapid growth in tourism in London.

1153 BRITISH TOURIST AUTHORITY (1978) Survey among visitors to London, Summer, 1978. British Tourist Authority, 57p.

One of a series of surveys whose aim is to provide a detailed picture of the characteristics of visitors to London, their activities and their reactions to the city.

1154 WHITE, D. (1978) Are tourists really a blight? *New Society*, Vol. 43, pp. 661–3.

A general review of the growth of tourism particularly in London, and the economic problems and benefits which tourists bring to specific areas.

1155 BRITISH TOURIST AUTHORITY and ENGLISH TOURIST BOARD (1979) Tourist growth and London accommodation: report on a conference held in London (5 December 1978). British Tourist Authority, 11p.

Looks at the problems, as well as opportunities, particularly for the accommodation sector, presented by the continuing growth of the London tourist trade.

1156 GOODEY, B. (1979) London's heritage deserves interpretation. *London Journal*, Vol. 5, pp. 256–61.

Reviews the *English Heritage Monitor* publication of the English Tourist Board, especially the tourist pressure on a few central, international buildings. Ways of redistributing visitors and their spending power, and the need for a new generation of urban interpreters to aid tourist awareness of place are suggested.

1157 LIPSCOMB, D, and WEATHERITT, L. (1979) Some economic aspects of tourism in London. *Greater London Intelligence Journal*, Vol. 42, pp. 15–7.

Presents certain data on tourism in London and discusses the impact that tourist spending has on London's economy and employment.

Section 6 Transport

London is the focus of south-east England's surface transport system and the chief centre for national air transport. Since transport in London is so closely linked to national and regional transport systems, it has often been difficult to isolate material for inclusion, and much which is relevant has had to be excluded from the bibliography.

An enormous amount has, however, been written on transport in London, particularly since the early 1960s. Preference has been given to literature above the local borough level concerned with metropolitan problems and policy or with major issues such as the motorway box, the third London airport, the decline of the docks, or traffic restraint. Many of the general planning documents, such as the *Greater London development plan* (GREATER LONDON COUNCIL, 1969) and the *Strategic plan for the South East* (SOUTH EAST JOINT PLANNING TEAM, 1970) which are included in section 8 (planning) are very much concerned with transportation policy. Planning material in this section is specifically related to transport planning. Publications which discuss the redevelopment of the docklands have been included in section 8, except where they are restricted to a consideration of transport planning in the area.

Much of the literature on transport in London is by economists, planners, engineers and others such as railway enthusiasts. The majority of authors work for various government organizations, both national (the Department of Transport and its predecessors, the Civil Aviation Authority, and various Royal Commissions) and local (the Greater London Council, London Transport and the 32 London boroughs). The Greater London Council, which has overall strategic planning authority, is now the most powerful body concerned with transport in London and is responsible for most of the research in this area. A large number of research memoranda are produced every year by the Greater London Council's Department of Planning and Transportation, and many are included here. Since the late 1960s an increasing number of local and national pressure groups concerned with various transport issues have been created. Bodies such as the Noise Abatement Society, the London Motorists' Association, the London Amenity and Transport Association, and various local action groups have been very active in their contribution to the literature.

This section has been subdivided according to mode of transport with general historical (6.2) and policy (6.3) subsections at the beginning. Material concerned with freight transport has been scattered according to mode, with general freight policy documents in section 6.3. The changing pattern of transport in London is reflected in the size and nature of the publications included in the various subsections. Much current writing on transport in London is concerned with road traffic problems. The third London airport controversy has also generated a great deal of research and publication over the past 20 years. The decline of the canals, local railways and, more recently, the docks, has meant that these modes of transport have tended to become the concern of the historian rather than the planner and economist. Few individual bibliographies of transport in London have been produced. Probably the best source of further references is *Urban Abstracts* (1974–) and its predecessor *Planning and Transportation Abstracts* (1969–74). The *Departments of Environment and Transport Library Bulletin* is also very useful.

6.1 Bibliographies and statistics

1158 LONDON TRANSPORT (19?–) Statistics. London Transport. (annual)

Detailed statistics of London Transport operations including information about rolling stock and buses in service.

1159 GREATER LONDON COUNCIL (1973–) Transport facts and figures. Greater London Council, No. 1– (irregular).

Basic data concerning London's transport system are given in the form of maps and diagrams.

1160 GREATER LONDON COUNCIL Department of Planning and Transportation (1973) Transportation in London: a select list of references. 2nd ed. Greater London Council, 14p. (GLC Research Bibliography, No. 16)

A brief listing of publications dealing with all aspects of transport in London.

1161 LINDSEY, C. F. (1973) Underground railways in London: a select bibliography. The author, 14p.

A brief list of references, mainly books, on the history and current organization of the London underground. There are no annotations.

1162 THOM, B. (1973) Thames Bridges in the Greater London area. Greater London Council, 22p. (GLC Research Bibliography, No. 45)

An annotated bibliography of 159 items concerned with various aspects of the road and rail bridges of London.

1163 Handlist of books in Guildhall Library relating to underground railways in London. (1974) *Guildhall Studies in London History*, Vol. 1, pp. 192–209.

The list is confined to printed books, and there are no annotations. It is divided into a general section and separate sections on individual lines and schemes.

1164 MUNBY, D. L. (1978) London's transport. In: MUNBY, D. L. Inland transport statistics Great Britain 1900–1970, Vol. 1. Oxford: University Press, pp. 457–658.

An explanatory chapter is followed by detailed statistics of London's public passenger transport. All aspects of London's railways, trams and buses are covered including revenue, employment, journeys, fares and assets.

6.2 Historical development

1165 SEKON, G. A. (1938) Locomotion in Victorian London. Oxford: University Press, 212p.

An account covering all aspects of transport in Victorian times from pedestrians to steamboats. The section on London's railways comprises a large portion of the book due to the great developments during this period.

1166 BARKER, T. C. and ROBBINS, M. (1963) A history of London transport: passenger travel and the development of the metropolis, Volume 1: The nineteenth century. George Allen and Unwin, 412p.

Covers the period between the development of the omnibus and the search for new forms of traction at the end of the nineteenth century. The construction of the underground railway, the horse tramways and the Victorian railways and roads are all described in relation to the growth of the city.

1167 HALL, P. G. (1964) The development of communications. In: COPPOCK, J. T. and PRINCE, H. C. (eds) Greater London. Faber and Faber, pp. 52–79.

An account of the geography of communications in London concentrating on two aspects: the morphology of the street pattern and the pattern of railways and rail traffic as affected by London's physical and economic geography.

1168 BARKER, T. C. (1965) London's third traffic crisis. *New Society*, Vol. 7, pp. 15–6.

Three traffic crises are identified: in the 1850s, the 1890s and in the 1940s. The first two were met by investment in overground and then underground railways; but the third did not stimulate the growth of railways since it was felt that the motor car might solve the problem.

1169 KELLETT, J. R. (1969) London. In: KELLETT, J. R. The impact of railways on Victorian cities. Routledge and Kegan Paul, pp. 244–83.

An analysis of the arrival of railways in London, examining the pattern of property ownership, the development of the railways and competition between the companies. The veto on surface lines crossing the capital led to a unique situation of a central area ringed with termini.

1170 PUDNEY, J. (1972) Crossing London's river: the bridges, ferries and tunnels crossing the Thames tideway in London. J. M. Dent, 176p.

Contemporary sources are used to trace the crossings over and under the 69 mile tidal river from old London Bridge to the Dartford Tunnel.

1171 BARKER, T. C. and ROBBINS, M. (1974) A history of London transport: passenger travel and the growth of the metropolis. Volume 2: The twentieth century to 1970. George Allen and Unwin, 554p.

All aspects of transport in London during the twentieth century are described from the rise of the electric tramways, the electrification of the underground and the beginnings of the motor bus to the various issues of present day London transport.

1172 KLAPPER, C. F. (1976) Roads and rails of London 1900–1933. Ian Allan, 191p.

The history of London transport during a period of great change witnessing the mechanization of London public transport by bus, tram and cab, and a rapid increase in the numbers of cars and lorries. The book ends with the creation of the London Passenger Transport Board in 1933.

1173 LONDON TRANSPORT (1979) Chronology. London Transport, 13p.

A brief chronology of the main events in the history of London Transport services.

6.3 Policy and surveys

1174 ROYAL COMMISSION ON LONDON TRAFFIC (1905–6) Report. HMSO, 8 vols. (Cd. 2597) Chairman: D. M. Barber

A large number of suggestions are made regarding the relief of congestion on roads and railways in London. A Traffic Board for London is proposed, a number of major improvement schemes recommended for the central area, and other improvements such as road widening, standard road widths and uniformity of building laws are called for.

1175 ADVISORY COMMITTEE FOR LONDON TRAFFIC (1920) Report. HMSO, 191p. (Cmd. 636)

The establishment of a London Traffic Authority is recommended and its constitution, duties and powers outlined.

1176 LONDON TRANSPORT EXECUTIVE (1950) London travel survey 1949. London Transport Executive, 48p.

The purpose of the survey was to establish the extent to which Londoners make use of public transport, particularly for the journey to and from work, and to discover also the purposes for which they travel. In addition, information was obtained on the social background of the Londoner in terms of the size of his household, classified according to his income, and the sex, age and status of the members of his household.

1177 LONDON TRANSPORT EXECUTIVE (1956) London travel survey 1954. London Transport Executive, 64p.

A repetition on a larger and more comprehensive scale of the first *London travel survey* (LONDON TRANSPORT EXECUTIVE, 1950). Many more households were covered by the interviews, and information obtained comprised a detailed record of all journeys made during the previous seven days, irrespective of the form of public or private transport used.

1178 DAVIES, E. (1962) Transport in Greater London. George Allen and Unwin, 15p. (Greater London Papers, No. 6)

A survey of transport organization and planning in London and the extent to which the recommendations of the *Royal Commission on Local Government in Greater London* (1960) might improve the situation. A number of suggestions are made as to how the proposals might be amended.

1179 FREEMAN FOX AND PARTNERS (1964) London traffic survey, Vol. 1. Existing traffic and travel characteristics in Greater London. London County Council, 214p.

The first volume of the survey is concerned with the collection and analysis of data about current London traffic and travel. Data have been collected by means of a variety of methods including home and roadside interviews, a commercial vehicle study, questionnaires and traffic counts.

1180 FREEMAN FOX AND PARTNERS (1966) London traffic survey, Vol. 2. Future traffic and travel characteristics in Greater London. Greater London Council, 242p.

The second phase of the *London traffic survey* (FREEMAN FOX AND PARTNERS, 1964) is concerned with forecasts of future travel demand in London. It includes the preparation of estimates of future traffic demands likely to develop by 1971 and 1981, generalized analyses of the adequacy of the present road improvement programmes for the 1962–71 period, and preliminary studies of alternative concepts of co-ordinated road and rail systems for 1981.

1181 MINISTRY OF TRANSPORT (1966) Transport policy. HMSO, 36p. (Cmnd. 3057)

A discussion of urban transport problems in general, and the special difficulties of London. Short-term policies are described.

1182 FREEMAN FOX, WILBUR SMITH AND ASSOCIATES (1967) London transportation study, phase 3. Freeman Fox, Wilbur Smith and Associates, 4 vols.

The third phase of the comprehensive survey of London's traffic pattern and future transportation needs which was started in 1962, the results of the two earlier phases having been published as the *London traffic survey* (FREEMAN FOX AND PARTNERS 1964, 1966). The main objectives of the final phase were to develop more advanced analytical methods that could be used to simulate the conditions expected in London in the future, to provide data on costs, and to make trial examinations of alternative networks (both for road and public transport systems) and land use patterns. All three phases of the survey are now known as the *London transportation study*, and findings have been incorporated in the *Greater London development plan* (GREATER LONDON COUNCIL, 1969).

1183 GREATER LONDON COUNCIL (1968) Generation of business traffic in Central London. Greater London Council, 54p. (GLC Planning Research Paper, No. 3)

A description of a sample survey of central area business premises which was undertaken in an attempt to provide a comprehensive record of business travel by all modes of transport. The results might contribute to the formulation of appropriate parking standards for new buildings, indicate the volumes of traffic generated and attracted by premises of different kinds, and provide information about the degree of business linkage that attracted firms to central London locations.

1184 MINISTRY OF TRANSPORT (1968) Transport in London. HMSO, 86p. (Cmnd. 3686)

A major reorganization of transport in London is recommended in that the GLC is to become the overall transport planning authority for London. In addition, the work of London Transport and the role of the Minister of Transport are reviewed.

1185 YATES, L. B. (1968) Traffic surveys in London. Greater London Council, 10p. (GLC Research Memorandum, No. 95)

The major and regular surveys of traffic movement in London are summarized. The overall programme is outlined together with future developments.

1186 GREATER LONDON COUNCIL (1969) Movement in London: transport research studies and their context. Greater London Council, 185p.

The findings of the *London traffic survey*, the *London transportation study* (FREEMAN FOX AND PARTNERS, 1964–7), and transportation research undertaken directly by the GLC are summarized. The effects of these studies on the development of LCC and GLC policy are assessed.

1187 JORDAN, D. (1969) Commuting into London. *Journal of the Town Planning Institute*, Vol. 55, pp. 72–3.

The 1966 Census work-place and transport tables are used in an investigation of employment patterns around London. The pattern of commuting from the outer metropolitan area and new towns to Greater London is outlined.

1188 BAYLISS, D. et al. (1970) 'Movement in London': proceedings of seminars. Greater London Council, 52p. (GLC Research Memorandum, No. 240)

Four papers on the *London traffic survey*, the *London transportation study* (FREEMAN FOX AND PARTNERS, 1964–7) and other research are presented together with summaries of the discussion which took place during the seminars.

1189 BAYLISS, D. (1970) The scale of travel demand in London 1962 – 1981–1991. Greater London Council, 7p.

Results of several travel surveys and forecasts, including data from the *London traffic survey* and *London transportation study* (FREEMAN FOX AND PARTNERS, 1964–7). Data are presented for these three years in a compatible form against a background of car ownership and planning assumptions.

1190 GREATER LONDON COUNCIL (1970) The future of London Transport: a paper for discussion. Greater London Council, 52p.

In 1970 responsibility for London Transport was transferred from the Ministry of Transport to the GLC. This paper outlines a number of problems to be faced by the new management and seeks to stimulate public discussion of key issues.

1191 RIDLEY, T. M. and TRESIDER, J. O. (1970) The London transportation study and beyond. *Regional Studies*, Vol. 4, pp. 63–71.

A description of the methodology, results and implications of the *London transportation study* (FREEMAN FOX AND PARTNERS, 1964–7).

1192 LONDON MOTORWAY ACTION GROUP and LONDON AMENITY AND TRANSPORT ASSOCIATION (1971) Transport strategy in London. London Motorway Action Group, 300p.

A report presented as evidence at the public inquiry into the *Greater London development plan* (GREATER LONDON COUNCIL, 1969). The proposals for inner motorways in London are attacked and an alternative strategy put forward.

1193 LONDON TRANSPORT EXECUTIVE (1971–) Annual report and accounts, 1970– . London Transport Executive (annual).

The report section gives a brief survey of developments during the year with regard to bus and underground services. There is also a general discussion of transport planning. Before the creation of the London Transport Executive, similar reports were issued by the London Passenger

Transport Board (1933–47), London Transport Executive (1948–62) and London Transport Board (1963–69).

1194 WACHER, T. (1971) Public transport and land use: a strategy for London. *Chartered Surveyor*, Vol. 104, pp. 16–26.

The problems of developing an effective public transport policy are discussed and related to land use. Various European approaches to the problem are discussed, and a number of recommendations are made.

1195 DAVIDSON, K. B. (1973) Relationships between land use and accessibility. Greater London Council, 39p. (GLC Research Memorandum, No. 378)

An analysis of accessibility to activities in the London Transportation Study area and their relationships to land use densities. Four measures of accessibility are defined and are related to population density and employment.

1196 GREATER LONDON COUNCIL (1973) Public transport in London: a regional approach. Greater London Council, 36p.

A review of the progress of the GLC in finding a solution to London's transport problems. The financial and organizational problems are discussed together with general principles adopted.

1197 TRANSPORT 2000 London and Home Counties Committee (1973) A transport strategy for London. Transport 2000, 20p.

Seven recommendations are made for the development of transport in the capital: improved pay and conditions for transport workers, more money to be spent on public transport rather than roads, more restrictions on private cars and commercial vehicles, better bus and tube services, a London regional transport authority, and an overall transport plan.

1198 CENTRAL LONDON PLANNING CONFERENCE (1974) Transport. Central London Planning Conference, 140p. (Advisory Plan for Central London Topic Paper, No. 14)

The current situation and trends in London transport are analysed. Alternative goals and policies are suggested within different facets of the transportation system.

1199 COLLINS, M. F. and PHAROAH, T. M. (1974) Transport organisation in a great city: the case of London. George Allen and Unwin, 660p.

The responsibilities of the various authorities involved with passenger transport in London are described, together with various factors such as planning, investment, subsidies and coordination. Fifteen case studies covering different transport problems are outlined.

1200 DORLING, N. et al. (1974) Transport management for London. *Journal of Transport Economics and Policy*, Vol. 8, pp. 152–60.

The price mechanism has been neglected in transport planning. An adjustment in user costs could arrest the decline in public transport in London.

1201 GREATER LONDON COUNCIL (1974–) Transport policies and programme 1975–80. Greater London Council, (annual).

A brief review of the five year rolling investment programme is followed by an examination of various services and issues.

1202 COLLINS, M. F. (1975) London's public transport: retrospect and prospect? *London Journal*, Vol. 1, pp. 267–74.

A review of T. C. Barker and M. Robbins *A history of London Transport Vol. 2. The twentieth century to 1970* (1974) and the GLC's *London rail study* (1974, 1975). Some of the main issues raised in both publications are then discussed.

1203 DOCKLANDS JOINT COMMITTEE (1975) The docklands spine – tube, bus or train. Docklands Joint Committee, 32p.

Existing public transport in docklands is reviewed and alternative rapid transit systems are discussed in terms of cost, service and other factors. Either a busway or an underground railway would seem to be the best solution.

1204 DOCKLANDS JOINT COMMITTEE (1975) Transport. Docklands Joint Committee, 56p.

A consultation paper outlining an integrated transport system for the redeveloped docklands. Proposals cover roads, buses, railways and water transport.

1205 GREATER LONDON COUNCIL (1975) Freight in London. Greater London Council, various paging.

The role and impact of freight in London are discussed. Various methods such as improved vehicle design, lorry routeing, cars and lorry parks are suggested as options for lessening the deterioration of the environment.

1206 LONDON FREIGHT CONFERENCE (1976) Freight in London: freight policy background information. Greater London Council, 172p.

A technical support document for the GLC's policy statement on freight in the capital. Full details of measures to be taken are given.

1207 STANDING CONFERENCE ON LONDON AND SOUTH EAST REGIONAL PLANNING (1976) Development of the Strategic Plan for the

South East: transport studies: issues and findings. Department of the Environment, 145p.

An account of the transport studies which were undertaken in order to update the *Strategic plan for the South East* (SOUTH EAST JOINT PLANNING TEAM, 1970). Major projects are reviewed in relation to the various plans, and there is also an analysis of transport funding.

1208 GREATER LONDON COUNCIL London Transport Committee (1977) London transport: a new look: consultation document. Greater London Council, 28p.

The new London Transport Committee recommends a variety of changes in London Transport's operations. Issues covered include productivity, buses, tubes, concessionary travel, and consumer protection.

1209 THOMAS, R. (1977) Commuting flows and the growth of London's new towns 1951–1971. Open University Press, 35p.

New towns around London are compared with other towns in the Outer Metropolitan Area. Commuting is discussed in relation to land use, community self-containment and the London labour market.

1210 THOMSON, J. M. (1977) London. In: THOMSON, J. M. Great cities and their traffic. Victor Gollancz, pp. 269–87.

A brief survey of the general structure and different elements of the transport system in London. Past surveys and proposals are discussed together with future prospects for the transport planner.

1211 FRYER, J. A. (1978) Travel to work in Greater London: selected results from the 1971 Census and the GLTS household survey 1971–72. Greater London Council, 19p. (GLC Research Memorandum, No. 551)

Data concerning travel to work are presented accompanied by a brief commentary. Tables are taken from the 1971 Census (10 per cent sample) and the Greater London Transportation Survey 1971–2.

1212 GREATER LONDON COUNCIL (1978) Freight policy in London. Greater London Council, 18p.

Past, present and future problems of freight in London are summarized. The GLC's policies concerning rail, port and river, air and road freight are outlined together with special measures to protect the environment.

1213 GREATER LONDON COUNCIL London Transport Committee (1978) London transport: a new look: the next steps. Greater London Council, 17p.

A statement of future developments in London Transport. Issues covered include commuter services, buses, tubes, the Jubilee Line, and fare systems.

1214 *Greater London Intelligence Journal*, (1978) No. 40 (whole issue).

The development of policies for freight transport in London is discussed. Road, rail, water and air freight systems are examined and an overall strategy outlined.

1215 HASELL, B. B. et al. (1978) Freight planning in London, 1. The existing system and its problems. *Traffic Engineering and Control*, Vol. 19, pp. 60–3.

A discussion of the development of the GLC's 1977 freight policy and the successes and failures of this policy.

1216 HASELL, B. B. (1978) Freight planning in London, 2. Assisting efficient freight operation. *Traffic Engineering and Control*, Vol. 19, pp. 126–9.

An outline of the main ways in which the GLC has influenced freight movement in London. Road improvement, traffic management, improved depots and retail areas, and vehicle design are the main topics considered.

1217 WISTRICH, E. (1978) Transport in Greater London. *Political Quarterly*, Vol. 49, pp. 1–12.

A discussion of a number of transport planning issues in London. The debate over the ringway, traffic restraint, public transport and forward planning are covered.

1218 LONDON TRANSPORT (1979) Basic facts. London Transport, (annual).

The organization, history and future development of London Transport's services are outlined.

1219 MOGRIDGE, M. J. H. (1979) Changing spatial patterns in the journey-to-work: a comparison of the 1966 and 1971 Census data in London. *Urban Studies*, Vol. 16, pp. 179–90.

The 1966 and 1971 journey to work data for London are analysed. Changes in modes of transport and trip length are noted.

6.4 Road

1220 ORDISH, T. F. (1911) Roads out of London: being photographic reprints extracted from Ogilby's 'Britannia' 1675 with so much of his text as relates to them. London Topographical Society, 17p.

Fourteen of Ogilby's strip road maps are reproduced along with his terse route information.

1221 ORDISH, T. F. (1913) History of metropolitan roads. *London Topographical Record*, Vol. 8, pp. 1–92.

A detailed 'official' history of the roads of London and district, originally published as a report of the London Traffic Board of the Board of Trade (1910). The main emphasis is on the Turnpike Trust roads and the improvements made by the Metropolis Road Board, created by Act of Parliament in 1826, and whose first task was the construction of the Finchley Road.

1222 PROUDLOVE, J. A. (1960) A traffic plan for London. *Town Planning Review*, Vol. 31, pp. 53–73.

A discussion of the proposals submitted in 1959 by J. A. Proudlove for the solution of London's traffic problems. The proposals won first prize in the Roads Campaign Council's competition for the design of a long-term plan of highway development in the London area.

1223 BUCHANAN, C. M. (1970) London road plans 1900–1970. Greater London Council, 52p. (GLC Research Report, No. 11)

A history of attempts made during the twentieth century to plan London's roads, concluding with the *Greater London development plan* (GREATER LONDON COUNCIL, 1969) road proposals.

1224 EYLES, D. and MYATT, P. (1970) Road traffic and urban environment in inner London (a study of LTS zone 277). Greater London Council, 157p. (GLC Research Memorandum, No. 250)

The environmental effects of traffic on the residents of local and secondary roads in an area of Wandsworth have been investigated. The study includes measurements of air and noise pollution, in addition to the attitudes of local residents.

1225 MOGRIDGE, M. J. H. and ELDRIDGE, D. (1970) Car ownership in London. Greater London Council, 68p. (GLC Research Report, No. 10)

Projections for car ownership in London in 1981 and 1991 have been made as a result of detailed analysis of the car ownership values found in the *London traffic survey* (FREEMAN FOX AND PARTNERS, 1964, 1966).

1226 THOMSON, J. M. (1970) Motorways in London. Duckworth, 194p.

The report of a London Amenity and Transport Association working party criticizing GLC proposals for motorways in London. The growth of London's transport problem is described together with the impact of the present proposals.

1227 WILLMOTT, P. and YOUNG, M. (1970) How urgent are London's motorways? *New Society*, Vol. 16, pp. 1036–8.

A report on the results of two surveys undertaken to discover the attitudes of Londoners to a variety of objects of expenditure which are the joint responsibility of central and local government. In the main survey, of a random sample of Londoners, building new motorways inside London came lowest in order of priority; in the smaller surveys of managing directors, this option was slightly more popular overall.

1228 JENKINS, S. (1973) The politics of London motorways. *Political Quarterly*, Vol. 44, pp. 257–70.

A review of the politics and history of plans for urban motorways in London which culminated in the *Greater London development plan's* (GREATER LONDON COUNCIL, 1969) proposals for the ringway. The GLC is criticized for the reasoning behind the plans and its powers as a strategic authority are revealed in the light of developments which have contributed to the plans not having been implemented.

1229 COTTLE, R. W. et al. (1974) City of London traffic study. Greater London Council, 35p. (GLC Research Memorandum, No. 428)

The results of a roadside origin-destination interview survey carried out jointly by the GLC and the Corporation of London. The data are used for network analysis, and to test various objectives.

1230 GREATER LONDON COUNCIL (1974) Bike-ways in urban areas. Greater London Council, 7p. (GLC London Topics, No. 1)

The pressure for bikeways and the current situation regarding bikeways in London are discussed. There is a list of concerned organizations and 29 references.

1231 GREATER LONDON COUNCIL Department of Planning and Transportation (1974) Routes for heavy vehicles in London: consultation paper. Greater London Council, 13p.

The progress and objectives of lorry routeing in London are discussed. Proposals regarding the scale of the routeing network and the size of lorries are presented.

1232 HASELL, B. B. (1974) Greater London Transportation Survey (GLTS): road traffic entering London in 1971: a preliminary analysis. Greater London Council, 55p. (GLC Research Memorandum, No. 304)

Data concerning traffic using roads crossing the Greater London Transportation Study external traffic cordon are analysed and compared with similar data for 1962 and with 1981 GLC traffic predictions.

1233 LONDON BOROUGH OF ISLINGTON Planning Department (1974) Highway planning and transportation. London Borough of Islington, unpaged.

A detailed account of public transport services and finance in Islington, including plans for future development.

1234 LONDON BOROUGH OF ISLINGTON Planning Department (1974) Public transport in Islington: a review of passenger facilities. London Borough of Islington, 40p.

Bus, tube and rail transport in the borough are outlined. Details of administration and finance are also given.

1235 FAIRHURST, M. H. (1975) The influence of public transport on car ownership in London. *Journal of Transport Economics and Policy*, Vol. 9, pp. 193–208.

Data from the *London transportation study* (FREEMAN FOX AND PARTNERS, 1964–7) are used to test two models developed to explain variations in car ownership. The main variables affecting ownership are household income, household size, access to public transport and residential density.

1236 GREATER LONDON COUNCIL (1975) East London river crossing study: report of the Steering Group. Greater London Council, 32p. East London river crossing study: studies volume. Greater London Council, 113p.

A new river crossing at Thamesmead with road links, north to South Woodford and south to Falconwood, is proposed. Full details of existing traffic patterns in the area are given, and the scheme is felt to be consistent with strategic transport policies.

1237 GREATER LONDON COUNCIL (1975) Greater London Transportation Survey: Vols 1–9, Greater London Council.

Vol. 1, Home interview, hotel and hostel survey report.
Vol. 2, External cordon and screen line survey report.
Vol. 3, Goods vehicle survey report.
Vol. 4, Home interview, hotel and hostel survey technical manual.
Vol. 5, External cordon and screen line survey technical manual.
Vol. 6, Goods vehicle survey technical manual.
Vol. 7, Address zone coding manual.
Vol. 8, Survey operation report.
Vol. 9, Structure report.

This survey comprised a number of large-scale studies aimed at compiling a picture of the way in which people, goods and vehicles travel within the Greater London area. The nine reports document the design, organization and conduct of the survey.

1238 GREATER LONDON COUNCIL, Department of Planning and Transportation (1975) An approach to the environmental assessment of traffic: progress report. Greater London Council, 73p.

An attempt to assess the environmental improvement brought about by measures such as cordon restraint, speedbuses, bus priority and pedestrianization. A more sensitive model is needed to take into account a wide range of factors.

1239 HOILE, J. P. C. (1975) Central London's pedestrian streets and ways. *Greater London Intelligence Quarterly*, No. 33, pp. 16–27.

The advantages and disadvantages of pedestrianization in central London are discussed, and the findings of a survey are presented which lend support to increased pedestrian areas. GLC policy and legal issues are also raised.

1240 TREVELYAN, P. (1975) Bicycle planning in London. *Planner*, Vol. 61, pp. 225–7.

The 1971 Census indicates that only 2.2 per cent of the population still cycle to work in Greater London. Existing bicycle policies in London are examined and it is suggested that boroughs develop a network of routes for cyclists.

1241 HART, A. (1976) Strategic planning in London: the rise and fall of the primary road network. Oxford: Pergamon, 238p.

A study of the relationship between strategic road planning and the concept of urban order during the period 1943–73. The development during the 1960s and final abandonment in 1973 of the proposals for a primary road network of urban motorways for London are discussed.

1242 LONDON AMENITY AND TRANSPORT ASSOCIATION (1976) Bicycling in London. London Amenity and Transport Association, 14p.

Increasing use of bicycles is examined in the light of facilities available. Boroughs are urged to provide routes enabling cyclists to reach all parts of London safely.

1243 BUCHANAN, C. D. (1977) London traffic inquiry. London Motorists' Association, 67p.

An independent inquiry into the GLC's traffic policies. It is particularly concerned with the effects of traffic and parking restraint on industrial and commercial life in the city.

1244 EVELYN, V. et al. (1977) How to build London's ringways at low cost and almost no environmental pain. *London Journal*, Vol. 3, pp. 3–23.

Some of London's railway lines should be converted into bus routes, and thus improve the inferior road network of the capital.

1245 ADAMS, J. (1978) Transport in the capital. In: CLOUT, H. (ed.) Changing London. University Tutorial Press, pp. 27–37.

A brief survey of planners' attempts to solve London's traffic problems. Various current issues and problems are examined.

1246 GREATER LONDON COUNCIL Department of Planning and Transportation (1978) Road for London. Greater London Council, 6p.

Contains descriptions and plans of the London road network, together with new projects in the 1978–83 development programme.

1247 HASELL, B. B. (1978) The control of goods vehicle movements in urban areas: the London experience. *ITE Journal*, Vol. 48, pp. 31–6.

A history of attempts to control road freight in London over the past ten years. Future policy is considered.

1248 MUNT, P. W. (1979) Traffic characteristics of Greater London's roads. Greater London Council, 50p. (GLC Research Memorandum, No. 554)

A variety of characteristics of London's roads are shown by means of maps and tables. Information given includes busiest roads and junctions, high speed roads, congested areas, and regular counting sites which carry the greatest number and highest percentage of various vehicle types during different periods of the day.

1249 MUNT, P. W. (1979) Traffic monitoring review for 1978. Greater London Council, 36p. (GLC Research Memorandum, No. 555)

The GLC's traffic monitoring programme is described and the recent trends and present state of road traffic in Greater London are outlined by means of short notes and tables.

6.5 Rail and underground

1250 COURSE, E. (1962) London railways. B. T. Batsford, 282p.

The author concentrates mainly on the early history of London's railways: how and why they were laid out, who designed them, who provided the money, and what survivals remain. A short chapter covers the contribution of the railways to the growth of the capital, and the post-war developments.

1251 JACKSON, A. A. and CROOME, D. F. (1962) Rails through the clay: a history of London's tube railways. George Allen and Unwin, 410p.

A history of the growth of London's underground system from the 1860s to the early 1960s, together with an analysis of its effect on the life of the capital.

1252 ROBBINS, M. (1969) The North London railway. 6th ed. Oakwood Press, 30p.

The history of the North London line from the passing of the first Act in 1836 to the end of the nineteenth century.

1253 JACKSON, A. A. (1969) London's termini. Newton Abbot: David and Charles, 368p.

An account of the development of London's 16 existing mainline termini. After an introductory survey, a chapter is devoted to each station or station group, beginning with Euston, the first London terminus for long distance traffic.

1254 WHITE, H.P. (1971) A regional history of the railways of Great Britain, Vol. 3. Greater London. 2nd ed. Newton Abbot: David and Charles, 228p.

After a brief definition of London, the author describes the growth of the railways in different sectors of the city: the building of the southern termini and approach lines, the rivalry between different companies, and the extensions across the inner suburbs and out beyond them. He also deals with central London and the dominant role played by the underground, and with the policies of the Great Eastern which helped the earliest spread of the working class suburbs.

1255 THOMAS, R. H. G. (1972) London's first railway: the London and Greenwich. B. T. Batsford, 270p.

A comprehensive account of London's first local line, the London and Greenwich Railway, from the formation of the company in 1831 to the twentieth century.

1256 CROWTHER, G. L. et al. (1973) A new ring rail for London: the key to an integrated public transport system. Just, 67p.

The need for a ring rail is assessed, and its operation and social and economic effects explained.

1257 LLEWELYN-DAVIES, WEEKS, FORESTIER-WALKER AND BOR (1973) S.E. London and the Fleet Line: a study of land use potential. Llewelyn-Davies, Weeks, Forestier-Walker and Bor, 172p.

The report carried out for London Transport Executive suggests how the various planning authorities could legislate for progressive urban development in order to maximize the economic and social benefits which the new Fleet Line could bring. The need for an integrated land use/public transport policy is emphasized.

1258 LONDON TRANSPORT (1973) A report on the traffic implications of the Victoria Line north of Victoria. London Transport, 44p.

The impact of the Victoria Line between 1969 and 1971 is assessed by means of two questionnaire surveys and a household survey. Its effect on property values and land use patterns, as well as on traffic flows, is measured.

1259 WILLIAMS, R. A. (1973) The London and Southwestern Railway, Vol. 2. Growth and consolidation. Newton Abbot: David and Charles, 383p.

A study of the twentieth century growth of the London and Southwestern railway and its impact on the area which it served.

1260 DAY, J. R. (1974) The story of London's underground. 6th ed. London Transport, 160p.

A concise history of the development of London's underground, from the construction of the Metropolitan line in 1863 to the Victoria line 100 years later.

1261 London rail study (1974) Greater London Council and Department of the Environment. Part 1, 43p.; Part 2, 117p. (Chairman: Sir David Barran)

The report of a study team set up in 1973 to review arrangements for passenger travel by rail in Greater London. The problems of both London Transport's underground and British Rail's London commuter network are discussed and various proposals suggested. Part 1 comprises the general arguments and conclusions; Part 2 presents a more detailed discussion.

1262 BAYLISS, D. (1975) London's railways in the future. *Greater London Intelligence Quarterly*, Vol. 31, pp. 31–7.

The main findings of the 1974 *London rail study* are summarized. Changes in the organization and planned extension of London's railways are discussed.

1263 ROBBINS, M. (1975) London railway stations. *London Journal*, Vol. 1, pp. 240–62.

The character and development of the British Rail and London Transport stations in the Greater London area are discussed. The main line terminals are individually described, and general comments are provided on underground stations and suburban stations.

1264 KLAPPER, C. F. (1976) London's lost railways. Routledge and Kegan Paul, 140p.

The decline of London's overground railway system during the twentieth century is recounted. Reasons for this decline and methods which might be employed for the rejuvenation of the railway are suggested.

1265 ROBBINS, M. (1977) The first London railways. *Transactions of the London and Middlesex Archaeological Society*, Vol. 28, pp. 292–304.

Specifically concerned with those railways constructed for London's own traffic, rather than those connecting London with places outside it. Two lines, the London and Greenwich and the London and Blackwall Railways, are considered in detail.

1266 JACKSON, A. A. (1978) London's local railways. Newton Abbot: David and Charles, 384p.

A survey of the history of almost 50 local railway lines in London, grouped in chapters under function. A discussion of unfinished lines is also included.

6.6 Sea and port

1267 PORT OF LONDON AUTHORITY (1911–) Annual report and accounts. Port of London Authority.

In addition to financial details, brief accounts of developments in the various docks are given. Trade, industrial relations, organizational changes and developments affecting the port such as the proposed Maplin Airport and the Thames Flood Barrier are also covered in the more recent annual reports.

1268 BROODBANK, J. G. (1921) History of the Port of London. O'Connor. Vol. 1, 270p.; Vol. 2, 516p.

A detailed account of the history and development of the port of London from pre-Saxon times until 1920.

1269 JONES, R. L. (1931) The geography of London river. Methuen, 184p.

Describes the physical setting of the Port of London, and the natural and artificial evolution of the estuary. The economic development of the port and its commercial structure in 1930 are analysed.

1270 BELL, A. (1934) Port of London 1909–1934. Port of London Authority, 74p.

Gives a brief history of the evolution of London's docks and trading pattern, and the crises that led to the creation of the Port of London Authority in 1909. The improvements made to the docks after the First World War are listed.

1271 BIRD, J. (1957) The geography of the Port of London. Hutchinson, 208p.

An account of the changing geography of the port from the twelfth century onwards including the rise of trade, the development of the dock system, and the nature of the river channel itself.

1272 BIRD, J. (1964) The Port of London. In: COPPOCK, J. T. and PRINCE, H. C. (eds) Greater London. Faber and Faber, pp. 202–24.

A brief account of the growth of the port and its trade up to 1962. The chief characteristics of the five London dock systems are described.

1273 HOWGEGO, J. L. (1967) Docks on the North Bank: a nineteenth century transformation. *East London Papers*, Vol. 10, pp. 75–108.

An account of the construction and enlargement of the London docks from the two docks planned at Wapping in the late eighteenth century, to the West India docks, St. Katherine's docks, Victoria docks and, finally, the Tilbury docks which were opened in 1846. Economic decline has since led to their gradual closure.

1274 REES, R. (1967) The Port of London and economic change. *East London Papers*, Vol. 10, pp. 109–24.

The structure and organization of the Port of London are discussed in relation to the growth of bulk cargoes. New port facilities at Tilbury are described.

1275 PUDNEY, J. (1975) London's docks. Thames and Hudson, 192p.

The growth of the various London docks is related from 1800 to the early 1970s, covering the change from sail to steam, the beginnings of organized labour, the Port of London Authority and the containerization revolution.

1276 JARVIS, R. C. (1977) The metamorphosis of the Port of London. *London Journal*, Vol. 3, pp. 55–72.

The historical and geographical development of the port is described from earliest times. Its decline and possible future are examined in the light of changing economic conditions.

1277 BROWN, R. D. (1978) The Port of London. Lavenham, Suffolk: Terence Dalton, 202p.

A description of the history of the Port of London from Roman times up to 1978.

1278 PORT OF LONDON AUTHORITY (1978) Your Port of London: the challenge of the future. Port of London Authority, 14p. (PLA Information Paper, No. 1)

A brief outline of the economic problems of the Port of London – a result of overmanning, insufficient investment and the retention of the loss-making Upper Docks.

1279 PORT OF LONDON AUTHORITY (1978) Your Port of London: the challenge of decision now. Port of London Authority, 14p. (PLA Information Paper, No. 2)

An analysis of government and union response to PLA Information Paper No. 1. As a result of this response, the PLA suggest modified proposals whereby the Royal Docks will be closed and the India and Millwall Docks retained.

1280 SPEARING, N. (1978) London's docks: up or down the river? *London Journal*, Vol. 4, pp. 231–44.

An account of the decline and contraction of the London Docks during the 1960s and 1970s. The various options open to the Port of London Authority are discussed.

1281 TAYLOR, R. (1978) The great PLA disaster. *Management Today*, Nov. pp. 74–81, 164.

An assessment of the enormous problems facing Sir John Cuckney, the Chairman of the Port of London Authority. The main issue is the huge financial loss resulting from overmanning, decline in traffic and poor management.

1282 PORT OF LONDON AUTHORITY (1979) Port of London Authority five year strategic plan 1979–1983. Port of London Authority, various paging.

A programme of development for the Port of London for the next five years. Issues covered include manpower, organization, dock operations and capital development.

1283 PORT OF LONDON AUTHORITY (1979) Your Port of London: planning the next five years. Port of London Authority, 10p. (PLA Information Paper, No. 3)

A number of different proposals for the future of the Port of London Authority are discussed, based on forecasts of various traffic levels.

1284 PORT OF LONDON AUTHORITY (1979) Your Port of London: a port on probation. Port of London Authority, 14p. (PLA Information Paper, No. 4)

A brief review of the objectives of the PLA's *Five year strategic plan* (1979) and the problems which lie ahead.

6.7 Canals and waterways

1285 VINE, P. A. L. (1968) London's lost route to Basingstoke. Newton Abbot: David and Charles, 212p.

The history of the Basingstoke canal which was opened in 1794 in an attempt to link London with the English Channel.

1286 VINE, P. A. L. (1973) London's lost route to the sea: an historical account of the inland navigations which linked the Thames to the English Channel. 3rd ed. Newton Abbot: David and Charles, 267p.

The story of the Wey and Arun Junction Canal and the navigations which formed the link between London and the sea at Portsmouth.

1287 GREENBERG, S. (1974) Canals in London, Greater London Council, 8p. (GLC London Topic, No. 3)

A review of the literature is followed by over 50 references concerned with various aspects of London's canals.

1288 GREATER LONDON COUNCIL and TRANSPORT PLANNING LTD. (1975) Waterborne freight in London: a report of studies. Transport Planning, 131p.

The value of the Grand Union Canal as a transporter of freight is assessed.

1289 PARKER, J. and DUNCAN, J. (1975) The Regents Canal. *Greater London Intelligence Quarterly*, Vol. 30, pp. 19–28.

Recent improvements of the Regent's Canal and the land alongside it are reviewed. Much of the area has been improved and three and one half miles of canal-side walk created.

1290 GREATER LONDON COUNCIL London Canals Consultative Committee (1976) London's canals. 2nd ed. Greater London Council, 20p.

Recent improvements to London's canals and their environs are reviewed, together with future plans.

1291 SPENCER, H. (1976) London's canal: the history of Regent's Canal, 2nd ed. Lund Humphries, 70p.

A history of the planning, construction and use of the Regent's Canal from the early nineteenth century to the present day.

1292 DENNEY, M. (1977) London's waterways. B. T. Batsford, 192p.

A historical account of London's various canals and waterways. Each waterway is dealt with in turn, from its earliest development to its current use.

1293 BURNBY, J. G. L. and PARKER, M. (1978) The navigation of the river Lee (1190–1790). Edmonton Hundred Historical Society, 26p.

An analysis of the organizational and physical difficulties in making the Lea navigable.

1294 FAIRCLOUGH, K. (1979) A Tudor canal scheme for the river Lea. *London Journal*, Vol. 5, pp. 218–27.

Describes an attempt to bring a canal from the lower river Lea to the edge of the City, to improve the supply of agricultural products to the City. The canal was never built, but riparian landholders effected a whole range of minor improvements along the river with the result that later City authorities were convinced that the Corporation had been responsible.

6.8 Air

1295 MINISTRY OF CIVIL AVIATION (1953) London's airports. HMSO, 8p. (Cmnd. 8902)

Government proposals for the rationalization of air traffic in the London region. Use of smaller airports around London is to cease and traffic is to be concentrated on Heathrow and Gatwick.

1296 DOUBLE, R. J. (1958) London airport: the planning of an international airport. *Town Planning Review*, Vol. 29, pp. 79–90.

A discussion of the choice of the Heathrow site for London airport, its layout and future requirements.

1297 BOARD OF TRADE and MINISTRY OF HOUSING AND LOCAL GOVERNMENT (1967) Third London Airport. HMSO, 34p. (Cmnd. 3259)

An account of the government's reaction to the public inquiry into the proposed development of Stansted as the third London airport. Stansted is regarded as the best site, and it is concluded that this airport will be needed by the mid 1970s.

1298 NOISEMENT ABATEMENT SOCIETY (1968) The third London airport. Noise Abatement Society, 78p.

The report of a study commissioned in order to find a feasible alternative to Stansted. It largely concentrates on Foulness which is regarded as a more suitable site, particularly with regard to the noise problem.

1299 FORDHAM, R. C. (1970) Airport planning in the context of the third London airport. *Economic Journal*, Vol. 80, pp. 307–22.

An exploration of the problems facing the Roskill Commission and possible approaches to them. The commission's interpretation of its terms of reference is criticized.

1300 ADAMS, J. G. U. (1971) London's third airport: from TLA to airstrip one. *Geographical Journal*, Vol. 137, pp. 468–504.

A review of the major issues in the airport debate. Assumptions behind the various arguments are examined, and forecasts compared.

1301 COMMISSION ON THE THIRD LONDON AIRPORT (1971) Report. HMSO, 275p. Index, 35p. (Chairman: Lord Roskill)

The report of a commission set up in 1968 to make a recommendation regarding the site for a third London airport. Several possible sites are discussed but, with the exception of Professor Colin Buchanan, the commission recommends that the airport be sited at Cublington in Buckinghamshire.

1302 LICHFIELD, N. (1971) Cost benefit analysis in planning; a critique of the Roskill Commission. *Regional Studies*, Vol. 5, pp. 157–83.

The methodological weaknesses of the work of the Roskill Commission on the Third London Airport are examined. Better use of cost benefit analysis might have led the commission to conclude that Foulness was the best site.

1303 BOURNE, W. R. P. (1972) Foulness and the third London ariport. *Journal of Environmental Planning and Pollution Control*, Vol. 1, pp. 55–64.

Argues against the siting of the third London airport at Foulness on the grounds that it should be preserved for the sake of its wildlife.

1304 CIVIL AVIATION AUTHORITY (1973) Forecasts of air traffic capacity at airports in the London area. Civil Aviation Authority, 36p.

Forecasts of air traffic and capacity at airports in the London area are reviewed up to 1985. It has been assumed that Maplin airport will be open in 1980.

1305 McKIE, D. (1973) A sadly mismanaged affair: a political history of the third London airport. Croom Helm, 256p.

The political side of airport planning in south-east England is examined during the period from 1920 to the rejection of the Roskill enquiry recommendations.

1306 BRITISH AIRPORTS AUTHORITY (1974) Gatwick Airport, London: master plan report. British Airports Authority, 13p.

The history of the airport is discussed together with developments necessary to increase capacity to 16 million passengers per year.

1307 BROMHEAD, P. (1974) The abandonment of the third London airport: some problems and solutions. *Town and Country Planning*, Vol. 42, pp. 373–7.

A number of methods of reducing the load on London's airports by the 1980s are suggested. These include more direct international flights to regional airports and the creation of a European transport plan by the EEC countries.

1308 DEPARTMENT OF TRADE (1974) Maplin: review of airport project. HMSO, 84p.

Traffic forecasts for the London area are reassessed. The effects of the Maplin project on airport noise, safety, urbanization and transport in the South East are examined.

1309 FOSTER, C. D. et al. (1974) Lessons of Maplin: is the machinery for government decision making at fault? Institute of Economic Affairs, 58p. (IEA Occasional Paper, No. 40)

The environmental and political aspects of the Maplin decision are discussed. Alternatives to the Maplin site for the third London airport are suggested.

1310 HALL, J. M. (1974) Swan song for the Maplin white elephant. *Geographical Magazine*, Vol. 46, pp. 721–2.

A brief review of the rise and final abandonment of plans to construct London's new third airport at Maplin. The author criticizes the government for the lack of a national integrated transport policy.

1311 DEPARTMENT OF TRADE (1975) Airport strategy for Great Britain. Part 1: the London area: a consultation document. HMSO, 78p.

This publication has been produced to serve as a basis for consultation with various concerned authorities about future development of air traffic around London. Possible diversion of air traffic to outside this area, the future of the four existing airports and the environmental and economic impact of developments are discussed.

1312 Take off points. (1975) *Town and Country Planning*, Vol. 43, pp. 452–4.

The expansion of London's two existing airports and the development of regional airports are discussed in the light of the decision to abandon Maplin as the third London airport.

1313 BRITISH AIRPORTS AUTHORITY (1976) Heathrow Airport, London: master development plan report, March 1976. British Airports Authority, 18p.

Development proposals for Heathrow Airport up to 1985 are described. Plans include the construction of a fourth terminal.

1314 STANDING CONFERENCE ON LONDON AND SOUTH EAST REGIONAL PLANNING (1976) South East airports. Standing Conference on London and South East Regional Planning, 49p.

An account of research undertaken since the abandonment of the Maplin project. Short and long term expansion of the existing airports in the South East is examined in relation to regional planning.

1315 HALL, P. G. (1977) Stansted – why not? *New Society*, Vol. 39, pp. 62–3.

The arguments for and against the selection of Stansted as the site for London's third airport are presented.

1316 HEATHROW AREA WORKING PARTY (1977) The planning implications of the further expansion of Heathrow: second report of the Working Party. Heathrow Area Working Party, 53p.

A consideration of the short and long-term implications of the development of Heathrow Airport, particularly for the surrounding areas. The effects which expansion to a four terminal airport would have on the local environment, economy and transport system are examined.

1317 DEPARTMENT OF TRADE (1978) Airports policy. HMSO, 48p. (Cmnd. 7084)

A review of government policy in connection with air transport and airports. The national strategy is outlined and policies for the individual London and regional airports are discussed.

1318 ADVISORY COMMITTEE ON AIRPORTS POLICY (1979) The need for a third London airport. HMSO, 58p. (Chairman: G. C. Dick)

A continuing rise in traffic at the London airports is predicted up to the year 2000, although figures are lower than previous forecasts. It is recommended that the extra demand be met by one of six short-listed sites.

1319 STANDING CONFERENCE ON LONDON AND SOUTH EAST REGIONAL PLANNING (1979) A third London airport. Standing Conference on London and South East Regional Planning, 8p.

A statement to the government of the Conference's views on the need for and location of a third London airport. A new airport site, preferably at Maplin, is supported, and either a fifth Heathrow terminal or second Gatwick runway are opposed.

1320 STUDY GROUP ON SOUTH EAST AIRPORTS (1979) Possible sites for a third London airport. HMSO, 162p. (Chairman: G. C. Dick)

Six short-listed sites are considered as contenders for a two runway airport. Stansted is demonstrated to have clear advantages.

Section 7 Social patterns and processes

The growing awareness during Victorian times of the appalling housing problems and deprivation amongst the working classes inspired many concerned tracts and exposés (subsection 7.3) and eventually led to the massive survey *Life and labour of the people in London* (BOOTH, C., 1902–3). A second great socio-economic survey of London followed in 1930 from the London School of Economics and Political Science: *The new survey of London life and labour* (LONDON SCHOOL OF ECONOMICS AND POLITICAL SCIENCE, 1930–5). Despite these two major early studies, the great bulk of publication on the social conditions and processes in London has taken place since the 1960s. Apart from the censuses there have been no surveys of the whole of London to compare with this early work. Interest has largely concentrated on particular areas of London such as Lansbury, Thamesmead, docklands, and more recently Lambeth, or on specific issues.

There have been several important social issues which have attracted the attention of geographers, planners, sociologists, anthropologists and economists writing about London during the 1960s and 1970s. First, there is the city's changing social composition (subsection 7.5), including the gentrification of certain areas and the social decline of others. Secondly, there are the interrelated problems of the decrease in job opportunities, unemployment, and the migration of people from the inner city (7.4). A third issue has been the arrival of immigrants, particularly from the New Commonwealth, and the associated problem of ghetto formation (7.6). Linked with all these issues has been the very basic problem of the perennial shortage of housing in the capital (7.7). During the 1970s a number of other concerns have attracted increasing research: how people in London spend their leisure time and the provision of recreational facilities for them (7.8), and their perception of and attitudes to their urban environment (7.9). Since so many of the social problems of the inner city are closely linked with changing economic activity, and are of major importance to the planner, a number of the items listed in sections 5 (Economic structure and patterns) and 8 (Planning the metropolis and beyond) would also be of interest to the student of London's social patterns and processes.

Much of the publication in this area has emanated from the Greater London Council, particularly in the field of housing. Central government has also been concerned in setting up the Committee on Housing in Greater London in the early 1960s, and subsequently, the Standing Working Party on London Housing, and the Department of the Environment Action Group on London Housing. Increasing realization of the problems of the inner city and the recognition of stress areas by central government in the mid 1970s resulted in the inner city study of part of Lambeth by the consultants SHANKLAND COX PARTNERSHIP (1974–8) and in the *Policy for the inner cities* (DEPARTMENT OF THE ENVIRONMENT, 1977). The social problems of London are also of obvious concern to the various borough planning and social services departments. Only a limited number of publications concerned with issues at borough level have, however, been included here. The majority of them will be found through *Urban Abstracts*.

Aside from government publications, London's social processes have been the concern of a number of monographs from anthropologists and sociologists. Since the 1950s Willmott, Young and Firth have published several kinship studies of different parts of London. The absorption of various ethnic groups, particularly of the West Indians, has been the subject of considerable research by sociologists. More recently, concerned immigrant and community groups such as the Runnymede Trust, SHAC, Shelter, and smaller local groups such as the Barnsbury People's Forum, have been producing publications concerned with various social issues.

The Greater London Council has produced a number of useful research bibliographies and topic papers on London's social problems, particularly housing. An extremely valuable source is the *Social atlas of London* (SHEPHERD, J. et al., 1974). Apart from the census volumes for London, the most useful source of statistical information is the *Annual Abstract of Greater London Statistics* (1966–).

7.1 Bibliographies and statistics

1321 MORREY, C. R. (1971) Quarterly birth and death statistics for the London boroughs. Greater London Council, 14p. (GLC Research Memorandum, No. 328).

Five sets of tables give the following data both for the London boroughs and for inner and outer London: (1) quarterly live births by legitimacy; (2) annual totals of births; (3) a comparison of provisional and final data on births for 1967 and 1968; (4) quarterly death statistics; (5) annual death statistics.

1322 JOHNSON, S. (1972) Perception of the environment, with special reference to the urban environment. Greater London Council, 8p. (GLC Research Bibliography, No. 36)

A select list of 87 references concerned with environmental perception.

1323 PEACOCK, A. (1972) Homelessness. Greater London Council, 26p. (GLC Research Bibliography, No. 34)

A list, with occasional annotations, of 211 references to books, periodicals and newspaper articles concerned with the problems of and policies towards homelessness in London.

1324 SCHLUTER, A. (1972) Outdoor recreation in London. Greater London Council, 18p. (GLC Research Bibliography, No. 39)

A select list of 88 references on outdoor recreation in London published since 1963.

1325 ARMSTRONG, W. and DUGMORE, K. (1973) 1971 Census county report for Greater London: selected topics and historical comparison, 1. Greater London Council, 70p. (GLC Research Memorandum, No. 415)

The 1971 Census report is briefly analysed, and age, sex, marital status, birthplace, households, institutional population and housing occupancy figures are compared with those of the 1961 and 1966 county reports. The population figures are compared with the projected population and the *Greater London development plan* (Greater London Council, 1969).

1326 GREATER LONDON COUNCIL (1973–) Housing facts and figures. No. 1– Greater London Council (irregular).

Basic facts about London's housing situation presented in a series of maps and diagrams.

1327 FIELD, A. M. et al. (1974) 1971 Census data on London's overseas-born population and their children. Greater London Council, 119p. (GLC Research Memorandum, No. 425)

The 1971 Small Area Statistics on Greater London's overseas born population are assessed. The authors then outline the distribution, concentration and various other demographic characteristics of the New Commonwealth ethnic groups.

1328 LONDON BOROUGH OF HILLINGDON (1974) The population of Hillingdon. London Borough of Hillingdon, 58p.

Basic population data for the borough are followed by various social statistics under the heading 'social malaise'.

1329 SHEPHERD, J. et al. (1974) A social atlas of London. Oxford University Press, 128p.

An atlas of the Greater London area with substantial text covering social development, housing, demographic and ethnic structure, socio-economic groups, employment, transport, environment and other aspects of socio-economic life. Most maps cover the whole city and have been constructed on the basis of electoral wards using 1971 Census and other data.

1330 GRAYSON, L. (1975) The single homeless. Greater London Council, 13p. (GLC London Topics, No. 2)

The nature and scale of the problem are outlined, and organizations concerned with the homeless listed. A bibliography of over 40 references is given.

1331 MILLER, C. (1975) Housing in London. 2nd ed. Greater London Council, 78p. (GLC Research Bibliography, No. 28)

An annotated bibliography, with a subject index, of references relating to various aspects of housing in London. Organized under headings such as statistics, housing policy, housing land and housing finance.

1332 DEPARTMENT OF THE ENVIRONMENT (1976) Census indicators of urban deprivation: Greater London. Department of the Environment, 11p.

1971 Census data for enumeration districts are used to locate areas of deprivation in London. The level of deprivation is compared with that of other British conurbations, and its spatial concentration is analysed.

1333 GARTON, B. et al. (1976) 1971 Census county report for Greater London: selected topics and historical comparison, II. 2nd ed. Greater London Council, 59p. (GLC Research Memorandum, No. 421)

A brief analysis of topics concerning housing. Figures are given for the 1961, 1966 and 1971 Censuses, and the problems of changes of definition and comparisons between censuses are discussed.

1334 FIELD, A. M. (1977) Index to published and unpublished work of the Population Studies Section. 3rd ed. Greater London Council, 18p. (GLC Research Bibliography, No. 53)

A compilation of 169 of the major published and unpublished reports of the GLC's Population Studies Section. Most items are concerned with the population of London and are listed alphabetically with a subject index.

1335 HAMLYN, P. (1977) Housing management in London. Greater London Council, 22p. (GLC Research Bibliography, No. 80)

A revision of part of Research Bibliography No. 28 *Housing in London* (MILLER, C. 1975). It contains 100 annotated items, mostly journal references and reports.

1336 HAMLYN, P. (1977) Housing the elderly and disabled in London. Greater London Council, 8p. (GLC Research Bibliography, No. 87)

A revision of part of Research Bibliography No. 28 *Housing in London* (MILLER, C. 1975). An annotated list of 47 references.

1337 SCOTT, G. and PENNELL, H. (1977) The building process. Greater London Council, 13p. (GLC London Topic, No. 18)

An account and bibliography of the local authority house building process, with particular reference to the GLC.

1338 FIELD, A. M. and GREEN, H. (1979) The 1978 guide to non-census sources of population studies data. Greater London Council, 146p. (GLC Research Memorandum, No. 552)

A guide to the major sources of population statistics other than the national censuses. Information is limited to data from 1965 to 1977 for Greater London and the south east England region.

1339 LONDON BOROUGH OF HILLINGDON Planning Department (1979) Population. London Borough of Hillingdon, 25p.

The level of population in the borough and various demographic characteristics are discussed. Future trends are forecast.

7.2 General

1340 LONDON SCHOOL OF ECONOMICS AND POLITICAL SCIENCE (1930–5) The new survey of London life and labour. P. S. King, 9 vols.

A vast socio-economic survey of London conducted by the LSE in 1928. Chapters on all aspects of London life are included and there is particular emphasis on the changes which have taken place since Charles Booth's inquiry into London life and labour undertaken 1886–1903.

1341 DURANT, R. (1939) Watling: a survey of social life on a new housing estate. P. S. King, 128p.

An account of community life and the role of the community centre in the London County Council housing estate constructed in Watling in the late 1920s. Data were collected during several intensive studies of local events, records, groups and organizations.

1342 JEFFREYS, M. (1964) Londoners in Hertfordshire: the South Oxhey Estate. In: CENTRE FOR URBAN STUDIES (ed.) London: aspects of change. Macgibbon and Kee, pp. 205–55.

The results are presented of a 1952–5 survey, based largely on interviews, of a London County Council out-county estate. Issues examined include use of health facilities, social activities, employment and attitudes to South Oxhey.

1343 WESTERGAARD, J. H. and GLASS, R. (1964) A profile of Lansbury. In: CENTRE FOR URBAN STUDIES (ed.) London: aspects of change. Macgibbon and Kee, pp. 159–206.

The results are described of a social survey undertaken in 1951 of Lansbury in the East End, which was the first outstanding example of comprehensive reconstruction in the metropolis. These results are compared with those of a second investigation in 1958.

1344 WESTERGAARD, J. H. (1964) The structure of Greater London. In: CENTRE FOR URBAN STUDIES (ed.) London: aspects of change. Macgibbon and Kee, pp. 91–144.

An outline analysis of the economic and social structure of London based largely on the 1951 Census results. Population growth and structure, socio-economic zones, journeys to work and housing are the main topics covered.

1345 PAHL, R. E. (1965) Urbs in rure: the metropolitan fringe in Hertfordshire. London School of Economics and Political Science, 84p. (Geographical papers, No. 2)

A study of the effects of urbanization on a rural area of Hertfordshire. Changes in economy, modes of life and attitudes have been measured to demonstrate the extent of metropolitan influence.

1346 WALLIS, C. P. and MALIPHANT, R. (1967) Delinquent areas in the county of London: ecological factors. *British Journal of Criminology*, Vol. 7, pp. 250–84.

The distribution of delinquency in the LCC area is related to a variety of ecological factors in an attempt to define the attributes that distinguish delinquent areas. Large correlations are identified between delinquent rates and factors such as overcrowding, education, social class and unemployment.

1347 EVERSLEY, D. E. C. (1972) Old cities, falling populations and rising costs. *Quarterly Bulletin of the Intelligence Unit of the Greater London Council*, No. 18, pp. 5–17.

Some of the economic and social problems of big cities and some of the successes of British planners with new town and green belt policies are described. Present difficulties, especially those of London with its declining population, are discussed in terms of the urban economic base, investment in declining areas and the needs of the people.

1348 LEE, R. K. (1972) Planning and social change in east London. *East London Papers*, Vol. 14, pp. 25–43.

The social consequences of physical planning are discussed, with particular reference to council house provision. Comprehensive housing redevelopment of the docklands is considered unwise.

1349 EVERSLEY, D. E. C. (1973) Problems of social planning in inner London. In: DONNISON, D. and EVERSLEY, D. E. C. (eds) London: urban patterns, problems, and policies. Heinemann, pp. 1–50.

The various social issues facing the London planner in the early 1970s are reviewed. The author questions the traditional approaches to problems such as low incomes, unemployment, housing and migration from inner city areas.

1350 GARDINER, G. (1973) The changing life of London. Tom Stacey, 169p.

The various problems of London and attempts which have been made to solve them are discussed. Particular emphasis is given to environmental issues, housing and transport.

1351 HATCH, S. and SHERROTT, R. (1973) Positive discrimination and the distribution of deprivations. *Policy and Politics*, Vol. 1, pp. 223–40.

An examination of the evidence about the degree of concentration of deprivations in areas of acute social need which were identified in the 1968 Urban Programme. A range of descriptive data, some derived from the census and others from local authority records, is analysed for the two London boroughs of Southwark and Newham.

1352 REDPATH, R. U. and CHILVERS, D. J. (1974) Swinbrook: a community study applied. *Greater London Intelligence Quarterly*, No. 26, pp. 5–17.

A report on the results of a census carried out in the Swinbrook Road area of north Kensington in an attempt to measure the degree of community attachment in the area. The importance of personal attachments which emerged from the study has influenced the GLC architects in their design of a phased redevelopment scheme.

1353 SHANKLAND COX PARTNERSHIP and INSTITUTE OF COMMUNITY STUDIES (1974–1978) Inner area study – Lambeth. Department of the Environment.

Project report. 39p. (IAS/LA/1)
Completes the initial stage and outlines the scope of the study. The programme, along with others in other towns, is to provide a base for general conclusions on policies and actions, within the sphere of the Department of the Environment, for such inner city residential areas which present a combination of poor or declining housing, bad environment and a concentration of social problems.

Changes in socio-economic structure. 31p. (IAS/LA/2)
Data from the 1951, 1961 and 1966 Censuses and the *Household survey* for 1973 have been used to detect changes in the socio-economic structure of the population of Lambeth. Various trends such as gentrification, polarization and migration are discussed.

Interim report on local services. 38p. (IAS/LA/3)
The availability and use of public services in Lambeth are studied, and the possibility of their reorganization discussed.

Labour market study. 78p. (IAS/LA/4)
An examination of the problems of unemployment, low incomes and inadequate job opportunities in the Lambeth Study Area. A number of suggestions are made regarding the causes and possible solution of these problems.

People, housing and district. 102p. (IAS/LA/5)
A study of the population structure of the area, and the housing, lives and attitudes of its residents. Based partly on a household survey, but also draws on census and other data.

Housing stress. 54p. (IAS/LA/6)
An assessment of the attitudes of local residents in an area of poor quality housing, and their reasons for remaining in the area.

Policies and structure. 72p. (IAS/LA/7)
An analysis of the various policies of the GLC and the London Borough of Lambeth which affect land use.

Housing and population projections. 8p. (IAS/LA/8)
A series of population projections have been developed to measure the effect of various housing policies on the population of the area.

Local services: consumers sample. 14p. (IAS/LA/9)
An examination of the accessibility of various health and social services to the local residents.

Poverty and multiple deprivation. 30p. (IAS/LA/10)
An interim report, based on the household survey, on the extent and forms of multiple deprivation, and the kinds of families which are most prone to it.

London's inner area: problems and possibilities. 53p. (IAS/LA/11)
A discussion paper, based on the findings of earlier projects by the consultants, which concentrates on the presentation of policies to tackle inner city decline.

The implications of social ownership. 70p. (IAS/LA/12)
A consideration of the practical implications of social ownership of housing within the study area.

Schools project. 50p. (IAS/LA/13)
An attempt to involve schools, children and parents in the inner area study. Material gathered about the area was made available to schools, and the perception of the area by children was assessed.

Multi-service project. 34p. (IAS/LA/14)
A number of suggestions are made regarding the improvement of public service provision to local residents.

Second report on multiple deprivation. 70p. (IAS/LA/15)
A final report on deprivation, based on household interviews.

Local employers' study. 88p. (IAS/LA/16)
An attempt to discover how and why job opportunities in the area are limited and changing. Based mainly on an in-depth interview survey of a sample of about 40 medium and large employers in inner south London.

The Groveway Project: an experiment in salaried child-minding. 86p. (IAS/LA/17)
An account of an action project to recruit and train child-minders and employ them on a monthly salaried basis.

Housing management and design. 99p. (IAS/LA/18)
A report of a survey of 18 housing estates in the study area. Various

problems on the estates are identified and suggestions are made as to how to deal with them.

Multi-space project. 14p. (IAS/LA/19)
An account of an attempt to improve the appearance of the study area by improving selected derelict sites and buildings.

Study of intending migrants. 31p. (IAS/LA/20)
A report of a survey concerned with would-be movers from the study area. It is recommended that more local authority housing be provided in outer London and that it be made easier for people in public housing to move.

1354 WILLMOTT, P. (1974) Population and community in London. *New Society*, Vol. 30, pp. 206–10.

Reasons for the changing geography of social class in London are suggested. Council housing policy should aim at preserving a stable community life.

1355 MYERS, M.(1975) Urban deprivation and the GLC. *Greater London Intelligence Quarterly*, No. 31, pp. 25–30.

GLC projects to solve the problem of urban deprivation in Spitalfields in Tower Hamlets and the Hanley Road area in Islington are described.

1356 ST CLAIR, S. and DINES, A. (1975) Thamesmead social survey: 1. Greater London Council, 73p. (GLC Research Memorandum, No. 465)

Data are presented which derive from a social survey of Thamesmead which was undertaken in March, 1972. Information covers housing, environment, various facilities and the move to Thamesmead.

1357 WILLMOTT, P. (1975) Whatever's happening to London? An analysis of changes in population structure and their effects on community life. London Council of Social Service, 14p.

A study of changes in London's population, housing and employment, based on an analysis of 1966 and 1971 Census data. A number of suggestions are made as to how certain detrimental changes might be halted.

1358 DAVIS, A. et al. (1977) The management of deprivation: final report of Southwark Community Development Programme. Polytechnic of the South Bank, 100p.

The Southwark Community Development Programme was set up in 1970 as an experiment to examine the causes of deprivation in the Newington ward of Southwark. A number of recommendations are made regarding the organization and provision of social services.

1359 SHANKLAND COX PARTNERSHIP (1977) Inner London: policies for dispersal and balance: final report of the Lambeth Inner Area Study. HMSO, 246p.

A series of reports have been produced on all aspects of socio-economic life in Lambeth (SHANKLAND COX PARTNERSHIP and INSTITUTE OF COMMUNITY STUDIES, 1974–8). This final report summarizes the research and makes a number of recommendations for the improvement of deprived inner city areas.

1360 BAKER, J. (1978) The neighbourhood advice centre: a community project in Camden. Routledge and Kegan Paul, 310p.

The development and aims of the neighbourhood advice centre are described. An assessment is made of its impact on the community.

1361 HOWE, G. M. (1979) Death in London. *Geographical Magazine*, Vol. 51, pp. 284–9.

An examination of some aspects of the medical geography of London. Standard mortality ratios for deaths from all causes, suicide, ischaemic heart disease, cancer of trachea, lung and bronchus, and chronic bronchitis have been calculated for the London boroughs, and variations are discussed.

1362 RHODES, G. (1979) Research in London 1952–1977. *London Journal*, Vol. 5, pp. 57–86.

There has been a great growth in research in the social sciences in relation to London, particularly since the early 1960s. The author demonstrates to what extent this has benefited the capital and what role it might have in the future. There are 88 references to research projects on housing, traffic and transport, population and employment, planning, local government, and the inner city.

7.3 Historical social patterns

1363 BEAMES, T. (1852) The rookeries of London, past, present and prospective. 2nd ed. Bosworth, 309p. (Reissued by Frank Cass, 1970.)

A vivid description of the slum tenements of London as observed in the 1840s. The evolution and condition of individual rookeries are described, the causes perpetuating the poverty analysed, and remedies suggested.

1364 HOLLINGSHEAD, J. (1861) Ragged London in 1861. Smith Elder, 338p.

A collection of articles describing and exposing poverty and deprivation in London, arranged by districts. Long appendices provide additional evidence and statistics.

1365 MAYHEW, H. (1861–2) London labour and London poor: a cyclopedia of the condition and earnings of those that will work, those that cannot work and those that will not work. Griffin Bohn, 4 vols.

A long and detailed account of working class poverty in London, with particular emphasis on street sellers, street buyers, labourers, prostitutes, thieves and beggars.

1366 HILL, O. (1883) Homes of the London poor. 2nd ed. 95p. MEARNS, A. (1883) The bitter cry of outcast London. 24p. (Reissued as one volume, Frank Cass, 1970.)

An account of Octavia Hill's social experiment in providing homes for the poor on a small profit making basis. Mearns's tract was an exposé of the despicable housing situation in London and his pamphlet was a force behind the creation of the 1890 Housing of the Working Classes Act.

1367 MEARNS, A. (1883) The bitter cry of outcast London. Edited with a new introduction by A. S. Wohl, 1970. Leicester: University Press, 155p.

Includes Mearns's exposé of housing conditions together with leading articles from the Pall Mall Gazette of 1883, plus articles by Lord Salisbury, Joseph Chamberlain, and Forster Gozier.

1368 BOOTH, C. (1902–3) Life and labour of the people in London. Macmillan, 18 vols.

An immense social investigation of the condition and occupations of the inhabitants of London at the end of the nineteenth century. The work is arranged as three series of volumes: (1) the series on poverty includes details of house-to-house surveys (with key maps of the status of the inhabitants) displaying an amazing situation of deprivation and iniquitous occupational structure; (2) the industry series provides detailed accounts of the main manufacturing and service industries in London; (3) the religious series is an area-by-area survey of churches and religious institutions and their influence on the neighbourhood.

1369 QUENNELL, P., (ed.) (1949) Mayhew's London. Pilot Press, 569p. (also Spring Books, 1956)

A selection from Henry Mayhew's *London labour and London poor* (1861–2), with an introductory essay on Mayhew. (Quennell has also edited other selections from Mayhew, including *London's underworld* (1950) and *Mayhew's characters* (1951).)

1370 DYOS, H. J. (1955) Railways and housing in Victorian London. *Journal of Transport History*, Vol. 2, pp. 11–21; 90–100.

Assesses the impact of railway construction on working class housing in London. Demolition of houses caused waves of local dispersion and overcrowding, a problem partly compensated by the advent of cheap workers' fares allowing some workers to live in the suburbs.

1371 MITCHELL, R. J. and LEYS, M. D. R. (1958) A history of London life. Longman, 302p.

Uses a variety of sources to examine different types of social, economic, religious, cultural, legal and working class life, arranged by periods.

1372 FRIED, A. and ELMAN, R. M., (eds) (1969) Charles Booth's London: a portrait of the poor at the turn of the century drawn from his 'Life and labour of the people in London'. Hutchinson, 342p.

General introductions to the character of Booth and his social environment are followed by selections from his great work (BOOTH, C., 1902–3).

1373 DYOS, H. J. (1967) The slums of Victorian London. *Victorian Studies*, Vol. 11, pp. 5–40.

Defines a slum and charts the increasing awareness of slum conditions in Victorian London. The process of slum creation is analysed using examples.

1374 TARN, J. N. (1968) The Improved Industrial Dwellings Company. *Transactions of the London and Middlesex Archaeological Society*, Vol. 22, pp. 43–59.

Outlines the history of the company set up by Sir Sydney Waterlow in 1864 to build houses for the skilled working class of London on a commercial basis. The architectural development of the large number of blocks of flats erected by the company is described and compared with similar blocks erected by the Peabody Trustees.

1375 WOHL, A. S. (1968) The bitter cry of outcast London. *International Review of Social History*, Vol. 13, pp. 189–245.

Examines the social background that led to the publication of Andrew Mearns's pamphlet *The bitter cry of outcast London* (MEARNS, A., 1883). The impact of the pamphlet in stirring social conscience and the outcome of the Royal Commission on the Housing of the Working Classes is analysed in detail.

1376 JONES, G. S. (1971) Outcast London: a study in the relationship between classes in Victorian Society. Oxford University Press, 424p.

Argues that the market for unskilled labour in London was so great that a huge population of casual poor congregated in the centre and East End. Housing was grossly inadequate and, as time passed, the middle and respectable working classes viewed the poor as a threat, without being able to solve the problem.

1377 THOMPSON, E. P. and YEO, E. (eds) (1971) The unknown Mayhew. Merlin Press, 489p.

Henry Mayhew published 82 lengthy articles in the Morning Chronicle 1849–50, only part of which was used in his *London labour and London poor* (MAYHEW, H., 1861–2). This book selects articles describing the more skilled trades of mid-Victorian London and examines Mayhew's role as a social investigator.

1378 WOHL, A. S. (1971) The housing of the working classes in London, 1815–1914. In: CHAPMAN, S. D. (ed.) The history of working class housing. Newton Abbot: David and Charles, pp. 13–54.

Argues that the demand for working class housing in central London was so great that even high rents obtained only substandard and insanitary accommodation. Cheap workmen's tickets led to the development of working class suburbs after 1883, but jerry building and poor sanitation only repeated the problem there at a later date. Model dwelling blocks and council housing slightly alleviated the shortage, but by 1914 there was still a great need to be met.

1379 DYOS, H. J. and REEDER, D. A. (1973) Slums and suburbs. In: DYOS, A. J. and WOLFF, M. (eds) The Victorian city: images and realities. Routledge and Kegan Paul, pp. 359–86.

Discusses the creation of slums in Victorian London as a result of poor quality building, gross demand and unscrupulous landlords. The flight of the middle classes to the suburbs was directly and indirectly at the expense of the working classes, though the poor were also found in middle class areas in pre-existing nuclei and in otherwise unlettable houses.

1380 STEFFEL, R. V. (1973) The slum question: the London County Council and decent dwellings for the working classes, 1880–1914. *Albion*, Vol. 5, pp. 314–25.

Surveys the slum problem and the developing housing policies of late nineteenth century London, with an analysis of the contribution made by the LCC in trying to inspect slums and bring pressure on landlords, and eventually in building its own houses.

1381 TARN, J. N. (1973) Five per cent philanthropy: an account of housing in urban areas between 1840 and 1914. Cambridge: University Press, 211p.

Examines the growth of working class housing in the nineteenth century, and the philosophy behind the developing housing movement. Philanthropic and governmental schemes are discussed, and though the book examines Britain as a whole, the greatest housing problems and the best known housing schemes were located in London.

1382 TARN, J. N. (1974) French flats for the English in nineteenth-century London. In: SUTCLIFFE, A. (ed.) Multi-story living: the British working class experience. Croom Helm, pp. 19–40.

Outlines the movement for building blocks of flats in London between 1840 and 1890. Although flatted model dwellings for the working class were successful, flats for the middle and upper classes eventually lost to the suburban house and garden movement.

1383 MALCOLMSON, P. E. (1975) Getting a living in the slums of Victorian Kensington. *London Journal*, Vol. 1, pp. 28–55.

Examines the occupational pattern of men and women in four distinctive slum communities in northern Kensington, mainly derived from census data for 1851–71. Male employment as pig-keepers, brick makers, gas workers and nurserymen was largely unskilled and seasonal, while female employment was dominated by laundry work for the many rich households of Kensington.

1384 WOHL, A. S. (1977) The eternal slum: housing and social policy in Victorian London. Edward Arnold, 386p.

A detailed discussion of the housing problem of Victorian London and how philanthropic societies, the Metropolitan Board of Works and the LCC attempted to cope with the situation. The migration to the developing suburbs is also discussed.

1385 CROSSICK, G. (1978) An artisan elite in Victorian Society: Kentish London 1840–1880. Croom Helm, 306p.

Shows how Greenwich, Woolwich and Deptford were separate communities dominated by an elite of tradesmen and merchants. Social groups tended to be well defined with little social mobility, while many sons followed their father's occupations.

7.4 Population and migration

1386 PRICE-WILLIAMS, R. (1885) The population of London, 1801–1881. *Journal of the Statistical Society of London*, Vol. 48, pp. 347–440.

A long and detailed analysis of the structure of London's population, concentrating mainly on the period 1841–81. Two detailed maps show population density and rates of increase and decrease in the LCC area, (and some adjoining areas) for the period 1871–81. Much data are summarized in tables and graphs.

1387 SHANNON, H. A. (1935) Migration and the growth of London 1841–91: a statistical note. *Economic History Review*, Vol. 5, pp. 79–86.

Analyses birthplaces of London inhabitants using census returns. In the period 1851–61 immigration into London was greater than the natural increase in population, but by 1881–91 immigration had fallen behind natural increase though the absolute figure was considerably higher than in 1851–61.

1388 WILLMOTT, P. (1963) The evolution of a community: a study of Dagenham after forty years. Routledge and Kegan Paul, 154p.

A study of the London County Council's estate in Dagenham, which was built in the 1920s and 1930s to rehouse people from the East End of London. The author traces the relationships of the people of Dagenham with relatives, friends and neighbours, and then examines their attitudes to each other, to politics and to social class.

1389 COPPOCK, J. T. (1964) Dormitory settlements around London. In: COPPOCK, J. T. and PRINCE, H. C. Greater London. Faber and Faber, pp. 265–91.

The distribution and nature of the dormitory settlements around London are discussed. The growth of such settlements is illustrated by the example of Radlett, whose history since the 1890s is described.

1390 HERAUD, J. B. (1966) The new towns and London's housing needs. *Urban Studies*, Vol. 3, pp. 8–21.

Movement of families from London to the new town of Crawley is examined, and it is discovered that a comparatively small proportion originated from areas of housing shortage. The government is urged to change its policy on new towns so that recruitment is from the most needy areas.

1391 LEVIN, P. H. (1968) Population trends, housing, and the overspill programme. *Quarterly Bulletin of the Research and Intelligence Unit of the Greater London Council*, No. 5, pp. 49–54.

A discussion of the planned overspill programme and the growth of the new and expanding towns in the light of London's declining population.

1392 TEPER, S. (1968) Patterns of fertility in Greater London. Greater London Council, 32p. (GLC Occasional Research Paper, No. 1)

A short descriptive study of fertility in the Greater London area during the early 1960s. The differences in fertility between London and England and Wales are examined in some detail, together with an analysis of London in relation to other conurbations.

1393 GRYTZELL, K. G. (1969) County of London: population changes 1801–1901. Lund, Sweden: Gleerup for the Royal University of Lund, Dept. of Geography, 129p. (Lund Studies in Geography, Series B, No. 33)

A detailed analysis of population changes presented in maps and tables. The main part comprises a decade by decade analysis of changes in population density and urban growth. London is also compared with England and Wales, and with 12 surrounding counties, for the period 1700–1911.

1394 JOHNSON, R. J. (1969) Population movements and metropolitan expansion: London 1960–1. *Transactions of the Institute of British Geographers*, No. 46, pp. 69–91.

A discussion of patterns of population mobility and movements within Greater London. Several models of intra-urban migration are examined.

1395 THOMPSON, E. J. (1969), Some data on past population changes in London and south east England, 1. *Quarterly Bulletin of the Research and Intelligence Unit of the Greater London Council*, No. 6, pp. 43–55.

Tables and diagrams, together with a brief explanatory text, relate to intercensal population changes in London and the South East. Additional data concern post-war births, deaths and natural increase in Greater London and migration to and from London before the 1961 and 1966 Censuses.

1396 RODERICK, W. P. (1971) The London new towns: origins of migrants from Greater London up to December, 1968. *Town Planning Review*, Vol. 42, pp. 323–41.

An attempt to discover a pattern in the movement of people from the Greater London boroughs to the surrounding ring of eight new towns established between 1946 and 1949. The London boroughs closest to the various new towns along the radial lines of communication tend to be those that contribute the greatest number of immigrants.

1397 THOMPSON, E. J. (1971) A review of population projections for Greater London issued by government departments (1963–70). Greater London Council, 20p. (GLC Research Memorandum, No. 281)

The six projections of the future level of London's population which were issued between 1963 and 1970 by government departments are compared and discussed.

1398 UNGERSON, C. (1971) Moving home: a study of the redevelopment process in two London boroughs. G. Bell, 99p.

A study of the strains imposed on people living in urban redevelopment areas. Various policies are suggested which might help remove the potential insecurities of those about to be moved.

1399 GEE, F. A. (1972) Homes and jobs for Londoners in new and expanding towns. HMSO, 112p.

The report of a survey into the working of the Industrial Selection Scheme, the official machinery for helping people find work in new and expanding towns. The survey was based on questionnaires sent to people registering in the scheme.

1400 SKUSE, C. M. (1972) Thamesmead: survey of ingoing tenants. Greater London Council, 54pp. (GLC Research Memorandum, No. 335)

Presents basic socio-economic and demographic data for about the first 1,000 ingoing households to Thamesmead. The data were collected voluntarily by interview for the Planning and Transportation and Housing Departments.

1401 BLOWERS, A. (1973) London's out-county estates: a reappraisal. *Town and County Planning*, Vol. 41, pp. 409–14.

The successes and failures of the 13 estates which were constructed in London's green belt as part of the post-war housing programme are discussed.

1402 MORREY, C. R. (1973) The changing population of the London boroughs. Greater London Council, 56p. (GLC Research Memorandum, No. 413)

The population of the present inner London boroughs is traced back to 1801, and that of the present outer boroughs to 1901. The methodology and results are discussed.

1403 THOMPSON, E. J. (1974) Population projections for metropolitan areas. *Greater London Intelligence Quarterly*, No. 28, pp. 5–10.

A description of the methods that can be used to make population projections for urban areas. Four approaches are discussed, and it is explained how the method chosen will depend not only on the data available but also on how the projection is to be used.

1404 DUGMORE, K. (ed.) (1975) The migration and distribution of socio-economic groups in Greater London – evidence from the 1961, 1966 and 1971 Censuses. Greater London Council, 123p. (GLC Research Memorandum, No. 443)

Changes in the distribution of men of different socio-economic groups in Greater London between 1961 and 1971 are assessed in an attempt to discover more about migration patterns in the capital.

1405 GILJE, E. and HOLLIS, J. (1975) Demographic projections for Greater London and the London Boroughs, 1974. Greater London Council, 254p. (GLC Research Memorandum, No. 455)

Four population projections have been produced using different fertility and migration assumptions. The assumptions behind the projections are described, and tabulations of population changes up to 1991 are given for Greater London, the borough groups and individual boroughs.

1406 GILJE, E. K. (1975) Migration patterns in and around London. Greater London Council, 30p. (GLC Research Memorandum, No. 470)

Presents an analysis of migrants to and from London, their characteristics, historical comparisons, reasons for moving and the effect of planned versus voluntary migration. The data used are the latest available from the 1971 Census, the National Movers Survey and the Greater London Transportation Survey.

1407 LOMAS, G. M. (1975) Population trends and housing needs. *Built Environment Quarterly*, Vol. 1, pp. 24–7.

London's housing problems are attributed to a mismatch of households and dwellings rather than to a shortage of accommodation. The author stresses the differential development of population and household structure and the stock of dwellings, and suggests some solutions.

1408 OGILVY, A. A. (1975) Bracknell and its migrants – twenty one years of new town growth. HMSO, 119p.

The planning and development of Bracknell together with inward and outward migration are examined. There is also a discussion of the role of the Industrial Selection Scheme in the town's growth.

1409 BIRD, H. (1976) Residential mobility and preference patterns in the public sector of the housing market. *Transactions of the Institute of British Geographers*, New Series, Vol. 1, pp. 20–33.

Requests to move by tenants of the Greater London Council and Newcastle Corporation have been analysed. Reasons for moving, distances and directions of would-be moves, and mobility rates are discussed.

1410 FERNANDO, E. and HEDGES, B. (1976) Moving out of Southwark. Social and Community Planning Research, 28p.

Part of a two-part survey to examine the reasons for migration from Southwark (*see* PRESCOTT-CLARKE, P. and HEDGES, B., 1976). Reasons for leaving are examined, in particular the role of the various facilities and services available in Southwark.

1411 DEAKIN, N. D. and UNGERSON, C. (1977) Leaving London: planned mobility and the inner city. Heinemann, 194p.

This research into the problems of the inner city and the attitudes of inhabitants falls into three sections: (1) a study of two deprived wards in Islington; (2) an examination of the working of the new and expanding towns scheme, (3) the experience of migrants from Islington to new or expanded towns is explored.

1412 FIELD, A. M. and CROFTS, C. (1977) Some aspects of planned migration to new and expanding towns. Greater London Council, 40p. (GLC Research Memorandum, No. 527)

Various data concerning the movement of population and industry to new and expanding towns are examined. The inadequacy of the industrial statistics is lamented.

1413 HUSAIN, J. (1977) Londoners who left. *New Society*, Vol. 41, pp. 17–8.

The results of an urban research project in Bury St Edmunds are outlined. Recent migrants from London have been interviewed concerning their attitudes to the move.

1414 CONGDON, P. et al. (1978) 1977–8 round of demographic projections for Greater London, Part 1: Methodology and assumptions. Greater London Council, 56p. (GLC Research Memorandum, No. 538)

The first in a group of three reports concerned with demographic projections for Greater London until 1991. A description of methodology and assumptions is followed by a commentary on the broad results.

1415 CONGDON, P. et al. (1978) 1977–8 round of demographic projections for Greater London, Part 2: population projections. Greater London Council, 153p. (GLC Research Memorandum, No. 539)

Three detailed tabulations of population projections up to 1991.

1416 CONGDON, P. et al. (1978) 1977–8 round of demographic projections for Greater London, Part 3: economically active and household projections. Greater London Council, 170p. (GLC Research Memorandum, No. 540)

An introductory note is followed by five detailed tabulations of the economically active, and household projections up to 1991.

1417 LEAN, B. (1978) A population policy for inner London. *Greater London Intelligence Journal*, No. 41, pp. 14–8.

Policies are proposed for halting the decline of London's population. A key solution would be the attraction of skilled labour to the inner city.

1418 STONE, P. A. (1978) The implications for the conurbations of population changes (with particular reference to London). *Regional Studies*, Vol. 12, pp. 95–123.

The level of population in British conurbations has fallen whilst other areas have increasing numbers. The causes and effects of such changes are analysed with particular reference to London.

1419 CONGDON, P. and HOLLIS, J. (1979) Demographic projections for the counties of south east England, 1978. Greater London Council, 132p. (GLC Research Memorandum, No. 557)

The results are given of a set of four projections of total population, economically active population and households for the counties of the south-east England region up to the year 1991. Figures for central, inner and outer London are included.

7.5 Social structure

1420 YOUNG, M. and WILLMOTT, P. (1957) Family and kinship in east London. Routledge and Kegan Paul, 232p.

This study of family and community life in Bethnal Green and a new housing estate in Essex, to which local residents were moving, reveals a number of interesting contrasts between the old district and the new.

1421 WILLMOTT, P., and YOUNG, M. (1960) Family and class in a London suburb. Routledge and Kegan Paul, 172p.

A study of family and community life in the largely middle class London suburb of Woodford, based on interviews with residents. Family and social relationships, the care of old people, and attitudes to social class are all examined and contrasted with attitudes prevalent in working class Bethnal Green.

1422 WHITEHAND, J. W. R. (1967) The settlement morphology of London's cocktail belt. *Tijdschrift voor Economische en Sociale Geografie*, Vol. 58, pp. 20–7.

An investigation into the form and origin of the settlements around London which comprise the cocktail belt. The morphology of the towns, their relation to railway lines and the nature of their housing are discussed.

1423 FIRTH, R. et al. (1970) Families and their relatives: kinship in a middle class sector of London. Routledge and Kegan Paul, 476p.

The authors have studied two areas of north London in an attempt to demonstrate that in industrial societies, families are not isolated from their kin and that kinship has positive functions. The survey is based largely on interviews.

1424 DALY, M. (1971) Characteristics of 12 clusters of wards in Greater London: an analysis using 11 general variables from the 1966 Census. Greater London Council, 36p. (GLC Research Report, No. 13)

Eleven variables from the 1966 Sample Census have been chosen to represent a general social, economic and demographic description of wards in Greater London. Cluster analysis techniques have been used to produce a classification of these areas.

1425 HARRIS, M. and LYONS, J. (1971) Social polarization. Greater London Council, 37p. (GLC Research Memorandum, No. 324)

Previous work on the subject of social polarization is reviewed with particular reference to London. A methodology for measuring this polarization is explained using calculations based on Greater London census data.

1426 KELLY, F. (1971) Classifications of the London boroughs. Greater London Council, 48p. (GLC Research Report, No. 9)

A number of classifications of London boroughs have been produced using data from the 1966 Sample Census. The classifications are derived from various socio-economic variables using cluster analysis techniques.

1427 HARRIS, M. (1973) Some aspects of social polarization. In: DONNISON, D. and EVERSLEY, D. E. C. (eds) London: urban patterns, problems and policies. Heinemann, pp. 156–89.

Recent work on social polarity and social polarization in London is reviewed. It is concluded that research does not suggest that there has been a tendency for social polarity to increase in Greater London, and that, in any case, the relationship between social polarity and social problems is not self evident.

1428 ROWLAND, J. (1973) Community decay. Harmondsworth: Penguin Books, 152p.

Surveys and case studies of the poor economic and social conditions in north Islington are described. A number of proposals are made for reviving community spirit in this area.

1429 WILLMOTT, P. and YOUNG, M. (1973) Social class and geography. In: DONNISON, D. and EVERSLEY, D. E. C. (eds) London: urban patterns, problems, and policies. Heinemann, pp. 190–214.

Special analyses of 1951 and 1966 Census data have been undertaken in order to compare the social class distribution in Greater London, and examine the changes which have taken place. The differential migration of social classes is viewed in the light of the physical and economic geography of the city.

1430 YOUNG, M. and WILLMOTT, P. (1973) The symmetrical family: a study of work and leisure in the London region. Routledge and Kegan Paul, 398p.

In a discussion of the results of a sociological survey in the London region, it is argued that a new style of family life has emerged. With wives increasingly working outside the home, and husbands inside it, the new family is under great pressure.

1431 WEBBER, R. (1974) Social area analysis of Greater London. Planning Research Application Group, 65p. (PRAG Technical Paper, No. 7)

A method of identifying social areas within London is described whereby spatial variation can be assessed by three factors which measure social status, employment structure and aspects of household stability. The pattern of social areas in London is discussed.

1432 BLAKE, J. (1975) The Lomas Report. *Town and Country Planning*, Vol. 43, pp. 294–8.

The economic and social problems of inner London are not as bad as have been suggested. Unemployment is less than the national level and polarization of social groups is less marked than in other great cities.

1433 CROSS, P. (1975) Population analysis in Hackney. *Greater London Intelligence Quarterly*, No. 30, pp. 38–46.

Various aspects of the population of the London Borough of Hackney are considered. Changes between 1951 and 1971 are analysed in relation to population structure, migration and social groups.

1434 WHITE, D. (1975) Newham: an example of urban decline. *New Society*, Vol. 34, pp. 201–4.

The decline of industry in Newham has led to the moving away of the middle classes and the decay of the community. Poorer, largely immigrant groups have moved into the borough.

1435 BERTHOUD, R. (1976) Where are London's poor? *Greater London Intelligence Quarterly*, No. 36, pp. 5–12.

An attempt to analyse variations in income between and within wards and between households in Greater London to test the hypothesis that there are areas of deprivation as suggested in the Plowden Report. The conventional definition of an inner ring of deprivation is re-examined and an alternative eccentric ring concept is advanced.

1436 HAMNETT, C. (1976) Social change and social segregation in inner London, 1961–1971. *Urban Studies*, Vol. 13, pp. 261–271.

Census data have been used in this examination of the differential decline of population across various socio-economic groups in London.

1437 HILL, S. (1976) The dockers: class and tradition in London. Heinemann, 252p.

A survey of the behaviour and attitudes of London's dockers and their place in the British class system.

1438 DEAKIN, N. D. et al. (1977) Assessing London's social problems. Greater London Council, 18p.

A discussion about the definition of deprivation, its measurement, and the location of areas of deprivation. How best to establish an objective programme for the alleviation of deprivation is considered.

1439 PITT, J. (1977) Gentrification in Islington. Barnsbury People's Forum, 36p.

The recent gentrification of Islington is discussed, together with policies necessary to stem its growth.

1440 CONGDON, P. (1978) Classifications of the London boroughs using 1971 Census data. Greater London Council, 60p. (GLC Research Memorandum, No. 550)

The London boroughs are classified using various sets of social, demographic and housing indicators from the 1971 Census. Results are compared with earlier classifications based on the 1966 Sample Census.

1441 CHANDLER, S. E. (1979) The incidence of residential fires in London – the effect of housing and other social factors. Building Research Establishment, 2p. (BRE Information Paper, 20/79)

Data on the incidence of domestic fires in Greater London have been analysed in relation to a number of parameters. Non-owner occupation, a high proportion of children in care, high population density and lack of basic household amenities all correlated with high fire incidence.

1442 HAMNETT, C. and WILLIAMS, P. R. (1979) Gentrification in London, 1961–71: an empirical and theoretical analysis of social change. University of Birmingham, 30p. (University of Birmingham Centre for Urban and Regional Studies, Research Memorandum, No. 71)

A study of the nature and scale of the problems of gentrification and consequent working class displacement in London. One of the main effects is the reduction of housing stock.

7.6 Ethnic structure

1443 GLASS, R. and POLLINS, H. (1960) Newcomers: the West Indians in London. Centre for Urban Studies, 278p (CUS Report, No. 1)

The characteristics and distribution of West Indians in London are analysed from data gathered from the Migrant Services Division of the West Indies Commission. Various problems of adjustment and attitudes are then reviewed.

1444 PATTERSON, S. (1963) Dark strangers: a sociological study of the absorption of a recent West Indian migrant group in Brixton, South London. Tavistock, 470p.

An in-depth survey of West Indian life in Brixton. Three main aspects of life are covered: economic life, housing and social relationships, together with the development of forms of social organization within the immigrant group.

1445 JACKSON, J. A. (1964) The Irish. In: CENTRE FOR URBAN STUDIES (ed.) London: aspects of change. Macgibbon and Kee, pp. 293–308.

An account of Irish immigration to London is followed by a discussion of their absorption into the economic and social life of the capital.

1446 PATTERSON, S. (1964) Polish London. In: CENTRE FOR URBAN STUDIES (ed.) London: aspects of change. Macgibbon and Kee, pp. 309–42.

Post-1939 Polish settlement in London is examined. Most aspects of Polish life in London are covered: kinship, social organization, political and religious attitudes, and absorption into British society.

1447 FITZHERBERT, K. (1967) West Indian children in London. G. Bell, 111p. (Occasional Papers on Social Administration, No. 19)

The family system in the West Indies is examined together with the ways in which it has been modified by immigrants in England. This is followed by a report based on case studies of West Indian children in care, and a field work report by a child care officer.

1448 LEECH, K. (1967) The role of immigration in recent east London history. *East London Papers*, Vol. 10, pp. 3–18.

The history of immigration of various groups into east London is traced from the arrival of the Huguenots in the seventeenth century, the Irish and Jews in the nineteenth and early twentieth centuries, to the New Commonwealth immigrants after 1945. Levels of integration and the out-movement of the existing population are analysed.

1449 NG, C. K. (1968) The Chinese in London. Oxford: University Press, 92p.

A history of Chinese immigration into Britain is followed by a survey of the economic and social background of the Chinese in London. Occupations, allegiances and absorption into British society are discussed.

1450 DOHERTY, J. (1969) The distribution and concentration of immigrants in London. *Race Today*, Vol. 1, pp. 227–31.

An analysis of the distribution and concentration of five immigrant groups in Greater London, using data from the 1966 Census. This shows the distribution of Irish, West Indians, Asians and all immigrant groups by ward.

1451 LEES, L. H. (1969) Patterns of lower class life: Irish slum communities in nineteenth century London. In: THERNSTROM, S. and SENNETT, R. (eds) Nineteenth-century cities. Yale University Press, pp. 359–85.

Re-examines the perception of Irish lower class life in London as described by contemporary observers, and argues that the implicit picture of chaos and disorder needs reconsideration. The poor Irish were little different from the poor British and their physical movements and slow social mobility followed a recognizable pattern.

1452 HADDON, R. F. (1970) A minority in a welfare state society: the location of West Indians in the London housing market. *New Atlantis*, Vol. 2, pp. 80–133.

An attempt to develop a framework for the analysis of the position of West Indians in the London housing market. 1966 Census data are used.

1453 LOMAS, G. M. and LUM, C. S. (1972) The housing of immigrants in London. Greater London Council, 36p. (GLC Research Memorandum, No. 336)

The relative concentration and diffusion, tenure and socio-economic status of immigrants in the London housing market are reviewed. This is followed by an examination of the policies and practices of the various bodies involved with the problem.

1454 DALTON, M. and SEAMAN, J. M. (1973) The distribution of New Commonwealth immigrants in the London Borough of Ealing, 1961–6. *Transactions of the Institute of British Geographers*, Vol. 58, pp. 21–40.

1961 and 1966 Census data are used to determine the overall increase in New Commonwealth immigrants, together with changes in their spatial distribution. The trend appears to be towards thickening of established clusters of immigrant settlement and hence increasing segregation from the host population.

1455 DEAKIN, N. D. and UNGERSON, C. (1973) Beyond the ghetto: the illusion of choice. In: DONNISON, D. and EVERSLEY, D. E. C. (eds) London: urban patterns, problems and policies. Heinemann, pp. 215–47.

The difficulties encountered by blacks in the housing and employment markets are examined, and the possibility of ghetto formation in London is considered. Various attempts to legislate against discrimination are considered.

1456 LOMAS, G. M. (1974) The inner city: a preliminary investigation of the dynamics of current labour and housing markets with special reference to minority groups in inner London. London Council of Social Service, 38p.

The worsening housing and unemployment problems of blacks in inner London are examined. Lomas suggests that coping with these problems requires a multiple approach affecting incomes, housing, employment and transport.

1457 BERMANT, C. (1975) Point of arrival: a study of London's East End. Methuen, 292p.

A study of the immigrant groups who arrived at London's East End, with special emphasis on Jews. Argues that the East End was attractive as an area of cheap housing, while religious and co-operative practices kept the groups intact for some time until cultural assimilation and wealth permitted them to move to other areas.

1458 DENCH, G. (1975) Maltese in London: a case study in the erosion of ethnic consciousness. Routledge and Kegan Paul, 302p.

The level and nature of Maltese immigration into London is examined, and the reasons for the bad public image of the Maltese discussed. The author concludes that the migratory process has led to disorganization of community life.

1459 KOSMIN, B. A. and GRIZZARD, N. (1975) Jews in an inner London borough: a study of the Jewish population of the London Borough of Hackney, based upon the 1971 Census. Board of British Jews Research Unit, 40p.

A survey of the demographic characteristics, social patterns, economic and community life of Jews in Hackney. Data are taken from the 1971 Census, the electoral register and other sources.

1460 RUNNYMEDE TRUST (1975) Race and council housing in London. Runnymede Trust, 11p.

The provision of council housing to people whose parents were born in the New Commonwealth is reviewed. Both the proportion of blacks in council housing and the types of estate in which they live is examined, and it is concluded that housing provision for blacks is inferior to that for whites.

1461 PARKER, J. and DUGMORE, K. (1976) Colour and the allocation of GLC housing: the report of the GLC lettings survey, 1974–75. Greater London Council, 94p. (GLC Research Report, No. 21)

A social investigation involving a random sample of households becoming tenants of the GLC over a one year period, from June 1974, is described and analysed. It is concluded that GLC housing allocations appear to perpetuate and reinforce the geographical concentration of immigrant families in the housing stress areas of Inner London. The causes of this discrimination are discussed.

1462 GREATER LONDON COUNCIL Policy Studies Administration (1977) Ethnic minorities and disadvantage. Greater London Council, 18p.

An examination of the distribution and problems of ethnic minorities in London. Implications for GLC policies are also considered.

1463 LEE, T. R. (1977) Race and residence: the concentration and dispersal of immigrants in London. Oxford: University Press, 194p.

An examination of the current levels of residential concentrations and segregation of the main immigrant groups in London, with particular emphasis on the position of the West Indians. The trends in the pattern of immigrant dispersal over the period 1961–71 are analysed mainly on the basis of census data, including special unpublished tabulations, which provide specific information on immigrant groups.

1464 TAPER, T. (1977/8) The allocation of Islington housing to ethnic minorities. *New Community*, Vol. 6, pp. 41–4.

The results of a questionnaire survey of households allocated local authority housing in Islington in 1975/6. It is hoped that improved treatment of ethnic minorities will result from the adoption of a number of recommendations.

7.7 Housing

1465 JOHNSON, J. H. (1964) The suburban expansion of housing in London 1918–1939. In: COPPOCK, J. H. and PRINCE, H. C. Greater London. Faber and Faber, pp. 142–66.

A discussion of the nature of and reasons for the expansion of housing outside the administrative county of London during the interwar period. Issues such as transport improvements, the changing distribution of population, and the provision of local authority housing are highlighted.

1466 COMMITTEE ON HOUSING IN GREATER LONDON (1965) Report. HMSO, 450p. (Cmnd. 2605) (Chairman: Sir Milner Holland)

The housing situation in Greater London has been surveyed with particular reference to the use, maintenance and management of both private and public rented accommodation, and to the relations between the occupiers of rented accommodation and private landlords.

1467 GLASS, R. and WESTERGAARD, J. H. (1965) London's housing needs: statement of evidence to the Committee on Housing in Greater London. Centre for Urban Studies, 108p. (CUS Report, No. 5)

This publication comprises three complementary reports concerned with aspects of the housing situation in Greater London. These are concerned with the main factors in housing needs, their geographical patterns, and the use of space in public and private housing.

1468 STANDING WORKING PARTY ON LONDON HOUSING (1967) The housing role of the Greater London Council within London. HMSO, 10p. (Standing Working Party on London Housing Report, No. 1)

The first report of a working party set up in 1965 'to consider ways and means by which central and local government can co-operate in implementing measures to deal with the problems reported on by the Milner Holland Committee (COMMITTEE ON HOUSING IN GREATER LONDON, 1965) and generally to improve the housing situation in Greater London and to keep under review any arrangements made for these purposes'. The role of the GLC is defined and the need for cooperation between the boroughs and the GLC is stressed.

1469 BARBOLET, R. H. (1969) Housing classes and the socio-ecological system. Centre for Environmental Studies, 47p. (CES University Working Paper, No. 4)

An examination of the relationship between spatial structure (measured in terms of commuting time), and social structure, which results in an imbalance between the housing market and the job market for some categories of workers. The report contains an analysis of records of first owners of all new houses built in Kent from 1966 by one private developer.

1470 CRAVEN, E. A. (1969) Private residential expansion in Kent 1956–64: a study of pattern and process in urban growth. *Urban Studies*, Vol. 6, pp. 1–16.

Data from planning applications in Kent demonstrate a trend towards larger sites and more medium density house types, closely linked with the growing importance of large non-local developers. Implications for land use planners are discussed.

1471 STANDING WORKING PARTY ON LONDON HOUSING (1969) Rehousing obligations of Local Authorities. HMSO, 3p. (Standing Working Party on London Housing Report, No. 2)

A consideration of the rehousing obligations of the local authorities under the Housing Acts, and their operation under London's conditions of housing pressure. A number of recommendations are made.

1472 CHAPMAN, J. S. and WHELLER, B. (1970) The characteristics of London's households. Greater London Council, 116p. (GLC Research Report, No. 5)

The second part of the GLC housing survey investigates the pressure for accommodation from households wishing to subdivide or to move, and the relationship between cost and income for various types of dwelling.

1473 CHAPMAN, J. S. and BRACEWELL, I. B. (1970) The condition of London's housing: a survey. Greater London Council, 164p. (GLC Research Report, No. 4)

Describes the first part of a survey undertaken in 1967 to provide information for the preparation of the *Greater London development plan* (GREATER LONDON COUNCIL, 1969). Over 100,000 properties were inspected and details of their condition, age, tenure, facilities and costs of improvement or conversion were collected and analysed.

1474 GLASS, R. (1970) Housing in Camden. *Town Planning Review*, Vol. 41, pp. 15–40.

A summary of the results of the Camden housing study which was completed in 1968. This full-scale housing survey of both public and private housing sectors of the borough was designed to provide information for a review of housing policy and to lead to relevant recommendations.

1475 STANDING WORKING PARTY ON LONDON HOUSING (1970) London's housing needs up to 1974. Ministry of Housing and Local Government, 53p. (Standing Working Party on London Housing Report, No. 3)

This report comprises mainly a projection of housing needs to 1974. The London boroughs and the GLC are urged to act over the worsening pressure on housing in inner London.

1476 GREVE, J. et al. (1971) Homelessness in London. Edinburgh: Scottish Academic Press, 304p.

The results of a short-term research project into the scale of homelessness and the characteristics of the homeless are described. A number of recommendations are made, including more co-ordination between the London boroughs, more central guidance on policy and a fairer distribution of social service resources in relation to needs.

1477 STANDING WORKING PARTY ON LONDON HOUSING (1971) The work of housing associations in Greater London – zoning. Department of the Environment, 4p. (Standing Working Party on London Housing Report, No. 4)

A consideration of the sort of work on which housing associations might best concentrate and how they and the housing authorities might co-operate. It is recommended that individual associations concentrate in comparatively small areas or zones.

1478 DEPARTMENT OF THE ENVIRONMENT Action Group on London Housing (1972) First and second interim reports. Department of the Environment, 44p.

A discussion of the results of a survey which has been undertaken to determine the amount of potential building land available in London.

1479 MANSFIELD, F. (1972) What next for housing? The Greater London problem. *Municipal Journal*, Vol. 80, pp. 1469–74.

According to the author, London's housing shortage can only be resolved by increased pressure on the London boroughs to release land for housing or by the establishment of an overall London housing authority.

1480 ADAMS, B. (1973) Furnished lettings in stress areas. In: DONNISON, D. and EVERSLEY, D. E. C. (eds) London: urban patterns, problems, and policies. Heinemann, pp. 354–82.

Data collected for the 1970 Committee on the Rent Acts are considered in this review of furnished lettings in inner London. Various proposals for the improvement of the housing situation in stress areas are discussed.

1481 COUNTER INFORMATION SERVICES (1973) The recurrent crisis of London: a CIS anti-report on the property developers. Counter Information Services, 68p.

The boom in office property values in London has led to increasing pressures on local government from property developers and has produced a decrease in the property available for housing. This report examines the schemes of one development company.

1482 DEPARTMENT OF THE ENVIRONMENT Action Group on London Housing (1973) Third interim report. Department of the Environment, 21p.

A report on various aspects of housing problems in London including the availability of land and levels of local housing need and provision. Assistance in areas of housing stress and the improvement of older houses are also discussed.

1483 GRIGSON, W. S. (1973) The obsolescence and ageing of London's housing. *Quarterly Bulletin of the Greater London Council*, No. 24, pp. 17–28.

Historical data concerning the rates of house building in London in the nineteenth century are analysed, and it is concluded that, contrary to a widely held view, the obsolescence of the housing stock is going to increase only slowly over the next few decades.

1484 HAMNETT, C. (1973) Improvement grants as an indicator of gentrification in inner London. *Area*, Vol. 5, pp. 252–61.

The pattern of improvement grants in London is examined and explained in terms of housing quality, social class and location.

1485 HARLOE, M. et al. (1973) The organizational context of housing policy in inner London: the Lambeth experience. In: DONNISON, D. and EVERSLEY, D. E. C. (eds) London: urban patterns, problems, and policies. Heinemann, pp. 313–53.

An examination of the roles played by various agencies in the housing market in Lambeth: local authorities, housing associations, private landlords, estate agents, building societies and private developers. The attempt by Lambeth Borough Council to mount a comprehensive attack on the housing problems in the borough is reviewed.

1486 JACKSON, A. (1973) Semi-detached London: suburban development, life and transport, 1900–1939. George Allen and Unwin, 381p.

A detailed survey of the development of London's suburbs in the first part of the twentieth century. A wide variety of aspects is examined, including speculators, building methods, finance and transport. Middle class and working class housing is compared, while case studies cover Ilford, Golders Green, Edgware and Stoneleigh.

1487 LOMAS, G. M. (1973) London's housing needs – current problems and prospects. *Housing Review*, Vol. 22, pp. 129–35.

The metropolitan housing situation is reviewed in detail and statistics given for building, overspill, population decline and levels of assistance. A more balanced pattern of home ownership and fairer allocation of council housing across the whole of London is advocated.

1488 LOVETT, A. F. and NIGHTINGALE, J. R. (1973) Factors affecting the role of private sector in the supply of new housing. Greater London Council, 33p. (GLC Research Memorandum, No. 396)

The amount of land available for house building in London and ways of increasing it are discussed. Household incomes are then considered in relation to house prices.

1489 PARKER, B. J. (1973) Some sociological implications of slum clearance programmes. In: DONNISON, D. and EVERSLEY, D. E. C. (eds) London: urban patterns, problems and policies. Heinemann, pp. 248–73.

A review of the problems caused by the destruction of slum communities and the rehousing of residents in suburban housing estates. It is suggested that the break-up of working class communities by slum clearance has no more than a minor and short-term effect on most people.

1490 POPE, E. (1973) Trends in the growth of London's housing 1971–1991. Greater London Council, 35p. (GLC Research Memorandum, No. 409)

The housing stock projection model has been used to project the growth of the housing stock of Greater London up to 1991. The most significant trend to emerge is the near disappearance of the privately rented sector from the housing market.

1491 SHANKLAND COX AND ASSOCIATES (1973) Private housing in London: people and environment in three Wates housing schemes. Shankland Cox and Associates, 79p.

Residents' attitudes to the design and layout of estates in Croydon, Blackheath and near Hampton Court are studied. Particular aspects examined include open space, landscaping, garage provision and privacy.

1492 DEPARTMENT OF THE ENVIRONMENT Action Group on London Housing (1974) Use of housing land in Outer London: fourth report to the Minister for Housing and Construction. Department of the Environment, 45p.

A report of results of an investigation into land available to the outer London borough councils for housing purposes, and their progress in finding more land.

1493 GREATER LONDON COUNCIL (1974) A strategic housing plan for London. Greater London Council, 76p.

Both the private and the public sector are included in this plan to improve London's housing situation. The programme covers new housing, conversion, greater control of private rented property and improved management of local authority housing.

1494 GREATER LONDON COUNCIL (1974) Strategic housing plan: report of studies. Greater London Council, 88p.

Various aspects of London housing are covered including household size, income, tenure, stock, land and housing finance. The formulation of a strategic plan and possible housing programmes are discussed.

1495 HARLOE, M. et al. (1974) The organisation of housing: public and private enterprise in London. Heinemann, 190p.

The authors describe how two London borough councils (Sutton and Lambeth) tried, in different ways, to influence and control housing associations, landlords, building societies, estate agents and developers to achieve, in conjunction with the councils' own building programmes, a comprehensive housing policy.

1496 KINGHAN, M. (1974) Squatting in London. *New Society*, Vol. 28, pp. 254–5.

Homelessness in London since 1969 is examined together with the approach of London boroughs to the situation. The role of the Family Squatting Advisory Service is outlined.

1497 KRISHNA-MURTY, R. (1974) Incomes and housing expenditure in Greater London. Greater London Council, 44p. (GLC Research Memorandum, No. 432)

Information on housing expenditure and income in Greater London households is analysed and compared with similar data for the rest of the country.

1498 McDONALD, G. (1974) Metropolitan housing policy and the stress areas. *Urban Studies*, Vol. 11, pp. 27–37.

An examination of London's housing policies and problems in the late 1960s and early 1970s. Results from a social survey of North Islington are used to highlight the effects of housing policy.

1499 MARKS, S. (1974) Planning for housing: a strategy for London. *Municipal Journal*, Vol. 82, pp. 1338–44.

The GLC's *Strategic housing plan for London* (1974) is discussed in the context of previous attempts at planning for housing in the capital.

1500 MINNS, R. (1974) Who builds more? *New Society*, Vol. 28, pp. 184–6.

The housing records of the various London boroughs are compared and related to the political compositions of the councils during the period 1967 to 1973. No significant difference between the performance of the two parties is noted.

1501 BALL, N. R. and CLARK, M. J. (1975) Private sector housing in south east England. *Tijdschrift voor Economische en Sociale Geografie*, Vol. 66, pp. 75–83.

The results are examined of a survey undertaken to assess the distribution of property types and the cost of housing around London. The existence of a zonal ordering of social classes within the private sector housing market is tested.

1502 CONDLIFFE, J. (1975) Formulating a housing development programme for Greater London. *Housing Review*, Vol. 24, pp. 9–12.

London's housing problem is described in terms of shortage, quality and distribution between different sectors. A model has been developed in an attempt to balance supply and demand.

1503 CROFTON, B. (1975) The allocation of public housing. *Housing Review*, Vol. 24, pp. 41–6.

A number of recommendations are made concerning housing shortage and overcrowding in London. A city-wide public sector building programme and allocation system for rehousing are proposed.

1504 DEAKIN, N. D. (1975) Some constraints on innovation in social policy in London. *Policy and Politics*, Vol. 3, pp. 61–73.

An assessment of the GLC's approach to the London housing problem. The GLC has been hampered by problems of administration and resource allocation, which might be ameliorated by a reordering of national priorities.

1505 DOCKLANDS JOINT COMMITTEE (1975) Housing. Docklands Joint Committee, 44p.

The current housing situation in docklands is reviewed. Future plans for building houses are described together with new schemes for creating more socially mixed areas and an alternative to the usual form of local authority tenure.

1506 HARLOE, M. (1975) Housing in London. *Housing Review*, Vol. 24, pp. 7–9.

The market system for housing and land and the political organization of

London housing are blamed for current problems of housing in London. The GLC's *Strategic housing plan* (1974) is reviewed in relation to these factors.

1507 LOMAS, G. M. (1975) London: the never ending crisis. *Housing Review*, Vol. 24, pp. 19–24.

Housing policies should reflect social or economic trends, such as movements of the labour market.

1508 LONDON BOROUGH OF SOUTHWARK (1975) Housing for the poor: council housing in Southwark 1925–1975. London Borough of Southwark, 57p.

The history of council housing in Southwark is described and possible changes in present policy are examined. Various improvements such as alternatives to council housing, grading of tenants and changes in the social structure of new estates are suggested.

1509 NEVITT, D. A. (1975) Towards a Greater London housing strategy. *London Journal*, Vol. 1, pp. 135–42.

A highly critical discussion of the GLC's 1974 consultation document: *A strategic housing plan for London*. It is claimed that the GLC has taken a very passive approach to London's housing problems largely because of its weak constitutional position.

1510 SHOULTS, T. and HENNEY, A. (1975) Homelessness in Haringey. *Greater London Intelligence Quarterly*, No. 30, pp. 29–37.

A sample study of homeless households in Haringey made in an attempt to discover the causes of homelessness and the types of household particularly at risk. Various data are presented and suggestions are made for the improvement of the situation.

1511 DUGMORE, K. (1976) Social pattern in GLC housing. *Greater London Intelligence Quarterly*, No. 35, pp. 26–33.

Census data have been used to provide information about the social characteristics of tenants on 72 GLC estates. Cluster analysis has been used to demonstrate that there is a relationship between the social characteristics of the estates and their age and physical type.

1512 GREATER LONDON COUNCIL (1976) Improving London's housing. Greater London Council, 36p.

A history of post-war schemes for modernization of London's housing and rehabitation.

1513 GREATER LONDON COUNCIL (1976) A strategic housing plan for London: second interim report. Greater London Council, various paging.

A summary of joint GLC/London Boroughs Association consultations on the *Strategic housing plan for London* (GREATER LONDON COUNCIL, 1974). Issues considered are targets for new building and rehabitation at borough level, and a co-ordinated allocation system.

1514 LEVER, J. (1976) Home sweet home: housing designed by the London County Council and Greater London Council architects, 1888–1975. Academy Editions, 111p.

Trends in London's public housing are described from the first flats and suburban cottage style estates to the more recent high rise schemes. Based on an exhibition at the ICA in 1973 entitled 'Home Sweet Home'.

1515 LYALL, S. (1976) GLC inner city housing. *RIBA Journal*, December, pp. 504–17.

An outline of the organizational structure of GLC housing and of various schemes and developments. A number of plans and illustrations of sites are included.

1516 STEFFEL, R. V. (1976) The Boundary Street estate. *Town Planning Review*, Vol. 47, pp. 161–73.

The Boundary Street estate in Bethnal Green was built between 1889 and 1914 and was the LCC's first major renewal scheme. The estate is described in the light of the development of public housing policy.

1517 WATES, N. (1976) The battle for Tolmers Square. Routledge and Kegan Paul, 232p.

The story of the long conflict which began in 1957 over the future of Tolmers Square in north London. Pressurized by local tenants, Camden Council finally purchased the land from property developers wishing to redevelop the Tolmers area for offices rather than housing.

1518 WILLIAMS, P. R. (1976) The role of institutions in the inner London housing market: the case of Islington. *Transactions of the Institute of British Geographers*, New Series, Vol. 1, pp. 72–82.

The influences of building societies and estate agents on the gentrification of Islington are discussed. It is concluded that they have an important effect on the rate of social change in residential areas.

1519 HAMMOND, C. (1977) Housing in London: the continuing crisis: indicators of housing stress. SHAC, 28p.

A discussion of the shortage of housing in London in 1977. Unless policies are changed the situation will have reached crisis proportions in the 1980s.

1520 KINGHAN, M. (1977) Squatters in London. Shelter, 89p.

The results of a 1975 survey into the problems of squatters in six London boroughs. A number of suggestions are made with particular emphasis on provision for special groups such as the childless and non-English speakers.

1521 LONDON BOROUGHS ASSOCIATION Housing and Works Committee (1977) Report on the regeneration of inner London. London Boroughs Association, 25p.

An examination of the problems of inner London and the policies which have been adopted for their alleviation.

1522 MARKS, S. (1977) In London's housing trap. *New Society*, Vol. 41, pp. 488–9.

A critique of the housing policy of the newly Conservative GLC. Sale of council houses, devolution of housing responsibility and the needs of the inner city are the major issues.

1523 O'MALLEY, J. (1977) The politics of community action: a decade of struggle in Notting Hill. Nottingham: Bertrand Russell Peace Foundation, 180p.

An account of community politics in Notting Hill between 1966 and 1974. The setting up of the Notting Hill People's Association together with its fight against planning decisions, and particularly against poor housing in the area, are described in detail.

1524 DERBY, C. (1978) Greater London housing: matching people's aspirations. *Local Government Chronicle*, No. 5793, pp. 411–4.

The directors of housing of the GLC and the London boroughs of Sutton and Enfield have been interviewed concerning the problems of finding suitable accommodation for Londoners.

1525 GREATER LONDON COUNCIL (1978) Homesteading: an introduction to the GLC's homesteading scheme. Greater London Council, 24p.

An outline of the homesteading programme which has been introduced to help solve London's housing shortage by encouraging people to restore old houses.

1526 GREATER LONDON COUNCIL (1978) A new housing policy for London: inner London must live. Greater London Council, 35p.

London's housing problems and policies are discussed and the new reorganized policy of the GLC explained. The new housing strategy, whereby the GLC will act as a co-ordinator rather than a direct spending agent, is explained.

1527 JACHNIAK, D. (1978) House prices in the GLC area, 1939–1971. University of Reading Department of Geography, 79p. (University of Reading Geographical Papers, No. 62)

An empirical study of temporal and spatial aspects of house prices in the GLC area in 1939, and between 1967 and 1971. Various trends are identified and reasons for them suggested.

1528 PINCH, S. (1978) Patterns of local authority housing allocation in Greater London between 1966 and 1973: an inter-borough analysis. *Transactions of the Institute of British Geographers*, Vol. 3, pp. 35–54.

Multivariate techniques have been used to demonstrate levels of housing need in the London boroughs. The geographical variations in the extent to which housing provision meets these needs are discussed.

1529 TAYLOR, G. H. (1978) A review of housing in London 1966–1976. Greater London Council, 72p. (GLC Research Memorandum, No. 534)

The development and impact of the GLC housing policy are summarized. Data from various housing surveys and recommendations of various bodies are discussed.

1530 WHITEHEAD, C. (1978) Private landlords in London: who stays, who goes? *CES Review*, Vol. 4, pp. 48–53.

An examination of the effects of the 1974 Rent Act on rented accommodation available in the capital. It is suggested that landlords of furnished tenancies are the least likely to take their properties off the market.

1531 YOUNG, K. and KRAMER, J. (1978) Local exclusionary policies in Britain: the case of suburban defense in a metropolitan system. In: COX, K. R. (ed.) Urbanization and conflict in market societies. Methuen, pp. 229–51.

Attempts by the GLC in the 1960s and 1970s to open up the suburbs in the interests of the inner city poor are described. Local suburban politics are shown to have been responsible for preventing the large-scale influx of urban working class residents.

1532 YOUNG, K. and KRAMER, J. (1978) Strategy and conflict in metropolitan housing: suburbia versus the GLC 1965–1975. Heinemann, 306p.

An examination of the GLC policy of opening up the suburbs to house the inner city poor between 1964–75. The authors argue that the metropolitan government has not achieved its aims, largely due to suburban resistance.

1533 BALCHIN, P. N. (1979) Housing improvement and social inequality, case study of an inner city. Saxon House, 261p.

A discussion of the economic and social aspects of housing improvement on parts of west London. The roles of various agents such as landlords, developers and local authorities are examined.

1534 BONNAR, D. M. (1979) Migration in the south east of England: an analysis of the interrelationship of housing, socioeconomic status and labour demand. *Regional Studies*, Vol. 13, pp. 345–59.

Socio-economic and tenurial features of long and short distance migration, and the relationship between migration and changes in the spatial pattern of labour demand are considered. The housing market was found to exert a significant frictional effect on the response of migration to labour demand changes in the south east.

1535 HAMNETT, C. (1979) The flat break-up market in London: a case study of large scale disinvestment – its causes and consequences. In: BODDY, M. Land, property and finance. University of Bristol School for Advanced Urban Studies, pp. 35–55. (University of Bristol Working Paper series, No. 2)

An account of the continuing decline of the privately rented sector of the housing market, with particular reference to flats. The renting out of flats provides a very low yield on investment, with the result that individual flats in blocks are gradually being sold off.

7.8 Recreation

1536 GREATER LONDON COUNCIL Planning Department (1970) Surveys of the use of open spaces, Vol. 1. Greater London Council, 115p.

A summary is presented of questionnaire surveys of 160 visitors in 16 London parks in 1964. The pattern of demand for open space highlighted by these surveys was to be taken into account in the formulation of the *Greater London development plan* (GREATER LONDON COUNCIL, 1969).

1537 LAW, S. and PERRY, N. H. (1970) Countryside recreation for Londoners: a preliminary technical study. Greater London Council, 48p. (GLC Research Memorandum, No. 267)

From data on facilities, social area composition and accessibility, a number of standards and indices of pressure on the countryside were computed in an attempt to illustrate the relative provision of existing countryside recreation facilities accessible to Londoners.

1538 LAW, S. (1972) Surveys of the use of open spaces, Vol. 2. Greater London Council, 234p. (GLC Research Memorandum, No. 381)

Full tabulations of results of surveys into the use of open space in London, which are described in volume 1 (GREATER LONDON COUNCIL Planning Department, 1970).

1539 GREATER LONDON COUNCIL (1973) Leisure for Londoners. Greater London Council, 55p.

A discussion paper setting out the main choices open to the GLC with regard to provision for the increasing demand for leisure activity in the London area.

1540 CITY OF WESTMINSTER Department of Architecture and Planning (1975) Recreation and leisure. City of Westminster, 104p.

Current provision for recreation in Westminster is described. Problems of recreation planning in the borough are discussed and various alternative solutions proposed.

1541 GREATER LONDON COUNCIL (1975–1976) Greater London recreation study. Greater London Council, 3 vols. (GLC Research Report, No. 19)

The first two parts deal with data gathered during a household interview survey concerning how people spend, and would like to spend, their leisure in recreational activities away from home. They also analyse demand by various demographic and socio-economic characteristics. The third part is an inventory of facilities and resources in Greater London, and provides a brief commentary.

1542 NICHOLLS, M. (1975) Recreationally disadvantaged areas in Greater London. Greater London Council, 25p. (GLC Research Memorandum, No. 467)

A description of the methods used to identify areas of recreational disadvantage in London with reference to sports and active recreation. The problems arising are discussed and, where possible, alternative methods which might be used in future projects are suggested.

1543 BERRY, J. G. (1976) Countryside inside London's fringe: manager's view. *Countryside Recreation Review*, Vol. 1, pp. 21–5.

Bad management and insufficient funds have led to the deterioration of London's commons. The situation is illustrated by an account of attempts to improve Mitcham Common.

1544 FITTON, M. (1976) The urban fringe and the less privileged. *Countryside Recreation Review*, Vol. 1, pp. 25–34.

An attempt to assess the needs of the less privileged for countryside recreation and the scope for providing for them in the urban fringe around London. The problems of accessibility, type of facilities and promotion are examined.

1545 STANDING CONFERENCE ON LONDON AND SOUTH EAST REGIONAL PLANNING (1977) Informal recreation in the South East. Standing Conference on London and South East Regional Planning, 20p.

The results of surveys into recreational habits in the south-east region. The report emphasizes the demand for informal countryside recreation and discusses to what extent this demand is being met.

1546 FERGUSON, M. J. and MUNTON, R. J. C. (1979) Informal recreation sites in London's green belt. *Area*, Vol. 11, pp. 196–208.

The provision and location of sites for recreation in the green belt are discussed. Certain areas appear to be deficient in such provision.

1547 SHOARD, M. (1979) Metropolitan escape routes. *London Journal*, Vol. 5, pp. 87–112.

It is argued that planners have failed to provide Londoners with sufficient opportunities for countryside recreation. National parks should be designated close to London in an attempt to bring recreation provision for the capital into line with that of other major British cities.

7.9 Attitudes and perception

1548 EYLES, J. D. (1968) The inhabitants' images of Highgate village, London: an example of a perception measurement technique. London School of Economics and Political Science, 31p. (LSE Graduate Geography Department Discussion Paper, No. 15)

A total of 89 mental maps of Highgate village were obtained from a sample of residents of the area around the village, and were related to the age, social class, and length and location of residence of the respondents. The technique used is discussed and suggestions made for further research.

1549 HARLOE, M. (1971) Inner London, *Official Architecture and Planning*, Vol. 34, pp. 363–5.

The author argues that a change in social attitudes would go a long way towards solving the inter-related social, economic and physical problems of the inner city.

1550 LONDON BOROUGH OF CAMDEN (1975) The London Borough of Camden: a survey of residents' attitudes. London Borough of Camden, 162p.

Results of a questionnaire survey to aid development of the borough plan. Issues covered include housing, health and recreational facilities, shopping and transport.

1551 EYLES, J. D. (1976) Environmental satisfaction and London's docklands: problems and policies on the Isle of Dogs. Queen Mary College Department of Geography, 31p. (QMC Department of Geography Occasional Paper, No. 5)

Levels of dissatisfaction amongst the residents on the Isle of Dogs are assessed. The likely success of the Isle of Dogs Planning Study, the Dockland Study and the Dockland Joint Committee's strategy in improving the area's quality of life is discussed.

1552 PRESCOTT-CLARKE, P. and HEDGES, B. (1976) Living in Southwark. Social and Community Planning Research, 207p.

The results of a survey investigating people's attitudes to the services and facilities in Southwark. The importance of housing, the environment, employment, shopping, transport and recreation to the lives of local residents is analysed.

1553 HOLLIS, G. E. and BURGESS, J. A. (1977) Personal London. *Geographical Magazine*, Vol. 50, pp. 155–61.

University College students participated in two projects to test their perception of certain parts of London. Their mental maps of their surroundings and their choice of key words to summarize four London walks are presented.

1554 RESEARCH SURVEYS OF GREAT BRITAIN (1977) The London project: commentary. Research Surveys of Great Britain, 41p.

A discussion of the results of a survey designed as an aid to planners, into the feelings of Londoners about their environment and lifestyles.

1555 BYRD, M. (1978) London transformed: images of the city in the eighteenth century. Yale University Press, 202p.

Contrasts the unembellished image of London presented in the writings of Boswell and Johnson, with the more dramatic descriptions of Defoe, Pope, Wordsworth and Blake.

Section 8 Planning the metropolis and beyond

The literature of planning in London is very recent compared with that in other sections of this bibliography. The first town planning Act (1909) concerned itself only with the development of new suburbs, for which preparation of a planning scheme was made optional. Despite the recommendations of the Royal Commission on Local Government (1923), no regional planning body was set up until 1927 when the Greater London Regional Planning Committee was established. This body was, however, purely advisory and was dissolved ten years later. The rapid and haphazard growth of the London region during this period led to the establishment of the Barlow Commission whose report in 1940 (ROYAL COMMISSION ON THE DISTRIBUTION OF THE INDUSTRIAL POPULATION, 1940) marked the first serious attempt to regulate London's growth. The destruction caused by the Second World War created the first opportunities for metropolitan planning, and the early 1940s saw the preparation of three advisory plans upon which post-war reconstruction would be based: the *County of London plan* (LONDON COUNTY COUNCIL, 1943), the *Greater London plan* (ABERCROMBIE, P., 1945) and the *City of London plan* (CORPORATION OF LONDON, 1944).

The period since the 1940s has witnessed a tremendous growth in the literature of planning. The expansion of London and the creation of the new towns have meant that the problems of the capital have become central to the planning of the south-east region (8.4), which is the responsibility of the South East Joint Planning Team within the Department of the Environment. The Greater London Council is responsible for strategic planning of the metropolitan area. The planning departments of the City of London and the 32 London boroughs are preparing local plans within the broad guidelines laid down in the *Greater London development plan* (GREATER LONDON COUNCIL, 1969). Local planning documents have generally been excluded here. They are listed in the excellent bibliography by COCKETT, I. (1977) and in *Urban Abstracts*. Much of the literature of planning in London has, however, focused on particular parts of the city such as Covent Garden, Thamesmead or docklands which are being redeveloped or have particular problems. A selection of the more important publications concerned with such areas has been included. The inner city study (SHANKLAND COX PARTNERSHIP, 1974–8) has centred attention on Lambeth. Some of the more general documents to emerge from this study have been included in this section. Others, which are more concerned with the social problems of the area, are found in Section 7.

The planning process has become increasingly complex during the 1960s and 1970s, as illustrated by the history of the *Greater London development plan* (GREATER LONDON COUNCIL, 1969). The rise in public participation has resulted in the production of numerous consultation papers from local, metropolitan and national government, and reports and propaganda from various pressure groups, as in the case of the redevelopment of Covent Garden and docklands. Bibliographical control of much of this local literature is poor since it is frequently unpublished. A considerable amount of publication in the field emanates from academic planners, economists and geographers writing in journals such as *East London Papers*, *New Society*, *Town Planning Review* and *Town and Country Planning.* In addition, a number of planning studies have been commissioned by the various planning authorities from consultants such as Colin Buchanan and Partners, and Shankland Cox Partnership.

The aim of this section is to list references concerned with the overall planning of the city or areas within it. Material on social, economic or transport planning will be found elsewhere in the relevant sections of the bibliography. Obviously there are many common themes since all the major planning documents contain detailed studies of the individual functions of the area under scrutiny. Literature concerned primarily with conservation rather than planning is listed in subsection 9.2 (Conservation of the man-made environment). Material relating to early attempts at planning, such as the construction of the Hampstead Garden Suburb, which was begun before the 1909 Town Planning Act, is included in section 4.

The best sources for further references relating to the planning of London are the GLC bibliographies, *Urban Abstracts*, *Geo Abstracts* and the *Departments of the Environment and Transport Bulletin.*

8.1 Bibliographies and statistics

1556 LONDON COUNTY COUNCIL (1963) The London County Council: a select bibliography. London County Council, 13p. (LCC Booklist No. 5)

A bibliography of works bearing on the constitution, policy and main services of the LCC from 1889–1963, largely published by the Council itself.

1557 BEER, J. and HECTOR, S. (1970) Planning surveys. Greater London Council, 123p. (GLC Research Memorandum, No. 242)

Details are given of planning surveys or studies completed and in progress or prepared by the London boroughs. Information is arranged according to ten subject categories, for example, industry and commerce, and leisure.

1558 JOHNSON, S. (1973) Postwar planning in London. 2nd ed. Greater London Council, 16p. (GLC Research Bibliography, No. 15)

A select annotated list of 93 references to the plans for London since 1943, the reorganization of London government in 1963 and plans for south-east England. Arranged under various headings such as postwar reconstruction and reorganization of London government.

1559 SKINNER, I. (1975) Thamesmead. Greater London Council, 20p. (GLC Research Bibliography, No. 62)

104 annotated references covering the development of Thamesmead over the past ten years. Social and economic as well as planning issues are covered.

1560 SKINNER, I. (1976) Docklands. Greater London Council, 39p. (GLC Research Bibliography, No. 73)

An annotated bibliography of 162 references on planning in London's docklands and other British dockland areas.

1561 COCKETT, I. (1977) Local plans in London. Greater London Council, 48p. (GLC Research Bibliography, No. 88)

An annotated bibliography of 199 local plans for the London boroughs. Arranged under the names of the boroughs with a subject, and district and area plans index.

1562 SCOTT, G. (1977) The literature of London government: a summary and review. Greater London Council, 10p. (GLC London Topics No. 20)

Summarizes and reviews recent literature on the development and operation of government in Greater London.

8.2 General

1563 ROYAL COMMISSION ON THE DISTRIBUTION OF THE INDUSTRIAL POPULATION (1940) Report. HMSO, 320p. (Cmd. 6153) (Chairman: Sir Montague Barlow).

Report of an inquiry into the causes which influenced the geographical distribution of the individual population of Great Britain, to consider the social, economic or strategical disadvantages arising from industrial concentrations and any remedial measures which should be taken in the national interest.

1564 CHILDS, D. R. (1962) Counterdrift: a programme to combat spreading congestion in the metropolitan region. *Journal of the Town Planning Institute*, Vol. 48, pp. 215–25.

A paper and discussion on the pattern of increasing concentration of population and industry in the South East. The author suggests a counterdrift plan based on new urban centres in declining areas and a new national and international highway system.

1565 HALL, J. M. (1976) London: metropolis and region. Oxford: University Press, 48p. (Problem Regions of Europe)

A study of the current problems of London at the local, regional and national scale. Attempts by planners to provide solutions are discussed, together with the future of the metropolis.

1566 DEPARTMENT OF THE ENVIRONMENT (1977) Policy for the inner cities. HMSO, 33p. (Cmd. 6845)

A statement of the government's plans for aid to the inner city in Great Britain. The nature of the problem is described, together with government proposals for action, resources and legislation. Special partnerships are to be offered to authorities responsible for certain inner city areas, which include Lambeth, Docklands, Hackney and Islington.

8.3 Government

1567 SHARPE, R. R. (1894/5) London and kingdom. 3 vols. Longman.

The political history of the City of London in its relations with the Crown and Parliament, written by the Records Clerk of the Corporation.

1568 ROYAL COMMISSION ON LONDON GOVERNMENT (1923) Report. HMSO, 282p. (Cmd. 1830) (Chairman: Lowther, J. W., Viscount Ullswater)

The Ullswater Commission report on London government inquired into the alterations needed to secure greater efficiency and economy into the administration of local government services in the London County Council and surrounding districts.

1569 HARRIS, P. A. (1931) London and its government. J. M. Dent 261p.

A description of the government of London in the late 1920s under various headings – education, health, housing, traffic and so on.

1570 HAWARD, H. (1932) The London County Council from within: forty years official recollections. Chapman and Hall, 437p.

A record of recollections of some of the principal activities of the London County Council, with special reference to finance.

1571 GIBBON, L. G. and BELL, R. W. (1939) History of the London County Council, 1889–1939. Macmillan, 696p.

A factual account of the history of the London County Council, its development and achievements.

1572 ROBSON, W. A. (1948) The government and misgovernment of London. 2nd ed. George Allen and Unwin, 484p.

A history of government in London from 1835, the problems of governing the metropolis in the late 1930s and a look at the future of London's government.

1573 CORPORATION OF LONDON Common Council (1950) The Corporation of London: its origins, constitution, powers and duties. Oxford University Press, 251p.

An account of the origin and long history of municipal government in the City and of the powers, duties and responsibilities of the Corporation of London.

1574 ROYAL COMMISSION ON LOCAL GOVERNMENT IN GREATER LONDON (1959/60) Minutes of evidence, from government departments. HMSO, 4 vols.

The full minutes of evidence taken before the Royal Commission between 5 March 1959 and 2 February 1960.

1575 ROYAL COMMISSION ON LOCAL GOVERNMENT IN GREATER LONDON, 1957–60. (1960) Report. HMSO, 382p. (Cmd. 1164) (Chairman: Sir Edwin Herbert)

Report of the Herbert Commission advocated a new series of authorities (population, 100,000–250,000) to be called 'Greater London Boroughs', responsible for the majority of local government functions, and a new directly elected authority, the Council for Greater London, to perform these functions 'which can only be or can be better performed over a wider area'.

1576 SHARPE, L. J. (1960) The politics of local government in Greater London. *Public Administration*, Vol. 38, pp. 157–72.

A study of aspects of the politics of Greater London (excluding the City of London) with special reference to the local government representative and the electorate.

1577 ROBSON, W. A. (1961) The Greater London boroughs. London School of Economics and Political Science, 16p. (Greater London Papers No. 3)

Examines and develops the proposals from the *Royal Commission on Local Government in Greater London* (1960) with special reference to the creation of the Greater London Boroughs.

1578 ROYAL COMMISSION ON LOCAL GOVERNMENT IN GREATER LONDON (1960) Wirtten evidence from local authorities, miscellaneous bodies and private individuals. HMSO, 5 vols.

Vol. 1, The Administrative County of London. 488p.
Vol. 2, Middlesex. 762p.
Vol. 3, Metropolitan Essex, metropolitan Hertfordshire and local authority associations. 695p.
Vol. 4, Metropolitan Kent and metropolitan Surrey. 513p.
Vol. 5, Miscellaneous bodies and private individuals. 867p.

1579 SELF, P. J. O. (1962) Town planning in Greater London. London School of Economics and Political Science, 23p. (Greater London Papers No. 7)

Examines the aims and organization of town planning in London during the post-war period and explores the conditions which must be met if effective planning is to be done.

1580 SHARPE, L. J. (1962) A metropolis votes: the London County Council election of 1961. London School of Economics and Political Science, 96p. (Greater London Papers No. 8)

Describes and interprets the London County Council election of 1961, with a detailed case study of Clapham.

1581 JACKSON, W. E. (1965) Achievement: a short history of the London County Council. Longman, 304p.

An account of the London County Council from 1939 to 1964.

1582 SMALLWOOD, F. (1965) Greater London: the politics of metropolitan reform. Indianapolis: Bobbs-Merrill, 324p.

An analysis of the political response to the reorganization plan for the creation of a new Greater London government in the early 1960s.

1583 BARR, J. (1967) How great a London? *New Society*, Vol. 236, pp. 495–8.

A discussion of the problems of the government of London. It is suggested that conflicts are likely to arise between the Greater London Council and the 32 London boroughs particularly in the spheres of transport, housing and planning.

1584 COX, K. R. (1968) Suburbia and voting behavior in the London metropolitan area. *Annals of the Association of American Geographers*, Vol. 58, pp. 111–27.

Analyses voting behaviour in the suburbs of Greater London using voting data for the General Elections of 1950 and 1951.

1585 SELF, P. J. O. (1969) What's wrong with the GLC? *Town and Country Planning*, Vol. 37, pp. 194–7.

A critical review of the Greater London Council policy statement on the *Greater London development plan* (1969).

1586 RHODES, G. (1970) The government of London: the struggle for reform. George Weidenfeld and Nicolson, 320p.

Describes and analyses the events leading up to the introduction of the Greater London Council in 1965, with special reference to the period from the setting up of the Royal Commission in 1957.

1587 RUCK, S. K. and RHODES, G. (1970) The government of Greater London. George Allen and Unwin, 197p.

An account of the working of the new system of government in Greater London, introduced in 1965, set against its historical background.

1588 ROWLEY, G. (1971) The Greater London Council elections of 1964 and 1967: a study in electoral geography. *Transactions of the Institute of British Geographers*, No. 53, pp. 117–31.

Examines aspects of the electoral geography of the Greater London area as revealed in the 1964 and 1967 Greater London Council elections. Those changes in political behaviour which occurred between the results of 1964 and 1967 have spread equally through the system and the basic spatial variation remains constant.

1589 FOLEY, D. L (1972) Governing the London region: reorganization and planning in the 1960s. Berkeley: University of California Press, 223p.

Examines the background to the creation of the Greater London Council (1965) and focuses on plans and policies introduced to solve the problems of the metropolis in the 1960s.

1590 REGAN, D. E. (1972) London. In: ROBSON, W. A. and REGAN, D. E. Great cities of the world: their government, politics and planning, Vol. 2. George Allen and Unwin, pp. 505–72.

Describes local government in London and the politics and planning of the Greater London Council area, with an emphasis upon the changes of the 1960s.

1591 RHODES, G. (ed.) (1972) The new government of London: the first five years. George Weidenfeld and Nicolson, 562p.

An evaluation of the new system of government in Greater London, established in 1965, over its first five years of operation. The author's main aim is to demonstrate changes in the performance of various functions, particularly planning, education, housing, social and health services and transportation.

1592 DEARLOVE, J. (1973) The politics of local government. Cambridge: University Press, 287p.

A study of decision-making in Kensington and Chelsea.

1593 DONNISON, D. (1973) Micro-politics of the city. In: DONNISON, D. and EVERSLEY, D. E. C. (eds) London: urban patterns, problems and policies. Heinemann, pp. 383–404.

Examines recent administrative and political developments and processes at the sub-borough scale in London and presents practical proposals to formulate and explore some of the political and professional problems discussed in the chapter.

1594 ELKIN, S. R. (1974) Politics and land use planning: the London experience. Cambridge: University Press, 196p.

This study of politics and planning, under the London County Council and later the Greater London Council, is largely based on interviews with London officials and the files of the London County Council. Two case studies are described to illustrate the planning process: the development of housing at World's End, Chelsea, and office construction at Centre Point and St Giles Circus.

1595 SHANKLAND, G. (1975) London planning now – precedent or warning? *Journal of Planning and Environment Law*, Vol. 1, pp. 5–15.

Looks at London's experience of multi-tier government to see what lessons it can offer the newly reorganized local authorites.

1596 TODD, N. (1975) The uses of contemporary suburban history (1918–50). *Local Historian*, Vol. 11, pp. 285–9.

Argues the case for serious historical study of the suburbs of large English cities. Suburbs can possibly make two distinctive contributions to contemporary political history; as mainly middle class areas they cast considerable light on right wing politics and because they are new and have changing populations, they are a source of information on the effects of population movement on political organizations. The article illustrates these trends with reference to suburban history at Bexley.

1597 MARSHALL, F. (1978) The Marshall inquiry on Greater London: report to the Greater London Council. Greater London Council, 134p.

Examines London government in terms of the relationship between central government, the GLC, other local authorities and statutory bodies in order to define the future direction that the government of London should take. Covers topics of planning, transportation, business and employment, housing, education, social services, health services, water services, the political and operational services, and makes recommendations.

1598 FREEMAN, R. (1979) The Marshall plan for London government: strategic role or regional solution? *London Journal*, Vol. 5, pp. 160–75.

Examines the future for London government, following the Marshall inquiry. Looks at the alternatives of a strategic role or a regional government and sees a strong representative regional assembly as the best hope for the metropolis.

8.4 Regional planning

1599 BEST, R. H. (1964) New towns in the London region. In: COPPOCK, J. T. and PRINCE, H. C. (eds) Greater London. Faber and Faber, pp. 313–32.

A brief history of the new town idea is followed by an account of the development of the eight new towns around London: Basildon, Bracknell, Crawley, Harlow, Hatfield, Hemel Hempstead, Stevenage, and Welwyn Garden City.

1600 THOMAS, R. (1969) London's new towns: a study of self-contained and balanced communities. Political and Economic Planning, 474p. (PEP Broadsheet, No. 510)

The aims of the new town planners with regard to people, jobs and journeys to work are compared with the current situation in London's new towns. The main themes covered are self-containment, independence from London, social balance, office employment. A more detailed study of Basildon is included.

1601 TOWN AND COUNTRY PLANNING ASSOCIATION (1961) The London region and the development of south east England: 1961 to 1981. *Town and Country Planning*, Vol. 29, pp. 225–34.

The TCPA advocate increased government control of London's growth. Greater development of the outer ring around London, new regional centres some distance from London, and increased economic development of East Anglia are the main measures suggested.

1602 MINISTRY OF HOUSING AND LOCAL GOVERNMENT (1964) The South East study: 1961–1981. HMSO, 146p.

A rapid population growth in south-east England is predicted for this 20 year period. The problems which this will create in the areas of employment, transport and housing are discussed, and a number of proposals put forward.

1603 HALL, P. G. (1967) Planning for urban growth: metropolitan area plans and their implications for south east England. *Regional Studies*, Vol. 1, pp. 101–34.

Five metropolitan development plans are examined and summarized. The application of the main principles to the whole of south-east England is considered.

1604 SOUTH EAST ECONOMIC PLANNING COUNCIL (1967) A strategy for the South East. HMSO, 100p.

This first report of the council considers the development of London and the South East until the end of the century, with particular emphasis on planning to 1981. The major emphasis of the report's proposals is on areas of planned expansion outside London and greater dispersal of employment away from the capital.

1605 COLLINS, M. P. (1970) Strategic planning for the South East. *Area*, Vol. 4, pp. 21–4.

A criticism of the methodology and recommendations of the *Strategic plan for the South East* (SOUTH EAST JOINT PLANNING TEAM, 1970). A comprehensive national framework with a series of physical, economic and social objectives is needed for future regional studies.

1606 SOUTH EAST JOINT PLANNING TEAM (1970) Strategic plan for the South East. HMSO, 110p.

This report was commissioned in 1968 by the government, local planning authorities and the South East Planning Council in order to provide a framework for long-term decisions on investment and economic and social policy. The main recommendations are for the development of major growth areas outside London, redevelopment in the capital, and the preservation of extensive areas of open country.

1607 KEEBLE, D. E. (1971) Planning and south east England. *Area*, Vol. 3, pp. 69–74.

Examines the proposals in the *Strategic plan for the South East* (SOUTH EAST JOINT PLANNING TEAM, 1970) and suggests that, despite certain limitations, it seems likely to represent a major step forward in the diagnosis and treatment of regional planning problems in south-east England.

1608 SOUTH EAST JOINT PLANNING TEAM (1971) Strategic plan for the South East: studies. HMSO, 6 vols.

Vol. 1, Population and employment.
Vol. 2, Social and environmental aspects.
Vol. 3, Transportation.
Vol. 4, Strategies and evaluation.
Vol. 5, Report of Economic Consultants Ltd.
Vol. 6, Index.

These studies volumes record part of the research and evaluation work done by the South East Joint Planning Team. They contain the background material concerning economic, social and environmental issues in the South East on which the *Strategic plan for the South East* (SOUTH EAST JOINT PLANNING TEAM, 1970) was based.

1609 HUDSON, D. M. (ed.) (1974) The future of the London town expansion programme. Regional Studies Association, 44p. (RSA Discussion paper, No. 5)

The history of the 1952 Town Development Act is examined, together with the current progress of town development around London. Various papers at a conference on the future of the programme are then reproduced.

1610 MOOR, N. (1974) Planning brief: south east England. Newman Books, 144p.

A regional survey of the development opportunities and constraints in the south-east region. The first section examines the region as a whole identifying its special characteristics, and this is followed by sections on each of the counties within the region, dealing with topics from residential development to tourism.

1611 STANDING CONFERENCE ON LONDON AND SOUTH EAST REGIONAL PLANNING (1974) London and south east England: regional planning, 1943–1974. Standing Conference on London and South East Regional Planning, 10p.

A chronological account of regional planning in south-east England from the Barlow Report (ROYAL COMMISSION ON THE DISTRIBUTION OF THE INDUSTRIAL POPULATION, 1940) to the early 1970s.

1612 GREATER LONDON COUNCIL Policy and Resources Committee (1975) Planned growth outside London. Greater London Council, 30p.

The nature of, and reasons behind, the migration of people and industry from London are discussed. There is a distinction between planned overspill and migration due to other factors such as industrial change.

1613 DEPARTMENT OF THE ENVIRONMENT (1976) Development of the strategic plan for the South East: interim report. Department of the Environment, 76p.

Interim findings are presented on various economic and social trends which have implications for the development of the plan. Particular attention is paid to the decline of population and industry in London.

1614 HANSON, M. (1976) Is London needed in the South East Plan? *Municipal Journal*, Vol. 84, pp. 357–60.

Points to basic conflicts within strategic planning policies in the South East, for example, the attitudes towards the effects of London's declining population and industry. It is concluded that without a prosperous business economy in London, there will not be the financial resources needed for environmental improvements.

1615 SOUTH EAST JOINT PLANNING TEAM (1976) Strategy for the South East, 1976 review: report of the Housing Group. Department of the Environment, 110p. Appendices, 274p.

Housing in the South East since 1946 is examined. Demand and supply, changes in the use of the dwelling stock and various constraints on the market are analysed.

1616 SOUTH EAST JOINT PLANNING TEAM (1976) Strategy for the South East, 1976 review: report of the Land Group. Department of the Environment, 122p.

An attempt to determine what urbanization demands are likely to be felt in the region over the next ten years in terms of type, quality and location of land, and whether there would be sufficient land to meet these demands in each of the planning areas.

1617 SOUTH EAST JOINT PLANNING TEAM (1976) Strategy for the South East, 1976 review: report of the Population Group. Department of the Environment, 71p.

Recent changes in population trends and projections for the South East are presented, and their implications for regional planning are discussed.

1618 SOUTH EAST JOINT PLANNING TEAM (1976) Strategy for the South East, 1976 review: report of the Resources Group. Department of the Environment, 83p. Annex 1–13.

An examination of the level of public and private financial resources available in the South East, and the way in which this could influence the implementation of the *Strategic plan for the South East* (SOUTH EAST JOINT PLANNING TEAM, 1970). More detailed analyses are contained in the annexes.

1619 SOUTH EAST JOINT PLANNING TEAM (1976) Strategy for the South East, 1976 review: report of the Transport Group. Department of the Environment, 158p.

Major transport projects being planned in the region are reviewed in the context of their impact upon county structure plans and the overall objectives of the *Strategic plan for the South East* (SOUTH EAST JOINT PLANNING TEAM, 1970). Major freight traffic generators and transport expenditure are examined.

1620 SOUTH EAST JOINT PLANNING TEAM (1976) Strategy for the South East, 1976 review: report with recommendations by the South East Joint Planning Team. HMSO, 70p.

The report of a study commissioned in 1974 to update the *Strategic plan for the South East* (SOUTH EAST JOINT PLANNING TEAM, 1970) in the light of recent economic and demographic changes. The problems of London are seen as particularly acute, and the Team recommends that urgent measures be taken to arrest its decline.

1621 DEPARTMENT OF THE ENVIRONMENT (1978) Strategic plan for the South East, review: government statement. HMSO, 38p.

Since the publication of the *Strategy for the South East, 1976 review* (SOUTH EAST JOINT PLANNING TEAM, 1976) the government has produced new legislation concerning the inner city, new towns, transport and other issues which have implications for the region. This statement clarifies the government's position with regard to strategic planning in the South East.

8.5 Metropolitan planning (including the green belt)

1622 LONDON COUNTY COUNCIL (1943) County of London plan. Macmillan, 188p.

This survey, prepared for the LCC by J. H. Forshaw and P. Abercrombie, and concerned with the post-war reconstruction and development of the county, was the first of the three advisory plans for London proposed during the 1940s. It combines short term plans for reconstruction and rehousing with long term proposals to control the growth of the capital.

1623 RICHARDS, J. M. (1943) London plans. *Geographical Magazine*, Vol. 16, pp. 1–13.

A critique of the various plans proposed for rebuilding London and, in particular, one presented by the Royal Academy. The author urges a thorough statistical survey and a master plan based on the survey.

1624 The County of London Plan (1944) *Architectural Review*, Vol. 96, pp. 77–86.

The attitude and proposals of the plan are commented on by five American planners. This is followed by a summary of the proposals and comments from the *Architectural Review*.

1625 ABERCROMBIE, P. (1945) Greater London plan, 1944. HMSO, 222p.

One of the three war-time advisory plans prepared to guide post-war development policy for London. Social, economic, and industrial patterns of the Greater London area are surveyed and possible solutions proposed. Two key proposals were the establishment of eight new towns and a metropolitan green belt.

1626 CARTER, E. and GOLDFINGER, E. (1945) The County of London Plan explained. West Drayton: Penguin Books, 80p.

A description and justification of the proposals contained in the *County of London plan* (LONDON COUNTY COUNCIL, 1943). Illustrated with maps and photographs.

1627 PURDOM, C. B. (1946) How should we rebuild London? 2nd ed. J. M. Dent, 308p.

A discussion of the principles which the author believes should influence the replanning and rebuilding of London after the war. The destruction of much of the city is seen as an opportunity for improving the housing conditions, transport and facilities available to Londoners.

1628 WOOLDRIDGE, S. W. (1945) Some geographical aspects of the Greater London regional plan. *Transactions of the Institute of British Geographers*, No. 11, pp. 1–20.

The author argues that planning for the region would be more effective if the physical geography and geology were taken into account.

1629 LONDON PLANNING ADMINISTRATION COMMITTEE (1949) Report. HMSO, 21p.

The report of a committee whose terms of reference were to advise on the appropriate mechanism for 'securing concerted action in the implementation of a Regional Plan for London as a whole'.

1630 LONDON COUNTY COUNCIL (1951) Administrative County of London development plan: analysis. London County Council, 325p.

A survey of environmental and economic conditions in post-war London which was used as a basis for the statement of proposals contained in the development plan.

1631 LONDON COUNTY COUNCIL (1951) Administrative County of London development plan: statement. London County Council, 208p.

A detailed statement of planning proposals to be undertaken over the next 20 years by the London County Council. The proposals cover housing, open space, schools, transport, hospitals, as well as eight schemes for the comprehensive development of certain areas.

1632 MIDDLESEX COUNTY COUNCIL (1952) County of Middlesex development plan 1951: report of the survey. Middlesex County Council, 239p.

An account of a broad survey of Middlesex undertaken in an attempt to analyse factors affecting the Middlesex County Council's planning policy which is based on the 1944 *Greater London plan* (ABERCROMBIE, P., 1945). Physical conditions, open space, transport and the whole social and economic background are covered.

1633 MIDDLESEX COUNTY COUNCIL (1952) County of Middlesex development plan 1951: written statement. Middlesex County Council, various paging.

An indication of the broad intentions of the county council for the future development of Middlesex and the various stages of development expected. Written statements concerning a number of comprehensive development areas are included.

1634 TOWN PLANNING INSTITUTE (1956) Report on planning in the London region. Town Planning Institute, 60p.

The report of the London Regional Planning Committee which was established by the Town Planning Institute in 1953 'to examine and report on the present situation of planning in Greater London'. The problems of implementing the three London advisory plans of the 1940s are discussed.

1635 LONDON COUNTY COUNCIL (1960) Administrative County of London development plan: first review. 1960, London County Council, 248p. (LCC County Planning Report, Vol. 1)

This is the first review of a plan for London published in 1951 and approved in 1955. Policies are reviewed in the light of changing circumstances since 1951 which have been assessed by means of a survey. Proposals for altering the plan have been made.

1636 POWELL, A. G. (1960) The recent development of Greater London. *Advancement of Science*, Vol. 17, pp. 76–86.

A review of economic and social changes which have taken place in the London region between the publication of the Barlow Report (ROYAL COMMISSION ON THE DISTRIBUTION OF THE INDUSTRIAL POPULATION, 1940) and the Abercrombie proposals (ABERCROMBIE, P., 1945) and the late 1950s. A revised regional plan is urgently required.

1637 THOMAS, W. (1961) The growth of the London region. *Town and Country Planning*, Vol. 29, pp. 185–97.

A review of the Abercrombie plan (ABERCROMBIE, P., 1945) and the trends in population, employment and housing in London during the 1950s.

1638 WISE, M. J. (1961) The crisis for British planning. *Town and Country Planning*, Vol. 29, pp. 179–84.

A discussion of the problems faced by planners trying to cope with the rapid growth of population in the London area. Greater co-ordination of economic planning with land use planning, and a stronger emphasis on a regional approach are recommended.

1639 CARTER, E. (1962) The future of London. Harmondsworth: Penguin Books, 198p.

A discussion of the various physical, social and administrative problems of present-day London and the various attempts that have been made to overcome them.

1640 MINISTRY OF HOUSING AND LOCAL GOVERNMENT (1962) The green belts. HMSO, 30p.

The origins of the green belts and the various legislation affecting them discussed. Problems associated with the maintenance of the green belt are reviewed.

1641 FOLEY, D. L. (1963) Controlling London's growth: planning the Great Wen, 1940–1960. Berkeley; University of California Press, 224p.

Primarily concerned with the advisory plans for London prepared in the 1940s and with the social policies which were incorporated in these plans. The extent to which these social policies have been carried out in the face of subsequent development forces is also discussed.

1642 THOMAS, D. (1963) London's green belt: the evolution of an idea. *Geographical Journal*, Vol. 129, pp. 14–24.

A history, with several maps, of the development of the London green belt. Legislation intended to extend and preserve the belt is reviewed.

1643 COPPOCK, J. T. (1964) The future of London. In: COPPOCK, J. T. and PRINCE, H. C. (eds) Greater London. Faber and Faber, pp. 358–80.

Discusses the problem of how to control London's growth while planning for lower living densities and preserving the countryside. New and expanded towns are seen as playing a major role in future developments.

1644 THOMAS, D. (1964) The green belt. In: COPPOCK, J. T. and PRINCE, H. C. (eds) Greater London. Faber and Faber, pp. 292–312.

The history of London's green belt and its effects on land use around the capital are traced. The author claims that green belt design has suffered as a result of confusion amongst local authorities over the true aims of the policy.

1645 LEAN, W. (1967) London regional planning strategy? *Journal of Town Planning Institute*, Vol. 53, pp. 50–4.

The government's policy for dealing with the projected increase of the Greater London area is criticized. The relative merits of various methods of catering for the increasing population and industrialization of the capital are then discussed.

1646 GREATER LONDON COUNCIL (1969) Greater London development plan: report of studies. Greater London Council, 327p.

A summary of research undertaken in the preparation of the plan. A land use survey, a housing survey and an employment survey were made, and the *London transportation study* (FREEMAN FOX, WILBUR SMITH AND ASSOCIATES, 1967) completed.

1647 GREATER LONDON COUNCIL (1969) Greater London development plan: statement. Greater London Council, 78p.

The first statutory development plan made for the area of Greater London as it was constituted in the 1963 London Government Act. There are sketch maps and diagrams, and chapters cover general strategy, population, housing, employment, transport, town and townscape and central London.

1648 GREATER LONDON COUNCIL (1969) Tomorrow's London: a background to the Greater London development plan. Greater London Council, 128p.

A discussion of the aim, philosophy and financial implications of the *Greater London development plan* (GREATER LONDON COUNCIL, 1969).

1649 HALL, P. G. (1969) London 2000. 2nd ed. Faber and Faber, 288p.

Trends in employment, housing, population and travel are predicted to the year 2000, and a number of suggestions are made as to how the London of the future should be planned and administered.

1650 GODDARD, J. B. (1970) Greater London development plan: central London a key to strategic planning. *Area*, Vol. 3, pp. 52–4.

A criticism of the policy for central London as laid down in the *Greater London development plan* (GREATER LONDON COUNCIL, 1969). More detail is required regarding appropriate economic activity and its location, and a detailed structure plan for the area should be drawn up.

1651 SELF, P. J. O. et al. (1970) London under stress: a study of the planning policies proposed for London and its region. Town and Country Planning Association, 80p.

These papers from a conference entitled *Whither London?* cover a number of issues from the significance of the *Greater London development plan* (GREATER LONDON COUNCIL, 1969) to the inner city, administration, employment, leisure and transport.

1652 THOMAS, D. (1970) Land use changes on London's fringe. *Journal of the Town Planning Institute*, Vol. 56, pp. 435–7.

Changes in and on the edges of the green belt round London between 1960 and 1966 are considered. The author urges careful scrutiny of any schemes likely to create more pressure on the green belt.

1653 THOMAS, D. (1970) London's green belt. Faber and Faber, 248p.

The history of London's green belt is examined, together with the effect of present controls on the green belt and their influence on landscape and land use in the entire zone surrounding the metropolis. The effects of differing degrees of control on the urban fringes of other world cities are described for comparative purposes.

1654 EVANS, H. (ed.) (1971) Region in crisis. Charles Knight, 54p.

A critical appraisal of the *Greater London development plan* (GREATER LONDON COUNCIL, 1969) presented as evidence by the Town and Country Planning Association to the public inquiry into the plan, and covering population, housing and employment, transport, and strategy.

1655 HILLMAN, J. (ed.) (1971) Planning for London. Harmondsworth: Penguin Books, 150p.

Contributions to this work cover different aspects of life and planning in London. They include discussions of the *Greater London development plan* (GREATER LONDON COUNCIL, 1969), housing, employment, shopping, transport, tourism, and various other factors affecting the planning and conservation of the capital.

1656 SELF, P. J. O. (1971) Metropolitan planning: the planning system of Greater London. George Weidenfeld and Nicolson, 54p. (Greater London Papers, No. 14)

The effects of the 1963 local government reorganization on the system of planning in London is examined, with particular reference to the *Greater London development plan* (GREATER LONDON COUNCIL, 1969). The new system is analysed and evaluated.

1657 ASH, M. (1972) A guide to the structure of London. Bath: Adams and Dart, 104p.

The author argues for a new approach to the problems of London such as whether urban motorways should be built or whether its population decline should be arrested. Social processes rather than architecture are the key to an understanding of the capital.

1658 DEPARTMENT OF THE ENVIRONMENT (1973) Greater London development plan: report of the panel of inquiry. HMSO, Vol. 1 Report; Vol. 2 Appendices. (Chairman: F. H. B. Layfield)

This report, also known as the 'Layfield Report', presents the findings of a panel set up in 1970 to examine the proposals of the *Greater London development plan* (GREATER LONDON COUNCIL, 1969). A number of weaknesses of the 1969 Plan are pointed out and suggestions are made towards improvement of the whole planning process.

1659 FOSTER, C. D. and WHITEHEAD, C. M. E. (1973) The Layfield Report on the Greater London development plan. *Economica* Vol. 40, pp. 442–54.

Concentrates on three issues raised by the report – the nature of an urban economic policy, an optimum size for London and housing policies for

London. Criticizes the report for its lack of a theoretical framework in which to assess its recommendations.

1660 GREATER LONDON COUNCIL (1973) London: the future and you. Greater London Council, 12p.

A review of the current problems of London life, housing, employment, transport, social services, shopping and population, followed by a discussion of the GLC's plans for meeting them.

1661 HALL, P. G. (1973) London's western fringes. In: HALL, P. et al. (eds) The containment of urban England, Vol. 1. George Allen and Unwin, pp. 447–83.

A detailed history of urban growth in some of the metropolitan areas of London's outer fringes. The unplanned growth between the wars, the effects of the post-1945 planning controls and the problems of regional planning are described.

1662 CLARK, D. J. (1974) London's green belt. *Greater London Intelligence Quarterly*, Vol. 29, pp. 5–17.

The history of the green belt is discussed, together with changes in attitudes and legislation which have affected its development.

1663 COLLEGE OF ESTATE MANAGEMENT (1974) The future of the green belt. Reading: College of Estate Management, 32p. (Occasional Papers in Estate Management, No. 5)

Four contributors discuss the establishment and future use of the green belt, mineral extraction, urban growth and improved green belt policies.

1664 ASH, M. (1975) What Mr Crosland should do about London. *Town and Country Planning*, Vol. 43, pp. 148–51.

The *Greater London development plan* (GREATER LONDON COUNCIL, 1969) is, in effect, dead according to Ash. He sugests that the Layfield Report (DEPARTMENT OF THE ENVIRONMENT, 1973), with its suggestion of the development of perhaps six new urban centres within the London area, may have much to be said for it. After a discussion of possible new centres, the issue of dockland is raised and the need for new river crossings to help break the cycle of poverty in the East End is stressed.

1665 DEPARTMENT OF THE ENVIRONMENT (1975) Modified Greater London development plan. Greater London Council, 164p.

This document contains the government's draft written statements of modifications to the *Greater London development plan* (GREATER LONDON COUNCIL, 1969). It is based largely on the recommendations of the Layfield Report (DEPARTMENT OF THE ENVIRONMENT, 1973).

1666 DROVER, G. (1975) London and New York: residential density planning policies and development. *Town Planning Review*, Vol. 46, pp. 165–84.

Post-war residential development in New York and London is examined in relation to the density planning policies of both cities.

1667 SHANKLAND, G. (1975) London planning now – precedent or warning? *Journal of Planning and Environmental Law*, Vol. 1, pp. 5–15.

A look at the planning process in London and the role of public participation. It is argued that the opening up of the decision making process can lead to inertia.

1668 DUNHAM, J. et al. (1976) Planning in London study. 2nd ed. Royal Town Planning Institute, 17p.

A consideration of the planning system in London and some of its failures. A number of proposals for improvement are made.

1669 GREATER LONDON COUNCIL (1976) Greater London development plan, approved by the Secretary of State for the Environment, on 9 July 1976: notice of approval, written statement, roads map, key diagram, urban landscape diagram. Greater London Council, 133p.

The final version of the plan incorporating various modifications recommended by the Layfield Panel of Inquiry (DEPARTMENT OF THE ENVIRONMENT, 1973).

1670 GREATER LONDON COUNCIL (1976) Modified Greater London development plan: Greater London Council amendments and detailed comments. Greater London Council, 44p.

Various amendments to and comments on the *Greater London development plan* (GREATER LONDON COUNCIL, 1969) made since the report of the Layfield Panel (DEPARTMENT OF THE ENVIRONMENT, 1973) and the Secretary of State for the Environment at the end of 1975.

1671 HALL, J. M. (1976) A mighty maze but not without a plan. *London Journal*, Vol. 2, pp. 117–26.

A history of the *Modified Greater London development plan* (GREATER LONDON COUNCIL, 1976) from 1965. Prospects for future planning are discussed.

1672 LOCK, D. (1976) Planned dispersal and the decline of London. *Planner*, Vol. 62, pp. 201–4.

A discussion of the changing policies of the GLC from planned overspill to increasing concern over the industrial and social decay of the capital.

1673 STANDING CONFERENCE ON LONDON AND SOUTH EAST REGIONAL PLANNING (1976) The improvement of London's green belt: report by the Green Belt Working Group. Standing Conference on London and South East Regional Planning, 36p.

A study of land use planning and the various pressures on the green belt. The local boroughs are urged to protect and improve the areas of green belt within their boundaries.

1674 HALL, P. G. (1977) London. In: HALL, P. G. The world cities. 2nd ed. George Weidenfeld and Nicolson, pp. 23–52.

A survey of the growth of London and its problems, particularly in comparison with other large world cities. Attempts to plan the city's growth are described.

1675 POOLEY, F. (1977) London after Abercrombie: planning for reorganisation. *Architects Journal*, Vol. 165, pp. 969–73.

The personal views of the GLC's Controller of Planning and Transportation on the simplification of London's planning process and the improvement of its transport system.

1676 WILCOX, D. and RICHARDS, D. (1977) London: the heartless city. Thames Television, 172p.

An examination of London's planning problems. Housing, transport, unemployment and economic planning are discussed.

1677 HEALEY, P. and UNDERWOOD, J. (1978) Professional ideals and planning practice: a report on research into planners' ideas in practice in London borough planning departments. *Progress in Planning*, Vol. 9, No. 2, pp. 73–127.

A report of a number of case studies of London planners. The relationship between the ideas of these planners and planning policies is explored.

1678 JENKINS, J. (1978) Population decline: some problems for London's strategic planning. *Greater London Intelligence Journal*, No. 41, pp. 9–13.

The decline in London's population has caused a number of problems for the planner. The problems of housing, education, ethnic minorities and the elderly are outlined.

1679 MUNTON, R. J. C. (1978) London's green belt. In: CLOUT, H. (ed.) Changing London. University Tutorial Press, pp. 99–109.

A brief account of the history of the metropolitan green belt and green belt policies. Land use changes 1964–74, and agriculture in the green belt are examined.

1680 THAMES TELEVISION (1978) Planning to the people: report of the London Looks Forward project. Thames Television, 128p.

A report of a Jubilee Year project on London planning.

1681 GREATER LONDON COUNCIL (1979) 1988 Olympic games: feasibility study. Greater London Council, 173p.

Sports facilities and tourist accommodation available and required for staging the Olympic Games in London are discussed, and a number of options compared. An evaluation is made of amenity and other benefits and of the financial and economic risks of holding the games.

8.6 Planning of areas within London – docklands

1682 GREATER LONDON COUNCIL (1967) Thamesmead: a riverside development. Greater London Council, 32p.

A short brochure describing the future development of the new community of Thamesmead planned for the area beside the Thames between Woolwich and Erith.

1683 New hope for London's other half (1971) *Official Architecture and Planning*, Vol. 34, pp. 572–7.

A comprehensive solution to the redevelopment of the docklands is urged, in the hope that east London will be revitalized.

1684 EVERSLEY, D. E. C. (1972) The docklands: an exercise in geopolitics. *East London Papers*, Vol. 14, pp. 51–64.

The historical reasons behind the deprivation of the docklands are given, and the complex redevelopment problems discussed.

1685 HALL, J. M. (1972) East London's future: visions past and present. *East London Papers*, Vol. 14, pp. 5–24.

The history of various attempts to plan east London is described, culminating in the 1971 *London docklands study*. Various key issues in the docklands, together with the growing interest in neighbourhood studies, are described.

1686 The Docklands Study reviewed (1973) *East London Papers*, Vol. 15, pp. 5–36.

Four papers discuss the GLC's dockland study (LONDON DOCKLAND STUDY TEAM, 1973). The proposals are first summarized and then criticized from the planning, social and local residents' viewpoints.

1687 LONDON DOCKLAND STUDY TEAM (1973) Docklands: redevelopment proposals for east London. Greater London Council. Vol. 1 Main report; Vol. 2 Appendices.

Comprises a comprehensive study of the area of Thames-side between London docks and Barking Creek, which was commissioned in 1971 by the Department of the Environment and the Greater London Council. A number of alternative plans for the area, with widely differing aims, are drawn up and discussed.

1688 GREATER LONDON COUNCIL (1974) Thamesmead. Greater London Council, 8p.

A short introduction to Thamesmead and its future development.

1689 LONDON BOROUGH OF SOUTHWARK Planning Department (1974) Southwark's Thames-side: a strategy plan. London Borough of Southwark, 60p.

Movement of the docks has left large derelict areas along Southwark's river frontage. This plan considers the need for redevelopment over the whole area.

1690 OWEN, J. and TYRRELL, B. (1974) Docklands study home interview survey, 1973, Greater London Council, 17p. (GLC Research Memorandum, No. 446)

An account of the questionnaire design, fieldwork and data processing connected with a survey of public opinion regarding the future development of dockland. The survey was also designed to monitor the effectiveness of publicity about the proposals.

1691 THAMES SIDE CONFERENCE (1974) Tower Bridge to Tilbury – an examination of the strategic possibilities. Greater London Council, 265p.

Redevelopment possibilities of the Thames-side area downstream from the docklands are discussed. Issues considered include housing, employment, recreation, transport and the environment.

1692 TURTON, A. R. (1974) Analysis of public reaction to redevelopment plans for London Docklands. Greater London Council, 62p. (GLC Research Memorandum, No. 440)

Public reaction to redevelopment plans for the docklands has been monitored through information centres, a home interview survey, a self completion questionnaire, and views expressed at public meetings and in letters. The results are discussed, and then summarized under the following headings: housing, employment, recreation, use of dock areas, public transport, and shopping.

1693 DOCKLANDS JOINT COMMITTEE (1975) A strategy for docklands: setting the scene: a working paper for consultation. Docklands Development Team, 36p.

An outline of the main issues in docklands, and the assumptions which have been made as a basis for preparing the strategic plan.

1694 EVERSLEY, D. E. C. (1975) The re-development of London docklands – a case study in sub-regional planning. Regional Studies Association, 19p.

A summary of the author's evidence to the House of Commons Select Committee on Expenditure. The advantages and disadvantages of comprehensive redevelopment, its administration and implications for the local community are discussed.

1695 HALL, P. G. (1975) Whose docklands? *New Society*, Vol. 31, pp. 519–21.

The problems facing the Docklands Joint Committee are examined. It is argued that given the vast investment of public money, the benefits of the redeveloped docklands must be shared with the rest of the capital.

1696 DOCKLANDS JOINT COMMITTEE (1976) London docklands strategic plan. Docklands Development Team, 118p.

A framework of guidelines for the development of the docklands, which was modified and finally approved after three months of public debate. Transport and employment are the most crucial issues, but the plan also encompasses social problems such as housing, education, health, welfare and recreation.

1697 HALL, J. M. et al. (1976) Rebuilding the London docklands. *London Journal*, Vol. 2, pp. 266–85.

A review of recent attempts to plan the docklands is followed by a discussion of the draft *London docklands strategic plan* (DOCKLANDS JOINT COMMITTEE, 1976). A planner, an academic and a local resident all give their views on the plan.

1698 JOINT DOCKLANDS ACTION GROUP (1976) Docklands – the fight for a future. Joint Docklands Action Group, 36p.

A response to the Docklands Joint Committee's draft strategy which was published in March 1976. Finance and implementation are discussed and a number of proposals made.

1699 RYAN, M. C. and ISAACSON, P. (1976) Planning and participation: London docklands. *Local Government Studies*, January, pp. 47–56.

The history of public participation in the docklands redevelopment scheme is traced. The contrasting views of the public and local government are examined.

1700 VIGARS, L. (1976) The redevelopment of London's docklands. *Port of London*, Vol. 51, pp. 70–2.

The history of the docklands area redevelopment is described, and the Docklands Joint Committee's strategic plan (1976) discussed.

1701 BYRNE, D. F. (1978) Developing London's docklands. *Town and Country Planning*, Vol. 46, pp. 361–5.

A discussion of plans for social development in the docklands. Community support and liaison, facilities and research are regarded as the four main requirements for a programme to help the local population adjust to rapid large-scale changes in the environment.

1702 CLOUT, H. (1978) The dockland dilemma. In: CLOUT, H. (ed.) Changing London. University Tutorial Press, pp. 61–9.

The problems of the East End and the decline of the docks during the 1960s and 1970s are reviewed. A brief account is given of the various plans for the area together with the controversial redevelopment at St Katharine's Dock.

1703 DOCKLANDS JOINT COMMITTEE (1978) London docklands operational programme 1978–1982. Docklands Joint Committee, 86p.

An account of dockland redevelopment proposals to be implemented over the next five years. The schemes cover a large number of activities and will be financed by a variety of agencies.

1704 DOCKLANDS JOINT COMMITTEE (1978) A review of progress: London docklands, 1978. Docklands Joint Committee, 10p.

A review of progress made towards implementation of the *London docklands strategic plan* (DOCKLANDS JOINT COMMITTEE, 1976). Land has been acquired, roads and services are being provided, and industrial and housing sites are being developed.

1705 JOINT DOCKLANDS ACTION GROUP (1978) Docklands: two years on. Joint Docklands Action Group, 32p.

A review of progress toward the implementation of the *London docklands strategic plan* (DOCKLANDS JOINT COMMITTEE, 1976). Insufficient government investment and release of land for development have, amongst other factors, inhibited the improvement of the area.

1706 BEARD, N. (1979) London docklands: an example of inner city renewal. *Geography*, Vol. 64, pp. 190–5.

The many problems of the London docklands are described together with the aims of the *London docklands strategic plan* (DOCKLANDS JOINT COMMITTEE, 1976).

1707 CHAPMAN, P. (1979) New town for docklands. *Architects Journal*, Vol. 170, pp. 1082–4.

A review of the background of the development of the London docklands over the past decade. The author outlines what needs to be done with special reference to the new Urban Development Corporation on the new town model, and looks at the problems it will face.

1708 LEITH, D. (1979) The new face of St Katharine's Dock. *Port of London*, Vol. 54, pp. 54–7.

The redevelopment of St Katharine's Dock is described. The area now has new offices, housing, a hotel, and accommodation for yachts.

8.7 Planning of areas within London – the inner city

1709 CORPORATION OF LONDON Improvements and Town Planning Committee (1944) Report on the preliminary draft proposals for post-war reconstruction in the City of London. B. T. Batsford, 64p.

This plan for the post-war reconstruction of London covers land use, buildings and amenities, traffic and the problems of legislation. The recommendations of the *County of London plan* (LONDON COUNTY COUNCIL, 1943) are taken into consideration.

1710 SMAILES, A. E. and SIMPSON, G. (1958) The changing face of east London. *East London Papers*, Vol. 1, pp. 31–46.

Outlines the growth and urban morphology of east London (Bethnal Green, Stepney and Poplar) and how post-war redevelopment schemes are transforming the urban texture.

1711 McEWEN, A. (1960) The Lansbury story. *East London Papers*, Vol. 3, pp. 67–86.

Lansbury was part of the Victorian expansion of Poplar but by the Second World War it had attained poor social and physical environments and suffered extensive war damage. This article describes the history of the scheme for planning Lansbury as a Comprehensive Development Area, commencing in the 1950s.

1712 ROBSON, W. A. (1965) The heart of Greater London: proposals for a policy. London School of Economics and Political Science, 40p. (Greater London Papers, No. 9)

A plan for the protection and preservation of the central area of London and for its future development is suggested.

1713 GREATER LONDON COUNCIL et al. (1968) Covent Garden's moving: Covent Garden area draft plan. Greater London Council, 118p.

A report presenting the essential elements of an outline redevelopment scheme for the area to begin after the removal of the market. The main emphasis is on the basic principles and major projects around which the scheme as a whole is organized.

1714 BROWNE, K. (1971) West End: renewal of a metropolitan centre. Architectural Press, 112p.

An examination of the character of different areas of the West End such as Soho, Covent Garden and Mayfair, with the help of maps, drawings and photographs. The good and bad points of the various neighbourhoods are investigated, largely with the planner in mind.

1715 COLIN BUCHANAN AND PARTNERS (1971) Greenwich and Blackheath study. Colin Buchanan, 90p.

A study undertaken to advise the Greater London Council on measures required to deal with present and future demands for movement in a manner consistent with the environmental objectives for the area. It is concerned mainly with the reorganization of the road system to deal with longer distance traffic.

1716 GREATER LONDON COUNCIL Department of Planning and Transportation (1971) Recreational appraisal of Thames-side. Greater London Council, 30p.

Examines the use of the river and its banks for recreation and looks at the problems of recreational use of the Thames. Suggests public expenditure should be given to the creation of recreation centres, public boating centres and the improvement of public river boat services.

1717 GREATER LONDON DEVELOPMENT PLAN and COVENT GARDEN JOINT DEVELOPMENT COMMITTEE (1971) Covent Garden, the next step: the revised plan for the proposed Comprehensive Development Area. Greater London Council, 32p.

As a result of public discussion of the *Covent Garden area draft plan* (GREATER LONDON COUNCIL et al. 1968), a revised Comprehensive Development Area proposal was submitted to the GLC in 1971. This booklet provides a detailed explanation of the proposal.

1718 FERRIS, J. (1972) Participation in urban planning, the Barnsbury case: a study of environmental improvement in London. G. Bell, 96p. (Occasional Papers on Social Administration, No. 48)

The influence of public participation particularly in the form of various local pressure groups, is discussed in relation to planning policy in Barnsbury during the period 1964–71. The main causes of controversy were the Barnsbury environmental study, and the experimental traffic scheme which was subsequently implemented.

1719 RICHARDS, J. (1973) Planning and redevelopment in London's entertainment area, with special reference to the theatre. Arts Council, 48p.

A survey of current planning and redevelopment procedure and practice in the West End of London, with reference to their effect on the London theatre. The author concludes that recent proposals for large-scale redevelopment do pose a threat to London's artistic and cultural amenities and should be resisted.

1720 CANTACUZINO, S. (1973) The Barbican development, City of London. *Architectural Review*, Vol. 918, pp. 66–90.

The historical influences, particularly that of Le Corbusier, on the Barbican scheme are examined and the finished product criticized. Despite a number of defects, the author approves the overall design.

1721 CENTRAL LONDON PLANNING CONFERENCE (1974) Advisory plan for central London – four outline strategies. Central London Planning Conference, 52p.

The objectives of the advisory plan are outlined. Details are then given of proposals for economic growth, improvement in working conditions and housing, and conservation in central London.

1722 CENTRAL LONDON PLANNING CONFERENCE (1974–) Advisory plan for central London: topic papers 1–4. Central London Planning Conference.

These four papers attempt to analyse the major planning issues in central London: economic activity, population and housing, conservation and development, and transport. The subsequent advisory plan was produced based on these investigations into the functions and role of the central area.

1723 CENTRAL LONDON PLANNING CONFERENCE (1974) Planning for the future of central London. Central London Planning Conference, 24p.

Discussion on papers given at a seminar in March 1974. Economic activity, population, housing, conservation and transport are amongst topics covered.

1724 CHRISTIE, I. (1974) Covent Garden – approaches to urban renewal. *Town Planning Review*, Vol. 45, pp. 30–62.

The Covent Garden plans were prepared within the framework of a Comprehensive Development Area. This comprehensive redevelopment approach is criticized and an alternative suggested.

1725 COVENT GARDEN DEVELOPMENT TEAM (1974) Covent Garden local plan: report of survey. Greater London Council. (Discussion papers 1–6)

A series of papers covering different aspects, such as housing, conservation and transport, of the proposed new plan for Covent Garden. Comments are invited from the public and from local interest groups.

1726 WEIGHTMAN, G. (1974) The new Garden. *New Society*, Vol. 29, pp. 533–6.

The decision to move Covent Garden Market to the Nine Elms site is discussed in its historical context, and the future of the market at its new site is debated.

1727 GREATER LONDON COUNCIL (1975) The future of the South Bank – a report on planning principles and the scope for the provision of housing. Greater London Council, 40p.

The present situation of the South Bank Comprehensive Development Area is described. Three options for development are suggested.

1728 CENTRAL LONDON PLANNING CONFERENCE (1976) Advisory plan for central London: policy document. Central London Planning Conference, 26p.

The advisory plan policies lie within the structural framework of the *Greater London development plan* (GREATER LONDON COUNCIL, 1969) and are based on previous decisions of the Central London Planning Conference. They cover housing, employment, transport and the environment.

1729 CENTRAL LONDON PLANNING CONFERENCE (1976) Advisory plan for central London: technical document. Central London Planning Conference, 54p.

A justification of the policies advanced in the policy document. The present situation in central London is described together with the aims and approaches of the planners.

1730 GREATER LONDON COUNCIL (1976) Covent Garden: GLC action area plan, written statement. Greater London Council, 154p.

In 1973 Covent Garden became a Comprehensive Development Area to enable the GLC and the local planning authority to develop the area subsequent to the removal of the market to Nine Elms. This is the final plan for the area produced by the GLC after liaison with the local planning authorities and consultation with the public.

1731 HUTCHINSON, D. and WILLIAMS, S. (1976) South Bank saga. *Architectural Review* Vol. 160, pp. 156–62.

The history of the South Bank and the controversy it has aroused over 25 years are reviewed.

1732 GREATER LONDON COUNCIL, Department of Planning and Transportation (1977) Kings Cross/St Pancras Action Area: report of survey. Greater London Council, 85p.

The problems and possibilities for development in the Kings Cross area are examined. Population and housing, economic activity, social and environmental issues are included in the discussion.

1733 BURGESS, J. (1978) Conflict and conservation in Covent Garden. *L'Espace Géographique*, Vol. 7, pp. 93–107.

A review of developments in Covent Garden since 1968 including a discussion of redevelopment plans, public participation, land use and financial implications.

1734 LOMAS, G. M. (1978) Inner London's future: studies and policies. *London Journal*, Vol. 4, pp. 95–105.

Various studies of London's inner area and the government's inner cities policy are examined in relation to national and regional plans.

1735 CHRISTENSEN, T. (1979) Covent Garden: a struggle for survival. *Political Quarterly*, Vol. 50, pp. 336–48.

An account of the history of Covent Garden. GLC policies and the various objections to them and the involvement of the local community are discussed.

1736 CHRISTENSEN, T. (1979) Neighbourhood survival: the struggle for Covent Garden's future. Prism, 149p.

The story of the opposition to the Covent Garden Plan by local community groups, chiefly the Covent Garden Community Association. It is argued that neighbourhood conservationists have been a major force in the scaling down of urban renewal and in changing the system of local planning.

8.8 Planning of other areas within London

1737 CIVIC TRUST (1964) A Lea valley regional park: an essay in the use of neglected land for recreation and leisure. Civic Trust, 46p.

A survey, with maps and pictures, of a 20 mile stretch of the Lea valley from Ware to West Ham. It is proposed that the area be declared a regional park, and subsequently developed as an area for recreation, sport, entertainment and leisure for residents of the East End, north-east London, Hertfordshire and Essex.

1738 LEE VALLEY REGIONAL PARK AUTHORITY (1969) Report on the development of the regional park with plan of proposals. Lee Valley Regional Park Authority, 82p.

This is the report of an authority established in 1967 in order to 'develop, improve, preserve and manage' the Lea valley for recreational purposes. The report includes surveys of the geography, population, communications and other aspects of the region, as well as plans for a variety of recreational centres and activities.

1739 COLIN BUCHANAN AND PARTNERS (1970) North east London: some implications of the Greater London development plan. Colin Buchanan and Partners, 141p.

The transportation policies put forward in the *Greater London development plan* (GREATER LONDON COUNCIL, 1969) are examined in relation to the north-east sector of London. The problem of reconciling demands for increased mobility and a better environment is discussed in detail.

1740 GREATER LONDON COUNCIL Department of Planning and Transportation Intelligence Unit (1970) The Thames-side survey, 1967. Greater London Council, 22p. (GLC Reserach Report, No. 6)

A brief description of a survey undertaken in 1967 in order to obtain detailed information about the uses of riverside sites and the extent to which these are dependent on the river, and to help in the formulation of a comprehensive policy for the Thames as part of the *Greater London development plan* (GREATER LONDON COUNCIL, 1969).

Section 9 Environmental problems: conservation, pollution and control

A major outcome of the rapid and extensive growth of London is that the environment both within and without the metropolis has come under increasing pressure from the residential, industrial, commercial, recreational and other needs of the urban complex.

The riverine and estuarine environments of the Thames have long suffered from the use of the river for transport and docking facilities, and as an open sewer. The pollution of the Thames presented a major problem to environmental engineers, but since the Second World War there has been a major clean-up of the river and this is described in several articles in subsection 9.5. Even areas around London that may still appear 'natural' to the lay observer are heavily modified by man. Epping Forest, for example, has a long history of human interference and control, and even the present day pattern of conservation is essentially to preserve a man-made forest environment. References on the conservation of Epping Forest are thus included in subsection 9.2 on the conservation of the man-made environment.

The preservation of London's urban fabric is a continuing problem. The need to conserve the historic buildings in the Cities of London and Westminster has long been recognized, but the extensive areas of Georgian and Victorian development that surround the historic centre are under constant threat from general decay and neglect on the one hand, and redevelopment and road improvements on the other. The need to preserve the distinctive urban landscapes of London is now recognized throughout the metropolis and a guide to conservation areas in Greater London is provided by DENBIGH, K. (1978) *Preserving London*. The threat to archaeological sites is also now recognized throughout the Greater London area, as noted in *Time on our side? A survey of archaeological needs in Greater London* (1976).

Further environmental problems stem from the need to supply London with adequate quantities of water for household and industrial uses. London's early water conduits date from at least the thirteenth century, but the other aspect of water supply, the disposal of waste and unwanted water, was much slower in its provision, and London's main sewers date from the second half of the nineteenth century (*see* HUMPHREYS, C. W., 1930) *The main drainage of London*. A growing problem in Greater London is the disposal of refuse and other solid wastes, reflected in an expanding literature in such journals as *Solid Wastes* and *Public Health Engineer*.

Conservation, pollution and control are interlinked in London's flood hazard problem. Flooding can seriously damage the urban environment and cause grave pollution risks. The tributaries of the Thames have a history of flooding, further aggravated by the rapid run-off from urban areas, which in turn brings pollutants into the river system. Control schemes have been constructed on several tributaries (*see* BUTTERS, K. and LANE, J. J., 1975) but it is the risk of flooding from tidal surges in the Thames that poses by far the greatest threat. An investigation by Sir Herman Bondi, which reported in 1967 (unpublished), was the first step in the scheme that led to the proposal for a flood barrier at Woolwich (*see* GREATER LONDON COUNCIL Department of Public Health Engineering, 1969) and a considerable literature on the flood barrier is summarized in subsection 9.4.

Human activities have also caused considerable problems from air and noise pollution in the capital. Noise from air and road traffic is a continuing problem, but air pollution has decreased throughout the century. The infamous London fogs and smogs that gripped London about the turn of the century and again in 1952 seem to have disappeared, but whether owing to the influence of the Clean Air Act of 1956 or because of meteorological changes is a subject of some debate (*see* RUBINSTEIN, H. T., 1975).

Due to the technical nature of much of the current work on environmental problems, references often appear as technical monographs or scattered throughout a large number of specialized journals. Abstracts of environmental publications are regularly given in the *Departments of Environment and Transport Library Bulletin*.

9.1 Bibliographies and general

1741 WOODWARD, H. B. (1906) Soils and sub-soils from a sanitary point of view, with special reference to London and its neighbourhood, 2nd ed. HMSO, 82p. (Memoirs of the Geological Survey)

Examines the relations between the nature of the soils and sub-soils and the sanitary requirements of the community in London and its suburbs. Covers geology, soils, sub-soils, house foundations and sites, water supply and drainage, weather and aspect, and cemeteries.

1742 LAWTHER, P. J. and BONNELL, J. A. (1970) Some recent trends in pollution and health in London and some current thoughts. Washington: International Union of Air Pollution Prevention Associations, 22p.

Examines diminishing trends in the relationship between morbidity and mortality and air pollution in London. The authors think that the improvement in health may well be the result of a decline in pollution.

1743 GREATER LONDON COUNCIL (1971) The human habitat: a review by the GLC for the working party on the human habitat to the United Nations conference on the environment, Stockholm, 1972. Greater London Council, 44p.

Discusses the environmental problems and possibilities facing the GLC in improving conditions in which Londoners and its visitors live, work and move about. Sections are included on town and landscape, roads and environment, and the planning and development of improvement standards.

1744 GREATER LONDON COUNCIL (1971) Pollution: a review by the Greater London Council for the Working Party on Pollution to the United Nations conference on the environment, Stockholm, 1972. Greater London Council, 16p.

Presents the GLC's views on pollution. Sections included are as follows: air, freshwater and marine pollution, agriculture, industry, refuse disposal, radioactive waste disposal, noise, economic implications, administrative and fiscal policy, long-term dangers, and international agreements. The final section identifies priorities for enquiry and action.

1745 Handlist of books in Guildhall Library relating to the river Thames (1972). *Guildhall Miscellany*, Vol. 4, pp. 184–93.

A list of books published before 1900 on the following themes: general description; guides, dictionaries and atlases; literary and artistic studies; watermen; navigation and conservancy; purity, pollution and embankments; fishing and fisheries; frost fairs.

1746 GREATER LONDON COUNCIL (1973) London's river Thames. Greater London Council, 65p.

A general survey of the Thames and its banks in London. The changes taking place in economic activity and the landscape of the upper, central and lower reaches of the river are examined.

1747 MANN, R. (1973) The Thames and the Lea. In: MANN, R., Rivers in the city. Newton Abbot: David and Charles, pp. 109–37.

Explores the riverine systems of the Thames and the Lea within their environmental and historical contexts.

1748 GREATER LONDON COUNCIL (1974) London's environment: the implication for London of the United Nation's conference on the human environment, Stockholm, 1972. Greater London Council, 66p.

Reviews the conference papers and the subsequent action plan and principles raised, and how they could be applied to London. There are sections on resources, pollution, social and economic implications, environmental influences in planning, transport, housing and education.

1749 GREATER LONDON COUNCIL Environmental Group and Pollution Control Group (1974) London's environment: first report. Greater London Council, 19p.

Outlines the aims of the groups as they affect air pollution, noise, residential and community facilities and visual intrusion. An appendix describes the Council's activities in relation to London's environment.

1750 GREATER LONDON COUNCIL et al. (1974) The Thames tomorrow: report of the conference held in the Waterloo Room, Royal Festival Hall on 24 June 1974. Greater London Council, 47p.

Includes papers on the recreational and tourist functions of the river, the redevelopment of the dockland area and river banks, the Thames barrier, commerce and industry along the river, the role of the Thames in transport and the wildlife of the river.

1751 HOWARD, P. (1975) London's river. Hamish Hamilton, 258p.

An account of the Thames from Hampton to the sea. The concluding part of the book deals with the control of the river and its future.

1752 SOUTHWOOD, J. (1978) Flood control on the river Thames. Greater London Council, 6p. (GLC Research Bibliography, No. 95)

An annotated bibliography of 36 references on flood control and protection in London with particular reference to the Thames flood barrier.

9.2 Conservation of the man-made environment

1753 Bloomsbury; the case against destruction (n.d.) London Borough of Camden, 28p.

A photographic survey of the area of Bloomsbury threatened with redevelopment for the new British Library. A strong plea is presented for preservation of the area and siting the library elsewhere.

1754 London's squares and how to save them (c. 1930). London Society, 52p.

Originally produced to publicize the threat to London's squares as a result of building developments in Endsleigh Gardens and Mornington Crescent and a threatened take-over of Mecklenburgh and Brunswick Squares.

1755 FORTY, F. J. (1955) London Wall by St Alphage's Churchyard. *Guildhall Miscellany*, Vol. 1, Pt. 5, pp. 4–39.

A fully illustrated account of the exposure (as a result of war damage) and preservation of a part of the Roman and medieval town wall.

1756 CREASE, D. (1965) New towers of London. *Geographical Magazine*, Vol. 37, pp. 778–88.

A photographic essay on the impact of high buildings on the London landscape since the first multi-storey block of flats was built in 1888.

1757 GREATER LONDON COUNCIL Department of Architecture and Civic Design (1968). Thames-side environmental assessment: a study of environmental quality on Thames-side within Greater London. Greater London Council, 90p.

Presents an assessment of existing qualities and characteristics in the Thames-side environment between Dartford Creek and Hampton as a basis for the formulation of an environmental policy for Thames-side.

1758 JENKINS, S. (1970) A city at risk: a contemporary look at London's streets. Hutchinson, 190p.

A guide to the architecture of London's streetscapes, with a strong plea for preservation. Arranged by districts, concentrating on central and west London.

1759 HOBHOUSE, H. (1971) Lost London; a century of demolition and decay. Macmillan, 250p.

A pictorial catalogue of the most important buildings, docks, bridges and whole streets that have disappeared from the London scene owing to speculative redevelopment, enemy action and lack of planning control. An introductory chapter charts the emergence of conservation planning.

1760 SHANKLAND COX AND ASSOCIATES (1971) Hampstead Garden Suburb: a plan for conservation. Shankland Cox and Associates, 146p.

Outlines the history, architecture, and development threat to the suburb. Short-term and long-term conservation plans are proposed, covering house maintenance, landscape policy and traffic control.

1761 BIDDLE, M. and HUDSON, D. (1973) The future of London's past: a survey of the archaeological implications of planning and development in the nation's capital. Worcester: Rescue, 83p. and 8 large scale maps.

An assessment of the current state of knowledge shows how much remains to be discovered of Roman, Anglo-Saxon and medieval London. However at the present rate of redevelopment, only minor pockets will survive by c. 1990 unless a major programme of archaeological investigation can be implemented.

1762 BOOKER, C. and GREEN, C. L. (1973) Goodbye London: an illustrated guide to threatened buildings. Fontana, 160p.

A guide to the major and minor development schemes being planned in 1973 (mainly in central London) and the buildings that are threatened by these schemes.

1763 A handlist of books in Guildhall Library relating to the preservation of Epping Forest (1973–5). *Guildhall Studies in London History*, Vol. 1, pp. 35–44.

Epping Forest became the property of the City of London in 1878 so that Guildhall Library has an especially good collection of books, pamphlets and reports on the Forest. This list (briefly annotated) has a short appendix on general works on Epping Forest.

1764 Battle of Trafalgar (1974). *Architectural Review*, Vol. 934, pp. 334–6.

Briefly discusses the proposed redevelopment of the south-east corner of Trafalgar Square.

1765 HOUSING CENTRE TRUST (1974) Bloomsbury: the case against destruction. *Housing Review*, Vol. 23, pp. 82–6.

Reports on a survey of housing and population in the area proposed for the construction of the British Library. It was found that a real community lived in the blocks of flats and they appreciated their proximity to the centre of London.

1766 NOBLE, G. B. (1974) Conservation areas in the City of London. *Journal of the London Society*, No. 339, pp. 2–7.

Outlines the 1967 Civic Amenities Act and the need for conservation and describes the eight conservation areas designated by 1974.

1767 ROCK, D. and LEALE, M. (1974) Long live Covent Garden. *Building*, Vol. 227, pp. 112–9.

Analyses the factors which determine the special character of the area, such as its historic continuity and human scale, and how this should influence the conservation of the area.

1768 CITY OF WESTMINSTER (1974) Conservation in Westminster. Westminster City Council, 44p.

Provides a brief history of Westminster and describes and maps conservation areas. Policies for preservation and improvement are outlined.

1769 WHARTON, K. et al. (1974) London pride is falling down. *Architect*, No. 3, pp. 24–7.

A series of short articles looking at redevelopment projects in London from a critical viewpoint. Includes specific comment on Trafalgar Square and a new Guildhall in the City of London and brief details of some of the buildings and statues preserved by the GLC's Historic Buildings Board.

1770 DOCKLANDS JOINT COMMITTEE (1975) Conservation and the role of the river. Docklands Joint Committee, 48p. (DJC Working Paper for Consultation, No. 8)

A discussion of the possibilities of conserving existing buildings, such as warehouses, granaries, houses, wharves and other dock features in dockland development schemes. Buildings proposed for retention are listed and mapped.

1771 GREATER LONDON COUNCIL (1975) Historic buildings in London. Academy Editions, 128p. (London Architectural Monographs)

A borough-by-borough inventory of the 1,000 historic buildings and monuments owned by the GLC. Each building is briefly described, and many are illustrated. An introductory chapter traces the history of the GLC historic building ownership and conservation.

1772 LLOYD, D. (1975) The City: conservation or destruction? *Building*, Vol. 229, pp. 86–9.

Uses photographs to illustrate the traditional townscape of the City and how it is being altered. The worst devastations tend to occur in areas with specific City character, though not in designated conservation areas.

1773 LLOYD, D. (1975) The City of London – has its medieval past a future? *Built Environment*, Vol. 1, pp. 66–72.

A review of the development of the City since medieval times with special emphasis on the radical changes following the bombing of 1940 and the incursions of property developers in the 1960s. Argues that the area is Britain's greatest victim of comprehensive redevelopment and calls for policy to preserve what is left of the surviving evolutionary landscape.

1774 The Heritage Year campaign (1975) *Planner*, Vol. 61, pp. 2–30.

Looks at the conservation of historic buildings and areas, with general articles on conservation and specific articles on particular areas, including some in London.

1775 MANSELL, G. and HIRSCH, J. M., (eds) (1975) The living heritage of Westminster. City of Westminster Department of Architecture and Planning, 108p.

Issued as a contribution to European Architectural Heritage Year, this book examines the problems of conservation and development in Westminster. Examples of preservation, restoration, replication and renewal are given.

1776 CRAWFORD, D. (1976) Conservation in the dock. *Built Environment Quarterly*, Vol. 2, pp. 68–72.

Sees London's docklands as a huge test area for conservation policies, rather than a zone for redevelopment. Re-use of the area for recreation, industrial and housing conservation is suggested.

1777 Time on our side? A survey of archaeological needs in Greater London (1976). Department of the Environment, Greater London Council and Museum of London, 15p. and maps.

States that time is not on our side and that the resources available to deal with the threat to thousands of potential archaeological sites in Greater London are insufficient, and an increasing number of archaeological sites will be destroyed unless resources are increased. Eight detailed maps show the distribution of archaeological sites in Greater London and a final map locates the major development schemes and projects known in 1974/5.

1778 ADDISON, W. (1977) Portrait of Epping Forest. Robert Hale, 191p.

A history of the preservation and management of Epping Forest, mainly after the 1878 Act which gave the Corporation of London control of the Forest. Also covered is the ecology of the forest deer and a description of earthworks and other historical remains.

1779 BOOKER, C. (1977) An exercise in conservation. *Spectator*, Vol. 238, No. 7755, pp. 14–5.

Reports on the redevelopment of the St Katharine's Dock site since 1969. Despite an agreement that the fine nineteenth century warehouses would be preserved, one has been demolished and a second threatened.

1780 LESTER, A. W. (1977) Hampstead Garden Suburb: the care and appreciation of its architectural heritage. Hampstead Garden Suburb Design Study Group, 58p.

A practical guide to the conservation and sympathetic alteration of the houses in London's first garden suburb.

1781 DENBIGH, K. (1978). Preserving London. Robert Hale, 240p.

A descriptive guide to 25 conservation areas in Greater London, arranged by districts, but excluding Westminster and the City. A full list of conservation areas in London is given in the appendix.

1782 KUTCHER, A. (1978) Looking at London: illustrated walks through a changing city. Thames and Hudson, 128p.

Examines the impact of redevelopment and change on the urban landscape of central London. Sixty views that deserve protection have been identified in addition to the six mentioned in the *Greater London development plan* (GREATER LONDON COUNCIL, 1969). A chapter on London's future examines recent planning issues and battles.

1783 LLOYD, D. et al. (eds) (1979) Save the City: a conservation study of the City of London. 2nd ed. Society for the Protection of Ancient Buildings, 196p.

Divides the City into 12 sectors and analyses, with maps and photographs, the old and new townscapes of each sector, with recommendations for improvement and preservation. There are also general statements of the conservation issues associated with traffic, open spaces, the waterfront, pubs and restaurants, Roman remains and the economics of conservation.

1784 STANDING CONFERENCE ON LONDON AND SOUTH EAST REGIONAL PLANNING (1979) Restoration of sand and gravel workings. Standing Conference on London and South East Regional Planning, 33p.

This paper reports the intention of local planning authorities in the South East to work closely with the Sand and Gravel Association in an attempt to clear up the backlog of gravel workings left unrestored for many years.

9.3 Water supply, drainage and waste disposal

1785 RICHARDS, H. C. and PAYNE, W. H. C. (1899) London water supply, 2nd ed. Edited by J. P. H. Soper. P. S. King, 310p.

An administrative and legal history of the various metropolitan water companies, chiefly during the nineteenth century, and their subsequent purchase by the LCC.

1786 JEPHSON, H. (1907) The sanitary evolution of London. Fisher Unwin, 440p.

A detailed account of the existing local government difficulties that led to the creation of the Metropolitan Board of Works in 1855, and the subsequent administrative history of the creation of a unified sewage system for London.

1787 FOORD, A. S. (1910) Springs, streams and spas of London: history and associations. Fisher Unwin, 352p.

Provides historical details about Walbrook, Fleet, Tyburn and other streams, together with spas and wells north and south of the Thames. The conduit system of water supply is described, including a chapter on the New River. An appendix lists shallow or surface wells and pumps.

1788 BARROW, G. and WILLS, L. J. (1913) Records of London wells. HMSO, 215p. (Memoirs of the Geological Survey)

A catalogue of wells in London with an introduction setting out the physical background.

1789 DAVIES, A. M. (1913) London's first conduit system: a topographical study. *Transactions of the London and Middlesex Archaeological Society*, New Series, Vol. 2, pp. 9–58.

Uses documents and maps to trace the sources and courses of London's early water supply from the thirteenth century onwards.

1790 HUMPHREYS, G. W. (1930) The main drainage of London. London County Council, 242p.

A descriptive account, with maps and diagrams, of the growth of London's main drainage after the creation of the Metropolitan Board of Works in 1855. The adaptations and additions to the network made by the LCC are described.

1791 BUCHAN, S. (1938) The water supply of the County of London from underground sources. HMSO, 260p.

An account of the geological formations and underground water supply with special reference to the catchment, intake and flow areas of the Chalk together with a detailed catalogue of wells.

1792 MINISTRY OF HEALTH (1948) Report of the departmental committee on Greater London water supplies. HMSO, 15p.

Report of a committee set up to examine water supply administration in the Greater London area (within 40 miles of Charing Cross).

1793 BOSWELL, P. C. E. (1949) A review of the resources and consumption of water in the Greater London area. Staple Press for the Metropolitan Water Board, 23p. map, plan.

Reviews the water resources and water consumption patterns in the Greater London area, covering ground and river water.

1794 FLOWERDEW, L. J. and BERRY, G. C. (1953) London's water supply, 1903–1953. Staple Press for the Metropolitan Water Board, 368p.

A history of the financial, engineering and water treatment activities of the Metropolitan Water Board, set up in 1903. The earlier history of London's water supply is briefly reviewed.

1795 DICKINSON, H. W. (1954) Water supply of Greater London. Leamington Spa: Newcomen Society, 151p.

A detailed account by an engineer of London's water supply from the early use of streams, wells and conduits to the creation of the Metropolitan Water Board. Contains excellent illustrations of pipes, waterworks and pumping engines.

1796 LONDON COUNTY COUNCIL (1955) Centenary of London's main drainage, 1855–1955. London County Council, 30p.

Commemorates the creation of the Metropolitan Board of Works in 1855 whose primary duty was the construction and maintenance of sewers. Bazalgette's main scheme and its subsequent additions and adaptations are outlined.

1797 GREEN, A. F. (1956) The problem of London's drainage. *Geography*, Vol. 41, pp. 147–54.

A study of the evolution and problems of London's drainage system –

defined as the removal of water supplies without due offence – from earliest times to contemporary improvements of the plant at the northern and southern outfalls to the east of London.

1798 BONIFACE, E. S. (1959) Some experiments in artificial recharge in the lower Lee valley. *Proceedings of the Institution of Civil Engineers*, Vol. 14, pp. 325–8.

Describes artificial recharge operations at three neighbouring well stations in the London basin with details of the hydro-geology of the area and a brief history of abstraction from the aquifers. Experimental methods of recharge are examined and a groundwater account is attempted.

1799 NASH, J. E. (1959) The effect of flood-elimination works on the flood frequency of the river Wandle. *Proceedings of the Institution of Civil Engineers*, Vol. 13, pp. 317–38.

The time response of the catchment of the river Wandle to rainfall will be altered as a result of the proposed works. This paper forecasts the alteration, and develops a method by which the existing record of discharges can be transformed by matrix inversion and multiplication in accordance with the change in the unit hydrograph of the catchment. A routine frequency analysis of the return periods of flood flows is made on the transformed hydrograph.

1800 METROPOLITAN WATER BOARD (1961) The water supply of London. Metropolitan Water Board, 113p.

An account of the activities of the Board produced with the general reader in mind. Following an historical background the engineering features and the methods of purification and quality control are examined.

1801 GOUGH, J. W. (1964) Sir Hugh Myddleton, entrepreneur and engineer. Oxford: Clarendon Press, 155p.

Four chapters are devoted to Myddleton's role in the creation of the New River scheme which brought a domestic water supply from Amwell in Hertfordshire to London in the early seventeenth century.

1802 VICK, E. H. (1966) Sewage treatment works in Greater London. Greater London Council, 16p.

A review of the sewage treatment works serving the new Greater London Council area. A reduction of the number of sewage works is recommended.

1803 WATER RESOURCES BOARD (1966) Water supplies in south east England: report of a technical committee. HMSO, 133p.

This report deals with likely water demands and water resources in south-east England up to 2001, in the light of the major population growth forecast in the *South East study* (MINISTRY OF HOUSING AND LOCAL GOVERNMENT, 1964).

1804 REES, J. A. (1969) Industrial demand for water: a study of south east England. Weidenfeld and Nicolson, 94p.

Examines the factors which influence industrial demands for water in south-east England and uses them to explain the varying quantities of water taken by different firms.

1805 DARLINGTON, I. (1970) The London Commissioners of Sewers and their records. Phillimore, 75p.

The London Commissioners of Sewers was established by an Act of Parliament in 1531 and was responsible for all navigable and drainage water courses in London and Westminster. This book provides a brief account of the Commissioners' activities and lists the relevant documents.

1806 PEARSON, R. F. (1970) Study for disposal of digested sewage sludge from the Greater London Sewerage area into the North Sea by pipeline. *Proceedings of the Institution of Civil Engineers*, Vol. 48, pp. 375–98.

Outlines the feasibility study into the disposal of digested sewage sludge from the major treatment works of the GLC at Beckton and Crossness into the North Sea, by means of a 130 km long pipeline.

1807 REES, J. A. and REES, R. (1972) Water demand forecasts and planning margins in south east England. *Regional Studies*, Vol. 6, pp. 37–48.

A study of the forecasts from the Water Resources Board producing a set of projections of required expenditure on new supply capacity. Suggests the forecasts are well above the demand levels expected from the best estimate of past trends and will lead to considerable excess capacity by 1985 if implemented.

1808 WATER RESOURCES BOARD Central Water Planning Unit (1972) Artificial recharge of the London basin; 1. Hydrogeology. Reading: Water Resources Board, 49p.

Shows that over the past 150 years a dewatering of large areas of the Chalk and lower Tertiaries has taken place. Available data were analysed to produce a storage capability with conditions modelled by analogue. Pilot schemes for recharge are in process.

1809 SATCHELL, R. L. H. and WILKINSON, W. B. (1973) Artificial recharge in United Kingdom with special reference to the London basin. In: AMERICAN ASSOCIATION OF PETROLEUM GEOLOGISTS INC. Underground waste management and artificial recharge, Vol. 1, pp. 34–59.

Report of a hydrogeological study to assess the potential for recharge of groundwater levels beneath London over the last 170 years. Groundwater levels have fallen in some areas more than 250 feet creating a storage volume exceeding 200 billion gallons. Concludes that additional yield of more than 70 million gallons per day could be made available at low cost without the need of further surface storage.

1810 SHIELDS, L. (1973) The lesson of London. *Municipal Journal*, Vol. 81, pp. 1161–3.

Comments on the decision to separate the responsibility for refuse collection and disposal and reviews the legal and financial problems met by the GLC – the first regional disposal authority (1968). The new disposal authorities are likely to meet difficulties with the shortage of tip space and increased responsibility for the disposal of bulky materials.

1811 VENABLES, J. (1973) The main drainage of London: Part 1, The early years; Part 2, Adapting to modern times. *Underground Services*, Vol. 1, pp. 17–20; 25–8.

Reviews the progress of London's drainage from the fourteenth century to Sir Joseph Bazalgette's main sewers plan of 1856. Subsequent additions to that system and current improvements are outlined.

1812 WATER RESOURCES BOARD Central Water Planning Unit (1973) Artificial recharge of the London basin; II, Electrical analogue model studies. Reading: Water Resources Board, 38p.

Describes the construction and operation of both steady state and non-steady state electrical network analogue models of groundwater flows in the Chalk and Tertiary sands of the London basin. Experiments are described to examine the artificial recharge of groundwater levels in two areas.

1813 REES, J. A. (1973) The demand for water in south east England. *Geographical Journal*, Vol. 139, pp. 20–42.

The problems of managing and planning the use of water resources with special attention to the difficulties of equating the demand for and supply of water in south-east England.

1814 BUSBY, P. R. A. (1974) The evolution and development of London's main drainage. *Public Health Engineer*, Vol. 12, pp. 213–8.

A history of London's sewerage system showing how it emerged from pressure arising from intolerable conditions. Details of the plans devised by Joseph Bazalgette in the mid-nineteenth century, which forms the basis of the present system, are given.

1815 PENDRED, B. W. (1974) Boring into the Thames and Kennet: the plans for the first stage of a scheme to enhance the water supply of London. *Field*, Vol. 244, pp. 558–60.

Outlines the scheme for enhancing London's water supply through 40 boreholes in the valleys of the Lambourn, Loddon, Pang and Enborne.

1816 GREATER LONDON COUNCIL Public Services Committee (1975) Review of the solid wastes management service. Greater London Council Agenda, No. 5, pp. 1–5.

The GLC was the first regional solid waste disposal authority in this country and this article reviews the first eight years of operations. The advantages of taking disposal from numerous small authorities have included lower rates for bulk haulage contracts, better qualified staff, more favourable contracts for sale of waste paper, release of land formerly used for disposal plants, a regional disposal system for old cars and the introduction of modern technology such as the Edmonton incinerator and containerized linear trains.

1817 MILLBANK, P. (1975) Newham: landmark in solid wastes management. *Surveyor*, Vol. 146, pp. 34–6.

Provides details of the GLC's refuse transfer station under construction in Newham which forms part of the GLC's policy of establishing larger centres as foci for collection authorities so that transport to distant disposal centres can be rationalized, there being few landfill sites left in the Greater London area.

1818 MUKHOPADHYAY, A. K. (1975) The politics of London water. *London Journal*, Vol. 1, pp. 207–24.

Analyses three phases of London's fight for water – in order to elucidate the decision-making process and examine the growth of the administrative state in nineteenth century Britain. Also examines the inadequacy of the present system of management of London water supply.

1819 PATRICK, P. K. (1975) Solid waste disposal in Greater London: the first ten years. *Public Health Engineer*, Vol. 18, pp. 173–7.

Reviews the development of the first regional solid waste disposal organization in the UK with discussion of the problems involved in transferring control from the London boroughs to the GLC; the development of long-term plans for incineration, landfill and the transport of waste, and the design of transfer stations and treatment plant.

1820 PATRICK, P. K. (1976) Greater London's plan for rail haul of solid wastes. *Solid Wastes*, Vol. 66, pp. 150–7.

Describes in detail the design and operation of the council operated station under construction at Brentford, West London, to enable London's waste to be hauled out by rail to more distant landfill sites than can be reached by road haulage.

1821 PENN, C. J. (1976) Waste paper salvage in the London Borough of Enfield. *Solid Wastes*, Vol. 66, pp. 5–16.

Describes the waste paper baling plant established next to the Edmonton incineration plant serving much of north London. The plant sorts, bales and sells the processed paper.

1822 TOWNSEND, W. K. (1976) The Civic Amenities Act 1967; Section 18 and the disposal authority. *Solid Wastes*, Vol. 66, pp. 168; 181–4.

There were 49 civic amenities sites in the GLC area in 1975 and the operation in north-west London (14 boroughs) is described. Ten of the 22 sites are transfer stations where the public is admitted. Guidelines are given for deciding whether separate civic amenities sites should be set up with public use as laid down in the Act.

1823 ROFE, B. H. (1977) Storage and the Thames scheme. *Consulting Engineer*, Vol. 41, pp. 27–9.

An account of the main components of stage 1 of the scheme to utilize groundwater by pumping from aquifers into the river systems in periods of drought.

1824 INSTITUTION OF CIVIL ENGINEERS and THAMES WATER AUTHORITY (1978) Thames groundwater scheme; proceedings of a conference held at Reading University 12–13 April. Institution of Civil Engineers, 241p.

Eight papers concerned with the Thames groundwater scheme for stream flow augmentation, developed since 1949. Includes material on the historical, technical and environmental aspects of the scheme.

1825 STANDING CONFERENCE ON LONDON AND SOUTH EAST REGIONAL PLANNING (1978) Waste disposal in south east England. Standing Conference on London and South East Regional Planning, 43p.

Presents information on the generation of waste in the region and of its disposal, the location and nature of disposal sites, and costings.

1826 WACHER, J. S. (1978) The water supply of Londinium. In: BIRD, J. et al. Collectanea Londiniensia. London and Middlesex Archaeological Society, pp. 104–8.

Examines the evidence for the sources and methods of supplying water in Roman London, in particular wells and springs.

1827 HOLLIS, T. (1978) Water for London. In: CLOUT, H., (ed.) Changing London. University Tutorial Press, pp. 118–27.

Examines the demand for and supply of water to London. Sections on the historical setting and the water management structure are followed by elements in the changing scene. Three hazards are analysed – flooding, tidal inundation and rising nitrate levels; the Lea valley is presented as a sub-regional case study.

9.4 Flood hazard

1828 THAMES CONSERVANCY (1947) Report on the flooding of urban and agricultural districts in the Thames valley with special reference to the high flood of March, 1947. Thames Conservancy, 27p.

Includes a description of possible methods of flood control and works proposed or carried out to this end.

1829 ROBINSON, A. (1953) The sea floods around the Thames estuary. *Geography*, Vol. 38, pp. 170–6.

Looks at the impact of the storm floods of the 1 February 1953 upon the Thames estuary to the east of London, along the Essex and Kent coasts.

1830 ESCRITT, L. B. (1966) London stormwater. Greater London Council, 65p.

An overall investigation into rainfall run-off in London, the problems produced by storm overflows and how these problems may be solved.

1831 SPEARING, N. (1969) The Thames barrier-barrage controversy: a review of possibilities. Institute of Community Studies, 58p.

Compares the various solutions to the danger of flooding and suggests environmental improvements to be included in the flood prevention works.

1832 WOODLEY, G. M. (1969) Notes on the risk of flooding in London by the Thames, Parts 1 and 2. *Chartered Surveyor*, Vol. 101, pp. 331–2; 389–90.

The increased risk of flooding from the Thames in the capital over the past century is discussed, together with the reasons for this increase. Methods of flood prevention, including a possible Thames barrage, are reviewed.

1833 GREATER LONDON COUNCIL Department of Public Health Engineering (1969) Thames flood prevention: first report of studies. Greater London Council, 245p.

Describes the character and size of the Greater London area threatened by tidal flooding. The effect of the construction of various alternative defence works on the normal activities is discussed, and the report concludes that a drum gate barrier at Crayfordness and a drop gate tidal control structure at Woolwich should be subjected to detailed design study.

1834 GREATER LONDON COUNCIL Department of Public Health Engineering (1969) Thames flood prevention: second report of studies. Greater London Council, 3 vols. (Vol. 1 and 2 bound together, vol. 3 and appendix 6 bound separately)

Reports on further studies (to January 1971) on the exact location and type of flood barrier to be built on the Thames. The proposal for a drop gate or a drum gate is replaced with a proposal for a rising sector gate, reported to be more reliable and less costly.

1835 GREATER LONDON COUNCIL (1970) Taming the Thames: protecting London from flooding. Greater London Council, 4p.

Describes the possible solutions to the flood dangers to London, such as raising the walls, constructing a barrage or a barrier. The advantages of each are outlined.

1836 DUNHAM, K. C. and GRAY, D. A. (1972) A discussion on the problems associated with the subsidence of south eastern England. *Philosophical Transactions of the Royal Society*, Series A, Vol. 272, pp. 79–274.

Contains 17 contributions, nine of which are concerned with the evidence for subsidence and eight with the Thames barrier. Certain papers are individually annotated in this bibliography.

1837 GREATER LONDON COUNCIL Department of Public Health Engineering (1974) River Brent flood alleviation feasibility study. Greater London Council, 26p. and maps

Identifies existing and potential flooding problems in the catchment area of the Brent and outlines basic hydraulic requirements to control and alleviate the flooding.

1838 BUTTERS, K. and LANE, J. J. (1975) Flood alleviation works on some river Thames tributaries. *Journal of the Institution of Water Engineers and Scientists*, Vol. 29, pp. 67–94.

Examines urban land drainage in six hydrological catchment areas in the so-called London Excluded Area. Types of alleviation works and costs and benefits are outlined, with special reference to the river Brent and the Ravensbourne.

1839 HORNER, R. W. (1975) The river Thames flood defence barrier. *Public Health Engineer*, Vol. 3, pp. 120–4.

Discusses the area at risk from flooding, the cause of the North Sea surges, and the methods of flood defence. The type of flood barrier being constructed and its costs and benefits are examined.

1840 ALDOUS, T. (1976) If London floods. *Architects' Journal*, Vol. 164, pp. 826–8.

Looks at the likely results of the Thames flooding and overflowing London. The construction of the Thames barrier at Woolwich is described.

1841 HORNER, R. W. (1976) Keeping London above water: the story so far. *Surveyor*, Vol. 146, p. 7.

Reports on the progress of the construction of the flood barrier at Woolwich and on the defences along the river.

1842 HORNER, R. W. (1976) The Thames tidal flood prevention scheme. *Long Range Planning*, Vol. 9, pp. 78–83.

An outline of the technical reasons for the decision to develop the flood prevention scheme on the Thames estuary. Describes some aspects of the operation of the scheme programmed for 1980.

1843 PULLIN, J. (1976) Taking the Thames flood defences down into Kent. *Surveyor*, Vol. 147, pp. 22–3.

The Southern Water Authority is in the process of constructing a £36.5 million scheme of flood defence for the Thames between Dartford and the Isle of Grain. It includes the rebuilding of tidal defence walls and a small flood barrier at Dartford Creek.

1844 GUNNELL, B. and REINA, P. (1977) Thames barrier. *New Civil Engineer*, 29 September, pp. 25–38.

Describes the background to the flood prevention scheme and the construction story to date.

1845 HORNER, R. W. et al. (1977) London's stormwater problem. *Public Health Engineer*, Vol. 5, pp. 146–51.

Assessment of the drainage facilities divided into districts based on river catchments which have been modified by urban development. A 'standard storm' is used for the purpose of comparison of districts. Storm water overflow quality samples are reported and methods of dealing with the problem are discussed. A glossary of terms is provided and the effect of a storm in 1973 on the tidal Thames is appended.

1846 HEWAYES, N. (1978) London maps out flood defence network. *Contract Journal*, Vol. 281, pp. 22–3.

Reports on the Greater London Council's efforts to resume a programme (dropped in 1973) of flood alleviation for non-tidal rivers.

1847 INSTITUTION OF CIVIL ENGINEERS (1978) Thames barrier design: proceedings of the conference held in London on 5th October 1977. Institution of Civil Engineers, 202p.

Includes 16 contributions, with discussions from the conference on the Thames barrier project; three are on the history and location of the barrier, two on the site considerations and the rest on the design of the barrier itself.

1848 HORNER, R. W. (1979) The Thames barrier project. *Geographical Journal*, Vol. 145, pp. 242–53.

Describes the flood risk to London and the decision in 1972 to build a barrier across the river in the western half of Woolwich Reach. The design, construction and eventual operation of the barrier are described.

1849 MARK, R. (1979) The Thames barrier. Greater London Council, 7p. (GLC London Topics, No. 32)

Reviews the history and causes of flooding in the Thames river and estuary, and the progress in design and construction of the barrier. Cites 40 references.

9.5 Water pollution

1850 WILKINSON, R. (1956) Quality of rainfall-runoff water from a housing estate. *Institute of Public Health Engineering*, Vol. 55, pp. 70–84.

Provides a classic base line study of the effects of urban stormwater pollution arising from housing development in north-west London.

1851 MINISTRY OF HOUSING AND LOCAL GOVERNMENT (1961) Pollution of the tidal Thames. HMSO, 68p.

The report of the Pippard Committee appointed in 1951 to investigate the effects of heated and other effluents and discharges on the condition of the tidal reaches of the river Thames.

1852 CARTER, G. (1963) The river Thames and Lee – chloride content and hardness. *Proceedings of the Society of Water Treatment and Examination*, Vol. 12, pp. 226–9.

Graphical data are given showing variations in chloride content and hardness in the rivers Thames and Lea over the period 1951–60, and in chloride content over the period 1905–60. Chloride content of the Lea has doubled since 1905, probably as a result of urban development causing increased volume of sewage effluents.

1853 THAMES SURVEY COMMITTEE and WATER POLLUTION RESEARCH LABORATORY (1964) Effects of pollution discharges on the Thames estuary. HMSO, 609p. (Water Pollution Research Laboratory, Technical Paper, No. 11)

The major report of the Thames Survey Committee appointed in 1949 to study the Thames estuary with particular reference to its capacity to purify the sewage and industrial effluents discharged into it.

1854 GAMESON, A. L. H. and HART, I. C. (1966) A study of pollution in the Thames estuary. *Chemistry and Industry*, 17 December, pp. 2117–23.

Reports that the middle (and most polluted) reaches of the Thames estuary showed a marked improvement in 1964–5 due to the introduction of secondary treatment at the southern outfall sewage works. The lower reaches, however, showed no improvement.

1855 WHEELER, A. C. (1969) Fish-life and pollution in the lower Thames: a review and preliminary report. *Biological Conservation*, Vol. 2, pp. 25–30.

Reports on how the Thames is becoming less polluted as evidenced by increasing numbers of species of fish found in the cooling water intakes of five power stations in the lower Thames.

1856 POTTER, J. H. (1971) Pollution control and the Port of London. *Port of London*, Vol. 46, pp. 3–7.

Describes the growth of pollution on the Thames and the efforts of the Pollution Control Department of the Port of London Authority to improve the quality of the water using the maxims of consultation, coordination and control. A final section deals with plans for further improvements.

1857 POTTER, J. H. (1971) Pollution and its control in the tidal Thames. *Community Health*, Vol. 3, pp. 103–10.

Looks at the deterioration of the condition of the tidal river up to 1959 and shows how the Port Authority, the body with a statutory responsibility for pollution control, has taken action to cleanse the river to the extent that fish are returning in increasing quantities.

1858 BARRETT, M. J. and MOLLOWNEY, B. M. (1972) Pollution problems in relation to the Thames barrier. In: DUNHAM, K. C. and GRAY, D. A. A discussion on problems associated with the subsidence of south eastern England. *Philosophical Transactions of the Royal Society of London*, Series A, Vol. 272, pp. 213–21.

Outlines a theoretical study undertaken at the request of the Greater London Council to assess the effect on the condition of the water of the Thames estuary of operating the barrier on a regular basis in such a way as to prevent water levels landwards of it falling below Newlyn Datum.

1859 WOOD, L. B. (1973) The condition of London's rivers in 1971. *Quarterly Bulletin of the Research and Intelligence Unit of the Greater London Council*, Vol. 22, pp. 18–36.

Examines the quality of the water in the Thames and its tributaries in the GLC area during 1971, and compares it with that observed in the previous year. The sources of pollution and the methods of monitoring are described, including a series of experiments which produced a record catch of 352 fish.

1860 McCROW, B. J. (1974) The biological effects of pollution on a stretch of the river Wandle, 1970–1971. *London Naturalist*, Vol. 53, pp. 17–33.

A study of pollution on the river Wandle using samples taken at four points along the river downstream from Carshalton ponds. Concludes that the pollution of the Wandle must be controlled more strictly to restore the condition of the river.

1861 ELLIS, J. B. (1976) Sediments and water quality of urban storm water. *Water Services*, Vol. 80, pp. 730–4.

Investigates the Silk Stream catchment, the largest tributary of the river Brent, draining 3,323 hectares in north London. From the monitoring of the catchment a full record of the pollution loadings and patterns was obtained; hydrological details and the nature of urban stormwater sediment and heavy metals are described. The reactions between pollutant species and colloidal sediments determines the relative concentration of contaminants in solution and suspension.

1862 DOXAT, J. (1977) The living Thames: the restoration of a great tidal river. Hutchinson Benham, 96p.

An account of the decline of the river Thames as pollution increased with the growth of London, followed by a description of the cleansing of the river in the post-war period.

1863 FREEMAN, L. (1977) Old Father Thames clean up complete by 1980? *Water*, Vol. 13, pp. 2–6.

Looks at the river Thames, how it became polluted, how it was improved and how it is being maintained.

1864 GAMESON, A. L. H. and WHEELER, A. C. (1977) Restoration and recovery of the Thames estuary. In: CAIRNS, J., (ed.) Recovery and restoration of damaged ecosystems. Proceedings of an international symposium at the Virginia Polytechnic Institute, Blacksburg, Virginia, March 23–25, 1975. Charlottesville: University of Virginia Press, pp. 72–101.

Outlines the long history of pollution in the Thames and how the fish species and stocks began to decline over 150 years ago. Detailed scientific investigations were carried out from 1949–64, and since 1965 sewage pollution has been reduced and the restoration of the water quality has been accompanied by a remarkable ecological recovery.

1865 ASTON, K. F. A. and ANDREWS, M. J. (1978) Freshwater macro-invertebrates in London's rivers, 1970–77. *London Naturalist*, Vol. 57, pp. 34–52.

The fauna of the freshwater reaches of the Thames tideway and of its main tributaries in London has been studied regularly since 1970. The rivers Darent and Cray, which join the Thames 29 km below London Bridge, have been studied since 1974. More than 1,100 samples have been taken at 50 sites. The distribution of macro-invertebrates in the individual rivers is discussed in relation to the known pollution history.

1866 BRYCE, D. et al. (1978) Macro-invertebrates and the bio-assay of water quality: a report based on a survey of the river Lee. Nelpress, 44p.

A report on a general survey of the river that provides basic information on the spatial distribution of invertebrate fauna found in the valley of the Lea.

1867 HORNER, R. W. et al. (1978) London's stormwater problem. *Public Health Engineer*, Vol. 6, pp. 25–7.

A supplement to the earlier article by Horner et al. (1977) describing the effect of the storm of the 16 and 17 August when heavy rainfall combined with a rising tide led to flooding of several areas. There was a reduction in the oxygen content of the river causing fish to die, while sewage load discharge was of increased proportions.

1868 ELLIS, J. B. (1979) The nature and sources of urban sediments and their relation to water quality: a case study from north west London. In: HOLLIS, G. E. (ed.) Man's impact on the hydrological cycle in the United Kingdom. Norwich: Geobooks, pp. 199–216.

Uses evidence from the Silk Stream to examine the impact of pollution loadings from urban catchments on river channels with an originally rural catchment.

1869 WHEELER, A. C. (1979) The tidal Thames: the history of a river and its fishes. Routledge and Kegan Paul, 228p.

Tells the story of the return of fish to the tidal Thames (Teddington to the sea) as the river has been reclaimed from pollution. A final chapter examines the promise for the future of the river with specific reference to the threats to fish life in the tideway.

9.6 Air and noise pollution

1870 ABSOLOM, H. W. L. (1954) Meteorological aspects of smog. *Quarterly Journal of the Royal Meteorological Society*, Vol. 80, pp. 261–6; discussion pp. 272–8.

Provides general descriptive information about fog and the polluted variety known as smog. The causes and extent of the 1952 smog are described and mapped.

1871 MINISTRY OF HEALTH (1954) Mortality and morbidity during the London fog of December, 1952. HMSO, 60p. (Ministry of Health Report No. 95)

Presents the basic information necessary for an assessment of the increased mortality and morbidity associated with the London fog of 5–8 December 1952, and for an understanding of the contributory causes.

1872 WILKINS, E. T. (1954) Air pollution aspects of the London fog of December 1952. *Quarterly Journal of the Royal Meteorological Society*, Vol. 80, pp. 267–78.

Describes and tabulates the type and distribution of pollutants in London from 2–10 December 1952. There was a striking correlation between the mean daily concentration of smoke and sulphur dioxide and the total number of deaths in those days.

1873 GOLD, E. (1956) Smog, the rate of influx of surrounding clean air. *Weather*, Vol. 11, pp. 230–2.

Extends the discussion following the paper by WILKINS, E. T. (1954). The effects of incoming air replacing warmer and rising polluted air are assessed.

1874 WAINWRIGHT, C. W. K. and WILSON, M. J. G. (1962) Atmospheric pollution in a London park. *International Journal of Air and Water Pollution*, Vol. 6, pp. 337–47.

Reports on measurements of atmospheric sulphur dioxide concentration in Hyde Park taken during 1959–60. The rate of decrease of the concentration of sulphur dioxide with distance from the upwind edge was found to be closely related to the lapse rate but not to the wind speed.

1875 McKENNELL, A. C. (1966) Aircraft noise annoyance around London (Heathrow) airport. HMSO, 3 vols.

A detailed technical report of a survey made in 1961 into the causes and problems of aircraft noise in the vicinity of Heathrow airport.

1876 McKENNELL, A. C. and HUNT, E. A. (1966) Noise annoyance in central London. HMSO, 128p.

Presents the results of a survey of reaction to noise carried out in central London during the summer of 1961. Aircraft and traffic noise caused the most annoyance.

1877 COMMINS, B. T. and WALLER, R. E. (1967) Observations from a ten year study of pollution at a site in the City of London. *Atmospheric Environment*, Vol. 1, pp. 49–68.

Describes the results from samples taken over ten years at St Bartholomew's Hospital and medical school in the City of London. The figures show a reduction in the annual, mean and peak concentrations of smoke during the ten years, but not in the levels of sulphur dioxide. There was also a decline in the amount of polycyclic aromatic hydrocarbons recorded in the atmosphere.

1878 LUCAS, D. H. et al. (1967) The measurement of plume rise and dispersion at Tilbury power station. *Atmospheric Environment*, Vol. 1, pp. 353–65.

Describes the system of instrumentation that has been installed at Tilbury power station to continuously record sulphur dioxide emission, the size of the plume, and weather variables in the area.

1879 PLANK, D. (1970) Progress and effects of smoke control in London: a report to the General Purposes Committee of the London Boroughs Association. Greater London Council, 18p.

A review of the progress of smoke control in London, and, with a number of readily available indicators, estimates of its most important effects.

1880 OFFICE OF POPULATION CENSUSES AND SURVEYS (1971) Second survey of aircraft noise annoyance around London (Heathrow) Airport. HMSO, 193p.

Presents the results of a survey in 1967 of residents' reactions to aircraft noise. Suggests that the existing formula gives too much weight to the number of aircraft movements and too little to annoyance under take-off and approach paths.

1881 WARREN SPRING LABORATORY (1972) National survey of air pollution, 1961–71, Vol. 1. HMSO, 195p.

Presents the results of the national survey over the ten year period, and the conclusions to be drawn from them on a national and regional basis. Pollution by smoke and sulphur dioxide is described and analysed nationally for the UK, regionally for the South East, and locally for Greater London.

1882 BRUTON, D. M. (1973) Pollution levels at London (Heathrow) airport and methods for reducing them. In: ADVISORY GROUP FOR AEROSPACE RESEARCH AND DEVELOPMENT. Atmospheric pollution by aircraft engines. Paris: AGARD, 6p.

Exhaust pollution levels were measured and medical surveys of the interior of buildings were conducted to determine pollution levels. Methods for reducing the exhaust fume emission are described.

1883 DERWENT, R. G. and STEWART, H. N. M. (1973) Elevated ozone levels in the air of central London. *Nature*, Vol. 241, pp. 342–3.

Describes the monitoring of various pollutants in central London (Holborn) undertaken by the Warren Spring Laboratory. The concentration of ozone for 12–14 July 1972 is given with values at maximum above 100 p.p.b. – a value having been proposed as an air quality standard in the USA.

1884 GREATER LONDON COUNCIL Intelligence Unit (1974) Statistical review of progress and effects of smoke control in London. Greater London Council, 23p. (GLC Research Memorandum, No. 422)

Looks at the development of smoke control in London and, using a number of yardsticks, illustrates some of the more important effects and compares present levels with suggested goals.

1885 MASTERS, B. R. (1974) The City of London and clean air, 1273 AD to 1973 AD. *Clean Air*, Vol. 4, pp. 1–7.

Examines the problem of air pollution in the City of London from the thirteenth century to 1973. Special attention is paid to the attempts made to control pollution and the success of the legislation passed in 1956.

1886 VULKAN, G. H. (1974) Noise control in London. *Proceedings of Inter-Noise*, Vol. 74, pp. 35–40.

Reviews the noise problem in London stating that the main causes are traffic and aircraft. Discusses the implications of the Land Compensation Act and the roles of planning control and local noise control powers in reducing noise levels.

1887 ASHBY, E. (1975) Clean air over London; the second Sir Hugh Beaver Memorial Lecture. *Clean Air*, Vol. 5, pp. 25–30.

Examines the history of the campaign for clean air in London since 1800 through to 1956 when public concern (over the 1952 smog), economic and technical feasibility, and a firm balance of benefits over costs, combined to produce the necessary legislation.

1888 GREATER LONDON COUNCIL Environmental Sciences Group (1975) Concorde noise and its effect on London. Greater London Council, 58p. (GLC Research Memorandum, No. 478)

Reports and analyses the noise levels measured during the take-off and landing of Concorde and certain other aircraft between 7 July and 13 September 1975. Estimates the number of people exposed to noise from certain types of aircraft and likely to be disturbed by individual flights. Describes the methods of measurement and analysis used in the survey and discusses the results obtained.

1889 MILLER, C. (1975) Lead in the air. Greater London Council, 12p. (GLC London Topics, No. 10)

A review of recent literature, legislation and research on lead-based air pollution including comment on its medical effects, the monitoring of pollution and the provision of standards.

1890 London makes its own air standards (1975) *New Scientist*, Vol. 67, p. 154.

Reports how the GLC set its own standards for air quality in the face of delays over the establishment of national standards.

1891 SMITH, S. R. (1975) John Evelyn and London air. *History Today*, Vol. 25, pp. 185–9.

A review of *Fumifugium; or the inconvenience of the air and smoke of London dissipated* published in 1661. John Evelyn discusses the harmful effects of bad air on health (he attributed it to the commercial and industrial burning of sea coal) and proposed that all offending trades should be moved five or six miles from London.

1892 TEBRAKE, W. H. (1975) Air pollution and fuel crises in pre-industrial London, 1250–1650. *Technology and Culture*, Vol. 16, pp. 337–59.

Argues that the smoke from sea-coal fires was a general nuisance in London by the end of the thirteenth century and during the fourteenth century. Pollution was less of a problem in the succeeding period up to the second half of the sixteenth century, when a second pollution phase took place. The phases of pollution can be directly correlated with shortage of firewood forcing the inhabitants to burn sea-coal.

1893 BALL, D. J. (1976) Photochemical ozone in the atmosphere of Greater London. *Nature*, Vol. 263, pp. 580–2.

Presents details of the spatial and temporal variation of ozone concentrations experienced in Greater London during the summer of 1975.

1894 BALL, D. J. (1976) An air pollutant emission inventory for the Greater London area. *Clean Air*, Vol. 5, pp. 7–9.

Explains the strategy for the development of the Greater London emission inventory, the sources of sulphur dioxide and inventories for other pollutants.

1895 BALL, D. J. (1976) Photochemical oxidants in the atmosphere of Greater London. Greater London Council, 32p. (GLC Research Memorandum, No. 485)

Reports on a survey undertaken in 1975 which measured photochemical ozones at three sites in Greater London. Concentrations were found to exceed the GLC guidelines for considerable periods and on 15 per cent of summer days.

1896 COMMINS, B. T. and HAMPTON, L. (1976) Changing pattern in concentrations of polycyclic aromatic hydrocarbons in the air of central London. *Atmospheric Environment*, Vol. 10, pp. 561–2.

Reports the marked decline in the measured levels of concentrations of polycyclic aromatic hydrocarbons in central London.

1897 HICKMAN, A. J. (1976) Atmospheric pollution measurements in west London. (1973) Crowthorne: Transport and Road Research Laboratory, 15p. (TRRL Report, No. LR709)

Measurements taken at three different sites showed that concentrations of carbon monoxide, hydrocarbons and nitric oxide fell with increasing distance from the major road to about 30 per cent of the kerbside level at 50 metres.

1898 LANGDON, F. J. (1976) Noise nuisance caused by road traffic in residential areas, Part 1. *Journal of Sound and Vibration*, Vol. 47, pp. 243–63.

The first of a number of reports dealing with external noise nuisance. This paper considers a wide range of urban conditions with varying amounts of traffic flow. The area sampled was Greater London, where noise was measured at 53 sites and 2,933 respondents were interviewed. The survey concentrated on 24 sites where traffic flowed freely.

1899 STEWART, H. N. M. et al. (1976) Ozone levels in central London. *Nature*, Vol. 263, pp. 582–4.

Presents a statistical analysis of the relationship between ozone levels in central London and some meteorological parameters measured at the London Weather Centre (0.4 km from the sampling site).

1900 BALL, D. J. and HUME, R. (1977) The relative importance of vehicular and domestic emissions of dark smoke in Greater London in the mid-1970s. *Atmospheric Environment*, Vol. 11, pp. 1065–73.

Evidence is presented to support the hypothesis that dark smoke in central London derives mainly from vehicular sources. The implication of the finding for the system of air quality management currently practised in Britain is discussed.

1901 BRIMBLECOMBE, P. (1977) London air pollution, 1500–1900. *Atmospheric Environment*, Vol. 11, pp. 1157–62.

Documentary evidence for the changes in the air pollutant levels and climate of London are compared with the results obtained from a simple single box model for the annual mean sulphur dioxide and particulate levels in the London air.

1902 SAINSBURY, R. B. and CASWELL, R. (1977) Air pollution and pedestrianization: studies at the Upper Norwood triangle – Weston Hill traffic experiment. Greater London Council, 29p. (GLC Research Memorandum, No. 496)

Outlines the monitoring of carbon monoxide, smoke and lead at five sites around the Upper Norwood triangle before and during a traffic experiment. Describes the methodology developed for analyses of the air pollution data to assess the effects of a traffic experiment on the local environment.

1903 VULKAN, G. H. (1977) Noise disturbance in urban centres. *Greater London Intelligence Journal*, Vol. 37, pp. 22–7.

Discusses the problem of noise in urban areas (with reference to London) caused by road traffic, aircraft and industrial and construction processes.

1904 THORNES, J. E. (1977) Applied climatology: ozone comes to London. *Progress in Physical Geography*, Vol. 1, pp. 506–17.

Since recording began in 1972, there has been an upward trend in the peak maximum hourly concentration of ozone measured at County Hall in central London. High ozone concentrations have detrimental effects upon health. It is argued that the reduction of smoke levels may inadvertently have helped to generate the recent high ozone levels. The problems of legislation against ozone, taking London as the example, are examined.

1905 BALL, D. J. and BERNARD, R. E. (1978) Evidence of photochemical haze in the atmosphere of Greater London. *Nature*, Vol. 271, pp. 733–4.

A brief analysis of a problem caused by excessive ozone in the atmosphere over London.

1906 BALL, D. J. and BERNARD, R. E. (1978) An analysis of photochemical pollution incidents in the Greater London area with particular reference to the summer of 1976. *Atmospheric Environment*, Vol. 12, pp. 1391–1401.

Describes and analyses the measured ozone concentrations in Greater London, which in 1976 exceeded 20 parts per hundred million for the first time, with higher levels reported from a nearby rural location at Harwell.

1907 BRIMBLECOMBE, P. and WIGLEY, T. M. L. (1978) Early observations of London's urban plume. *Weather*, Vol. 33, pp. 215–20.

Examines changes in the public awareness of smoke plumes downward from London from an historical viewpoint stretching from the thirteenth century through to its disappearance from the vocabulary of south-east England in the late nineteenth century.

1908 BALL, D. J. and RADCLIFFE, S. W. (1979) An inventory of sulphur dioxide emissions to London's air. Greater London Council, 36p. (GLC Research Report, No. 23)

An overview of the sulphur dioxide emission patterns in Greater London in 1975/76 with a description of the methods employed in making this inventory, the problems encountered and the resources needed. The data are presented in tables, graphs and maps.

1909 THORNES, J. E. (1979) Applied climatology: the best practicable means of air quality management in the European Community. *Progress in Physical Geography*, Vol. 3, pp. 427–42.

Looks at British tactics to implement European Commission sulphur dioxide and smoke standards with special reference to air quality findings from the GLC undertaken in the 1970s. Sulphur dioxide levels were below European Commission standards in many of the inner London sites monitored by the GLC.

APPENDIX
Libraries in London containing collections on the London Region

This is a select list of libraries in Greater London which possess sizeable collections on the region. Before visiting any of them it is advisable to telephone to check opening hours and conditions of access. Some libraries allow researchers in only if they have made a prior appointment with the librarian.

Academic and special libraries

Bishopsgate Institute Library,
230 Bishopsgate,
London EC2M 4QH
Tel: 01-247 6844
Collection of London history with an emphasis on the City area; contains the London and Middlesex Archaeological Society library.

Department of the Environment Library,
2 Marsham Street,
London SW1P 3EB
Tel: 01-212 4847
Huge collection encompassing local government, housing, transport and all aspects of planning.

Directorate of Ancient Monuments and Historic Buildings Library,
Fortress House,
23 Savile Row,
London W1X 2AR
Tel: 01-734 6010
Possesses a special collection of books and prints on London (the Mayson Beeton collection).

Greater London Council History Library,
Room 114,
County Hall,
London SE1 7PB
Tel: 01-633 6759/7132
Collects material relating to London history and topography, local government and associated topics. Run in conjunction with the Greater London Record Office (recently merged with the Middlesex Record Office) which contains records of the Greater London Council and its predecessors, manorial, ecclesiastical, business, estate and private records relating to the London area.

Greater London Council Research Library,
Room 514,
County Hall,
London SE1 7PB
Tel: 01-633 6483/6068/7007/7169
Excellent collection including transportation, local government, planning, social services, housing, tourism and recreation.

Institute of Historical Research Library,
Senate House,
Malet Street,
London WC1E 7HU
Tel: 01-636 0272
Open to members only.

London Transport Museum Library,
39 Wellington Street,
London WC2E 7BB
Tel: 01-379 6344

Smallish collection on the history of London and the history of London transport. Open to the public by appointment only.

Museum of London Library,
London Wall,
London EC2Y 5HN
Tel: 01-600 3699
The emphasis of the collection is on London topography, but there are also special collections, including one of ephemera.

Port of London Authority Library,
London Dock House (South),
Thomas More Street,
London E1 9AZ
Tel: 01-476 6900
Collects material on the local history of London as well as on transport, docks, and port administration.

University College Library,
Gower Street,
London WC1E 6BT
Tel: 01-387 7050
Contains a special collection on the history of London.

Public libraries

The libraries listed all contain collections of material about their local areas, usually the borough or part of the borough in which they are located. Further details of these collections may be obtained from the *Guide to London local studies resources* which has ben produced by the Association of London Chief Librarians. Although this document has not been published, it is available for consultation in most public libraries in London.

London Borough of Barking Libraries,
Valence House Museum,
Becontree Avenue,
Dagenham,
Essex RM8 3HT
Tel: 01-592 2211

London Borough of Barnet Local History Library,
Hendon Catholic Social Centre,
Church Walk,
Egerton Gardens,
London NW4 4BA
Tel: 01-202 5625

London Borough of Bexley Libraries,
Local Studies Section,
Central Administrative Offices,
Hall Place,
Bourne Road,
Bexley
Kent DA5 1PQ
Tel: Crayford 26574

London Borough of Brent Libraries,
The Grange Museum,
Neasden Lane,
London NW10 1QB
Tel: 01-452 8311

London Borough of Bromley Local Studies Library,
Central Library,
Bromley,
Kent BR1 1EX
Tel: 01-460 9955

London Borough of Camden, Swiss Cottage Library,
88 Avenue Road,
London NW3 3HA
Tel: 01-278 4444

London Borough of Camden, Holborn Library,
32–8 Theobalds Road,
London WC1X 8PA
Tel: 01-405 2705

City of London, Guildhall Library,
Aldermanbury,
London EC2P 2EJ
Tel: 01-606 3030

Comprehensive collection on the City with coverage in depth of the inner London boroughs and in outline of the outer London area; the official repository for deposited records of the City.

London Borough of Croydon Local History Library,
Central Library,
Katherine Street,
Croydon CR9 1ET
Tel: 01-688 3627

London Borough of Ealing Central Library,
Walpole Park,
Ealing,
London W5 5EQ
Tel: 01-579 2424

London Borough of Enfield Local History Library,
Southgate Town Hall,
Green Lanes,
Palmers Green,
London N13 4XD
Tel: 01-886 6555

London Borough of Greenwich Local History Library,
'Woodlands',
90 Mycenae Road,
Blackheath,
London SE3 7SE
Tel: 01-858 4631

London Borough of Hackney Archives Department,
Rose Lipman Library,
De Beauvoir Road,
London N1 5SQ
Tel: 01-985 8262

London Borough of Hammersmith and Fulham Libraries,
Fulham Local Collection,
Fulham Library,
598 Fulham Road,
London SW6 5NX
Tel: 01-736 1127

London Borough of Hammersmith and Fulham Libraries,
Hammersmith Local Collection,
Hammersmith Central Library,
Shepherds Bush Road,
London W6 7AT
Tel: 01-748 6032

London Borough of Haringey Libraries,
Bruce Castle Museum,
Lordship Lane,
London N17 8NU
Tel: 01-808 8772

London Borough of Haringey, Hornsey Library,
Haringey Park,
London N8 9JA
Tel: 01-348 3351

London Borough of Harrow Central Reference Library,
The Civic Centre,
PO Box 4,
Station Road,
Harrow HA1 2UU
Tel: 01-863 5611

London Borough of Havering Central Reference Library,
St Edwards Way,
Romford,
Essex RM1 3AR
Tel: Romford 46040

London Borough of Hillingdon Libraries,
Local History Collection,
Uxbridge Library,
22 High Street,
Uxbridge,
Middlesex UB8 1JN
Tel: Uxbridge 50111

London Borough of Islington Central Reference Library,
2 Fieldway Crescent,
London N5 1PF
Tel: 01-609 3051

London Borough of Islington, Finsbury Library,
245 St John Street,
London EC1V 4NB
Tel: 01-609 3051

Royal Borough of Kensington and Chelsea Libraries,
Kensington Collection,
Central Library,
Hornton Street,
London W8 7RX
Tel: 01-937 2542

Royal Borough of Kensington and Chelsea Libraries,
Chelsea Collection,
Chelsea Library,
Chelsea Old Town Hall,
Kings Road,
London SW3 5EZ
Tel: 01-352 6056

London Borough of Kingston upon Thames Central Reference Library,
Fairfield Road,
Kingston upon Thames,
Surrey KT1 2PS
Tel: 01-549 0226/0227

London Borough of Lambeth, Minet Library,
52 Knatchbull Road,
London SE5 9QY
Tel: 01-274 5325

London Borough of Lewisham Libraries,
The Manor House,
Old Road,
London SE13 5SY
Tel: 01-852 0357

London Borough of Merton Reference Library,
Wimbledon Hill Road,
London SW19 7NB
Tel: 01-946 7979/7432

London Borough of Newham, Stratford Reference Library,
Water Lane,
London E15 4NJ
Tel: 01-534 4545

London Borough of Redbridge Reference Library,
112B High Road,
Ilford,
Essex IG1 1BY
Tel: 01-478 0017/0018/5680/8993

London Borough of Richmond Central Reference Library,
Little Green,
Surrey TW9 1QL
Tel: 01-940 0981/6857

London Borough of Richmond, Twickenham Reference Library,
District Library,
Garfield Road,
Twickenham,
Middlesex TW1 3JS
Tel: 01-892 8091

London Borough of Southwark, Newington District Library,
Local History Department,
Walworth Road,
London SE17 1RS
Tel: 01-703 3324/5529/6514

London Borough of Sutton Central Library,
St Nicholas Way,
Sutton SM1 1EA
Tel: 01-661 5050

London Borough of Tower Hamlets Central Library,
277 Bancroft Road,
London E1 4DQ
Tel: 01-980 4366

London Borough of Waltham Forest, Leyton Library,
High Road,
London E10 5QH
Tel: 01-539 1223

Westminster City Central Reference Library,
St Martins Street,
London WC2H 7HP
Tel: 01-930 3274

Westminster City, Marylebone Library,
Marylebone Road,
London NW1 5PS
Tel: 01-828 8070

Name index

This index comprises the names of all authors and editors, together with the names of most series and the titles of certain works which are either anonymous or are generally known by their titles. The numbers given refer to items and not to pages.

Subject index